AF894477

Xanthomonas

Xanthomonas

Edited by

J.G. Swings

Laboratorium voor Microbiologie
Universiteit Gent
Gent
Belgium

and

E.L. Civerolo

USDA-ARS
Beltsville
USA

SPRINGER-SCIENCE+BUSINESS MEDIA, B.V.

First edition 1993

© 1993 Springer Science+Business Media Dordrecht
Originally published by Chapman & Hall in 1993
Softcover reprint of the hardcover 1st edition 1993
Typeset in 10/12pt Palatino by Best-set Typesetter Ltd, Hong Kong

ISBN 978-94-010-4666-4 ISBN 978-94-011-1526-1 (eBook)
DOI 10.1007/978-94-011-1526-1

Apart from any fair dealing for the purposes of research or private study, or criticism or review, as permitted under the UK Copyright Designs and Patents Act, 1988, this publication may not be reproduced, stored, or transmitted, in any form or by any means, without the prior permission in writing of the publishers, or in the case of reprographic reproduction only in accordance with the terms of the licences issued by the Copyright Licensing Agency in the UK, or in accordance with the terms of licences issued by the appropriate Reproduction Rights Organization outside the UK. Enquiries concerning reproduction outside the terms stated here should be sent to the publishers at the London address printed on this page.

The publisher makes no representation, express or implied, with regard to the accuracy of the information contained in this book and cannot accept any legal responsibility or liability for any errors or omissions that may be made.

A catalogue record for this book is available from the British Library

Library of Congress Cataloging-in-Publication data available

Contents

Contributors

A. Alvarez
University of Hawaii at Manoa,
College of Tropical Agriculture and Human Resources,
Department of Plant Pathology,
3190 Maile Way,
Honolulu, HI 96822,
USA.

E.L. Civerolo
USDA-ARS,
Beltsville, MD 20705,
USA.

M.J. Daniels
The Sainsbury Laboratory,
John Innes Centre,
Norwich Research Park,
Norwich NR4 7UH,
UK.

E. Duveiller
Centro International de Mejozamiento de Maiz y Trigo (CIMMYT),
El Batán, Texcoco,
Mexico.

T.R. Gottwald
USDA-ARS,
2120 Camden Rd,
Orlando, FL 32803,
USA.

M.J. Hattingh
Department of Plant Pathology,
University of Stellenbosch,
Stellenbosch 7600,
South Africa.

A.C. Hayward
Department of Microbiology,
The University of Queensland,
Queensland 4072,
Australia.

S. Hokawat
Department of Plant Pathology,
Faculty of Agriculture,
Khon Kaen University,
Khon Kaen 40002,
Thailand.

B. Hoste
Laboratorium voor Mikrobiologie,
Universiteit Gent,
Ledeganckstraat 35,
B-9000 Gent,
Belgium.

K. Kersters
Laboratorium voor Mikrobiologie,
Universiteit Gent,
Ledeganckstraat 35,
B-9000 Gent,
Belgium.

M. Koizumi
Fruit Tree Research Station,
Okitsu, Shimizu,
Shizuoka 424-02,
Japan.

J.E. Leach
Department of Plant Pathology,
Kansas State University,
Manhattan, KS 66506,
USA.

F. Leyns
Plant Genetic Systems,
Jozef Plateaustraat 22,
B-9000 Gent,
Belgium.

B.Q. Manicom
ITSC,
Private bag X 11208,
Nelspruit 1200,
Republic of South Africa.

H. Maraite
Unité de Phytopathologie,
Faculté des Sciences Agronomiques,
Université Catholique de Louvain,
Place Croix du Sud, 2, Bte 3,
B-1348 Louvain-la-Neuve,
Belgium.

T.W. Mew
International Rice Research Institute (IRRI),
Los Banos,
The Philippines.

M.P. Natural
Department of Plant Pathology,
University of the Philippines at Los Banos College,
Laguna 4031,
The Philippines

W. Pagel
Geranien-Endisch,
Geranienweg,
D-6729 Hagenbach,
Germany.

O. Pruvost
CIRAD/FLHOR,
Station de Bassin Martin,
BP 180,
97455 Saint-Pierre Cédex,
La Reunion, France.

B. Rat
INRA,
Station de Pathologie Végétale et de Phytobactériologie,
Beaucouzé, 4900 Angers,
France.

M. Ridé
INRA,
Station de Pathologie Végétale et de Phytobactériologie,
Route de St. Clement,
Beaucouzé, 4900 Angers,
France.

E.J.A. Roebroeck
Bulb Research Centre,
PO Box 85, 2160 AB Lisse,
The Netherlands.

P. Rott
CIRAD-CA,
Station de Roujol,
97170 Petit Bourg,
Guadeloupe,
French West Indies.

K. Rudolph
Institut für Pflanzenpathologie und Pflanzenschutz,
Grisebachstrasse 6,
D-3400 Göttingen,
Germany.

N.W. Schaad
USDA-ARS,
Foreign Disease-Weed Science Research,
Fort Detrick, Building 1301,
Frederick, MD 21702,
USA.

R.E. Stall
Department of Plant Pathology,
University of Florida,
Gainesville, FL 32611,
USA.

I.W. Sutherland
Institute of Cell and Molecular Biology,
Division of Biology,
Edinburgh University,
Edinburgh EH9 3JH,
Scotland.

J. Swings
Laboratorium voor Mikrobiologie,
Universiteit Gent,
Ledeganckstraat 35,
B-9000 Gent,
Belgium.

J. van Doorn
Bulb Research Centre,
PO Box 85, 2160 AB Lisse,
The Netherlands.

L. Vauterin
Laboratorium voor Mikrobiologie,
Universiteit Gent,
Ledeganckstraat 35,
B-9000 Gent,
Belgium.

A.K. Vidaver,
Department of Plant Pathology,
University of Nebraska,
Lincoln, NE 68583-0722,
USA.

P. Yang
Laboratorium voor Mikrobiologie,
Universiteit Gent,
Ledeganckstraat 35,
B-9000 Gent,
Belgium.

A. Zomorodian
Institut für Pflanzenpathologie und Pflanzenschutz,
Grisebachstrasse 6,
D-3400 Göttingen,
Germany.

1

The hosts of *Xanthomonas*

A.C. Hayward
Department of Microbiology,
The University of Queensland

1.1 INTRODUCTION

Bacteria belonging to the genus *Xanthomonas* are, with the exception of *X. maltophilia* which is cosmopolitan in distribution and not known to be a plant pathogen (Swings *et al.*, 1983), plant-associated bacteria and not usually encountered in other environments. They are not invariably plant pathogens. Isolates resembling *X. campestris* pv. *campestris* have been found in asymptomatic association with plant tissue or as epiphytes (Maas *et al.*, 1985; Angeles-Ramos *et al.*, 1991) and there are other xanthomonads which appear to be limited to secondary invasion of plant tissue after infection by a primary bacterial pathogen (Gitaitis *et al.*, 1987b). An analysis of published information on the hosts of *Xanthomonas* (Leyns *et al.*, 1984) showed that *Xanthomonas* infections occur on at least 124 monocotyledonous and 268 dicotyledonous plant species. Among monocotyledons the range of hosts extends across 11 families comprising at least 70 genera, and there are hosts in 57 families of dicotyledons comprising more than 170 genera. No strong association of hosts with any of the subclasses of the angiosperm families has been found (Leyns *et al.*, 1984).

The list of species and pathovars of *Xanthomonas* in Table 1.1 is derived from several authoritative sources (Dye *et al.*, 1980; Bradbury, 1984; Leyns *et al.*, 1984; Young *et al.*, 1991b). *Xanthomonas oryzae*

Xanthomonas.
Edited by J.G. Swings and E.L. Civerolo.
Published in 1993 by Chapman & Hall, London.
ISBN 0 412 434202

Table 1.1 List of species of the genus *Xanthomonas*

Species	*Hosts*	*Families*
X. albilineans (Ashby 1929) Dowson 1943	*Saccharum officinarum* Other natural hosts are: *Brachiaria piligera, Coix lacryma – jobi, Imperata cylindrica* var. *major, Paspalum conjugatum.* Some other gramineous plants can be inoculated successfully, but are probably not natural hosts	Poaceae
X. axonopodis Starr and Garces 1950	*Axonopus affinis, A. compressus, A. micay, A. scoparius, Agropyron* sp.	Poaceae
Pathovars of *X. campestris*: (Pammel 1895) Dowson 1939		
X. campestris pv. *campestris* (Pammel 1895) Dowson 1939	*Brassica* spp., *Capsella bursapastoris, Lepidium sativum, Matthiola* spp., *Raphanus sativus, Rorippa armoracia, Boerhaavia erecta*	Brassicaceae Nyctaginaceae
X. campestris pv. *aberrans* (Knösel 1961) Dye 1978	*Brassica oleracea* var. *botrytis*	Brassicaceae
X. campestris pv. *alangii* (Padhya and Patel 1962) Dye 1978	*Alangium lamarckii*	Alangiaceae
X. campestris pv. *alfalfae* (Riker, Jones and Davis 1935) Dye 1978	*Medicago sativa, Melilotus indica, Pisum sativum, Phaseolus vulgaris, Trigonella foenum-graecum*	Fabaceae
X. campestris pv. *amaranthicola* (Patel, Wankar and Kulkarni 1952) Dye 1978	*Amaranthus* spp.	Amaranthaceae
X. campestris pv. *amorphophalli* (Jindal, Patel and Singh 1972) Dye 1978	*Amorphophallus campanulatus*	Araceae
X. campestris pv. *aracearum* (Berniac 1974) Dye 1978	*Xanthosoma sagittifolium*	Araceae

Table 1.1 *Continued*

Species	*Hosts*	*Families*
X. campestris pv. *arecae* (Rao and Mohan 1970) Dye 1978	*Areca catechu*	Arecaceae
X. campestris pv. *argemones* (Srinivasan, Patel and Thirumalachar 1961) Dye 1978	*Argemone mexicana*	Papaveraceae
X. campestris pv. *armoraciae* (McCulloch 1929) Dye 1978	*Armoracia rusticana, Brassica oleracea* var. *botrytis, B. oleracea* var. *capitata, Phaseolus vulgaris*	Brassicaceae Fabaceae
X. campestris pv. *arracaciae* (Pereira, Paradella and Zagatto 1971) Dye 1978	*Arracacia xanthorrhiza*	Apiaceae
X. campestris pv. *arrhenatheri* Egli and Schmidt 1982	*Arrhenatherum elatius*	Poaceae
X. campestris pv. *azadirachtae* (Desai, Gandhi, Patel and Kotasthane 1966) Dye 1978	*Azadirachta indica*	Meliaceae
X. campestris pv. *badrii* (Patel, Kulkarni and Dhande 1950) Dye 1978	*Xanthium strumarium, Pisum sativum*	Asteraceae Fabaceae
X. campestris pv. *barbareae* (Burkholder 1941) Dye 1978	*Barbarea vulgaris*	Brassicaceae
X. campestris pv. *bauhiniae* (Padhya, Patel and Kotasthane 1965) Dye 1978	*Bauhinia racemosa*	Fabaceae
X. campestris pv. *begoniae* (Takimoto 1934) Dye 1978	*Begonia* spp.	Begoniaceae
X. campestris pv. *betae* Robbs, Kimura and Ribeiro 1981	*Beta vulgaris*	Chenopodiaceae

Table 1.1 *Continued*

Species	*Hosts*	*Families*
X. campestris pv. *betlicola* (Patel, Kulkarni and Dhande 1951) Dye 1978	*Piper betle, P. hookeri, P. longum*	Piperaceae
X. campestris pv. *bilvae* Chakravarti, Sarma, Jain and Prasad 1984	*Aegle marmelos*	Rutaceae
X. campestris pv. *biophyti* (Patel, Chauhan, Kotasthane and Desai 1969) Dye 1978	*Biophytum sensitivum*	Oxalidaceae
X. campestris pv. *blepharidis* (Srinivasan and Patel 1956) Dye 1978	*Blepharis boerhaavifolia, B. molluginifolia*	Acanthaceae
X. campestris pv. *boerhaaviae* (Mathur, Swarup and Sinha 1964) Bradbury 1986	*Boerhaavia repens*	Nyctaginaceae
X. campestris pv. *brunneivaginae* Luo, Liao and Chen 1988	*Oryza sativa*	Poaceae
X. campestris pv. *cajani* (Kulkarni, Patel and Abhyankar 1950) Dye 1978	*Cajanus cajan*	Fabaceae
X. campestris pv. *cannabis* Severin 1978	*Cannabis sativa*	Cannabaceae
X. campestris pv. *cannae* Easwaramurthy, Kaviyarasan and Gnanamanickam 1984	*Canna* × *generalis*	Cannaceae
X. campestris pv. *carissae* (Moniz, Sabley and More 1964) Dye 1978	*Carissa congesta, C. carandas, Thevetia nerifolia, Cestrum nocturnum*	Apocynaceae Solanaceae
X. campestris pv. *carotae* (Kendrick 1934) Dye 1978	*Daucus carota*	Apiaceae

Table 1.1 *Continued*

Species	*Hosts*	*Families*
X. campestris pv. *cassavae* (Wiehe and Dowson 1953) Maraite and Weyns 1979	*Manihot* spp.	Euphorbiaceae
X. campestris pv. *cassiae* (Kulkarni, Patel and Dhande 1951) Dye 1978	*Cassia tora, C. occidentalis, Cicer arietinum, Pisum sativum*	Fabaceae
X. campestris pv. *celebensis* (Gaumann 1923) Dye 1978	*Musa* spp.	Musaceae
X. campestris pv. *centellae* Basnyat and Kulkarni 1979	*Centella asiatica*	Apiaceae
X. campestris pv. *cerealis* (Hagborg 1942) Dye 1978	*Agropyron* spp., *Avena* spp., *Bromus* spp., *Hordeum* spp., *Secale cereale, Triticum* spp.	Poaceae
X. campestris pv. *citri* (Hasse 1915) Dye 1978	*Aegle marmelos, Atalantia* spp., *Balsamocitrus paniculata, Casimiroa edulis, Chaetospermum glutinosa, Citropsis schweinfurthii, Citrus* spp. and hybrids, *Clausena lansium, Eremocitrus glauca, Evodia* spp., *Feronia* spp., *Feroniella* spp., *Fortunella* spp., *Hesperethusa crenulata, Limonia* spp., *Melicope triphylla, Microcitrus* spp., *Murraya exotica, Paramigyna longipedunculata, Poncirus trifoliata* and hybrids, *Severina buxifolia, Toddalia asiatica, Zanthoxylum* spp.	Rutaceae
X. campestris pv. *clerodendri* (Patel, Kulkarni and Dhande 1952) Dye 1978	*Clerodendrum phlomoides*	Verbenaceae
X. campestris pv. *clitoriae* Pandit and Kulkarni 1979	*Clitoria biflora*	Fabaceae

Table 1.1 *Continued*

Species	*Hosts*	*Families*
X. campestris pv. *convolvuli* (Nagarkoti, Banerjee and Swarup 1973) Dye 1978	*Convolvulus arvensis*	Convolvulaceae
X. campestris pv. *coracanae* (Desai, Thirumalachar and Patel 1965) Dye 1978	*Eleusine coracana*	Poaceae
X. campestris pv. *coriandri* (Srinivasan, Patel and Thirumalachar 1961) Dye 1978	*Coriandrum sativum, Foeniculum vulgare*	Apiaceae
X. campestris pv. *corylina* (Miller, Bollen, Simmons, Gross and Barss 1940) Dye 1978	*Corylus* spp.	Betulaceae
X. campestris pv. *cucurbitae* Bryan 1926) Dye 1978	*Citrullus vulgaris, Cucumis sativus, Cucurbita* spp.	Cucurbitaceae
X. campestris pv. *cyamopsidis* (Patel, Dhande and Kulkarni 1953) Dye 1978	*Cyamopsis tetragonolobus*	Fabaceae
X. campestris pv. *daturae* (Jain, Dange and Siradhana 1975) Bradbury 1986	*Datura metel*	Solanaceae
X. campestris pv. *desmodii* (Patel 1949) Dye 1978	*Desmodium diffusum*	Fabaceae
X. campestris pv. *desmodiigangetici* (Patel and Moniz 1948) Dye 1978	*Desmodium gangeticum*	Fabaceae
X. campestris pv. *desmodiilaxiflori* Pant and Kulkarni, 1976	*Desmodium laxiflorum, Tamarindus indica*	Fabaceae
X. campestris pv. *desmodiirotundifolii* (Desai and Shah 1960) Dye 1978	*Desmodium rotundifolium*	Fabaceae

Table 1.1 *Continued*

Species	*Hosts*	*Families*
X. campestris pv. *dieffenbachiae* (McCulloch and Pirone 1939) Dye 1978	*Aglaonema robellinii, Anthurium andraeanum, Dieffenbachia* spp., *Philodendron scandens, Dracaena fragrans*	Araceae Agavaceae
X. campestris pv. *durantae* (Srinivasan and Patel 1957) Dye 1978	*Duranta repens*	Verbenaceae
X. campestris pv. *erythrinae* (Patel, Kulkarni and Dhande 1952) Dye 1978	*Erythrina indica*	Fabaceae
X. campestris pv. *esculenti* (Rangaswami and Easwaran 1962) Dye 1978	*Hibiscus esculentus*	Malvaceae
X. campestris pv. *eucalypti* (Truman 1974) Dye 1978	*Eucalyptus citriodora, E. maculata*	Myrtaceae
X. campestris pv. *euphorbiae* (Sabet, Ishag and Khalil 1969) Dye 1978	*Euphorbia acalyphoides, E. nilli, E. pulcherrima*	Euphorbiaceae
X. campestris pv. *fascicularis* (Patel and Kotasthane 1969) Dye 1978	*Corchorus fascicularis*	Tiliaceae
X. campestris pv. *fici* (Cavara 1905) Dye 1978	*Ficus carica*	Moraceae
X. campestris pv. *glycines* (Nakano 1919) Dye 1978 (Syn. *X. phaseoli* var. *sojensis* (Hedges 1922) Starr and Burkholder 1942)	*Brunnichia cirrhosa, Dolichos uniflorus, Glycine* spp., *Phaseolus lunatus, P. vulgaris*	Polygonaceae Fabaceae
X. campestris pv. *graminis* (Egli, Goto and Schmidt 1975) Dye 1978	*Arrhenatherum elatius, Alopecurus pratensis, Agrostis* spp., *Bromus mollis, Dactylis glomerata, Festuca* spp., *Lolium multiflorum, L. perenne, Phleum pratense*	Poaceae

Table 1.1 *Continued*

Species	*Hosts*	*Families*
X. campestris pv. *guizotiae* (Yirgou 1964) Dye 1978	*Guizotia abyssinica*	Asteraceae
X. campestris pv. *gummisudans* (McCulloch 1924) Dye 1978	*Gladiolus* sp.	Iridaceae
X. campestris pv. *hederae* (Arnaud 1920) Dye 1978	*Hedera helix*	Araliaceae
X. campestris pv. *heliotropii* (Sabet, Ishag and Khalil 1969) Dye 1978	*Heliotropium aegypticum*, *H. sudanicum*	Boraginaceae
X. campestris pv. *holcicola* (Elliott 1930) Dye 1978	*Sorghum halepense*, *S. vulgare*	Poaceae
X. campestris pv. *hordei* (Hagborg 1942) Dye 1978	*Agropyron* spp., *Bromus inermis*, *Hordeum* spp.	Poaceae
X. campestris pv. *hyacinthi* (Wakker 1883) Dye 1978	*Hyacinthus orientalis*	Liliaceae
X. campestris pv. *incanae* (Kendrick and Baker 1942) Dye 1978	*Matthiola incana*	Brassicaceae
X. campestris pv. *ionidii* (Padhya and Patel 1963) Dye 1978	*Ionidium heterophyllum*	Violaceae
X. campestris pv. *juglandis* (Pierce 1901) Dye 1978	*Juglans* spp.	Juglandaceae
X. campestris pv. *khayae* (Sabet 1959) Dye 1978	*Khaya senegalensis*, *K. grandifoliola*	Meliaceae
X. campestris pv. *lantanae* (Srinivasan and Patel 1957) Dye 1978	*Lantana camara* var. *aculeata*	Verbenaceae
X. campestris pv. *laureliae* (Dye 1963) Dye 1978	*Laurelia novae-zelandiae*	Monimiaceae
X. campestris pv. *lawsoniae* (Patel, Bhatt and Kulkarni 1951) Dye 1978	*Lawsonia alba*	Lythraceae

Table 1.1 *Continued*

Species	*Hosts*	*Families*
X. campestris pv. *leeana* (Patel and Kotasthane 1969) Dye 1978	*Leea edgeworthii*	Vitaceae
X. campestris pv. *leersiae* (Fang, Ren, Chen, Chu, Faan and Wu 1957) Young *et al.* 1991b	*Leersia hexandra*	Poaceae
X. campestris pv. *lespedezae* (Ayres, Lefebvre and Johnson 1939) Dye 1978	*Lespedeza* spp.	Fabaceae
X. campestris pv. *maculifoliigardeniae* (Ark and Barrett 1946) Dye 1978	*Gardenia* spp., *Ixora coccinea*	Rubiaceae
X. campestris pv. *malvacearum* (Smith 1901) Dye 1978	*Gossypium* spp., *Thespesia lampas*, *Ceiba pentandra*	Malvaceae Bombacaceae
X. campestris pv. *mangiferaeindicae* (Patel, Moniz and Kulkarni 1948) Robbs, Ribeiro and Kimura 1974	*Anacardium occidentale*, *Mangifera indica*, *Spondias mangifera*	Anacardiaceae
X. campestris pv. *manihotis* (Berthet and Bondar 1915) Dye 1978	*Manihot* spp.	Euphorbiaceae
X. campestris pv. *martyniicola* (Moniz and Patel 1958) Dye 1978	*Martynia diandra*	Martyniaceae
X. campestris pv. *melhusii* (Patel, Kulkarni and Dhande 1952) Dye 1978	*Tectona grandis*	Verbenaceae
X. campestris pv. *melonis* Neto, Sugimori and Oliveira 1984	*Cucumis melo*	Cucurbitaceae
X. campestris pv. *merremiae* Pant and Kulkarni 1976	*Merremia gangetica*	Convolvulaceae
X. campestris pv. *mirabilis* (Durgapal and Trivedi 1976) Young *et al.* 1991	*Mirabilis jalapa*	Nyctaginaceae

Table 1.1 *Continued*

Species	*Hosts*	*Families*
X. campestris pv. *musacearum* (Yirgou and Bradbury 1968) Dye 1978	*Ensete ventricosum*, *Musa* spp.	Musaceae
X. campestris pv. *nakataecorchori* (Padhya and Patel 1963) Dye 1978	*Corchorus acutangulus*	Tiliaceae
X. campestris pv. *nigromaculans* (Takimoto 1927) Dye 1978	*Arctium lappa*	Asteraceae
X. campestris pv. *olitorii* (Sabet 1957) Dye 1978	*Corchorus olitorius*	Tiliaceae
X. campestris pv. *papavericola* (Bryan and McWhorter 1930) Dye 1978	*Meconopsis baileyi*, *Papaver* spp.	Papaveraceae
X. campestris pv. *passiflorae* (Pereira 1969) Dye 1978	*Passiflora edulis*	Passifloraceae
X. campestris pv. *patelii* (Desai and Shah 1959) Dye 1978	*Crotalaria juncea*	Fabaceae
X. campestris pv. *paulliniae* Robbs, Medeiros and Kimura 1982	*Paullinia* spp.	Sapindaceae
X. campestris pv. *pedalii* (Patel and Jindal 1972) Dye 1978	*Pedalium murex*	Pedaliaceae
X. campestris pv. *pelargonii* (Brown 1923) Dye 1978	*Geranium* spp., *Pelargonium* spp.	Geraniaceae
X. campestris pv. *pennamericanum* Qhobela and Claflin 1988	*Pennisetum americanum*, *Panicum miliaceum*	Poaceae
X. campestris pv. *phaseoli* (Smith 1897) Dye 1978	*Lablab purpureus* (syn. *Dolichos lablab*), *Lupinus polyphyllus*, *Phaseolus lunatus*, *Phaseolus vulgaris*	Fabaceae

Table 1.1 *Continued*

Species	*Hosts*	*Families*
X. campestris pv. *phlei* Egli and Schmidt 1982	*Phleum pratense*	Poaceae
X. campestris pv. *phleipratensis* (Wallin and Reddy 1945) Dye 1978	*Agropyron* sp., *Bromus* sp., *Phleum pratense*	Poaceae
X. campestris pv. *phormiicola* (Takimoto 1933) Dye 1978	*Phormium tenax*	Agavaceae
X. campestris pv. *phyllanthi* (Sabet, Ishag and Khalil 1969) Dye 1978	*Phyllanthus niruri*	Euphorbiaceae
X. campestris pv. *physalidicola* (Goto and Okabe 1958) Dye 1978	*Physalis alkekengi* var. *francheti*	Solanaceae
X. campestris pv. *physalidis* (Srinivasan, Patel and Thirumalachar 1962) Dye 1978	*Physalis minima, P. peruviana*	Solanaceae
X. campestris pv. *pisi* (Goto and Okabe 1958) Dye 1978	*Pisum sativum*	Fabaceae
X. campestris pv. *plantaginis* (Thornberry and Anderson 1937) Dye 1978	*Plantago* spp.	Plantaginaceae
X. campestris pv. *poae* Egli and Schmidt 1982	*Poa trivialis*	Poaceae
X. campestris pv. *poinsettiicola* (Patel, Bhatt and Kulkarni 1951) Dye 1978	*Euphorbia pulcherrima, E. milii, Manihot esculenta*	Euphorbiaceae
X. campestris pv. *populi* (ex de Kam 1984) Young *et al.* 1991b	*Populus alba* var. *nivea, P. balsamifera, P. balsamifera* × *deltoides, P. deltoides, P.* × *euramericana, P. nigra, P. tremula, P. trichocarpa, Salix* × *dasyclada*	Salicaceae

Table 1.1 *Continued*

Species	*Hosts*	*Families*
X. campestris pv. *pruni* (Smith 1903) Dye 1978	*Prunus* spp., *Sorbus japonica*	Rosaceae
X. campestris pv. *punicae* (Hingorani and Singh 1959) Dye 1978	*Punica granatum*	Punicaceae
X. campestris pv. *raphani* (White 1930) Dye 1978	*Brassica* spp., *Raphanus sativus*, *Capsicum annuum*, *Lycopersicon esculentum*, *Nicotiana tabacum*	Brassicaceae Solanaceae
X. campestris pv. *rhynchosiae* (Sabet, Ishag and Khalil 1969) Dye 1978	*Lupinus termis*, *Mucuna pruriens* (syn. *Stizolobium aterrimum*), *Rhynchosia memnonia*	Fabaceae
X. campestris pv. *ricini* (Yoshi and Takimoto 1928) Dye 1978	*Ricinus communis*	Euphorbiaceae
X. campestris pv. *secalis* (Reddy, Godkin and Johnson 1924) Dye 1978	*Agropyron* sp., *Bromus* sp., *Secale cereale*; *Hordeum* and *Triticum* spp. may be artificially infected	Poaceae
X. campestris pv. *sesami* (Sabet and Dowson 1960) Dye 1978	*Sesamum orientale*	Pedaliaceae
X. campestris pv. *sesbaniae* (Patel, Kulkarni and Dhande 1952) Dye 1978	*Sesbania aegyptiaca*	Fabaceae
X. campestris pv. *spermacoces* (Srinivasan and Patel 1956) Dye 1978	*Spermacoce hispida*	Rubiaceae
X. campestris pv. *syngonii* Dickey and Zumoff 1987	*Syngonium podophyllum*	Araceae
X. campestris pv. *tamarindi* (Patel, Bhatt and Kulkarni 1951) Dye 1978	*Caesalpina sepiaria*, *Tamarindus indica*	Fabaceae

Table 1.1 *Continued*

Species	*Hosts*	*Families*
X. campestris pv. *taraxaci* (Niederhauser 1943) Dye 1978	*Taraxacum bicorne*	Asteraceae
X. campestris pv. *tardicrescens* (McCulloch 1937) Dye 1978	*Belamcanda* sp., *Iris* spp.	Iridaceae
X. campestris pv. *theicola* Uehara and Arai 1980	*Camellia sinensis*	Theaceae
X. campestris pv. *thespesiae* Patil and Kulkarni 1981	*Thespesia populnea*	Malvaceae
X. campestris pv. *thirumalacharii* (Padhya and Patel 1964) Dye 1978	*Triumfetta pilosa*	Tiliaceae
X. campestris pv. *translucens* (Jones, Johnson and Reddy 1917) Dye 1978	*Hordeum* spp. and other cereals	Poaceae
X. campestris pv. *tribuli* (Srinivasan and Patel 1956) Dye 1978	*Tribulus terrestris*	Zygophyllaceae
X. campestris pv. *trichodesmae* (Patel, Kulkarni and Dhande 1952) Dye 1978	*Trichodesma zeylanicum*	Boraginaceae
X. campestris pv. *undulosa* (Smith, Jones and Reddy 1919) Dye 1978	*Agropyron* sp., *Bromus* spp., *Triticum* spp., *Secale cereale*, *Hordeum* spp. by inoculation	Poaceae
X. campestris pv. *uppalii* (Patel 1948) Dye 1978	*Ipomoea muricata*, *Tropaeolum majus*	Convolvulaceae Tropaeolaceae
X. campestris pv. *vasculorum* (Cobb 1893) Dye 1978	*Bambusa vulgaris*, *Brachiaria mutica*, *Coix lacryma-jobi*, *Panicum maximum*, *Pennisetum purpureum*, *Saccharum officinarum*, *Sorghum* spp., *Thysanolaena maxima*, *Zea mays*, *Cocos nucifera*, *Dictyosperma album*, *Roystonea regia*	Poaceae Arecaceae

Table 1.1 *Continued*

Species	*Hosts*	*Families*
X. campestris pv. *vernoniae* (Patel, Desai and Patel 1968) Dye 1978	*Vernonia cinerea*	Asteraceae
X. campestris pv. *vesicatoria* (Doidge 1920) Dye 1978	*Capsicum* spp., *Datura stramonium*, *Hyoscyamus* spp., *Lycium* spp., *Lycopersicon* spp., *Nicotiana rustica*, *Nicandra physalodes*, *Physalis minima*, *Solanum* spp., *Tinospora cordifolia*, *Argemone mexicana*	Solanaceae Menispermaceae Papaveraceae
X. campestris pv. *vignaeradiatae* (Sabet, Ishag and Khalil 1969) Dye 1978	*Lablab purpureus* (syn. *Dolichos lablab*), *Vigna radiata*	Fabaceae
X. campestris pv. *vignicola* (Burkholder 1944) Dye 1978	*Phaseolus vulgaris*, *Vigna pubigera*, *V. unguiculata*	Fabaceae
X. campestris pv. *vitians* (Brown 1918) Dye 1978	*Lactuca* spp., *Syngonium podophyllum*	Asteraceae Araceae
X. campestris pv. *viticola* (Nayudu 1972) Dye 1978	*Azadirachta indica*, *Phyllanthus maderaspatensis*, *Vitis vinifera*	Meliaceae Euphorbiaceae Vitaceae
X. campestris pv. *vitiscarnosae* (Moniz and Patel 1958) Dye 1978	*Vitis carnosa*	Vitaceae
X. campestris pv. *vitistrifoliae* (Padhya, Patel and Kotasthane 1965) Dye 1978	*Vitis trifolia*	Vitaceae
X. campestris pv. *vitiswoodrowii* (Patel and Kulkarni 1951) Dye 1978	*Vitis woodrowii*	Vitaceae
X. campestris pv. *zantedeschiae* (Joubert and Truter 1972) Dye 1978	*Zantedeschia aethiopica*	Araceae

Table 1.1 *Continued*

Species	*Hosts*	*Families*
X. campestris pv. *zingibericola* (Ren and Fang 1981) Bradbury 1986	*Zingiber officinale*	Zingiberaceae
X. campestris pv. *zinniae* (Hopkins and Dowson 1949) Dye 1978	*Zinnia elegans*	Asteraceae
Xanthomonas fragariae Kennedy and King 1962	*Fragaria* × *ananassa*, *F. vesca*, *Potentilla* spp.	Rosaceae
Xanthomonas maltophilia (Hugh and Ryschenkow 1961) Swings *et al.* 1983	*X. maltophilia* as far as known is not a plant pathogen. Most strains isolated from clinical specimens; opportunistic human pathogen, also found in water, milk and frozen food and from the rhizosphere of some cultivated plants	
Pathovars of *Xanthomonas oryzae* (Ishiyama 1922) Swings *et al.* 1990:		
Xanthomonas oryzae pv. *oryzae* (Ishiyama 1922) Swings *et al.* 1990	*Isachne globosa*, *Leersia* spp., *Leptochloa* spp., *Oryza sativa*, *Oryza* spp., *Phalaris arundinacea*, *Phragmites communis*, *Zizania aquatica*, *Bromus* sp., *Cyperus difformis*, *C. rotundus*	Poaceae Cyperaceae
Xanthomonas oryzae pv. *oryzicola* (Fang, Ren, Chen, Chu, Faan and Wu 1957) Swings *et al.* 1990	*Oryza sativa*, *Oryza* spp., *Leersia hexandra* (weak)	Poaceae

(Ishiyama 1922) Swings *et al.* 1990 is listed as a species distinct from *X. campestris* (Pammell 1895) Dowson 1939 according to Swings *et al.* (1990). There is a lack of consensus on the status of the different xanthomonads associated with disease on *Citrus* spp. Although it is known that the xanthomonads pathogenic for citrus constitute different biological entities (Vauterin *et al.*, 1991b) their interrelationships are not fully understood. There are five pathogenicity groups designated A, B, C, D and E of which Gabriel *et al.* (1989) proposed to reclassify strains of the former *X. campestris* pv. *citri* into *X. citri* (group A), *X. campestris* pv. *aurantifolii* (groups B, C and D) and *X. campestris* pv. *citrumelo* (Florida nursery strains: group E). However, in Table 1.1 only *X. campestris* pv. *citri* is listed and groups A to E are considered to be subdivisions within this pathovar, because the criteria on which the proposal of Gabriel *et al.* (1989) was based were considered to be defective in terms of current taxonomy and the standards for naming pathovars put forward by the International Society for Plant Pathology (Vauterin *et al.*, 1990; Young *et al.*, 1991a,b).

Some additional hosts of *Xanthomonas* are given in Table 1.2. The available evidence indicates that these are most probably all pathovars of *X. campestris* but either no formal proposal has been made, or the pathovar was not validly published (Young *et al.*, 1991b). Some of the hosts are those listed in the catalogue of the International Collection of Microorganisms from Plants (ICMP; Young, 1987).

The nomenclature of plants used in Tables 1.1 and 1.2 is that of Takhtajan (1980) as recommended by Leyns *et al.* (1984). Names of authors for each plant species have not been included following the convention adopted in Bergey's Manual (Bradbury, 1984).

The host range of each *Xanthomonas* species or pathovar is in most cases limited to a few species or sometimes a few genera of the same plant family, the name of the pathogen often being derived from the genus or species name of the host plant (Leyns *et al.*, 1984). Some xanthomonads are relatively catholic in host range, others specialized to one or more related hosts. For example, *X. campestris* pv. *campestris* is a common pathogen of plants belonging to the Brassicaceae and *X. c.* pv. *graminis* affects several genera of the Poaceae, whereas *X. c.* pv. *phlei*, *X. c.* pv. *poae* and *X. c.* pv. *arrhenatheri* are limited to one host (Egli and Schmidt, 1982). However, it is likely that in many cases our knowledge of the host range of *Xanthomonas* spp. is incomplete, because wild plant species of no economic importance have rarely been examined.

Bacteria belonging to the genus *Xanthomonas* cause a variety of disease symptoms including necrosis, gummosis and/or vascular or parenchymatous diseases on leaves, stems or fruits of many plants. Damage to fruits and vegetables is often sufficient to cause considerable loss in yield. The following sections have been chosen to represent some of the major diseases caused by xanthomonads with emphasis on symptoms, host range, isolation and detection methods, geographical distribution and economic importance as well as some salient features of disease biology including epidemiology and disease management.

Table 1.2 Undesignated pathovars of *Xanthomonas campestris*

Disease, host and locality	*Families*	*Authors*
Veinal necrosis of *Fittonia verschaffeltii*, Florida, USA	Acanthaceae	Blake *et al.* (1989)
Bacterial leaf disease on taro (*Colocasia esculenta*), Papua, New Guinea	Araceae	Tomlinson (1987)
Leaf spot on *Panax* sp., Hawaii, USA	Araliaceae	Buddenhagen (personal communication)
Foliar and stem disease on *Asclepias* spp., Nebraska, USA	Asclepiadaceae	Heise and Vidaver (1988)
Leaf spot on *Bellis perennis*, Iran	Asteraceae	Rahimian (1989b)
Leaf spot on *Alnus subcordata*, Iran and California, USA	Betulaceae	Rahimian (1989a) Bishop and Basarich (1989)
Leaf blight and dieback of *Lobelia erinus* cv. *Richardii*, Germany	Campanulaceae	Poschenrieder *et al.* (1990)
Leafspot on *Diospyros kaki*, New Zealand	Ebenaceae	Young (1987)
Leaf spot on onion (*Allium cepa*), Hawaii, USA and Brasil	Liliaceae	Alvarez *et al.* (1974)
Leaf streak on *Cordyline terminalis*, Hawaii, USA	Liliaceae	Alvarez *et al.* (1991)
Leafspot on *Magnolia stellata*, New Zealand	Magnoliaceae	Young (1987)
Bacterial disease on *Morus alba*, Korea	Moraceae	Choi *et al.* (1988)
Leaf streak[a] of maize (*Zea mays*), South Africa	Poaceae	Qhobela *et al.* (1990)
Leaf spot and twig blight on *Eriostemon myoporoides*, New Zealand	Rutaceae	Young (1987)
Leaf spot and blight of *Strelitzia reginae*, Florida, USA	Strelitziaceae	Chase and Jones (1989)
Leaf spot on *Pellionia* spp. and *Pilea* spp., Florida, USA	Urticaceae	Miller and Chase (1986)

[a] Pathogen designated as *X. campestris* pv. *zeae* but not validly published according to Young *et al.* (1991b).

1.2 *XANTHOMONAS CAMPESTRIS* PATHOVARS ON CASSAVA: CAUSE OF BACTERIAL BLIGHT AND BACTERIAL NECROSIS

H. MARAITE
Unite de Phytopathologie, Faculté des Sciences Agronomiques, Université Catholique de Louvain

1.2.1 Introduction

Cassava (*Manihot esculenta* Crantz) is the natural host of the non-pigmented *Xanthomonas campestris* pv. *manihotis* (Berthet and Bondar) Dye (1978), causing cassava bacterial blight (CBB), and of the yellow pigmented *X. c.* pv. *cassavae* (Wiehe and Dowson) Maraite and Weyns (1979), associated with cassava bacterial necrosis (CBN).

Apart from *Manihot esculenta* (syn. *M. utilissima* Pohl.), severe field infections by pv. *manihotis* occur only on *M. palmata* Pax, *M. aipi* Pax (Lozano and Sequeira, 1974a) and *M. glazovii*. With artificial infection, Dedal *et al.* (1980) induced similar symptoms on variegated ornamental cassava, *Euphorbia pulcherrima* and *Pedilanthus tithymaloides*, but not on 22 other plant species from families other than the Euphorbiaceae. Van den Mooter *et al.* (1987), however, observed no or only very restricted lesions on inoculation of *E. pulcherrima* with a wide range of pv. *manihotis* and pv. *cassavae* strains and did not confirm the infection of cassava by pv. *poinsettiicola*, as reported by Sabet *et al.* (1969). The limited area of distribution of pv. *cassavae* in Africa, its heterogeneity compared to pv. *manihotis* and its absence in South America, the centre of origin of cassava, suggest that cassava is not the primary host of this pathovar. Some overlap in the host range of *X. campestris* strains from Euphorbiaceae has been noticed (Sabet *et al.*, 1969; Van den Mooter *et al.*, 1987).

CBN was first described in Malawi (Wiehe and Dowson, 1953), but a similar disease had already been noticed in Uganda (Hansford, 1937). Since then it has only been detected in other East African countries (Rwanda, Kenya, Tanzania, Burundi) at altitudes exceeding 800 m, possibly related to the lower optimum temperature of pv. *cassavae*, compared to pv. *manihotis*. An overlap in the distribution of the two pathovars has been noticed in Kenya (Onyango and Ramos, 1979) and Burundi. Despite its restricted geographical origin, pv. *cassavae* appears to be more heterogeneous than pv. *manihotis* (Van den Mooter *et al.*, 1987).

CBB is a major constraint to cassava cultivation and losses can be extremely severe after introduction of the pathogen, or possibly of more aggressive strains, in a region where highly susceptible cultivars are grown. In 1973, one year after the first report of CBB in Nigeria, estimated yield losses were 75% (Ezelio, 1977). Crop losses as great as 90–100% were observed in some parts of Uganda, 2 years after it was first recorded (Otim-Nape, 1980). In Zaire, the epidemics between 1971 and 1973 in the Kasai and Bandundu provinces led to severe starvation reflecting the importance of cassava roots and leaves as staple foods in this area (Maraite and Meyer, 1975). In field experiments with CBB-free and CBB-infected stem cuttings, Otim-Nape (1983) observed a reduction in fresh tuber yield from 40.1 to 26.6 t/ha. By comparing the yield of susceptible clones with those resistant under natural infection in Colombia during 1974–82, Umemura and Kawano (1983) observed an 18–92% yield reduction, depending on locality, planting time and degree of simultaneous infection by *Elsinoe brasiliensis*. Under the selection pressure of severe CBB epidemics, the most susceptible clones are eliminated from the mixture of clones generally planted by the smallholders and overall severity decreases. Based on a CIAT survey of five major cassava-producing zones in Colombia, the estimated reduction in national production was, nevertheless, still 6.64% in 1973 (Lourido, 1974). Resistance to CBB appears to be due to polygenes, mainly with additive effects but to some extent also with non-additive effects, and it appears to be recessive to susceptibility (Hahn, 1979; Umemura and Kawano, 1983).

Yield losses also depend on environmental conditions. In Africa, CBB is more prevalent in the savannah and the forest–savannah transition zones than in the forest zones, or even the forest galleries within the savannahs (Maraite and Meyer, 1975; Persley, 1980). This may be due to a lower survival rate of the bacteria in moist conditions (Persley, 1979b), but is also due to increased susceptibility of cassava on the leached savannah soil. Indeed, an increase in the potassium content of the leaves by potassium fertilization tends to reduce disease severity and enhances yield (Odurukwe and Arene, 1980). In South America, temperature is considered a major factor affecting CBB severity. In regions such as the Amazon, with average minimum and maximum temperatures above 20 and 30°C, respectively, CBB is not an important disease, despite the high rainfall (Takatsu *et al.*, 1979). Fluctuating day/night temperatures exceeding 10°C during the rainy season are also found to enhance disease severity (Lozano, 1986; Joseph and Elango, 1991), which may also explain the low incidence in forest zones with buffered temperature variations.

Yield losses induced by CBN are much lower than those caused by CBB, but no precise data are available. As for CBB, prevalence is highest on poor, gravely sandy, and phosphorus- and potassium-deficient soils. Fertilizers retard development of this disease (Butare and Banyangabose, 1982).

1.2.2 Disease symptoms

CBB is characterized by a unique diversity of symptoms. Leaves become infected through stomata or wounds in epidermal tissues (Lozano and Sequeira, 1974a), often after an asymptomatic epiphytic phase (Persley, 1978; Daniel and Boher, 1985). The first symptoms are visible after 3–7 days as dark green to blue, watersoaked, angular spots (1–4 mm in diameter), limited by veinlets and irregularly distributed on the lamina. With time they frequently extend and coalesce along the veins or the edges of the leaf, the central portion turns brown and the watersoaked part often becomes surrounded by a chlorotic halo. The lesions appear as translucent spots when viewed against the light. Under a magnifying glass, small droplets of exudate oozing from

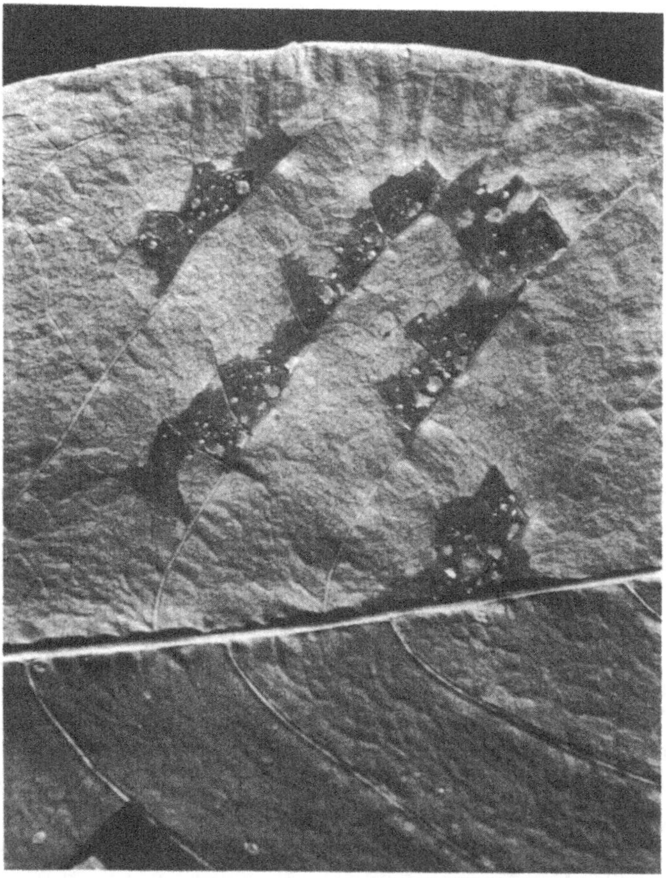

Figure 1.1 *X. c.* pv. *manihotis*: angular leaf spot symptoms on cassava under dry conditions.

the central portion of the lesion are visible on the lower surface of the leaves (Figure 1.1). The droplets, which first glisten creamy white and later yellow, are easily dissolved by rain or dew; on drying they form a thin scale. Under favourable conditions (young leaves, high soil and air humidities) pin-point watersoaked spots develop, scattered around young angular spots. The surrounding part of the lamina turns light brown and within 2–3 days, extensive areas of the leaflets become withered towards the tip or border of the leaflet, but also towards the base. The affected parts show light brown and green zonations as if they had been superficially burnt. These necrotic areas are not translucent, no bacterial exudate is observed and bacteria are absent or present in only very limited amounts at the border of the extending blight lesions.

The appearance and extent of the blight symptoms are associated with an increased leaf concentration of the blight-inducing toxin 3-(methylthio) propionic acid (Perreaux *et al.*, 1986) and other phytotoxic carboxylic acids, such as tiglic acid, phenylacetic acid and cyclopentanecarboxylic acid, produced by the bacteria (Ewbank, 1992). These toxins result from the induction of transamination and decarboxylation catabolism of methionine (Ewbank and Maraite, 1990) or other amino acids in the bacteria (Ewbank, 1992).

A severe attack leads to premature drying and shedding of the leaves. Under conditions of high humidity, infection may spread through the vascular bundles from the leaflets to the petiole and twigs or stems, with the formation of black and dark brown streaks as well as exudation drops along the pathway of progression. These parts may also become infected directly through wounds, for example due to removal of leaves for consumption or insect punctures. In Central and Western Africa, stem infections have often been found to originate from punctures by the bug *Pseudotheraptus devastans* Distant, an occasional vector of the bacteria (Maraite and Meyer, 1975). On the unlignified twig or stem, a dark green to black watersoaked area develops around the infection point. Large exudation drops appear some distance away from the infection point, in the axis of vascular bundles, and one or a small number of leaves, located on the same side, show a sudden loss of turgidity, followed by rapid wilting and shrivelling. Afterwards the base of the petiole collapses, but the dried leaves generally remain attached for some time. All leaves located above those showing the first symptoms wilt progressively. Finally the unlignified tip dies, appearing as a wick on the withered stem end, giving the 'candle' symptom, while new shoots grow out lower down on the stem. As the infection progresses towards the base of the stem these shoots often also become wilted leading to plant die back. In the infected shoots, xylem vessels are brownish. Under the microscope they appear obstructed by bacteria, tyloses and mucilaginous substances, including among others the bacterial extracellular polysaccharide (Ikotun, 1984).

Lytic pockets, due to bacterial multiplication in the intercellular spaces of parenchymatous tissues (Ikotun, 1978) and the possible expression of its ability to produce pectinases and cellulases (Starr and Nasuno, 1967; Maraite and Weyns, 1979; Dianese, 1985), develop around the protoxylem. The spread of these pockets causes rupture of the xylem ring, development of lytic pockets in the phloem, and later rupture of the fibre ring, in the cortical collenchyma. Externally the latter pockets become visible as dark green watersoaked spots and small black streaks, corresponding to altered laticifers. These spots swell, rupture and extrude a sticky white-yellow gum (Perreaux *et al.*, 1978).

In fully lignified stems or branches, only internal vascular browning is visible. Infection may spread more than 50 cm below any external visual symptom, but does not generally reach the roots, except in some very susceptible cultivars, whose swollen roots may show dry, rotted spots around the dead vascular strands (Lozano, 1986). On the green capsules, watersoaked expanding spots can also be observed. Heavily infected seeds from such fruit may be deformed with corrugation of the testa and necrotic areas on the cotyledons and endosperm.

Large variations in the predominance and severity of the various symptoms are observed according to the location, season, aggressiveness of the endemic bacterial strains and cassava cultivars. A variation in aggressiveness, but no clear-cut pathogenic specialization, is observed among *X. c.* pv. *manihotis* isolates from various countries, but also among those from a single country (Maraite *et al.*, 1981; Alves and Takatsu, 1984). Cultivar resistance is expressed for some cultivars by a reduced spread of the symptoms and occasionally by a hypersensitive reaction (Lozano and Sequeira, 1974b). Clearer differences are observed with regard to restriction of systemic invasion and tissue degradation in the stem, associated with the formation of a cork layer around the infected cortical or phloem tissues and increased primary fluorescence extending from the angles of the parenchyma cells surrounding the protoxylem to the pith (Lambotte and Perreaux, 1979).

On leaves, *X. c.* pv. *cassavae* induces watersoaked angular leaf spots, which often extend along the veins and are very similar to those induced by pv. *manihotis*, but without the formation of the small secondary spots evolving into blight areas. Larger necrotic areas only develop on the lamina when infection is located in the rib, or by coalescence of several angular spots. With ageing, the centre of the spots turns dark brown, covered with small yellow drops of exudate and surrounded by a narrow watersoaked line and a yellow halo.

Green stems become infected through wounds possibly caused by hail or wind-driven sand. After 2 weeks, pin-point dark green spots develop around the infection point. The lesion expands slowly, reaching 1–2 cm after one month and small yellow exudate droplets appear on the lesion. With the browning of the centre, cracks often develop lengthways in the bark, originating at the infection point. The spread and confluence of the lesion can lead to girdling of the shoot. Leaves located above this area may show yellowing, loss of turgidity during the day and finally permanent wilting. Otherwise, a cicatricial tissue develops around the lesion and secondary colonization by *Colletotrichum gloeosporioides* can be seen (Mostade and Butare, 1979). Sections through the darkened area reveal small lytic pockets, restricted to the cortex and surrounded by cicatricial tissue, limited browning of phloem laticifers and of cells surrounding xylem vessels, and increased primary fluorescence around the affected area. Systemic invasion of the xylem vessels is not observed (Maraite and Perreaux, 1979).

1.2.3 Isolation and disease diagnosis

The similarities of leaf symptoms in the early stages have raised the question of whether pv. *cassavae* is a yellow variant of pv. *manihotis* or an unrelated taxon. Because of the isolation in Brazil of a pv. *manihotis* strain with reduced virulence, inducing on the cultivar used for inoculation symptoms similar to pv. *cassavae*, Robbs *et al.* (1972) proposed considering pv. *cassavae* as a synonym of pv. *manihotis*. In several physiological, biochemical and pathogenicity tests, the strain used by Robbs *et al.* nevertheless behaved in a similar fashion to other strains of pv. *manihotis* but clearly unlike strains of pv. *cassavae* (Maraite and Perreaux, 1979; Maraite and Weyns, 1979). Numerical analysis of 267 phenotypic features, comparison of gel electrophoregrams of soluble proteins, mol%G+C determinations, DNA:DNA hybridizations, combined with virulence tests on 28 pv. *manihotis* and 14 pv. *cassavae* strains of various geographical origins, have demonstrated that pv. *manihotis* strains constitute a phenotypically and genetically homogeneous group of bacteria, clearly distinct from pv. *cassavae* (Van den Mooter *et al.*, 1987). Distinctive phenotypic features are: yellow pigment formation, growth on D-saccharic acid but not on DL-glyceric acid and the absence of Tween-60 hydrolysis for pv. *cassavae*, and the reverse for pv. *manihotis*. Pathovar *cassavae* strains also show slower growth on nutrient agar, a high thiamine requirement, more rapid formation of acid from maltose and a lower optimum and maximum temperature, compared to pv. *manihotis*. The optimum temperature for disease development is around 25°C for pv. *cassavae* and 30°C for pv. *manihotis* (Maraite and Perreaux, 1979). The pathovars *cassavae* and *manihotis* are not related serologically (Elango *et al.*, 1981; Ikotun, 1981a).

Xanthomonadin-producing strains of X. *campestris* have also been isolated at low altitude from cassava in Colombia. They were found to be serologically similar to pv. *cassavae* (Elango *et al.*, 1981), but electrophoretic profiles of soluble proteins (Kimura and Dianese, 1983) or of peptides from the cell envelopes (Santos and Dianese, 1985) have suggested that the Colombian strains are closer to pv. *manihotis* than to pv. *cassavae*. Van den Mooter *et al.* (1987) demonstrated that these strains are electrophoretically and genetically different from pv. *manihotis* and pv. *cassavae*, but similar to *X. campestris* pv. *poinsettiicola*, although being unable to infect *Euphorbia pulcherrima*, the natural host of this pathovar. A deviant pv. *cassavae* strain has recently been isolated from leaf spots on cassava in Niger (Janse and Defrancq, 1988).

1.2.4 Sources of inoculum and disease management

The distribution of infected vegetative planting material from diseased plantations has been the main means of disseminating pv. *manihotis* over long distances into Africa and Asia. It is also the source of primary infections in newly established plantations. This is due mainly to the lack of visible symptoms on lignified stems and to the ability of the pathogen to survive in the invaded tissue for a very long time (Lozano and Nolt, 1989). Production of pathogen-free planting material is based on visual inspection before and during harvesting of the cuttings in the field, as well as during multiplication by rooting shoot tips in greenhouses or by meristem cultures *in vitro*.

The possibility of transmission of pv. *manihotis* by botanical seed has been domonstrated by Persley (1979a) by a leaf infiltration technique, using as inoculum the supernatant from seeds soaked in sterile water at 30°C for 2–4 h. The threshold of sensitivity of this technique was 10^5 cells/ml. This very simple technique was also useful for the study of epiphytic survival on symptomless plants (Persley, 1978) or on debris (Persley, 1979b). By immunofluorescence (IF) and ELISA on enriched embryonic extracts, Elango and Lozano (1980) confirmed seed transmission, with embryo infection varying from 0 to 40% depending on the sampling period. Clear positive reactions were obtained with 10^5 cells/ml by IF and with 10^4 cells/ml by ELISA.

With leaf impression and direct leaf disc methods associated with IF, Daniel and Boher (1981) were able to study the spatial distribution of pv. *manihotis* on the leaves. Enrichment in nutrient broth, followed by infiltration of the enriched culture into detached cassava leaves, or by an Ouchterlony double diffusion test, enabled detection of very low populations of pv. *manihotis* in flood water (Joseph and Elango, 1991).

By using antiserum against the membrane protein complex of pv. *manihotis* in indirect IF staining, Leksomboon *et al.* (1989) were able to detect a minimum of 10^2 cells/cm^2 of leaf and 10^3 cfu/g of soil. These authors also developed a semi-selective medium showing 90–100% inhibition of soil saprophytes and having the same sensitivity threshold as the IF assay.

Ikotun (1981b) concluded from studies in Nigeria and Colombia that pv. *manihotis* does not survive naturally on alternate hosts in and around cassava plants and observed a hypersensitive reaction by inoculation of *Euphorbia repanda*. Marcano and Trujillo (1984), however, demonstrated the survival of a rifamycin-resistant isolate of pv. *manihotis* under high relative humidity, heavy rainfall and thick cloud cover but not under dry conditions on several plant species, and identified *Amaranthus* sp., *Panicum fasciculatum*, *Sida* sp., *Sorghum halepense* and several species belonging to the Euphorbiaceae as alternate hosts in Venezuela (Marcano and Trujillo, 1982).

CBB was first described in 1912 in Brazil (Bondar, 1912), and was suddenly found during the 1970s in most of the cassava-growing areas of Central and South America, the Caribbean, Africa and Asia (Bradbury, 1986). Some of the recent reports are due to increased attention being paid to cassava as a major staple food crop. Indeed, detailed field surveys in several countries have revealed widespread occurrence soon after it was first detected, indicating that it had been established long before the epidemic development was noticed (Maraite and Meyer, 1975; Persley, 1979a). In some areas, its recent introduction and rapid spread have been demonstrated (Kwaje, 1982; Joseph and Elango, 1991), justifying strict quarantine measures on exchange of germplasm. No relation between differences in aggressiveness and geographical origin have been demonstrated. Isolates of pv. *manihotis* from Africa could be differentiated from those of other origins by bacteriophage typing (Persley, 1980), as well as by their low amylase activity contrasting with the variable amylase production observed for strains from South America (Maraite *et al.*, 1981; Alves and Takatsu, 1984). This suggests that the epidemic development in Africa originated from a limited number of introductions.

1.3 *XANTHOMONAS CAMPESTRIS* PV. *MALVACEARUM*: CAUSE OF BACTERIAL BLIGHT OF COTTON

A. ZOMORODIAN and K. RUDOLPH
Institut für Pflanzenpathologie und Pflanzenschutz, Universität Göttingen

1.3.1 Introduction

Since the initial description of leaf symptoms by Atkinson (1891), bacterial blight has been reported in most cotton-producing areas of the world (Arnold and Brown, 1968; Cia *et al.*, 1973; Nayudu, 1973; Verma and Singh, 1975; Abdel-Fatakh and Safiyazov, 1976; Van Heerden, 1976; Zurini *et al.*, 1981; Zachowski and Rudolph, 1988; Zachowski *et al.*, 1989a; Allen and West, 1991). As a seed-borne bacterial disease and potentially very destructive it is particularly troublesome in Africa (Follin *et al.*, 1988; Poswal, 1988), although in Egypt it occurs too late in the season to cause crop loss.

It has recently become one of the more serious diseases of cotton in India and Pakistan (Hussain and Brinkerhoff, 1978; Verma *et al.*, 1980). This was primarily due to the increased substitution of the mostly resistant indigenous, diploid cottons, *G. arboreum* L. and *G. herbaceum* by the tetraploid and highly susceptible Upland and Egyptian cotton, *G. hirsutum* and *G. barbadense*. General reviews on bacterial blight have been published by Wickens (1953), Lagière (1960), Stapp (1966), El-Nur (1970), Brinkerhoff (1970), Brown (1976), Ebbels (1976), Innes (1983) and Verma (1986).

Objective estimates of the yield loss caused by *X. campestris* pv. *malvacearum* are difficult to obtain as many factors are involved. These include the stage at which the plant is infected, environmental conditions prevailing at the time of, and subsequent to, infection and the capacity of the plant to 'grow away' from the disease.

The disease is potentially very destructive in areas where wind-driven rain or sprinkler irrigation disseminate the pathogen (Brinkerhoff, 1963, 1970; Innes, 1983). Yield losses vary from 1%, when leaves are infected, to as much as 50% in epidemics where extensive leaf, stem, and boll infection occur. The disease is particularly severe in Africa, Asia, Australia, and Central and South America with estimated yield losses of 10–30% (Verma, 1986; Zachowski and Rudolph, 1988).

In the USA, bacterial blight caused an average yield loss of 1.0% annually from 1952 to 1984, with a range of 0.1–2.3% (El-Zik, 1986). In 1959, in the USA bacterial blight was estimated to have caused a loss of over 419 000 bales (Smith, 1960). Simpson (1956) has stated that in epiphytotic form losses of up to 50% occur and in one epidemic on 40 000 acres in the Pecos Valley of New Mexico Leyendecker (1950) estimated a reduction in yield of 35–50%. In 1976, the Cotton Disease Council of the United States estimated a loss of 73 000 bales as a result of the foliar and stem phases of bacterial blight (Bird *et al.*, 1981).

Very destructive outbreaks led to the ruination of the famous irrigation scheme in the Sudan Gezira in the early 1930s (Gaitskell, 1959). El-Nur (1970) has described results from trials in Sudan in which yield losses of the order of 20% were commonplace in irrigated varieties of cotton (*G. barbadense* L.); in one trial severe infection led to a 77% reduction in yield in partially resistant varieties. Knight (1949) recorded that in Sudan the resistant counterpart of a susceptible

Upland variety gave 64% more lint than the control, and attributed this difference almost entirely to the effect of bacterial blight.

In Australia where bacterial blight is present in most cotton-growing areas, the use of planting seed infested with *X. c.* pv. *malvacearum* was found to be a major factor contributing to severe epidemics of the disease. Inspection of the commercial cotton crops early in the 1984–85 season indicated that 3.3% of planting seed had been infested with the bacterial blight pathogen (Allen, 1986; Allen and West, 1987). The average incidence of bacterial blight on bolls in pure seed crops growing in Australia's major seed production area during the 1985–86 season was found to be 30.4% (Allen and West, 1987).

1.3.2 Disease symptoms

Xanthomonas campestris pv. *malvacearum*, the incitant of bacterial blight of cotton, can affect all aboveground parts of the plant through the growing season. The disease has thus acquired different names descriptive of the characteristic symptoms produced, such as seedling blight, angular leaf spot, leaf vein blight, black arm (on stems, monopodia and sympodia), bract spot and bacterial bollrot.

Infection arising from the seed results in seedlings with dull green flaccid areas extending from the periphery of the cotyledons (Wickens, 1956).

Black arm, the name by which the disease was originally known, refers to infection of stems and sympodial or monopodial branches. Infection may spread from cotyledons and leaves down to parenchymatous tissues of the petiole into the cortex of the stem, or may come from entry through wounded tissues. Lesions on hypocotyls are black, elongate cankers, which often girdle and kill the seedlings.

Angular leaf spot was the term used to describe the watersoaked lesions on cotton leaves (Atkinson, 1892). In infection of leaves, which the bacteria enter through the stomata, the disease manifests itself as watersoaked lesions on the lamina. These lesions increase in size in a susceptible variety up to 6–8 mm in diameter, are often bounded by veins and veinlets, thus angular in form (Figure 1.2), being at first translucent but drying later to a sooty black in Egyptian (*Gossypium barbadense* L.) types or a brownish-black in Upland cotton (*G. hirsutum* L). In severe cases the watersoaked lesions may follow the principal veins of the leaf blade resulting in the death of the plant's growing point or in the destruction of the terminal bud. Leaf shedding may occur under conditions favourable for disease development.

The watersoaked symptom is caused by bacterial slime containing extracellular polysaccharides which are hydrophilic and fill the intercellular spaces of the plant tissue (Rudolph *et al.*, 1989). Histological

Figure 1.2 *X. c.* pv. *malvacearum*: angular watersoaked spots on field-grown cotton.

studies have shown that exopolysaccharide production *in planta* and induction of watersoaking is light-dependent (Zachowski *et al.*, 1989b).

Bacterial bollrot (boll blight) occurs directly by extension of infection of the calyx, bracts or receptacle, or through the vascular system from the infection on a fruiting branch. Lesions on cotyledons and bolls are irregular, oval to round. On the bolls, these appear watersoaked at first (Figure 1.3), sunken and brown to black with age and, although they may dry later, they provide a point of entry for saprophytic organisms. Pathogen growth inside infected cotton bolls produces a yellow slime, where it may stain the lint and carpel wall. It also damages seed cotton and provides a means whereby the pathogen is carried over from season to season.

1.3.3 Isolation and disease diagnosis

The most common detection methods are based on the direct isolation of the pathogen from infected plant parts (i.e. hypocotyls, cotyledons, true leaves, bolls and seeds). In leaves and bolls the bacteria can

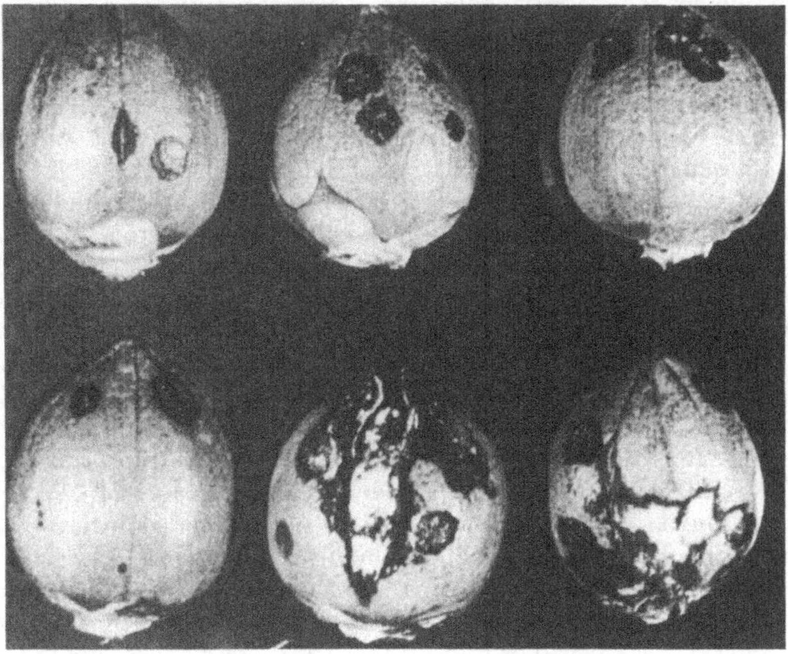

Figure 1.3 *X. c.* pv. *malvacearum*: several stages of boll blight.

be isolated easily from young watersoaked spots. Homogenates of infected plant parts in boiled tap-water are streaked on a starch-containing medium (10 g soluble starch; 1 g peptone; 5 g NaCl; 0.5 g KH_2PO_4; 20 g agar; and 1000 ml of distilled water; pH adjusted to 7.0) and incubated at 27°C. Single colonies can be isolated after 3–5 days (Zachowski and Rudolph, 1988).

The primary inoculum source of *X. c.* pv. *malvacearum* is infested seed. Therefore, even very low rates of seed infestation should be detected. To detect the bacteria in seeds the seed-soaking method is useful. Large volumes of seeds are soaked in boiled tap-water at room temperature for up to 24 h, so that the bacteria can diffuse out of the few infested seeds present. The concomitant infection of healthy seeds by this method improves the sensitivity of the test. All the soaked seeds are planted in sterile soil and observed for disease symptoms during germination. The disease can be detected on hypocotyls as watersoaked spots within 14 days after sowing (Brinkerhoff and Hunter, 1963; Halfon-Meiri and Volkani, 1977). We improved this method by shaking 100 seeds together with 50 g of sterile sand in 50 ml of sterile tap-water for 2 days at 25°C. The seeds were then germinated in sand and developed typical watersoaked spots on the cotyledons, from which the bacteria could be isolated easily (Zachowski and Rudolph, 1988).

Strains of X. c. pv. *malvacearum* show a high variability in virulence, which has not been reported for other *Xanthomonas* pathovars to the same extent. Brinkerhoff (1963) identified 12 races of X. c. pv. *malvacearum* on the basis of disease reactions on a set of differential lines of cotton. Subsequently, using an extended set of differentials, Hunter *et al.* (1968) identified a further three races. Verma and Singh (1970) described two more races, and Hussain and Brinkerhoff (1978) reported an additional one, race 18. Ruano and Mohan (1982) isolated a new strain of X. c. pv. *malvacearum* and proposed race 19. Recently, the so-called HV (highly virulent) strains have been isolated in Africa, which infect the long-established immune cotton cultivar '101–102B' (Bird, 1986; Follin *et al.*, 1988). Studies are underway to add differentials and assign a race number to the HV strains (Bird, 1986).

The occurrence of different virulent strains or races of X. c. pv. *malvacearum* is now well established in nearly all countries where bacterial blight is a problem. The highly virulent race 18 seems to predominate in Central America (Zachowski, 1989), but has not yet reached cotton-growing areas in the Near East, such as Turkey (Zachowski *et al.*, 1989a) or Iran. In these areas race 7 was shown to be predominant. Some bacterial strains show attenuated virulence after long-term storage on slants, for example race 7 changed into race 6 (Zachowski *et al.*, 1989a).

The mechanisms of pathovar specificity have not been elucidated. A phytotoxin of X. c. pv. *malvacearum* possessing an aromatic ring was reported by Pandian *et al.* (1982) to cause typical blight symptoms. Quantity and quality of exopolysaccharides may partially explain differences in virulence of races of X. c. pv. *malvacearum* (Zachowski, 1989).

1.4 *XANTHOMONAS ORYZAE* PATHOVARS ON RICE: CAUSE OF BACTERIAL BLIGHT AND BACTERIAL LEAF STREAK

T.W. MEW
International Rice Research Institute, Los Banos

1.4.1 Introduction

There are two bacterial diseases of rice with which xanthomonads are associated: bacterial blight (BB) caused by *X. oryzae* pv. *oryzae* (Xoo), and bacterial leaf streak (BLS) caused by *X. oryzae* pv. *oryzicola* (Xocola). For many years they were thought to occur only in Asia until Aldrick *et al.* (1973), Lozano (1977) and several others (Reckhaus, 1983; Mew, 1989) reported that BB was present in Australia, Latin America and

Africa, respectively. BB has also been observed in Texas and Louisiana, USA (Jones *et al.*, 1989). With these reports of occurrence, BB of rice has become a disease of global distribution, though of economic importance only in Asia. BLS may also be found outside Asia; however, confirmed reports are few.

BLS was reported to be present in the humid tropics but has not been found to occur in temperate regions such as Japan, Korea and North China. In South China, it is considered one of four major diseases of rice (Xie *et al.*, 1991b). It is common in lowland as well as upland rice during the rainy season, the bacteria infecting the rice plant at any growth stage of the crop.

The pathogen of BB, Xoo, attacks *Oryza sativa* and many related wild species of rice, such as *O. rufipogon*, *O. australiensis*, *O. nivara* and *O. glaberrima* (Mizukami and Wakimoto, 1969; Ou, 1985b; Mew, 1987). In addition, weeds also serve as alternate hosts. These weeds include *Leersia oryzoides* var. *japonica*, *L. oryzoides* and *Zizania latifolia*. There are other weed species (Ou, 1985) which are reported to be hosts of Xoo. However, no natural infection has been shown; only the results of artificial inoculation are available. Since there is no detailed description of the lesion and lesion development so far, one cannot make a comparison of the lesions on artificially inoculated weeds and lesions which appear on rice. Without such comparison, it is difficult to decide whether these weed species should be considered hosts of Xoo.

BB has become a major rice disease in tropical Asia in the last two decades because of the extensive use of nitrogen-responsive modern cultivars. BB usually occurs in fields which are prone to flood. The topography of the fields determines whether they are conducive to the disease. For instance, fields in mountain basins, valleys, or fields along the rivers, and lakes as well as fields of poor drainage, with stagnant water, have all been reported to be very conducive to the disease. Needless to say, BB is a post-typhoon disease. Reports of its occurrence and distribution in Asia are mostly in the rainy season and in typhoon-path countries such as China, Korea, Japan and the Philippines, and Indochina, India, Bangladesh and Burma. Because the monsoon season brings frequent rainstorms in countries like Indonesia, Malaysia and Thailand, BB also occurs more often.

Several epidemics in the region have occurred and caused severe damage to rice crops as well as yield losses (Mew, 1987). In 1978–79, in the Punjab and Haryana area, thousands of hectares were lost to BB due to a serious occurrence of kresek on a susceptible cultivar, Jaya, which was planted to large acreages in successive and intensive cropping. When hybrid rice was first introduced to the farmers in China, hundreds and thousands of hectares were infected with Xoo, and BB became widespread, because of lack of resistance to BB in the hybrids and also due to change of cultural practices. In the Philippines where the author has been working on this disease for the last 15 years, BB occurs every year in Central Luzon, considered the rice granary of the country, and other major rice-growing districts of the archipelago. This is because of a new race virulent to the cultivars planted since the mid-1970s.

Infection by BB causes reduction of total dry matter, grain weight of paddy rice, increase in the number of sterile grains and results in poor maturation and broken grains in milling. Generally, the extent of damage depends on the severity of the disease and the stage of the rice crop growth when infection took place. Mizukami and Wakimoto (1969) estimated that yield losses in Japan reached 20–30% in rice fields with moderate level of disease, while in severely infected fields yield loss was over 30%, even though this occurred in Japan only occasionally.

In the tropics, damage is more severe than in the temperate regions. This is because the occurrence of BB can take place at a very early stage of rice plant growth. However, most of the yield loss assessment was based on experiments done in controlled conditions. Ou (1985) reported a loss of 50% on very susceptible cultivars, 10% on moderately susceptible cultivars and the loss in resistant cultivars was insignificant under field conditions. In field experiments, Reddy *et al.* (1979) assessed infection initiated at flowering stage and showed that epidemics which began before panicle initiation significantly reduced grain yield, panicle fertility, and kernel weight. High nitrogen levels increased disease and reduced yield in BB-susceptible cultivars. BB-resistant cultivars were not affected by the application of high levels of nitrogen. Yield losses may be assessed by a field plot technique. Plot arrangement using an experimental design such as a randomized complete block is common in crop-loss work (Teng *et al.*, 1989). The plots are either paired-treatment or multiple-treatment, where treatments are the desired levels of disease or pathogen population. An important consideration in plot techniques is the plot size. Rice disease-loss experiments vary from 10 row × 6 m to 25 m × 25 m (Teng *et al.*, 1989). In BB, yield loss assessment was also performed using the 'single tiller/plant method'. In this method, hundreds of tillers are tagged, with care being made to select tillers which reflect wide ranges of disease severity, including zero and maximum disease. Fields are visited using a predefined survey procedure, and in each field, disease intensity is assessed and tillers are harvested. Each tiller represents a single point for regression analysis.

Economically BLS is not as important as BB. It appears that the rice plant compensates rapidly for the infection with new growth. During the rainy season this is especially true in conditions when long dry periods occur between rainstorms. Secondary infection ceases to spread without continuous rain. In severe cases of infection, however, yield losses may be in the range of a few percent to as much as 15 or 20% (Ou, 1985). Other features of the disease and transmission of the pathogen are very similar to those for BB. No detailed studies have compared ecological and epidemiological difference between the two diseases.

1.4.2 Disease symptoms

BB and BLS can be clearly distinguished by symptoms (Figure 1.4) reflecting differences in mode of infection. Xoo enters through the hydathodes, multiplies in the epitheme and moves to the xylem vessels

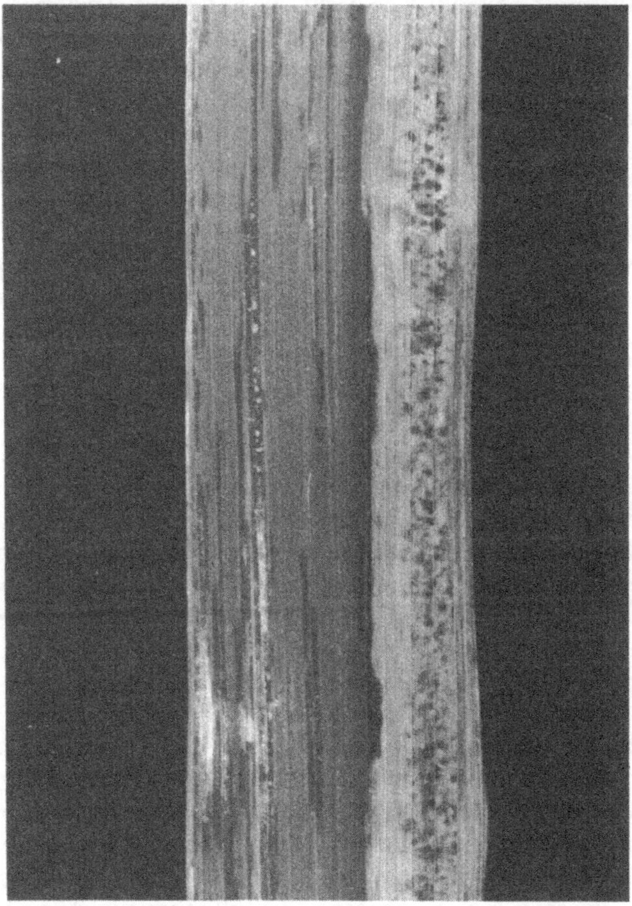

Figure 1.4 *X. oryzae* pvs *oryzae* (BB) and *oryzicola* (BLS): rice leaf infected by BB and BLS organisms.

resulting later in blight on the leaves. On the other hand, Xocola enters through the stomata and mutiplies in the parenchyma tissues of the leaves and causes linear leaf streak on leaves that coalesce to mimic BB. The transparent streaks differentiate leaf streak lesions from those of leaf blight which are opaque against the light. In extreme cases, Xocola is found in vascular tissues but apparently does not cause any visible symptoms.

BB is a vascular disease on rice, thus infection is systemic. Two symptoms of the disease are known: kresek and leaf blight. Although 'pale yellow' is considered by some scientists (Ou, 1985) as part of the syndrome of BB, in reality it is only a secondary effect as a result of leaf blight or seedling wilt (kresek; Mew, 1987). New leaves of kresek-infected plants turn pale yellow. Plants become wilted if infection occurs during the seedling or early tillering stage. Lesions of leaf blight enlarge in length and width and may have wavy margins at either one or both leaf edges. The lesion turns gradually from greyish-green to become straw coloured. At a later stage of lesion development, it usually turns chlorotic with clear margins.

In the tropics, the disease may appear at first as watersoaked stripes. Droplets of bacterial ooze can be observed early in the morning at the edges of leaf blades with no visible lesions. This is seen during the rainy season at or before the maximum tillering stage. In general, BB becomes noticeable at the heading stage both in temperate and tropical regions, although in the tropics it can also be observed at an earlier stage, or sometimes in the dry season if there are frequent showers. In a given field, however, BB usually occurs at the edge of a field in shady corners. Occasionally, its occurrence in a field appears to show several foci. Whether this is due to the initial infection originating from seedborne inoculum remains a matter of speculation.

On *Leersia* plants, watersoaked and discoloured lesions appear first at the edges of leaf tips. They then extend downwards along the vein, increase in size, and often form wedge-shaped streaks or large distinct lesions along the leaf edges. Margins of the lesions are orange-red, and the central parts of the lesions turn greyish-brown. The lesions sometimes enlarge on leaf sheaths as in the case of rice plants. On *Zizania latifolia*, lesions are initiated at the edges of leaf tips. They gradually enlarge in length and width and are bordered by a yellow wavy margin. Yellow spots sometimes appear along the leaf veins adjacent to main lesions and rapidly increase in size.

Earlier reports indicating confusion of BLS with BB may be attributed to the fact that both bacterial pathogens can infect the same rice leaves and the diseases can be observed simultaneously. The symptoms of BLS are distinct from those of BB (Figure 1.5). When a rice leaf is infected with Xocola, numerous tiny drops of bacterial exudate are

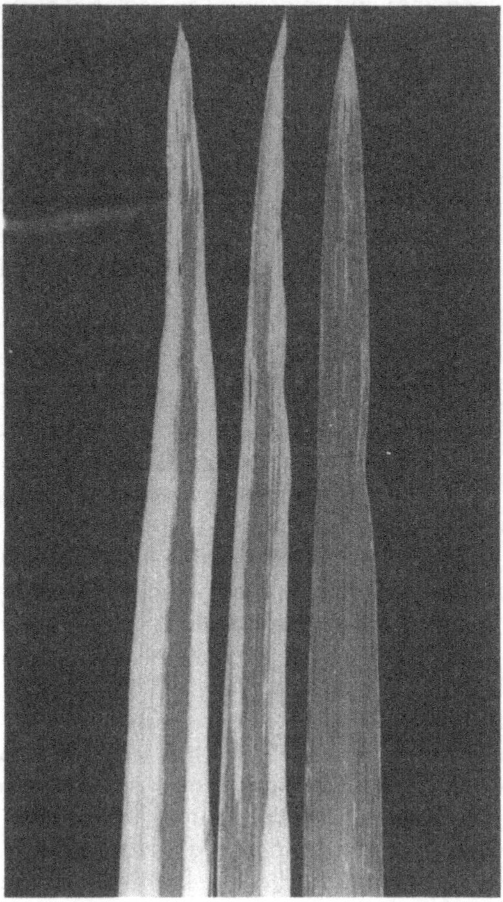

Figure 1.5 *X. oryzae* pv. *oryzicola* (BLS): symptoms on rice leaves at different stages.

always present on the surface of the lesions. However, when the streaks coalesce to cover the entire leaf which becomes greyish-white and then dries, the symptoms become indistinguishable from leaf blight. Xie *et al.* (1991b) compared the incidence of disease caused by *X. oryzae* pv. *oryzae* and *X. oryzae* pv. *oryzicola* using different inoculation methods, by spraying or by smearing a bacterial suspension with a brush onto the lower surface of rice leaves, and then leaving the plants either still or shaken every other day. In addition, other methods were used, such as introduction of the bacterial suspension to the guttation water drop in the evening with a fine needle infested with the bacterial suspension; clipping the leaf tips with a pair of scissors infested with the bacterial suspension; and injecting the bacterial suspension into the midrib of the leaves with a hypodermic needle. The results suggested that smearing the lower leaf surface produced 90% infection with Xocola, but only 10% with Xoo as compared to guttation water-drop infestation which produced 90% infection with Xoo but 2% with Xocola. Although both bacterial pathogens can be observed in xylem vessels or parenchymatous tissues, it is the pathogen that establishes and multiplies to a large degree that causes the respective diseases. There is an infection court specificity as suggested by Mew and Vera Cruz (1979).

1.4.3 Isolation and disease diagnosis

In general it is very difficult to isolate Xoo from dried diseased leaves because of the low cell numbers in the tissues. It has been observed that when the diseased leaves are in contact with rain-water, large numbers of bacteria stream out from the wounds/injury as bacterial ooze. The method of Wang *et al.* (1964) is to isolate Xoo from dried diseased leaves. Dried leaves are cut into 4 cm pieces and inserted vertically into washed sand placed in Petri dishes and moistened with distilled water. The Petri dishes are placed in moist chambers to provide 100% relative humidity at 25–28°C for 12–24 h. It is expected that mucoid substance will ooze out from the cut end. Using a sterile needle, some of the ooze is picked up, and inoculated to healthy leaves of a susceptible rice plant. Experience has shown that each leaf could accommodate 10–15 pricks applied gently to the leaf. If the ooze is Xoo, one would expect the lesion to develop within a week's time. Reisolation from these newly infected leaves is advisable for culture preservation. This could also be used for phage and serological tests and for reconfirmation of pathogenicity.

The pathogens causing BB and BLS of rice were previously regarded as pathovars of *Xanthomonas campestris* (Mew, 1987). However, recent studies have shown that *X. c.* pv. *oryzae* was differentiated into a phenon distinct from all *Xanthomonas* species based on 295 phenotypic properties (Vera Cruz *et al.*, 1984; Van den Mooter *et al.*, 1987). The fatty acid profiles of *X. c.* pv. *oryzae* and *X. c.* pv. *oryzicola* were similar, but there were differences in the presence of 12:0 iso 3OH, and 15:0 anteiso fatty acids. Both profiles were very different from those of all other *X. campestris* pathovars investigated (Stead, 1989). Similarly, the comparison of SDS–PAGE protein patterns together with other studies (Kersters *et al.*, 1989) allowed separation of the two pathovars from all others. Swings *et al.* (1990) concluded that both pathogens 'clearly belong to the genus *Xanthomonas* but cannot be assigned to the named species of *Xanthomonas*.' They therefore proposed to transfer both pathogens to a new species with two pathovars, namely, *X. oryzae* pv. *oryzae* and *X. oryzae* pv. *oryzicola*. In the absence of the host, several differentiating features are available to distinguish the two *Xanthomonas* bacterial pathogens of rice. These include a few phenotypic features: polyacrylamide gel electrophoresis protein fingerprints, fatty acid profiles, pathovar-specific monoclonal antibodies, and phage typing. The BB pathogen is susceptible to bacteriophages (Mizukami and Wakimoto, 1969). By using these phage strains, Xoo can be classified into five groups. In general, the phage strains are stable and specific to each bacterial group. However, sensitivity of the bacterial lysotypes may be changed after repeated phage passages through the host bacterial strains. The bacteriophages of Xoo in the tropics may be quite different from those in the temperate regions (Ou, 1985).

X. oryzae pv. *oryzae* varies not only in sensitivity to bacteriophages but also in virulence. Testing bacterial isolates on sets of differential rice cultivars, which differ in genes for resistance to the bacterial pathogen, has led to identification of races in the bacterial population (Mew, 1987). In the Philippines, six races have been identified based on six rice differential cultivars. When the differential system becomes more defined, it is expected that more races will also be identified. For the past 20 years, four (races 1, 2, 3 and 5) out of the six have often been detected. Besides race 2 which has a narrow spectrum of virulence, the other four races appear to have distinct geographical distribution (Mew *et al.*, 1992). Race 1 continues to be detected in lowlands where traditional cultivars are planted or where cultivars with the *Xa*-4 gene have not been cultivated extensively. Race 3, which was once predominant in the southern Philippines, is now present in low numbers, similarly with race 4 which was found only in Palawan. Race 5 exists as a highland race, and has dominated the bacterial population in Banaue, the mountain terraces of which are more than 5000 feet above sea level.

There was an apparent shift in frequency of virulence of the bacterial population since wide cultivation of modern cultivars with the *Xa*-4 resistance gene. Since its introduction before the mid-1970s, the majority (82%) of the isolates collected were virulent to traditional cultivars whose resistance to BB was unknown. Cultivars like TKM-6 and Sigadis carrying the *Xa*-4 gene were resistant. Gradually, the frequency of race 2 virulent to the R-gene *Xa*-4 increased with time as these modern cultivars were continually grown over about 90% of the rice area. This group dominated the bacterial population in areas planted with modern cultivars (Mew *et al.*, 1992). Our knowledge of the extent and distribution of pathogenic variation is improved by recent advances in molecular biology. DNA analysis has been used to measure the genetic variation in Xoo. The DNA types of Xoo were determined using *Pst* I analysis and RFLP analysis using the probe pJEL101, with bacterial isolates collected from single fields planted to rice cultivars with different resistance (Leach *et al.*, 1992). Race 1 isolates appear to be dominant in fields where rice cultivars possessing no *Xa*-4 gene for resistance were planted. In contrast, race 2 became prevalent in fields planted to rice cultivars possessing the *Xa*-4 gene.

1.4.4 Sources of inoculum and disease management

The questions that have been repeatedly asked are: 'Where does Xoo overwinter/summer, and how does it spread to cause new infections and develop new disease in the field?' This has become a key issue in the control of the disease. In China, many investigations and observations have shown that the bacterial pathogen can be detected and isolated from seed. Chinese scientists have further proved that the bacteria can survive in rice seed. Seedlings were shown to be infected if derived from seeds harvested from a diseased crop. Many Japanese scientists believe that weed hosts such as *Leersia oryzoides* var. *japonica*, *L. oryzoides* and *Zizania aquatica* are major sources of inoculum for the spread and overwintering of Xoo, from which rice leaves can be infected. It is clear that rice straw and diseased leaves can also serve as sources of inoculum. For disease control measures, it is therefore essential to manage crop residues by preventing the seedlings from coming into contact with rice straw. It is advisable to remove as much straw as possible left from previous crops, especially if it was diseased. Accumulation of rice straw near the source of irrigation water or ditches of secondary or tertiary irrigation canals should also be avoided.

In Japan, bacteria easily overwinter and survive until spring in dried rice straw kept on farms. Usually, the bacteria are eradicated in 1 or 2 months when the straw is scattered in the field and ploughed into the soil. However, straw is often piled up outdoors, protected from rain and the pathogen in straw which has been protected from rain survives until the next spring.

Rice stubble is another source of inoculum. After harvesting the grain, most of the stubble in the tropics survives until the next season. The bacteria can easily survive in this stubble. In temperate regions where the winter is freezing cold, the stubble withers fast and provides little protection for the bacteria even if the stubble is left in open fields. In the tropics, volunteer rice, or ratoon rice regenerated from infected stubble are other possible sources of inoculum for the next season or crop.

Plant pathogenic bacteria do not possess many recognized survival structures. Saprophytic survival ability is in close association with the host. Seed provides a logical venue for Xoo to survive the period between crops and for short as well as long distance dissemination of the disease. However, the importance of seedborne inoculum in the disease cycle and in disease development is difficult to assess. A vast volume of literature confirms that Xoo can be detected in rice seed. Whether seed is the only venue for the pathogen to complete its life cycle, as well as the disease cycle, is a matter that needs to be examined in the context of rice ecosystems and climatic conditions where rice is cultivated. In temperate environments where only one rice crop is grown, and both the temperature and the relative humidity decline rapidly at the time of crop maturity, the chance of Xoo survival in rice seed also decreases. In the tropics the reverse is true; although seed may be a venue for survival, in reality there are rice plants growing in the fields all year around, thus the importance of seedborne inoculum may not be significant. Another important requirement to establish the seedborne nature of the rice BB pathogen, is a well-defined method to isolate the bacteria from rice seed. Direct isolation using common laboratory media may not be feasible due to the slow-growing nature of Xoo. Indirect detection with bacteriophage or other methods has several limitations. The third problem in relation to the detection of Xoo is that the literature often does not mention the condition of the seed source, i.e. the infection level of the mother plants, the stage of crop growth when the infection took place, the length of the period between plant infection and seed harvest, and the status of seed being used for such detection; for example, the conditions (temperature and seed moisture levels) of seed storage and duration of storage with reference to seed harvest, are often not given in the studies published in the literature. It is therefore difficult to repeat the experiments. Later, when the phage technique was applied to detect Xoo in rice seed harvested from infected fields, the results were positive, thus proving that Xoo is seedborne. Because there are different views on the source of primary infection, the control measures also use different approaches. Reports from China indicate that seed treatment has resulted in both reduction of leaf infection and the number of focal centres. However, there were many cases which indicate that seed treatment did not produce the expected control of this disease. Positive evidence of the seedborne nature of Xoo was provided by Fang *et al.* (1956), who observed that the bacteria were present not only in the vascular systems of glumes but also occasionally in the endosperm. These authors considered seed as the primary source of inoculum for new infection. Fang and Hsu (1978) reported the epidemics of BB throughout China except in Qinhai, Gansu, Xinjiang and Inner Mongolia during the 1960s when seed transport was very intensive. There were other scientists who used the bacteriophage method to show that the bacterial pathogen survived in seed (Ou, 1985). The incidence of rice seeds carrying the bacterial inoculum appears to be very high, ranging from 3 to 100%. The critical problem related to seedborne Xoo is that much of the data and methods reported earlier by one scientist in one location could not be repeated by another in a different location. The duration of survival and the consequent transmission of the bacterial pathogen to cause new infection also varies. Reports on isolation of Xoo from seed often lack bacteriological tests or confirmation of pathogenicity by re-isolation from infected plants. The lesions caused by Xoo at the seedling stage due to direct transmission of the bacterial cells from seed are not well defined.

Unlike Xoo, the seedborne nature of Xocola has never been a controversial issue. Xie *et al.* (1991a) tested 230 seed lots from 11 counties of Zhejiang province using an immunoradiometric assay. The results showed that the percentage of positive samples, an indication that the seed carried the bacterial pathogen, was significantly correlated with samples coming from diseased fields, and the 19 seed lots from disease-free fields were all negative. They reported that Xocola could survive in the glume for up to one year. Quantitatively, the number decreases over time of storage after harvesting the seeds. Immature seed appears to harbour more bacteria than fully mature and healthy seeds.

1.5 *XANTHOMONAS CAMPESTRIS* PV. *PHASEOLI*: CAUSE OF COMMON BACTERIAL BLIGHT OF BEAN

A.K. VIDAVER
Department of Plant Pathology, University of Nebraska

1.5.1 Introduction

Xanthomonas campestris pv. *phaseoli* (Smith) Dye, causal agent of common bacterial blight (CB) of bean, was first described in 1897 and has undergone a number of taxonomic revisions over the years (Bradbury, 1986). Whether or not it warrants separate species status is a matter of debate (Young *et al.*, 1991b). While much is known about its biology, epidemiology, host range and other properties (Zaumeyer and Thomas, 1957; Schuster and Coyne, 1981; Saettler, 1989), and disease management (Coyne and Schuster, 1983; Saettler, 1991), it remains one of the most serious deterrents to bean production throughout the world. It is most serious in areas with warm and wet or humid conditions during the growing season. However, the minimal requirements for field infection, particularly temperature and humidity, have not been fully established (Saettler, 1989).

Currently, the disease occurs on all continents (Bradbury, 1986). Its wide distribution is presumed to be due to efficient seed transmission.

Economic losses due to CB result from reduced marketability of the seed, or even loss of seed production. Reliable data in monetary terms are difficult to ascertain but losses of thousands to millions of US dollars have been reported. Cultivars and land races vary in response to infection, even when conditions are conducive for disease development. Yield loss data have been reported to range from 10 to 45% (Saettler, 1989).

Natural host infection has been reported for *Calopogonium* sp., *Lablab purpureus* (syn. *Dolichos lablab*), *Macroptilium lathyroides* (syn. *Phaseolus lathyroides*), *Phaseolus lunatus, P. vulgaris* (common bean), *Pisum sativum* (pea), *Pueraria* sp., *Vigna aconitifolia* (*Phaseolus aconitifolius*), *V. angularis, V. mungo* (urd bean), *V. radiata* (*P. aureus*; mung bean), *V. umbellata* (*Phaseolus calcaratus*) and the weed, *Strophostyles helvola*. By artificial inoculation, several other hosts have been reported: *Helianthus annuus* (sunflower; Kore and Biradar, 1986); *Lupinus polyphyllus, Phaseolus acutifolius* var. *acutifolius* (tepary bean), *P. acutifolius* var. *latifolius*, and *P. coccineus* (scarlet runner bean). Some of these and other reports of infection by inoculation require confirmation (Bradbury, 1986; Saettler, 1989).

1.5.2 Disease symptoms

X. c. pv. *phaseoli* (*Xcp*), including the melanin-producing variant ('fuscans'), induce differentiable symptoms on leaves, stems, pods and seeds (Zaumeyer and Thomas, 1957; Saettler, 1989, 1991). Symptoms may be similar to those caused by *Pseudomonas syringae* pv. *phaseolicola*, causal agent of halo blight of bean, and thus require identification or isolation of the pathogen for confirmation.

Leaf symptoms initially are seen as watersoaked spots arising several days (4–10) after infection or inoculation. The spots enlarge, become flaccid and may coalesce with adjacent lesions. Necrosis then develops, usually, but not always, accompanied by a bordering zone of bright yellow tissue. Different strains and host cultivars, as well as environmental conditions, can affect the size of development of the yellow zone. Lesions may be in any area of the leaf. In severe infection, dead leaves may remain attached to the plant or defoliate. Stem girdling may also occur. Exudate may appear on the leaves, providing inoculum for secondary spread.

Although less commonly seen, stems may be infected. This may occur (a) via stomata of the hypocotyl and epicotyl, (b) through the vascular elements leading from the leaf and (c) from infected cotyledons (Zaumeyer and Thomas, 1957). Symptoms are similar to leaf symptoms, except that lesions are usually smaller and chlorosis is absent. Stem girdle or joint rot may develop at the cotyledonary node.

Pod symptoms also initially are watersoaked. Lesions enlarge with time and become dark brown or red-brown and slightly sunken. Lesion shape is oval to oblong. However, lesion size, shape and development depend on pod age and environmental conditions. Bacterial ooze with a yellow cast is often seen, especially under humid conditions.

On seeds, symptoms can only readily be detected as yellow to brown irregular spots on white seeds. Infection is obscured on coloured seed. The spots can be distributed throughout the seedcoat or restricted to the hilum area. If infection occurs early in development, infected seed may rot or shrivel, although water-soaking has not been reported. Seed infection can occur via the pedicel or pod vascular system, micropyle, or funiculus. Infected seed can be found in symptomless pods (Aggour *et al.*, 1989; Saettler, 1989), which raises concern about pathogen transmission.

1.5.3 Isolation and disease diagnosis

Presumptive diagnosis of CB can be based on symptoms, especially on leaves. However, confirmation of the disease and identification of *Xcp* require further tests. Techniques for detection of *Xcp* vary in complexity, ranging from agar plate dilution assays to immunofluorescence staining techniques (Sheppard *et al.*, 1989). The methods usually involve extraction of bacteria from suspect plant material and subsequent detection and identification of the extracted bacteria. If high concentrations of other microorganisms are present, low levels of the pathogen can be missed. While semi-selective media are available (Sheppard *et al.*, 1989) and used widely, a minority of *Xcp* strains do not grow on such media or with sufficiently high plating efficiency to detect them. Thus, non-selective media, such as yeast dextrose agar, remain useful in regulatory and research programmes. The growth medium procedure, as well as host inoculation, are time-consuming, labour- and cost-intensive. Plant tests require space and appropriate facilities. Immunoassays for *Xcp* have not yet been sufficiently specific or sensitive to be used routinely (Sheppard *et al.*, 1989; Wong, 1990).

A diagnostic probe, based on the presence of a common plasmid sequence in all 50 geographically representative strains tested (Gilbertson *et al.*, 1989), is promising for detection and identification. If it proves accurate in further testing and sufficiently sensitive in a non-radioactive form, usage will be enhanced.

Detection of *Xcp* on media may be complicated by primary isolation from beans of pectolytic xanthomonads that are non-pathogenic to bean (Gilbertson *et al.*, 1990; Angeles-Ramos *et al.*, 1991). These strains can be differentiated further by phenotypic and genotypic properties. Currently, the only reliable method of identification of *Xcp* is by host plant assays of trifoliate leaves of a susceptible host following artificial inoculation (Coyne and Schuster, 1983; Saettler, 1989). A detached leaf assay of a susceptible host also shows promise (Arunakamari and Vidaver, 1986).

Thus, there is no consensus on the best detection method. Furthermore, there is no indication of virulence or pathogenic determinants, as yet, that could assist in identification, except possibly for extracellular polysaccharide (El-Banoby and Rudolph, 1979). This finding awaits confirmation. Races have not been definitively demonstrated or generally recognized.

X. c. pv. *phaseoli*, and its melanin pigment variant, are currently in one taxon (Bradbury, 1986). This grouping is supported by protein patterns (Vauterin *et al.*, 1991a) of 27 strains (18 of 'fuscans' and nine of *Xcp*). Furthermore, the strains were highly related to other pathovars of legumes, viz. pv. *alfalfae*, pv. *vignicola*, pv. *cajani* and pv. *glycines* at about the 85% level. DNA–DNA hybridization of single strains showed similar relationships, albeit at a lower level (40–80%), except for a single strain of 'fuscans' tested against a single strain of *Xcp* (Hildebrand *et al.*, 1990). Other properties are as for the genus (Bradbury, 1984; Chapters 2 and 3), except that it is non-pectolytic, in contrast to other xanthomonads found on bean (Gilbertson *et al.*, 1990; Angeles-Ramos *et al.*, 1991) and some other xanthomonads.

1.5.4 Sources of inoculum and disease management

Host resistance has long been sought in edible beans. Compounding the difficulty of evaluating resistance or tolerance are the following: (a) leaf and pod reactions to the pathogen can be inherited independently (Coyne and Schuster, 1983; Park and Dhanvantari, 1987), (b) day length can affect susceptibility of some genotypes in different environments (Saettler, 1989), (c) immunity (no disease symptoms) has not been detected in *P. vulgaris* but only in related hosts from which it is difficult to transfer this property (Coyne and Schuster, 1983; Park and Dhanvantari, 1987), (d) wide differences occur in virulence of naturally-occurring populations of the pathogen (Schuster and Coyne, 1981; Saettler, 1989) and (e) large epiphytic populations can occur on tolerant plants (Saettler, 1989; Ishimaru *et al.*, 1991). Besides host resistance, other techniques that should be used are applicable to other plant pathogens. Seed that is inspected for the pathogen and found free of detectable infection should be used. Crop rotation, with at least 2 years between bean crops, minimizes disease carry-over. Fields should be kept free of weeds, volunteer beans and other alternate hosts (Saettler,

1991). Reports differ on the efficacy of antibiotics and copper compounds that are used as seed or foliar treatments. Disease may be reduced in some environments, but these treatments may not provide a consistently adequate level of control (Saettler, 1989).

1.6 *XANTHOMONAS CAMPESTRIS* PV. *GLYCINES*: CAUSE OF BACTERIAL PUSTULE OF SOYBEAN

S. HOKAWAT
Department of Plant Pathology, Faculty of Agriculture, Khon Kaen University

K. RUDOLPH
Institut für Pflanzenpathologie und Pflanzenschutz, Göttingen

1.6.1 Introduction

Bacterial pustule occurs in most soybean-growing areas of the world with warm temperatures and sufficient rainfall (Moffett and Croft, 1983; Sinclair, 1984; Bradbury, 1986). The disease has been observed in the USA since 1916 (Hedges, 1922) where it is now widespread. The causal organism was first described in Japan (Nakano, 1919) as *Pseudomonas glycines*. Bacterial pustule has since been reported from many countries, including Canada, Latin-America and Australia (Sinclair and Dhingra, 1975; Bradbury, 1986). More recently the disease has been described in Venezuela (Diaz, 1974), Thailand (Hokawat, 1978) and Cuba (Morales and Gutierrez, 1986). In contrast to bacterial blight caused by *Pseudomonas syringae* pv. *glycinea*, bacterial pustule of soybean is not limited by high temperatures and, therefore, occurs world-wide where warm temperatures and frequent showers prevail during soybean growing.

This disease is sometimes more important in tropical areas than bacterial blight caused by *P. syringae* pv. *glycinea* (Lelliott and Stead, 1987), and is one of the limiting factors in improvement of soybean production in Thailand (Hokawat, 1978; Prathuangwong, 1985). The disease causes premature defoliation that may decrease yield by reducing seed size and number (Laviolette *et al.*, 1976).

Studies of Hartwig and Johnson (1953) showed that moderately infected susceptible lines yielded 8–11% less than did the resistant lines. Mean yield, seed size, percentage protein and iodine number of the oil of susceptible lines were significantly lower than those of the resistant lines. The mean disease score for susceptible lines was 3, on a 1 to 5 rating scale, whereas resistant lines showed no well-defined pustule development. Weber *et al.* (1966) showed that, under a moderate artificial infection, the average seed yield reduction in the susceptible soybean lines was 4.3%, of which 86% was attributed to reduced seed number and 14% to reduced seed size. In tropical soybean-growing areas, such as Thailand, Prathuangwong (1985) reported that yield losses due to natural dissemination in the susceptible soybean cultivar 'SJ4' were 20–35%. The most susceptible age of 'SJ4' was between 6 and 7 weeks after emergence (at the flowering stage). Khundet (1989) also reported that yield losses due to bacterial pustule disease showed highest correlation with disease severity when evaluated at the growth stage of V7 and R2 of soybean plants. Evaluation of income from plots planted with various percentages of artificially infected soybean seeds demonstrated a break-even point, or income covering the cost of production, at 41% contaminated seeds (1 kg soybean seed costs *ca.* $0.25 in Thailand). Weeds such as red vine (*Brunnichia cirrhosa*) and *Dolichos biflorus* are also natural hosts (Bhatt and Patel, 1954; Jones, 1961) in addition to soybean. *Phaseolus vulgaris, P. lunatus* (Bhatt and Patel, 1954), *Lablab purpureus, Arachis hypogaea* and *Pachyrhizus erosus* (Prathuangwong, 1985) can be infected by artificial inoculation.

1.6.2 Disease symptoms

The disease occurs primarily on the foliage. Early symptoms are minute pale green spots with elevated centres on either or both leaf surfaces. Thereafter, a small, raised, light-coloured pustule forms in the centre, usually in lesions on the under surface of the leaf (Figure 1.6). These symptoms are sometimes confused with those of soybean rust.

Figure 1.6 X. *c.* pv. *glycines*: typical pustules on the lower leaf surface (artificial infection).

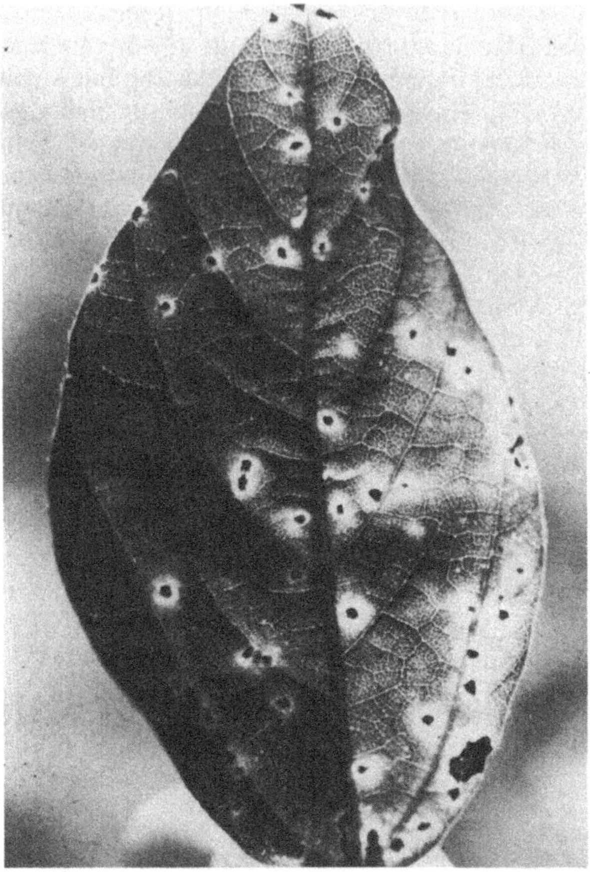

Figure 1.7 *X. c.* pv. *glycines*: mature lesions on the upper leaf surface after natural infection.

However, pustule lesions are characterized visually by small pustules surrounded by yellowing haloes, whereas rust disease forms tan or brown lesions in which uredospores are formed and released through a central pore.

Spots of bacterial pustule vary from minute specks to large, irregular, mottled brown areas that arise when smaller lesions coalesce. In later stages, dried, broken remnants of pustules may be seen on small brown necrotic areas surrounded by narrow yellowing haloes (Figure 1.7). Leaves become ragged when dead areas are torn away by the wind. Severe infection often causes yellowing and death of interveinal tissue resulting in premature defoliation. Leaf spots sometimes form without developing pustules. Small, reddish-brown, slightly raised spots may develop on stems and pods of susceptible cultivars. Since the pustule pathogen survives commonly in soybean seeds and also in soybean crop debris (Graham, 1953; Fett, 1979), first infection may occur on the primary leaves of seedlings (Hartwig and Lehman, 1951). The symptoms on seedlings in the greenhouse are initially white pustules later turning into reddish-brown spots which appear on hypocotyls, cotyledons and the primary leaves (Prathuangwong *et al.*, 1989). Pustules are mainly formed by hypertrophy, but hyperplasia may also occur (Hedges, 1924; Jones and Fett, 1987).

Symptoms on resistant soybean cultivars have also been reported, such as occasional small chlorotic spots, but no well-defined pustules (Hartwig and Lehman, 1951), or light green chlorosis and slight browning in greenhouse trials (Fett, 1984; Hokawat, 1988), or fewer and smaller pustules (Groth and Braun, 1986).

1.6.3 Isolation and disease diagnosis

The semi-selective medium of Mulrean and Schroth (1981) for *X. c.* pv. *juglandis* was improved by Groth and Braun (1989) by addition of 100 µg/ml cycloheximide to the brilliant cresyl blue-starch (BS) medium, and used for detecting epiphytic populations of *X. c.* pv. *glycines* on susceptible and resistant soybean cultivars. Blue colonies of 1 mm diameter were surrounded by a zone of starch hydrolysis. The semi-selective medium MXG of Khundet (1989) gave green convex-shaped colonies, 3–6 mm in diameter with smooth margin and fluidal character after 3 days, and was used effectively to detect the pathogen in infected seeds by using iodine solution to demonstrate the clear zone formation around infected seeds. The MXG medium consists of K_2HPO_4 (0.8 g), KH_2PO_4 (0.6 g), yeast extract (0.7 g), soluble starch (8.0 g), glucose (1.0 g), agar (15.0 g), distilled water (1000 ml), 1% methyl green aqueous solution (1 ml), cycloheximide (50 mg), bacitracin (10 mg), cephalexin (50 mg) and kasugamycin (10 mg).

X. c. pv. *glycines* is a motile Gram-negative rod with a single polar flagellum. Colonies on beef infusion agar, nutrient agar and Wakimoto's agar are pale yellow, circular and smooth with an entire margin. The pathogen produces abundant, slimy, yellow growth on sugar-containing media (Hokawat, 1978). Optimum temperature for growth on potato dextrose agar is 25–30°C, maximum 35°C and minimum 10°C (Bhatt and Patel, 1954).

The bacterium penetrates the soybean plant through natural openings and wounds (Sinclair, 1984) and multiplies intercellularly (Jones and Fett, 1985). The degree of attack was associated with the number of stomata on the under surface of leaves (Morales and Gutierrez, 1986). Strains of the pathogen showed large differences in their virulence

(Hokawat and Rudolph, 1991). Several factors which may contribute to the virulence of *X. c.* pv. *glycines* have been described, such as production of indoleacetic acid and cytokinin (Millar, 1955; Fett and Dunn, 1987), extracellular polysaccharides (Jones and Fett, 1985; Rudolph *et al.*, 1989), toxin (Paxton, 1972), bacteriocins (Fett *et al.*, 1987) or cellulases and protopectinases (Srivastava and Bais, 1986). The bacterium also secretes amylase, gelatinase and proteinase (Rudolph, unpublished). Extracellular polysaccharides prolonged the life of the pathogen under dry conditions and sunlight (Vorenkevich and Shmanenkova, 1968), increased survival rate in the laboratory, and supported growth of the pathogen in soybean tissue (Hokawat and Rudolph, 1988). Physiological races of the pathogen have not yet been identified (Jindal *et al.*, 1981; Fett, 1984; Hokawat, 1988).

1.7 *XANTHOMONAS CAMPESTRIS* PV. *CITRI*: CAUSE OF CITRUS CANKER

R.E. STALL
Department of Plant Pathology, University of Florida

E.L. CIVEROLO
USDA-ARS, Beltsville

1.7.1 Introduction

Citrus canker was noted on herbarium specimens of *Citrus medica* collected in India in 1827–31 (Fawcett and Jenkins, 1933). However, it was first described as a new disease after introduction into the USA where the causal agent was identified (Hasse, 1915). It now occurs in at least 30 countries around the world (Civerolo, 1991) where rainfall and high temperatures occur concomitantly (Peltier and Frederich, 1926). Citrus canker has been eradicated from Australia (Hill, 1918; Curry, 1989), New Zealand (Dye, 1969), South Africa (Doidge, 1929), and the USA (Dopson, 1964).

Since the original description of citrus canker, the citrus diseases caused by xanthomonads have been described as forms of citrus canker. However, the most widely distributed and severe form is the originally described Asiatic canker (Canker A). Canker A affects a wide range of rutaceous plants, but grapefruit, limes, and some sweet oranges are particularly susceptible (Lee, 1918). Cancrosis B occurs in Argentina, Paraguay and Uruguay and is most prevalent on lemon (Bitancourt, 1957). In Brazil cancrosis C was found on Mexican lime trees and the pathogen is serologically and physiologically distinct from other xanthomonads on citrus (Namekata and Oliveira, 1972).

A fourth form (Canker D) was identified in Mexico and is sometimes referred to as citrus bacteriosis. It is known to occur on Mexican lime and does not affect fruit (Rodrigues *et al.*, 1985). A fifth form, described in the USA in 1984, was called Canker E, but is also referred to as citrus bacterial spot. It differs from the others in appearance of lesions and only affects very young trees (Schoulties *et al.*, 1987). Presently, there is uncertainty as to whether B, C, D, and E are forms of citrus canker caused by *Xanthomonas campestris* pv. *citri* (Stall and Civerolo, 1991). Therefore, only Asiatic canker will be considered here.

1.7.2 Disease symptoms

Symptoms of Asiatic canker include lesions on leaves, stems, and fruit (Figure 1.8). Lesions begin as pin-point necroses slightly raised above the plant tissue surface. Lesions may expand to 9–10 mm in diameter in very susceptible leaves (Stall *et al.*, 1980). As the lesions enlarge, they become crateriform with raised margins and sunken centres. Usually the lesions are circular, but they may enlarge irregularly. Lesions 2 mm or larger usually have a watersoaked margin. When lesions reach 1–2 mm in diameter, a characteristic yellow halo forms around each lesion. Leaves are susceptible for a period of only 14–21 days near the time when they reach full expansion. Therefore, multiple lesions on leaves tend to be the same size because infections

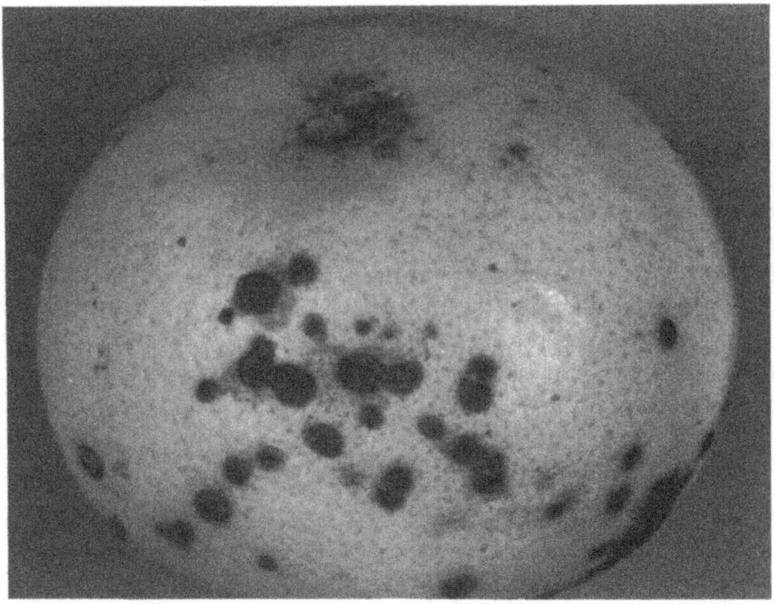

Figure 1.8 *X. c.* pv. *citri*: fruit lesions.

occur at nearly the same time. Fruit may be susceptible for 60–90 days after petal fall. Multiple lesions on fruit may be variable in size because more than one infection cycle can occur on fruit. Severely affected leaves or fruit abscise. The latter may occur because ethylene is produced in diseased tissue (Goto *et al.*, 1980). Lesions do not affect fruit quality and diseased fruit may be processed. Lesions may affect marketability of fresh fruit, however.

1.7.3 Isolation and disease diagnosis

Xanthomonas campestris pv. *citri* grows readily on most laboratory media and can be isolated easily from diseased tissue by standard techniques. Colonies appear on agar media after 2–3 days incubation at 28°C. Colonies have the distinctive yellow colour of xanthomonads and xanthomonadin pigment can be extracted from the cells. On high carbohydrate media cultures are mucoid. Pathogenicity can be confirmed by placing a drop of turbid inoculum on a young leaf and piercing the leaf through the drop. If the inoculated plants are kept at 28°C, raised, corky lesions appear in 10–14 days.

Inoculation by injection of high inoculum (10^8 cfu/ml) gives uncertain results. Hypersensitivity in tobacco is also not conclusive. Strains have the biochemical and physiological characteristics of *X. campestris* (Bradbury, 1984). Strains can be identified by fatty acid profiles (Stall and Hodge, 1989).

Canker A strains from many areas of the world are genetically very uniform based on DNA fingerprinting (Hartung and Civerolo, 1987) and restriction fragment length polymorphism (Gabriel *et al.*, 1988). In Japan, strains that differ in utilization of mannitol, mannose and lactose occur, and 12 lysotypes were differentiated with six virulent phages (Goto, 1962). About 80% of strains from Asiatic canker were lysed by citriphages, CP1, CP2, or both (Goto and Starr, 1972). These phages can be used to identify strains from Canker A, but should be used with caution. Strains from Asiatic canker are serologically similar based on polyclonal antisera (Bach *et al.*, 1976). A monoclonal antibody also demonstrated a high level of uniformity among strains from Asiatic canker (Alvarez *et al.*, 1991). Both polyclonal and monoclonal antibodies reacted with antigens of only a few other xanthomonads.

1.7.4 Sources of inoculum and disease management

Citrus trees are perennial plants and *X. c.* pv. *citri* can survive in the trees from year to year (Kuhara, 1978). Inoculum that oozes from lesions decreases as lesions age (Timmer *et al.*, 1991). Cells of *X. c.* pv. *citri* survive in soils and on surfaces of weeds for short periods (Goto *et al.*, 1975). Lesions have been associated with insect injuries, but survival in and transmission by insects have not been documented clearly. The citrus canker bacterium is dispersed short distances by wind-driven rain and mechanical equipment used in grove maintenance and care (Serizawa *et al.*, 1969; Stall *et al.*, 1980). Long distance spread is usually with diseased propagating material such as budwood, rootstocks and budded trees. Seed transmission of the pathogen has not been recorded. Potentially the pathogen could move long distances with diseased fruit, but there is no record of disease outbreaks as the result of marketed fruit. Low numbers of cells (10^2 cfu/ml) can be detected by development of lesions after infiltration, or multi-needle punctures of susceptible leaves with bacterial suspensions (Goto *et al.*, 1970).

Citrus canker does not occur in all areas where citrus grows and in some countries quarantines preclude importation of citrus for fear of importation of citrus canker. If citrus canker is detected early in disease progress, or the climate is not conducive for disease development, eradication of the disease may be feasible. Eradication requires accurate surveys for the disease, strict regulations to prevent spread, and rapid destruction of diseased trees (Stall *et al.*, 1986). In areas where citrus canker is endemic disease severity varies from year to year. Management of the disease requires a combination of resistant species and/or cultivars of citrus, integrated cultural practices and chemical sprays (Civerolo, 1984). Numerous citrus species and cultivars are relatively resistant to citrus canker, but mandarins and Valencia orange can be cultivated profitably in canker-infested areas. The severity of citrus canker can be reduced by the use of wind-breaks, such as bamboo, pine trees and plastic screens. The growth of vegetation between rows of trees may prevent injuries caused by wind-blown sand which facilitate ingress of bacteria. Cultivation practices should be avoided during periods when the trees are wet to prevent spread of the pathogen. Sprays of fixed copper control citrus canker if properly timed in relation to growth of new shoots and fruits. A forecasting system may be followed for optimum use of chemical sprays (Kuhara, 1978).

1.8 *XANTHOMONAS CAMPESTRIS* PV. *CAMPESTRIS*: CAUSE OF BLACK ROT OF CRUCIFERS

N.W. SCHAAD
USDA-ARS, Foreign Disease-Weed Science Research, Frederick

A. ALVAREZ
Department of Plant Pathology, College of Tropical Agriculture and Human Resources, University of Hawaii at Manoa

1.8.1 Introduction

Black rot, caused by *Xanthomonas campestris* pv. *campestris*, is found wherever crucifers are grown and is considered the most destructive disease of crucifers world-wide (Walker, 1952; Chupp and Sherf, 1960; Williams, 1980). Many species of cultivated crucifers are susceptible to infection, including cabbage, broccoli, cauliflower, Chinese cabbage, kale, rutabaga and Brussels sprouts (Walker, 1952; Chupp and Sherf, 1960). The causal agent of black rot was described in the USA (Iowa) in rutabaga in 1895 by Pammel (Walker, 1952). The pathogen was shown to be seedborne in 1904, and transmission of the pathogen by seed was demonstrated in 1921 (Walker, 1952). Black rot is widespread in crucifer fields in the USA (Walker, 1952; Chupp and Sherf, 1960; Hunter *et al.*, 1975; Schaad, 1976a; Alvarez *et al.*, 1985), Asia (Schaad and Thaveechai, 1983; Kim, 1986), India (Patel *et al.*, 1949) and Africa (Onsando, 1987) when the weather is favourable for disease. The pathogen grows best at relatively high temperatures (25–35°C) and humidities (80–100%) whereas the host favours a low to moderate temperature (10–18°C). In tropical and subtropical areas entire fields are often lost during the rainy season.

1.8.2 Disease symptoms

Symptoms of black rot vary considerably depending upon host genus, species, cultivar, plant age and the environment. Most cultivated crucifers express characteristic V-shaped yellow lesions on the margins of leaves. As the lesion enlarges the veins become black. Chlorotic spots, wilting, and black veins without yellowing also can occur (Schaad and Thaveechai, 1983). Atypical vein-delimited, tan necrotic tissue has been observed in tropical areas following water congestion and rapid collapse of the mesophyll (Alvarez *et al.*, 1987). Seedlings often are wilted and stunted and may die prior to formation of the second true leaf if the weather is hot. In contrast, when temperatures remain below 15–18°C (typical temperatures when crucifers are sown in temperate climates), symptoms seldom appear. Later, if temperatures rise, symptoms begin to appear after 4–6 weeks. Symptoms on cauliflower often consist of black flecks or scorched leaf margins. Internal parts of cauliflower curds are often black. Black veins often are faintly visible, especially in mature flowering plants. During periods of high temperatures when the plant is near maturity soft rot may occur. The name black rot originates from the occurrence of black, rotting plants; however, in many cases the soft rot is due to secondary invasion by *Pseudomonas* or *Erwinia* (Schaad and Thaveechai, 1983).

Although many cruciferous weed species are infected (Walker, 1952; Chupp and Sherf, 1960) many do not show typical symptoms (Schaad, 1976a; Schaad and Dianese, 1981; Schultz and Gabrielson, 1986; Alvarez *et al.*, 1987). Cruciferous weeds apparently have evolved with the disease and are highly resistant. For example, *B. nigra*, *B. campestris*,

Cardaria pubescens, Barbarea vulgaris, Raphanus sativus and *R. raphanistrum* may show diffuse yellowing or tan-brown lesions on the margins of older leaves, and the leaf often falls off prematurely. Small black veins sometimes occur. Species such as *Sisymbrium officinale, Lepidium virginicum,* and *Capsella bursa-pastoris* normally show no symptoms.

Symptoms on cultivated seed plants seldom occur making field diagnosis of black rot in seed plants difficult. Senescent leaves and opportunistic fungi can easily mask symptoms of black rot. Symptoms caused by *Fusarium* sp. also may be confused with black rot (Chupp and Sherf, 1960), although this is not normally a problem because most commercial cultivars are resistant to *Fusarium*.

1.8.3 Isolation and disease diagnosis

Visual detection of black rot is relatively easy in young plants of most commercial cultivars when the temperatures are between 25 and 32°C. Under cooler conditions, however, or when plants begin to bolt, symptoms can be confusing or lacking. In every case isolations (Schaad and Stall, 1988) or immunological assays (Alvarez *et al.*, 1985; Alvarez and Lou, 1985) should be made to confirm the presence of *X. c.* pv. *campestris*. Because the pathogen is a vascular invader (Walker, 1952), leaf tissue should be taken from healthy tissue adjacent to the lesion to avoid saprophytic bacteria associated with dead tissues. Detection is quite easy due to the large number of cells of *X. c.* pv. *campestris* in the vascular tissue. Semi-selective media (Schaad, 1989) are useful for isolating *X. c.* pv. *campestris* from plant tissues, but yeast dextrose–calcium carbonate agar is the best medium for observing the characteristic yellow pigmentation and mucoid colony morphology (Schaad and Stall, 1988). However, such presumptive identification must always be confirmed by pathogenicity tests.

1.8.4 Sources of inoculum and disease management

Plants normally become infected via seeds or leaf stomates and/or hydathodes. Although the pathogen quickly invades the xylem (Walker, 1952), most seeds become contaminated via infection of siliques during high humidity or overhead watering rather than infection via vascular tissue. Seeds are a major source of inoculum (Walker, 1952; Chupp and Sherf, 1960; Hunter *et al.*, 1975; Randhawa and Schaad, 1984; Schultz and Gabrielson, 1986). Pathogen-free seeds become even more important when transplants are used due to the favourable conditions for black rot development and spread in plant beds or greenhouses. Contaminated seeds are easily assayed by soaking samples in a solution of saline and Tween 20 (Schaad, 1989). Water should not be used because the death rate of *X. c.* pv. *campestris* in water is quite high (Schaad and Donaldson, 1980). Since the pathogen resides in the vascular system of the funiculus (Walker, 1952), cells ooze into the wash solution within minutes (Schaad, 1988). Several semi-selective media, based on starch hydrolysis, are available for isolating *X. c.* pv. *campestris* from seeds (Schaad, 1989). Because seed-associated bacteria can be antagonistic to *X. c.* pv. *campestris* on semi-selective media (Randhawa and Schaad, 1984) more than one medium should be used.

Other than seeds and crop residues, weeds are perhaps the most important source of inoculum for black rot (Chupp and Sherf, 1960; Schaad, 1976a; Schaad and Dianese, 1981; Schultz and Gabrielson, 1986). The pathogen not only infects many weeds but also survives as an epiphyte on symptomless leaves (Alvarez and Lou, 1985; Kuan *et al.*, 1986).

Detection of *X. c.* pv. *campestris* in the environment on leaf surfaces, decayed stem debris and in soil is much more difficult due to the relatively low number of *X. c.* pv. *campestris* in relation to numbers of other bacteria. Using semi-selective media, however, *X. c.* pv. *campestris* has been shown to survive for relatively long periods in plant residues in soil (Schaad and White, 1974; Schultz and Gabrielson, 1986; Alvarez *et al.*, 1987). Serological methods such as immunofluorescence and ELISA (Alvarez and Lou, 1985) can be useful for rapid identification of suspected colonies of *X. c.* pv. *campestris* isolated from infected plant material. Identification should, however, be confirmed with pathogenicity tests. Monoclonal antibodies are especially useful for identification of a particular strain in the field or determining the source and epidemiological fitness of a given strain (Alvarez *et al.*, 1987; Yuen *et al.*, 1987). Black rot has been shown to spread up to 30 m from infected weeds to cabbage (Schaad and Dianese, 1981), and inoculum is spread by wind-blown rain, aerosols (Kuan *et al.*, 1986) and insects (Williams, 1980).

Control of black rot requires use of pathogen-free seeds or transplants, rotation to avoid inoculum in soil debris, and elimination of cruciferous weeds (Chupp and Sherf, 1960; Williams, 1980; Schaad and Dianese, 1981; Alvarez *et al.*, 1987). Several seed treatments including hot water, antibiotics, antibiotics and sodium hypochlorite, hydrogen peroxide, and hot acidified cupric acetate or zinc sulphate are available (Walker, 1952; Chupp and Sherf, 1960; Humaydan *et al.*, 1980; Schaad *et al.*, 1980; Huang and Lee, 1988); however, none are totally effective. Either the pathogen is not completely eradicated by the seed treatment, such as hot water or antibiotics, or germination (Williams, 1980) and vigour of the cultivar are adversely affected (Patel *et al.*, 1949). Black rot is controlled best through sanitation and management practices that include a 2- to 3-year crop rotation, proper weed control, and use of assayed seed found to be free of *X. c.* pv. *campestris.*

1.9 *XANTHOMONAS CAMPESTRIS* PV. *GRAMINIS*: CAUSE OF BACTERIAL WILT OF FORAGE GRASSES

F. LEYNS
Plant Genetic Systems, Gent

1.9.1 Introduction

A bacterial disease of forage grasses was first described by Egli *et al.* (1975) in Switzerland. The causal organism was found to be a *Xanthomonas* species similar to '*Xanthomonas oryzae*', the organism that causes bacterial leaf blight of rice. Somewhat later, several reports suggested that the bacterium was present in most countries of west and central Europe, New Zealand and even the USA (Leyns *et al.*, 1981b; Roberts and Vargas, 1984; Falloon and Hume, 1988; Sletten, 1989). As several bacteria were being isolated with a difference in host range, a study was undertaken to test the pathogenicity of these bacteria on several grasses. The study showed that at least four different pathovars of *Xanthomonas campestris* were responsible for diseases on grasses. Some strains, now gathered under the name *X. c.* pv. *graminis sensu stricto* showed a broad spectrum host range. They showed pathogenicity towards several *Agrostis, Alopecurus, Arrhenatherum, Bromus, Festuca, Lolium, Phleum* and *Trisetum* species. *X. c.* pv. *phlei* causes severe symptoms on *Phleum* but seemed to be restricted to that host only. *X. c.* pv. *arrhenatheri* and *X. c.* pv. *poae* are restricted to *Arrhenatherum* spp. and *Poa* spp., respectively. *Poa annua*, however, showed resistance towards the latter bacterium in all trials (Egli and Schmidt, 1982; Paul and Birkenstaedt, 1989; Schmidt, 1989).

Since the initial report from Switzerland, France and Germany (Egli *et al.*, 1975), reports from England (Wilkins and Exley, 1977), Belgium (Van den Mooter *et al.*, 1981), Holland (Labruyere, 1975), the USA (Roberts and Vangas, 1984), New Zealand (Falloon and Hume, 1988) and Norway (Sletten, 1989) have shown that the disease is more widely spread than first suspected. The fact that, in many cases, the disease symptoms are sometimes not manifest and progress slowly in temperate climates may be the cause of the relative late discovery of the disease. In some cases it occurs together with fungal diseases (Falloon and Hume, 1988). A detailed study in northern Belgium showed that in over 80% of the investigated fields, diseased plants could be detected (Leyns *et al.*, 1981b). Most of the fields, however, contained some diseased plants per square meter only. In isolated cases, a severe disease aspect affecting most of the plants was observed in Belgium.

In western Scotland, half of the temporary hay and silage fields were infected with this pathogen in the summer of 1979. This figure reached 71% in 1980 (Channon and Hissett, 1984).

Little is known about the yield losses in naturally infected fields. Most data originate from small-scale experiments performed in the greenhouse or in artificially infested small field plots. Gondran and Betin (1989) found up to 47% loss after three cuts following inoculation of Italian ryegrass in a field plot. Losses were less severe with other grasses such as *Lolium perenne, Festuca arundinacea, F. pratensis* and *F. rubra*. Schmidt (1988) found a yearly yield reduction of 10% in fields seeded with westerwold ryegrass and up to 56% in greenhouse trials with cultivars of *Festuca pratensis*. In most experiments *Dactylis glomerata* seemed to be resistant to the disease (De Cleene *et al.*, 1981). In other grasses, yield losses of up to 91% in fresh weight were observed after 21 days following artificial inoculation of the grasses in the greenhouse (Leyns *et al.*, 1981a). Paul and Freudenstein (1989) found that five susceptibility classes can be found within Italian ryegrass. Tetraploid plants seem to be more resistant than others.

Assessment of the resistance by symptom scoring is not always reliable. Although most of the wilt scores and yield losses correlated satisfactorily, Gondran and Betin (1989) found that in a resistant Italian ryegrass cultivar 0% yield loss was observed whilst 10% of its leaves were wilted.

Standardization of the assessment procedures is necessary to obtain consistent results. An effort was made by Paul and Birkenstaedt (1989) to select methods for assessing the susceptibility of forage grass varieties. Injecting the bacteria directly into the tiller or clipping the plant with an infected pair of scissors results in rapid disease development useful in large screening programmes, whereas a slower disease development was observed after spraying intact plants with bacteria. These results are useful in more detailed resistance studies. Some controversy exists with the latter inoculation method; in many cases no disease development was observed macroscopically or microscopically (De Cleene, 1989).

1.9.2 Disease symptoms

X. c. pv. *graminis* causes wilting symptoms on forage and other grasses. Infection of a leaf or flowering tiller leads to a gradual drying out and necrosis of the leaf or tiller starting from the infection site and slowly progressing towards the leafbase or base of the plant. Yellowish stripes may appear near the vascular tissue. When the bacterium reaches vascular tissue, the disease spreads rapidly throughout the whole plant, causing wilt of several leaves. In some environmental conditions, the plant may die within 20 days after infection. Early in spring, a few diseased plants only can be detected in the field. The number of infected plants gradually increases during summer and after each mowing. In a few cases, especially in older fields, the whole field is affected and has a brownish-yellow aspect in late summer.

1.9.3 Isolation and disease diagnosis

A careful examination is sometimes necessary for the detection of diseased plants within a field. In some environmental conditions and in newly sown fields only a few diseased plants will be present per square metre. In mid-summer when the bacteria are allowed to grow extensively within the plant, squeezing the flowering tiller between the fingernails will result in extensive pale-yellowish slime, oozing out of the cut. Plating this slime onto GYCA (glucose 0.5%, yeast extract 0.5%, $CaCO_3$ 2%, agar) medium, supplemented with a fungicide, is a quick way to obtain pure cultures of the bacterium. The bacterium produces pale yellow, flat slimy colonies. Antisera have been developed for rapid identification of the causal organism by ELISA (Sletten, 1989) or double-diffusion tests (Samson *et al.*, 1989). Until now, the pathogen had not been isolated from within the seed, but was found on the glumes of seed, mowing material in diseased and symptomless plants, and plant debris (Leyns *et al.*, 1988; Schmidt, 1989; Sletten, 1989).

1.10 *XANTHOMONAS CAMPESTRIS* PV. *VESICATORIA*: CAUSE OF BACTERIAL SPOT OF TOMATO AND PEPPER

R.E. STALL
Department of Plant Pathology, University of Florida

1.10.1 Introduction

Bacterial spot of pepper (*Capsicum* spp.) and tomato (*Lycopersicon* spp.) probably occurs wherever those crops are grown commercially in the world. However, it is a severe disease only in climates where high rainfall and high temperatures occur concomitantly. Diseased plants have reduced growth and yields compared to healthy plants. The extent of the reductions depend upon the severity of the disease and the stage of plant affected. Infections on young plants reduce yields more than infections of mature plants (Pohronezny and Volin, 1983). Diseased fruit may not be saleable in some regions.

The hosts of this bacterium in addition to plants of *Capsicum* spp. and *Lycopersicon* spp. are not certain. Many other solanaceous plants have been listed as hosts (Elliott, 1951), but the methods of inoculation used for those determinations often would not have precluded confusion with a hypersensitive reaction. Also, pathogenic specialization within this bacterium may include interactions with intragenus variations among solanaceous plants around the world.

1.10.2 Disease symptoms

Symptoms of disease on tomato consist of numerous small (1–5 mm), circular lesions on leaves, stems and fruit. On pepper, necrotic spots and blotches on leaves can be of any size and shape, but on stems and fruit lesions are circular in shape and about the same size as spots on tomato. Peduncles and sepals of both crops are particularly susceptible to bacterial spot. Bacterial spot lesions can be distinguished from fungal leaf spots by their greasy appearance on the abaxial side of leaves. A general chlorosis develops in diseased leaves of both pepper and tomato, extending from the area of lesion appearance. The chlorosis is preceded by epinasty of leaves and followed by leaf abscission with pepper and complete necrosis of tomato leaflets. The chlorosis, leaf abscission, and death of leaves can be reproduced in healthy leaves by ethylene treatment. Sufficient ethylene is produced in diseased leaves to account for the latter symptoms (Stall and Hall, 1984; Ben-David *et al.*, 1986).

1.10.3 Isolation and disease diagnosis

Xanthomonas campestris pv. *vesicatoria* grows well on common laboratory media and can be isolated readily from diseased tissue. Colonies can be seen in 2–3 days, depending on the medium, at 28°C, and are yellow in appearance. Many saprophytic bacteria also occur in diseased tissue and may be in higher numbers than the pathogen. Some saprophytes may prevent the pathogen from growing after isolation onto media. Positive identification of *X. c.* pv. *vesicatoria* is most easily accomplished by determination of the presence of the xanthomonadin pigment and of pathogenicity to a susceptible plant. The former can be determined from the spectrophotometric absorption spectrum of a pigment extract, or by thin-layer chromatography of a pigment extract (Irey and Stall, 1982). Pathogenicity can be determined by the presence or absence of lesions after cells in a turbid suspension are spread lightly with a cotton applicator onto an immature leaf that was previously dusted with an abrasive material (carborundum). Hypersensitivity in tobacco is variable. An accurate identification of *Xanthomonas* can also be made from profiles of fatty acids extracted from cells (Roy, 1988). Pathogenicity to pepper or tomato should also be determined if the bacterium is identified by fatty acid profiles. Selective media have been described for the preferential isolation of the bacterium (McGuire *et al.*, 1986), but none is sufficiently selective for positive identification.

Numerous strains of the organism occur that have specific pathogenicity. Strains pathogenic to tomato and pepper, to pepper only, and to tomato only may occur in various regions. In addition, races of the pathogen have been distinguished based on cultigen resistances. Specific interactions appear to be inherited in both the pathogen and the plant in a gene-for-gene manner. Avirulence genes in the bacterium are involved in the host specificities as well as the cultigen specificities (Minsavage *et al.*, 1990b). Pathogenicity-specific strains are not distributed evenly world-wide (Cook and Stall, 1982). In addition, strains that are resistant to copper and streptomycin are commonly selected in fields sprayed with those materials (Stall and Thayer, 1962; Marco and Stall, 1983).

Most of the avirulence genes and genes for resistance to copper and streptomycin are located on plasmids in the bacterium (Stall *et al.*, 1986; Minsavage *et al.*, 1990a,b). Great diversity in plasmid content occurs in this bacterium. Among a population of nearly 500 strains, plasmids varied in size from 3 to 300 kb pairs. Thirteen size classes of plasmids were differentiated and 71 different plasmid profiles occurred among the strains (Canteros, 1990). The plasmid that contains the copper resistance genes is self-transmissible (Stall *et al.*, 1986).

Two groups of strains can be distinguished within *X. c.* pv. *vesicatoria* based on physiology. One group strongly hydrolyses starch and causes pits in a sodium polypectate medium. The other group is negative for those characteristics. Two serotypes were identified among a population of the pathogen based on polyclonal antibodies to heat-stable antigens (Charudattan *et al.*, 1973). Three serotypes were identified among another population based on polyclonal antibodies to ribosomal proteins (Schaad, 1976b). Those serotypes were not related to pathogenic specificity. However, Lovrekovitch and Klement (1965) noted serological differences between those strains pathogenic to pepper and those pathogenic to tomato. They also noted differences in phage sensitivities of the tomato and pepper strains. However, with a panel of seven bacteriophages 19 groups of strains were designated among a world-wide collection (Dye *et al.*, 1964).

1.10.4 Sources of inoculum and disease management

The pathogen may survive between seasons in lesions on volunteer plants and on seeds harvested from diseased plants. Survival on weeds and solanaceous plants, either as epiphytes or in restricted lesions, may occur in some areas of the world. However, the bacterium does not compete well with soil bacteria and disappears from soils as the vegetative material containing the bacterium decomposes (Jones *et al.*, 1986). There is no evidence for survival in or on insects. Transplants used to start some crops may be contaminated with *X. c.* pv. *vesicatoria*. Detection of *X. c.* pv. *vesicatoria* in low populations is not adequate for testing survival in all habitats.

The causal bacterium spreads readily from diseased to nearby healthy plants by wind-blown water (Volcani, 1969). Within-field dispersal occurs rapidly through management practices of thinning plants, transplanting, pruning, staking and harvesting. Dispersal occurs more frequently when those practices are done while plants are wet versus dry (Pohronezny *et al.*, 1990). Injuries, such as those caused by wind-blown sand or high-pressure sprays, increase the ingress of the pathogen (Vakili, 1967). The timing of stomatal openings with inoculation periods may influence the severity of disease and plant susceptibility (Ramos and Volin, 1987). Immature leaves and fruit are more susceptible than mature leaves and fruit (Nayudu and Walker, 1960).

A certification programme is sometimes used to aid in production of bacterial spot-free transplants. As part of the programme seeds are tested for the presence of *X. c.* pv. *vesicatoria*. Inspection of seedlings for bacterial spot is an integral part of the programme and rapid identification of the causal agent is needed (Gitaitis, 1991).

Numerous sources of resistance to bacterial spot occur in pepper (Sowell and Dempsey, 1977), and a source of resistance to bacterial spot of tomato was described recently (Scott and Jones, 1989). However, commercial cultivars of neither crop contain resistance to bacterial spot. Chemical sprays for control have been described, but control by that method has only been partially successful. Prevention of the disease is successful in some regions (Goode and Sasser, 1980).

1.11 *XANTHOMONAS CAMPESTRIS* PV. *PRUNI*: CAUSE OF PRUNUS BACTERIAL SPOT

E.L. CIVEROLO
USDA-ARS, Beltsville

M.J. HATTINGH
Department of Plant Pathology, University of Stellenbosch

1.11.1 Introduction

Prunus bacterial spot (PBS) disease was first described in 1902 in North America (Bradbury, 1986). The disease affects most commercially grown cultivars of several *Prunus* species. The causal agent of PBS disease is *Xanthomonas campestris* pv. *pruni*. PBS disease occurs in most countries where stone fruit are grown. However, it is most serious in areas with warm and wet or humid conditions during the growing season (Fahy and Persley, 1983; Bradbury, 1986; Du Plessis, 1988a). Currently, the disease is known to occur in all continents (Anonymous, 1978; Commonwealth Mycological Institute, 1987). *X. c. pruni* naturally infects all cultivated *Prunus* species and hybrids, including apricot (*P. armeniaca*), cherry (*P. avium*), almond (*P. amygdalus*), peach and nectarine (*P. persica* and *P. persica* var. *nectrina*), European plum (*P. domestica*), Japanese plum (*P. salicina*), flowering plum (*P. blireiana*) and sour cherry (*P. cerasus*) (Bradbury, 1986; Du Plessis, 1988b; Fahy and Persley, 1983). Several other *Prunus* species and *Sorbus japonicum* can be infected by artificial inoculation (Bradbury, 1986). A single report of natural infection of *Terminalia catappa* (Indian almond or myrobalan) has not been confirmed (Bradbury, 1986).

In the field, cultivated stone fruit species, cultivars and hybrids range from highly susceptible to highly resistant (Keil and Fogle, 1974; Du Plessis, 1987b; Topp *et al.*, 1989; Topp and Sherman, 1990). For example, no symptoms developed on Laetitia plum in South Africa (Du Plessis, 1987b) or on 17 plum cultivars and selections in Australia (Topp *et al.*, 1989) during several years, even though environmental conditions favoured development of PBS. *P. cerasifera*, *P. angustifolia*, and *P. salicina* might be major sources of resistance to PBS (Topp and Sherman, 1990).

Direct economic losses due to PBS disease result from the reduced marketability of fruit with unsightly blemishes. The entire crop may be affected when conditions are conducive for infection and disease development. In addition, twig and heavy leaf infection with subsequent defoliation devitalizes trees. Severely affected trees are susceptible to winter damage including terminal or extensive dieback. Chronically infected trees become unproductive.

1.11.2 Disease symptoms

Lesions on leaves develop generally as small, circular or irregular (0.5–1 mm diameter) watersoaked spots. These enlarge and become distinctly angular if restricted by smaller leaf veins. However, spots also occur along the main and secondary veins. In addition, developing lesions become dark red, purple, brown or black. The dark spots may be surrounded by grey to yellowish tissue. Several spots may coalesce to form larger affected areas. Affected leaves become yellow with varying degrees of chlorosis developing around individual and coalesced lesions, as well as in the distal portion of the leaf. As lesions mature, cracks develop through and around the spots. Severely affected leaves defoliate prematurely. The affected areas often break away from adjacent healthy or unaffected tissue and drop out, resulting in a shot hole effect. Atypical symptoms on peach leaves include a grey leaf spot on the upper surface, and greenish, yellow translucent appearance of areas infiltrated by bacteria.

Fruit lesions initially appear as small, circular watersoaked spots becoming grey to brown on the surface. These spots may be slightly depressed. As the spots develop, they become light, dark brown, purplish black or black. Small lesions, especially on young fruit, may be surrounded by a narrow zone of watersoaked tissue. Individual spots may be up to 15 mm in diameter depending upon the specific host. Individual spots may coalesce to affect large parts of the fruit surface. Pitting and cracking of these lesions occur as the fruit develops. Gum exudations are frequently associated with fruit lesions.

Watersoaked, circular to elliptical lesions 1–3 cm long develop around buds just below the attachment of the petiole and on green twigs in the spring and summer. Twig lesions that develop in the spring are usually slightly raised, while those that develop in the summer are slightly depressed. The lesions become brown, dark purple or black. The margins of twig lesions may also be watersoaked. As the lesions develop, the epidermis ruptures. Exudations containing masses of bacteria may be associated with these spring lesions. Spring and summer canker symptoms develop distinctly (Du Plessis, 1987c). Spring cankers develop after budbreak on shoots from the previous season's growth, whereas summer cankers appear on green shoots of the current season's growth. Both types of cankers are associated with systemic movement of the pathogen from infected leaf and shoot tissues (Du Plessis, 1984, 1986, 1987a,b, 1988a).

1.11.3 Isolation and disease diagnosis

Presumptive diagnosis of PBS disease can be based on symptoms on leaves, fruit and twigs. Gumming or bacterial ooze may be associated with lesions. However, confirmation of the disease and identification of *X. c. pruni* require additional tests.

X. c. pruni can be isolated on several commonly used bacteriological media containing glucose or sucrose. Colonies are generally wet, shiny, convex, slightly mucoid, and produce a yellow, water-insoluble pigment (Bradbury, 1984). A simple selective isolation medium (XPSM) has been developed for detection and identification of *X. c. pruni* in clinical samples (Civerolo *et al.*, 1982). The efficiency of plating of strains of *X. c. pruni* on XPSM is greater than 90%. *X. c. pruni* can also be isolated on general or selective media following selective enrichment in detached leaves (Randhawa and Civerolo, 1985).

Like other xanthomonads, *X. c. pruni* generally grows more slowly on nutrient agar medium than other bacteria commonly associated with diseased plant tissue. The refractive quality of *X. c. pruni* colonies that develop on nutrient agar spread with washings of peach leaves can be used to distinguish the pathogen from other yellow-pigmented bacteria (Gitaitis *et al.*, 1987a).

X. c. pruni is an aerobic, motile, Gram-negative rod-shaped (0.2–0.8 × 0.8–1.7 μm), non-sporing bacterium with a single polar flagellum. *X. c. pruni* produces xanthomonadin; other biochemical and physiological characteristics are described elsewhere (Bradbury, 1984).

The most reliable confirmation of the identity of *X. c. pruni* is development of characteristic symptoms in leaves of a susceptible host following artificial inoculation (Civerolo, 1975; Du Plessis, 1988a,b). This can be done by inoculating attached leaves on greenhouse-grown seedlings or plants by spraying inoculum at high pressure or by injection infiltration of inoculum into leaf tissue. Alternatively, detached leaves of a susceptible host can be inoculated to confirm pathogenicity of presumptive *X. c. pruni* strains (Randhawa and Civerolo, 1985).

Virulent bacteriophages with high specificity for *X. c. pruni* strains have been described (Eisenstark and Bernstein, 1955; Eisenstark *et al.*, 1955; Stolp and Starr, 1964; Civerolo, 1974, 1976; Du Plessis *et al.*, 1981). However, the use of bacteriophages to detect and identify *X. c. pruni* has been limited. *X. c. pruni* strains that have been examined to date contain one to four DNA plasmids in 10 size classes ranging from about 4.2 to 35.1 megadaltons (MDa). A 26.7 MDa plasmid occurs most frequently (Randhawa and Civerolo, 1987), and was not detected in 24 strains of other bacteria (Kado and Liu, 1981).

1.11.4 Sources of inoculum and disease management

Management of PBS disease is based on resistant cultivars and/or hybrids (Du Plessis, 1988a), integrated cultural practices (Du Plessis, 1988a), and treatment with chemicals (Keil and Carroll, 1967; Moffett, 1973; Keil, 1979; Du Plessis, 1983, 1987b). Several *Prunus* species, cultivars and hybrids are relatively resistant to PBS disease (Keil and Fogle, 1974; Du Plessis, 1987b; Topp *et al.*, 1989; Topp and Sherman, 1990). Precautions should be taken to prevent mechanical transmission of *X. c. pruni* from diseased to healthy trees during pruning (Goodman and Hattingh, 1988a). Also, *X. c. pruni* can be transmitted to plum and apricot nursery trees by budding (Goodman and Hattingh, 1986). Losses due to this type of transmission can be reduced by using buds cut early in summer from healthy trees (Goodman and Hattingh, 1988b). Chemical sprays of copper-containing compounds (e.g. oxychloride and cupric hydroxide plus lime) and antibiotics (e.g. streptomycin or oxytetracycline) can reduce the amount of leaf and fruit infection when applied at the proper time throughout the growing season (Keil and Carroll, 1967; Du Plessis, 1983, 1987a, 1988a). Injection of oxytetracycline into tree trunks can also reduce the incidence of PBS disease (Keil, 1979; Keil and Civerolo, 1979; Du Plessis, 1987a). Resistance to copper or antibiotics (streptomycin or oxytetracycline) has not been reported for *X. c. pruni* populations in orchards.

1.12 *XANTHOMONAS POPULI*: CAUSE OF BACTERIAL CANKER OF POPLAR

M. RIDÉ
INRA, Station de Pathologie Végétale et de Phytobactériologie

1.12.1 Introduction

Bacterial canker due to *Xanthomonas populi* is probably potentially the most damaging bacterial disease of poplar in Europe where it has been known for over 100 years (Koning, 1938). The causal agent was at first identified as *Aplanobacterium* (Ridé, 1958), then *Aplanobacter* (Ridé, 1963) but is now definitively classified as *Xanthomonas* (Ridé and Ridé, 1978a). Poplar canker continues to appear as an endemic disease because of the existence of infected plantations of very old susceptible cultivars such as *Populus euramericana* cvs. *regenerata, grandis* or *brabantica* and the inconspicuous presence of the disease in spontaneous populations of aspen (*Populus tremula*). Moreover the more recent distribution of polyclonal mixtures not tested against *X. populi* and of some clones whose susceptibility to bacterial canker was not sufficiently known (*P. euramericana* cv. Dorskamp, *P. trichocarpa* x. *P. deltoides* Barn) have created good conditions for new infection sources. Therefore bacterial canker remains a constant preoccupation of geneticists who have, in their breeding programmes, made use of North American provenances of *P. trichocarpa, P. deltoides* and *P. tremuloides,* often susceptible or highly susceptible to bacterial canker (*X. populi* is not known on *Populus* in North America).

Poplar bacterial canker occurs in Belgium, France, Germany, Czechoslovakia, the Netherlands, the UK, and probably also in Poland (where it may be confused with *Ceratocystis fimbriata* canker). Its status in Asia is doubtful.

Under the conditions of sanitation now prevailing in the main European poplar-growing countries, the disease is not present in nurseries. But it affects young industrial poplar plantations when susceptible cultivars are introduced near wild cankered aspen or old cankered plantations. One of the best examples occurred when *P. euramericana* cv. Blanc de

Poitou was introduced in 1958–60 into a high-risk area of northern France. In 1964, 30% of the trees were affected by *X. populi* and 10–28% of them had trunk cankers. The cankered trees suffered losses in commercial value of their timber of 5–60% according to the situation.

The response of the tree to *X. populi* in terms of inoculum dose, physiological stages of annual shoots and aggressiveness of bacterial strains gives a valuable indication of the degree of cultivar field resistance. Inoculation by dropping a bacterial suspension onto fresh leaf scars obtained by removing the leaves is suitable through the growing season, from the end of shoot formation to the cessation of growth of the main shoot terminal bud. Inoculation during that susceptible stage was proposed by Ridé and Steenackers and adopted by the FAO International Working Group on Poplar Diseases (1966) as a way of testing various clones and selected trees in European countries for their resistance (Ridé, 1963; Steenackers, 1966; Burdekin, 1972; Gremmen and Koster, 1972; Kechel, 1982).

1.12.2 Disease symptoms

The first signs of infection are observed in spring on the previous year's shoots, shortly before bud-burst: the tissues around infected buds swell, then burst, exuding a thick, whitish bacterial slime which later turns brown. The exudate (Figure 1.9) may also appear as droplets through the lenticels or cracks on the internodes of 1-year-old shoots as

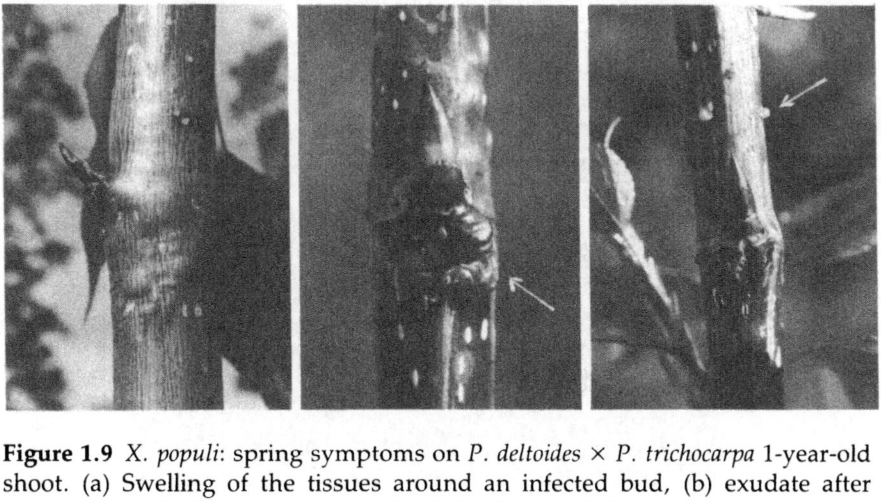

Figure 1.9 *X. populi*: spring symptoms on *P. deltoides* × *P. trichocarpa* 1-year-old shoot. (a) Swelling of the tissues around an infected bud, (b) exudate after stipule scar infection, (c) cracks and whitish bacterial slime appearing through the lenticels.

well as from the edges of reaction callus of perennial cankers. When the branch is stripped, the cambium, phloem and cortical parenchyma appear glassy due to the intercellular progression of *X. populi*; the medullar rays are also invaded. Mucous exudate may also appear later in the season (July and August) on some very susceptible clones, at the edges of developing lesions. The buds infected during the previous year are generally killed, but sometimes give leaves which may exhibit blackened or discoloured areas due to petiole invasion by the bacteria; many of the tender green shoots developing from these buds are killed. On the susceptible cultivars, lignified young twigs either die back at the end of the season or exhibit cracks which develop into perennial cankers. The final appearance of cankers depends on the species or clone of poplar; it may vary from closed cankers to open

Figure 1.10 *X. populi*: open canker on hybrid poplar.

Figure 1.11 *X. populi*: perennial canker on *P. deltoides*.

cankers (Figures 1.10 and 1.11) with more or less even callused margins to those with a hyperplasic pseudo-tumorous appearance (on *P. candicans* for example).

On susceptible poplar clones, active cankers develop year by year on twigs, main branches or trunks and may girdle and kill but also induce changes of tree form resulting from stem breakage and subsequent sprouting. Active cankers also cause a massive degradation of wood structure, even when they are healed.

1.12.3 Isolation and disease diagnosis

X. populi is relatively easily isolated from 1-year-old infected shoots during spring. Small pieces of suitable glassy cortical tissue are either deposited directly on YPGA [LPGA medium in French (Ridé, 1963): yeast extract Difco (5 g), bactopeptone Difco (5 g), glucose (10 g), agar (15 g), distilled water (1000 ml) pH 7–7.2] or macerated in sterile water and streaked on the same medium. At 23°C, smooth, mucoid, cream opaque bacterial growth develops around the pieces after 2–4 days, or from the macerate often after more prolonged incubation. Culture or colonies are suspended in sterile water and restreaked on YPGA to obtain single colonies. *X. populi* may also be isolated at all times of the year from very small nodules taken from canker reaction tissues: in this case, *X. populi* develops much later (10–15 days) around the nodules. It is more difficult to isolate *X. populi* from the exudate which is rapidly invaded by epiphytic microflora.

The following characters at 23°C are generally used for a presumptive diagnosis: aminopeptidase positive, oxidase negative, no growth on nutrient agar or in Hugh and Leifson medium. Hayward's (1964) liquid medium supplemented with 1% yeast extract is used for carbohydrate utilization and on a reciprocal shaker rapidly gives the following spectrum of utilization: L-arabinose –, D-glucose +, mannose +, galactose +, trehalose +, cellobiose – (Ridé and Ridé, 1978a), which is different from that of other members of the genus *Xanthomonas*. Virulent strains of *X. populi* give a hypersensitive reaction on tobacco leaves, avirulent strains do not.

A rapid presumptive diagnosis can also be made by using antisera specific to *X. populi* and the indirect immunofluorescence (IF) technique (Ridé *et al.*, 1975), which is also suitable for detection and enumeration, for example during epidemiological studies (Ridé *et al.*, 1978).

An entirely separate pathovar has been described on *Salix dasyclada* by De Kam (1981) as a subspecies '*X. p.* subsp. *salicis*'. Cross-infection between *X. populi* and *X. p.* subsp. *salicis* on their respective hosts does not occur and the two bacteria are also well separated by molecular methods. Moreover these bacteria cannot be confused with *X. campestris* pv. *populi*, the agent of bacterial bark necrosis of poplar (De Kam, 1984).

Few data are available concerning virulence factors of *X. populi*. Many *X. populi* strains have a plasmid the role of which is not known. A wide range of aggressiveness can be observed for different tested clone × isolate combinations (Ridé and Ridé, 1978b; Ridé *et al.*, 1985). British isolates are generally the most virulent compared to continental ones. Recent work shows a probable differential aggressiveness in some combinations but it is difficult to prove that specific clone ×

isolate interactions exist in terms of 'races'. This is probably a result of the polygenic nature of bacterial canker resistance suspected by Muhle-Larsen (1966) and Steenackers (1976) in a full-sib *P. trichocarpa* breeding programme.

1.12.4 Sources of inoculum

The main source of pathogenic bacteria is the slime produced in spring from active cankers and cracks on 1- to 2-year-old shoots infected during the previous season. The bacterial exudate is dispersed between adjoining trees by wind and rain, mainly in atlantic regions. The viability and persistence of inoculum in the slime and on the leaf surface depends on poplar clones, temperature, rainfall, activity of epiphytic microflora and probably on the release or non-release of inhibitory substances produced on the stipule or leaf surface (Ridé *et al.*, 1978).

Bacterial penetration mainly occurs on preformed and newly developing shoots through fresh stipule scars or fresh petiole scars arising from storm damage or early leaf disease (*Marssonina brunnea*), or else through wounds inflicted by hail or insects. The larvae of the cambium miner *Phytobia cambii* are important in spreading the bacterium within the tree. It is carried along the mine formed in spring and appears as cankers along the galleries the following year.

1.13 *XANTHOMONAS FRAGARIAE*: CAUSE OF ANGULAR LEAF SPOT OF STRAWBERRY

B. RAT
INRA, Station de Pathologie Végétale et Phytobactériologie

1.13.1 Introduction

Angular leaf spot of strawberry is a potentially serious disease which may be a limiting factor in establishment of a new strawberry cultivar (Hildebrand *et al.*, 1967). The disease was first reported in the USA (Kennedy and King, 1962a) and subsequently in New Zealand, Australia, France, Italy, Spain, Greece, Romania, Venezuela, Brazil and Argentina.

1.13.2 Disease symptoms

Small angular watersoaked spots appear first, exclusively on the lower surface of leaves. In the morning when the foliage is heavy with dew, a yellow-whitish exudate appears. After drying out it gives a glittering aspect to the spots. The bacteria are disseminated from the spots by droplets of water (rain or dew): new lesions appear frequently along the main veins of the leaf (Kennedy and King, 1962b). The size of these lesions increases progressively and subsequently the spots coalesce and become visible on the upper surface of the leaf. The death of the tissues invaded by the bacteria leads to reddish-brown irregular spots. The leaflets finally shrivel completely. No symptoms can be seen on the fruits. Similarly, in normal conditions, the bacteria never invade the vascular system of the plant. One report only describes a systemic infection, resulting in a vascular collapse (Hildebrand *et al.*, 1967). This probably takes place exclusively under damp nursery conditions and usually involves a restricted number of plants.

Both the oldest and youngest leaves are resistant to infection. Therefore symptoms essentially show up on leaves of 1–2 months of age. Severe damage may arise in newly planted plots; weakening of young plants results in poor crops.

1.13.3 Sources of inoculum and disease management

The disease spreads on plants collected in contaminated nurseries. The disease shows up on fresh young plants as well as after storage in the cold chamber. The progressive distribution of the disease to new countries probably reflects the ease with which the pathogen is carried on planting material.

The best control of the disease is provided by the use of young plants originating from healthy nurseries. When no symptoms are visible on the plants, the detection of the bacteria is very difficult because the growth of bacteria on agar plates gives rise to a low number of colonies and is often prevented by antagonism from other bacteria. Moreover, in France, saprophytic *Xanthomonas* associated with young strawberry plants have been regularly isolated, even after 6 months cold storage. This precludes the use of serological techniques in diagnosis because of cross-reactions between saprophytic *Xanthomonas* and *X. fragariae*.

1.14 *XANTHOMONAS ALBILINEANS*: CAUSE OF LEAF SCALD OF SUGARCANE

P. ROTT
CIRAD-CA, Station de Roujol, Guadeloupe

1.14.1 Introduction

Two species of bacteria pathogenic for sugarcane belong to the genus *Xanthomonas*. These are *X. albilineans*, the causal agent of leaf scald

(Ricaud and Ryan, 1989) and *X. campestris* pv. *vasculorum*, the causal agent of gumming disease (Ricaud and Autrey, 1989). Although both these pathogenic agents are a threat to sugarcane production, when one considers the geographic distribution and the damage caused, leaf scald is generally considered to be a more serious disease than gumming disease.

Leaf scald can cause serious damage to sugarcane. Whole fields planted with susceptible varieties can be destroyed in a few months or years. The highest yield losses occur during the acute phase of the disease (Martin and Robinson, 1961). The disease affects both the tonnage of cane and the quality of the juice extracted (Spence, 1957; Whittle, 1980).

Recently, in Taiwan, there was a drop in sugar content in mature sugarcane from 20.2% to 23.5% (Chen *et al.*, 1990). In recent years, apart from a few exceptional cases, losses caused by *X. albilineans* throughout the world have been insignificant. This situation has been brought about thanks to the production of healthy plant material and the use of resistant varieties (Egan and Sturgess, 1980). The disease is nevertheless still considered to be potentially highly dangerous and necessitating regular surveillance.

Leaf scald also causes indirect losses. Certain high-yielding varieties which are susceptible to the disease, such as Q63 in Australia, have to be avoided. The cost incurred by replanting areas destroyed by the pathogen, of producing healthy plant material (vitroculture or thermotherapy) and the selection of resistant varieties must be taken into consideration. Leaf scald also necessitates strict measures controlling the movement and exchange of plant material between sugar-producing countries. It is necessary for plant material to pass through a quarantine station such as at Montpellier (France) or Beltsville (USA) which are isolated from sugarcane-growing areas.

There is no precise method for evaluating production losses caused by leaf scald on commercially produced and naturally infected sugarcane. The incidence of the disease is often assessed in terms of the percentage of infection or of diseased plants. This type of assessment is, however, imprecise as leaf scald is a latent disease whose evolution seems to be closely linked to growth and environmental conditions (variety, soil, climate, etc.). This problem also occurs with other sugarcane diseases (Koike, 1981). In experimental conditions, yield losses due to *X. albilineans* can be evaluated in trials, including healthy control plants and artificially contaminated diseased plants. Various parameters can thus be analysed at harvesting time, such as yield, brix, sucrose percentage, etc. These trials are rarely performed because they are expensive and require large areas and many replications. For these reasons the incidence of the disease is often assessed by comparing the frequency and intensity of the symptoms to those of the known reference varieties. However, no method is universal because the systems of notation and the reference varieties vary between countries or research centres.

Sugarcane is practically the only cultivated host of *X. albilineans*. Leaf scald has been observed on maize in Cameroon and in Cuba, but all the other known artificial or natural hosts are wild members of the *Poaceae* (Rott *et al.*, 1988a).

As leaf scald is a latent disease, sugarcane can be contaminated for long periods without showing symptoms. Cuttings taken from the apparently healthy infected stalks represent an important source of primary inoculum and one of the principal means of dissemination of *X. albilineans*. This mode of dissemination is undoubtedly responsible for the wide geographical distribution of the disease.

Leaf scald of sugarcane has been identified in 55 countries up to 1991: in North America, Central America and the West Indies, South America, Africa, South Africa, Asia, and Australia and Oceania.

1.14.2 Disease symptoms

In symptomatology, two phases can be distinguished, chronic and acute, which sometimes appear independently, but generally there is a gradual evolution from the chronic phase to the acute phase (Martin and Robinson, 1961).

The chronic phase is characterized by the presence of lines which are parallel to the leaf veins. These are generally narrow 'pencil line' streaks which can also be several mm wide. Their colour varies from white to yellow on the leaf blade, and they often become violet as they continue along the sheath. On ageing, the lines can become reddish. This is the only external symptom which develops in resistant varieties. When the disease intensifies, a necrosis of the leaf tissue around the white line can be seen which extends progressively from the tip towards the base of the leaf (Figure 1.12). The lines also tend to widen and they become more diffuse on leaves reaching maturity. The fine central line which is characteristic of the disease can, however, always be seen in the centre of the lesion. The widening of the lines coincides with the chlorosis or the whitening of the leaf tissue (Figure 1.13). Chlorosis or whitening may be found on all the leaves. This discoloration of the lamina is accompanied by a drying of the leaf extremities which curl inwards giving the shoot a tapered aspect. This process, which resembles scalding, explains the name given to this disease.

On older stalks, the chronic phase is particularly characterized by the almost synchronous appearance and development of side shoots, without apical dominance. As a result, the shoots at the base of the stalk are the most developed, while the apical dominance is removed when

Figure 1.12 *X. albilineans*: stiff leaves with inward curling at the tip and side shooting.

the extremity of a healthy stalk is cut off. In this case, the lateral shoots which form show the same symptoms as the principal shoots. A longitudinal section of a diseased stalk shows a reddening of the vessels near the nodes and sometimes in the internodes (Figure 1.14). When the cane is very diseased, longitudinal lysigenous cavities can be seen in the stalk. The acute phase is characterized by a sudden wilting of mature stalks, without previous symptoms characteristic of the chronic phase. Apparently healthy sugarcane dies as if it had been killed by drought. This acute phase often occurs after a period of rain followed

Figure 1.13 *X. albilineans*: whitening of the leaf.

by a period of prolonged dry weather, but seems to be limited to highly susceptible varieties (Whittle, 1980).

Latency and eclipse phases can also be distinguished. In the case of many contaminated plants, the disease does not develop for a long period. These plants have either no symptoms or a few white lines on the leaves. This is the latency period which comes to an end for reasons which are as yet unknown (Hayward, 1974). The most commonly given reason is that of stress, particularly climatic or nutritional, which could encourage the development of the disease. Another phase, known as the eclipse phase, can take place at the same time as the

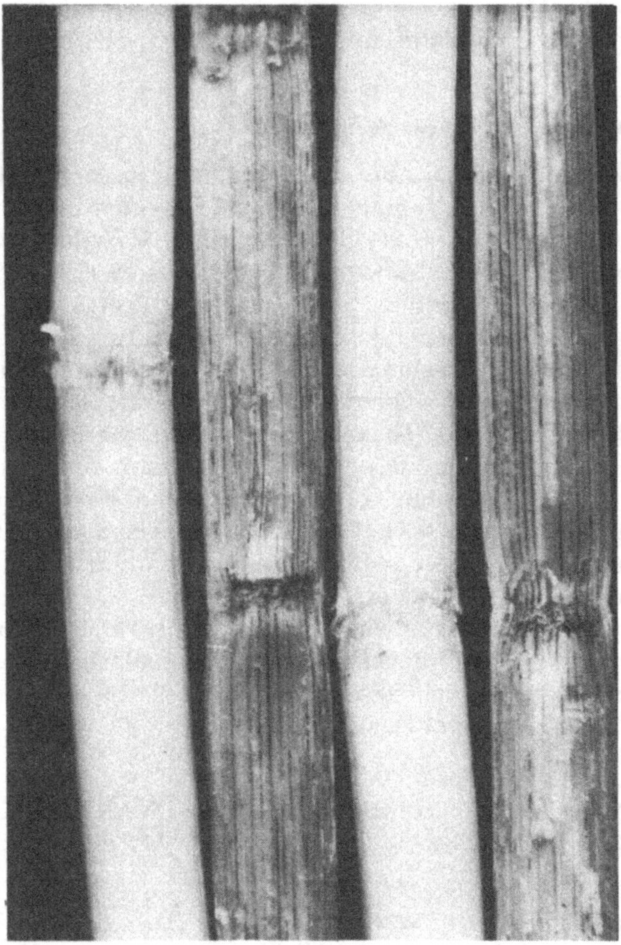

Figure 1.14 *X. albilineans*: diseased cane with dark-red necrosis of vascular bundles and healthy cane without necrosis.

latency phase: as the plant grows, white lines, whitening or necroses appear then disappear on the foliage. These symptoms are no longer visible after the senescence and death of the oldest leaves. Thus the newly formed leaves no longer show symptoms. Depending on the date of observation, a plant can thus be identified as being diseased or healthy. Several of the symptoms of leaf scald are common to other diseases of sugarcane whether they are caused by pathogens or not. *X. campestris* pv. *vasculorum* induces yellow streaks on the leaves and necroses similar to those caused by *X. albilineans*. In addition, the chlorosis or whitening on newly grown leaves is very similar for the two diseases. However, in the case of gumming disease, a gummy exudation occurs when the contaminated stalks are cut.

1.14.3 Isolation and disease diagnosis

Immunological techniques are most frequently used to detect *X. albilineans* in a host plant. Various techniques have been developed, such as immunofluorescence (Léoville and Coléno, 1976; Rott *et al.*, 1988b) and ELISA tests (Moffett and Croft, 1983; Autrey *et al.*, 1990). Three serovars have been identified in *X. albilineans* (Rott *et al.*, 1986). This variability must be taken into consideration during the devising and use of immunological tests. Effective selective culture media to isolate the pathogenic agent in the environment, for example from soils, have not yet been developed. The selective medium proposed by Rivera *et al.* (1985) inhibits the growth of numerous strains of *X. albilineans*. In certain cases large quantities of the pathogen are present in sugarcane stalks without symptoms. The bacteria can thus be detected by isolation on a non-selective culture medium by sufficiently diluting the sample to be analysed.

Apart from the causal agent of leaf scald, several phytopathogenic bacteria can be isolated from sugarcane, in particular *X. c.* pv. *vasculorum* which causes gumming disease, certain of the symptoms of which closely resemble those of leaf scald.

1.15 *XANTHOMONAS CAMPESTRIS* PATHOVARS ON CEREALS: CAUSE OF LEAF STREAK OR BLACK CHAFF DISEASES

E. DUVEILLER and H. MARAITE
Centro International de Mejozamiento de Maiz y Trigo, El Batán, Texcoco and Unité de Phytopathologie, Faculté des Sciences Agronomiques, Université Catholique de Louvain

1.15.1 Introduction

The genus *Xanthomonas* is associated with small-grain cereals (barley, rye, spelt, triticale, wheat) causing bacterial leaf streak (BLS), bacterial stripe or black chaff. BLS of cereals is widely distributed but is sporadic. It has been found in several areas of the USA, being more serious in the western states. In Mexico the disease is more severe at high elevations and warmer wheat-growing areas at sea level in the northwest are at high risk. In southern Brazil, where the area sown to cereal crops has expanded dramatically during the last decade, this pathogen is a potential threat. Optimal environments for its spread also exist elsewhere in South America. BLS has been observed in Africa (Duveiller *et al.*,

1991), Asia (Patel and Shekhawat, 1971), Australia (Moffett, 1983) and the Commonwealth of Independent States (Schneider and Iliukhina, 1983). BLS is currently considered to be absent from western Europe.

The disease is often considered unimportant but it can be devastating if the optimal combination of factors inducing epidemics is encountered. As it is seed transmitted, BLS is a constraint to germplasm exchange. Yield losses are difficult to evaluate because no effective treatment exists and thus healthy control plots cannot be used as a comparison. Shane *et al.* (1987) found that 50% disease severity on the flag leaf of spring wheat resulted in a 13–34% yield loss. A study over 2 years, using single tillers of genotype Alondra, with different degrees of disease severity, and labelled in naturally infected plots in Mexico, showed that 23% yield losses can be expected in a field when BLS incidence is 100% and when 50% of the flag leaf area is damaged (Duveiller, unpublished data). In triticale, losses of up to 43% have been reported (Cunfer, 1988).

1.15.2 Disease symptoms

Symptoms are evident on the leaf and begin as fine watersoaked translucent streaks evolving into elongated brown lesions, several centimetres long, extending between the veins. Lesions are initially distinct but coalesce, covering large areas eventually resembling a blight. Milky drops of bacterial exudate appear under humid conditions which dry to

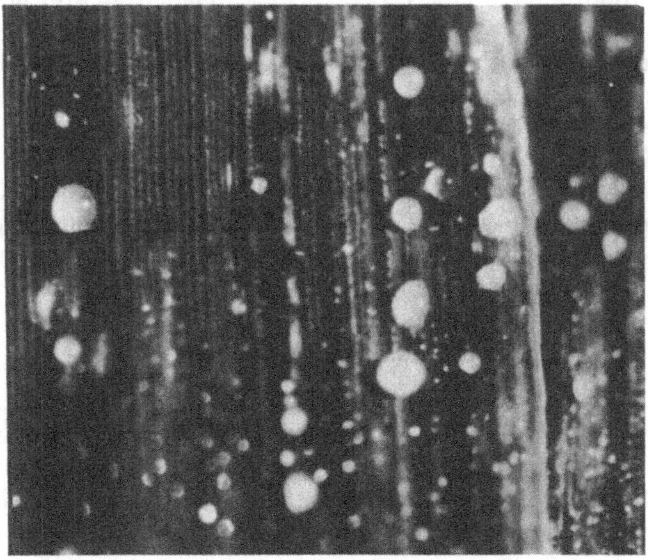

Figure 1.15 *X. c.* pv. *undulosa*: milky droplets of bacterial exudate.

form yellow beads or shiny scales on the leaf (Figure 1.15). In wheat, exudates are not always observed in commercial fields and the pathogen is not always easily identifiable unless it is isolated. Black chaff is characterized by dark brown streaks and blotches on the glumes.

1.15.3 Isolation and disease diagnosis

Semi-selective agar medium XTS has been proposed for isolating the pathogen from seed (Schaad and Forster, 1985). The recovery of the bacterium on XTS may, however, be difficult in some cases due to gentamycin used in the medium. Wilbrink's boric acid-cephalexin (WBC) medium (Duveiller, 1989) may represent an alternative. However, testing seed leachates and dilution plating on to an agar medium is laborious for routine certification. Monoclonal antibodies have been recently developed against *Xanthomonas* pathovars of cereals. Their use in seed detection is promising and might speed up seed testing procedures (Duveiller and Bragard, 1992).

Five pathovars associated with BLS are now included in the International Society of Plant Pathology (ISPP) list of approved names for plant pathogenic bacteria (Dye *et al.*, 1980): *X. c.* pv. *translucens, X. c.* pv. *cerealis, X. c.* pv. *undulosa, X. c.* pv. *secalis* and *X. c.* pv. *hordei.* Protein electrophoresis, fatty acid analysis and DNA–DNA hybridization studies of xanthomonads confirmed that these bacteria belong to the same group, exhibiting significant degrees of DNA binding (Vauterin *et al.*, 1992).

Many reports have failed to present a detailed identification of the bacterium causing BLS, and the pathovar name *translucens* has been used without confirmation of the host range (Bradbury, 1986). Recent studies of multiplication of *X. c.* pv. *secalis* and pv. *translucens*, in host and non-host plants, showed that these two pathogens are different (El-Banoby and Rudolph, 1989). Similarly, multiplication of *X. c.* pv. *undulosa* and pv. *hordei* in a set of wheat genotypes revealed differences in susceptibility and differences between pathotype strains, confirming that precise identification of the pathovars is justified despite overlapping host ranges (Duveiller *et al.*, 1991).

1.15.4 Sources of inoculum and disease management

Little quantitative information is available on the different factors related to the occurrence of field epidemics. Infected seed is the primary source of inoculum but BLS has a very low transmission rate and symptoms are rare on young plants. This makes the appraisal of effective bacteriocide seed treatments difficult. However, seed detection methods have to be specific but not highly sensitive. The pathogen survives poorly in soil and plant debris. The bacterium is ice-nucleation active (Kim *et al.* 1987) and BLS spreads more rapidly in frosted tissues (Azad and Schaad, 1988) but this mechanism is not essential to induce an epidemic. High humidity favours infection and the dissemination of the bacteria and consequently the number of lesions. Bacteria enter the plant through the stomata and multiply in large masses in the parenchyma. The high temperature prevailing in warmer environments boosts the growth of the organism, although the bacterium tolerates a wide range of temperatures (15–30°C). The disease can be observed in areas as diverse as sprinkler-irrigated fields in the USA and subtropical highlands characterized by marked diurnal temperature changes, or with a marked but transient temperature fall due to local thunder. The latter induces tissue congestion and facilitates penetration of the bacterium into the leaf. BLS is also seen in countries where warm and cloudy days may occur alternately. Since there is no effective seed treatment, efforts to control BLS are oriented to seed indexing and the search for resistant genotypes. Resistance to BLS is partial and genotypes may only be classified according to levels of susceptibility. In studies on bread wheat genotypes some *X. c.* pv. *undulosa* strains showed differences in aggressiveness and genotype specialization; for example, Tucano was more resistant than Alondra when inoculated with UPB480 strain from Pakistan but both genotypes were susceptible when inoculated with UPB513 from Mexico. Pavon was more resistant than Tucano and Alondra to both bacterial strains (Duveiller *et al.*, 1991). The appraisal of resistance to BLS in screening fields is performed at the flowering stage and can be based on the principles used for other foliar diseases in cereals (Duveiller, 1990). A disease severity scale based on percentage leaf area damaged has been proposed (James, 1971) but scales that better reflect the nature of BLS are required.

1.16 *XANTHOMONAS CAMPESTRIS* PV. *PELARGONII*: CAUSE OF BACTERIAL LEAF SPOT AND STEM ROT ON *PELARGONIUM* SPP.

W. PAGEL
Geranien-Endisch, Geranienweg

1.16.1 Introduction

Xanthomonas campestris pv. *pelargonii* is the causal agent of bacterial leaf spot and stem rot on *Pelargonium* spp., which is economically the most important disease of that ornamental plant. This is mainly attributable to the practice of vegetative propagation on cuttings, in which a latent infection may easily be transmitted from one generation to the next (Munnecke, 1954). In such a case the entire geranium stock of a nursery can break down within a few weeks. Exact figures concerning the amount of damage caused are not available; however, millions of plants may lose commercial value due to *Xanthomonas* attack each year. For example, during 1983 and 1984 more than 100 000 plants were lost in Minnesota (Kennedy and Pfleger, 1986). This pathogen is feared in all countries in which geraniums are cultivated and sold. This includes not only the European countries and North America but also Africa, Asia South America and Australia (Bradbury, 1986).

All *Pelargonium* spp. are natural hosts of *X. c.* pv. *pelargonii*. Cultivars of *P. graveolens*, *P. acerifolium* and *P.* x *domesticum* show resistance. *P. zonale* and *P. peltatum* hybrids, which are important for horticultural production, are highly susceptible, while *P. grandiflorum* hybrids are among the resistant geraniums (Munnecke, 1954; Knauss and Tammen, 1967; Wainwright and Nelson, 1972).

1.16.2 Disease symptoms

Hellmers (1952) was the first to show that both bacterial leaf spot and stem rot are caused by the same pathogen. Thirty years previously, Brown (1923) described the bacterial leaf spot caused by *X. c.* pv. *pelargonii*. The leaf spots can first be seen on the undersurface of leaves where very small dark yellowish green watery bubbles appear which, against the light, can be recognized as oily yellowish spots. As the spots enlarge, the rotted tissue becomes brownish-red and turns necrotic with collapse of the undersurface of the leaf (Figure 1.16). Against the light, the dark spots are surrounded by a greenish yellow halo. At a later stage, the affected intercostal region yellows and withers. Finally the leaf lamina and the petiole dries out by which time the bacteria have already entered the stem (Munnecke, 1954).

Stem rot is the consequence of an acute or latent infection of the stem. Under appropriate conditions, a rapid multiplication of the bacteria and their spread throughout the plant is the result. The first symptoms are of more or less leaf yellowing of the vein and swellings, followed by a withering of the leaf lamina. In dry soil and at low temperatures this stage may last for several weeks. The wilting of *P. peltatum* leaves is not so obvious as that of *P. zonale* leaves, due to their greater firmness. The first stem symptoms can often be recognized at the stem bases by brownish-black colouring of the tissue moving upwards along the stem. The first spots may also occur close to the point of entry of bacteria. The coloured stem areas gradually dry out. Later the plant bends (Hellmers, 1952; Munnecke, 1954).

At high humidity and with rapid multiplication of bacteria, dry rot in the stem does not occur first, but rather bacterial slime oozes out of the split stem (De Peuter, 1989). Following this, soft rotted tissue appears which later dries out.

Figure 1.16 *X. c.* pv. *pelargonium*: wilting of *Pelargonium zonale* after inoculation.

Cuttings infected with *X. c.* pv. *pelargonii* fail to root or show a delayed rooting. The typical stem rot symptoms may develop within 2–4 weeks (Munnecke, 1954). In other cases infected cuttings remain symptomless (Dougherty *et al.*, 1974).

1.16.3 Isolation and disease diagnosis

Diagnostic methods for *X. c.* pv. *pelargonii* have changed in recent years, but the need for reliable test methods remains the same. The two most important requirements are: the greatest possible sensitivity, as theoretically only one bacterium is sufficient for disease; and a high level of reliability of the results. The latter of these requirements not only presumes accurate recognition of the bacteria but must also take into consideration a possible discontinuity in the distribution of the pathogen in the stem when samples are taken (De Peuter, 1989).

The histofluorescence test according to Lemattre (1966) is an older method which is based upon the increased build-up of lignin in infected tissue. However, this elegant method has some disadvantages. Because of the discontinuous distribution of pathogens in the stem, several sections per plant must be investigated. In addition, the sensitivity of this method for the production of elite stock does not seem to be sufficient (Digat, 1978).

Many nurseries use an isolation method, where basispetal stem portions are ground in a buffer and aliquots of the extract are plated on a nutrient medium (YDC agar or PDA; Digat, 1978). As there is no selective medium for *X. c.* pv. *pelargonii* (Schaad, 1988), saprophytes and endophytes also grow on the nutrient medium. Colonies suspected to be of *Xanthomonas* therefore need to be checked for their pathogenicity, which can take several weeks. Since specific antisera are available these time-consuming tests can be replaced by serological methods so that a sure identification is now possible. Among the advantages of the isolation method is its potentially high sensitivity (Van Vuurde, 1987), although there are also disadvantages: results can only be obtained after a few days; the plants contain substances which could inhibit bacterial multiplication (Reuther and Sonneborn, 1983); and a high saprophyte–*Xanthomonas* ratio can reduce the sensitivity to 10^7 cells/ml because the fast-growing saprophytes overgrow the xanthomonads (Van Vuurde, 1987). The latter problem might be reduced by an immunoisolation technique where the *X. c.* pv. *pelargonii* cells are bound to a specific antibody and directly and specifically taken from the plant extract and spread onto a nutrient medium. Apart from elimination of the saprophytic bacteria there is a concentration of the target organism and an increase in sensitivity (Jones and Van Vuurde, 1989).

Direct proof of the presence of *X. c.* pv. *pelargonii* in plant extracts can be achieved by serological methods, and the IF test is most commonly used. It reaches a relatively high sensitivity (10^2–10^3 cells/ml). Apart from their fluorescence, the shape of the fluorescing cells can be used as a criterion to avoid misinterpretation (Digat, 1978; Van Vuurde, 1987).

The ELISA test is so far inadequate for identification of *X. c.* pv. *pelargonii*. The sensitivity of the currently available antisera is too low, as is the reliability of the results (Tuinier and Stephens, 1989; Anderson and Nameth, 1990). As only one criterion is used, false positive results are more easily obtainable than in the IF test (Van Vuurde, 1987). Should in the future a highly specific antiserum become available, the use of ELISA will increase, because this method can be automated. The same applies to the dot-blot method, where new techniques and blotting materials enable greater sensitivity of detection of *X. c.* pv. *pelargonii* (Anderson and Nameth, 1990; Jacquet, 1990).

1.16.4 Sources of inoculum and disease management

Because of the dire consequences of a *Xanthomonas* infection for the cultivation of geraniums, a special propagation system is used which differs from most other ornamental plants. Starting from a *Xanthomonas*-free nuclear stock, a mother stock of plants is built up by vegetative propagation over three to five generations, from which the cuttings are harvested for commercial use.

The basis for the yearly production of new nuclear stock consists of the best plants of the old nuclear stock, which, in the year before, had been examined for various signs of *Xanthomonas* and other diseases. Some young plant nurseries only use the meristem, or stem tips, of the best plants from which new plants grow within 3–5 months (Pillai and Hildebrandt, 1968; Reuther, 1975). Other nurseries use cuttings, all of which have their lower parts checked. Although time-consuming, meristem tip culture minimizes the risk of a dormant latent infection; repeated checks of nuclear plants must not be neglected.

For many years bacterial leaf spot and stem rot have not been associated with seed-propagated geraniums, because the level of seed transmission of *X. c.* pv. *pelargonii* was regarded as low (Koucheki, 1973). But with the growing popularity of geraniums in the USA, the disease has also been observed in plants grown from seed (Stephens and Tuinier, 1989). Therefore, the *Xanthomonas* problem is unlikely to become less important in the future.

1.17 *XANTHOMONAS CAMPESTRIS* PV. *HYACINTHI*: CAUSE OF YELLOW DISEASE IN *HYACINTHUS*

J. VAN DOORN and E.J.A. ROEBROECK
Bulb Research Centre, The Netherlands

1.17.1 Introduction

Yellow disease (Dutch 'Geelziek') of hyacinth was the first disease caused by a xanthomonad to be described and one of the first plant diseases recognized to be caused by a bacterium.

At the instigation of the hyacinth growers, the late Professor Hugo de Vries recommended that his pupil Wakker investigate a strange, very contagious disease among hyacinths, which endangered hyacinth culture in Holland in the 1870s (Beyer, 1966). The causative agent was named *Bacterium hyacinthi* by Wakker in 1881 (Wakker, 1887), later changed to *Pseudomonas hyacinthi*. In 1939 the name was changed to *Xanthomonas hyacinthi*. Finally it was designated a pathovar of *Xanthomonas campestris*: *Xanthomonas campestris* pv. *hyacinthi*.

Originally, the sole host was thought to be *Hyacinthus orientalis*, but later it was shown that some other, closely related bulbous Liliaceae can be infected as well. The closely related plant, *Scilla tubergeniana*, is apparently another natural host; *Scilla hispanica*, *Muscari armeniacum*

Table 1.3 Area of hyacinths cultivated in 1989

Country	*Area (ha)*
The Netherlands	1024
Japan	24
France	12
Australia	10
Poland	5

Data from the International Flowerbulb Centre, Hillegom, The Netherlands.

and probably also *Galtonia candicans* show symptoms only when infected artificially with the pathogen (Janse and Miller, 1983; Beyer, 1972).

Reports on the occurrence of *X. c.* pv. *hyacinthi* (*Xch*) in the field are (with a single exception) restricted to The Netherlands (Huang, 1986) where by far the greater part of hyacinth culture is situated (Table 1.3). In 1989, about 200 million hyacinth bulbs with a production value of 50 million Dutch guilders were produced in The Netherlands. According to data of the Dutch Flowerbulb Inspection Service for 1982–89 between 0.7 and 2.4% of hyacinth bulbs were rejected annually because of yellow disease. Yield losses resulting from control measures involving mechanical or chemical destruction of diseased plants in the field and heat treatment of bulbs are hard to assess.

1.17.2 Disease symptoms and course of infection

The typical yellow disease symptoms of leaf and bulbs of hyacinths will be demonstrated by describing the course of infection of this particular example of xanthomonad bacteriosis. As a primary infection, yellow disease is dispersed by wind and water droplets. A certain sequence of weather conditions is particularly associated by experienced growers with a high rate of spread of yellow disease among the plants.

Wet and windy or misty weather, followed by dry, warm periods may give rise to a rapid development of the disease. There is an increased risk of infection following damage inflicted by a frost or hail storm. The prevailing southwesterly winds in Holland result in typical infection patterns of yellow disease in a northeasterly orientation in the field.

In February or March the first hyacinth sprouts emerge in the fields. Shoots from infected bulbs will carry *Xch* and act as foci of infection. Primary infection occurs via hydathodes, stomata and wounds on hyacinth shoots (Saaltink, 1966; Beyer, 1972; Huang, 1986).

Figure 1.17 *X. c.* pv. *hyacinthi*: leaf top infected near the hydathodes ('spetters').

Wet conditions will give the bacteria the opportunity to spread via droplets on the leaves, possibly aided by flagellar locomotion. Guttation droplets, excreted under humid conditions by hydathodes at the top of leaves can be colonized by *Xch*. When in springtime morning mists are dispersed by the sun, guttation droplets with *Xch* are sucked into the air chambers beneath hydathodes or stomata. There *Xch* can multiply and cause localized lesions. Macroscopically these infections appear as small, dark green and watersoaked spots ('spetters') (Figure 1.17). These spots may spread over the leaves and coalesce to form stripes of watersoaked tissue, especially near the edges of the leaves. The affected tissue finally becomes brown to black; this symptom is known among hyacinth growers as black rim ('zwartrand') (Figure 1.18).

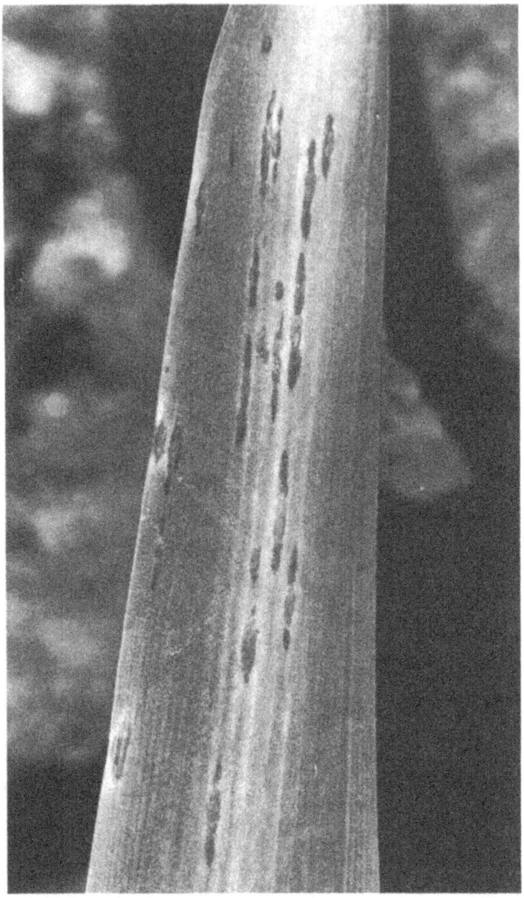

Figure 1.18 *X. c.* pv. *hyacinthi*: leaf with black rim symptoms.

Within the leaves the bacteria will spread under favourable conditions in the intercellular spaces between parenchyma cells. In some cases the bacteria may enter the vessels, leading to rapid development of infection and symptoms. A streak-like symptom may develop particularly from the marginal leaf spots, whereas in the white leaf part at subsoil level the affected tissue is visible as bright-yellow lines (Figure 1.19).

Besides humidity, temperature is the most important physical parameter that influences the course of infection. A high humidity, followed by hot, dry weather can give the characteristic symptoms in less than 1 week. In the more susceptible cultivars the shortest incubation period was found to be only 6–7 days at 25°C. At temperatures in the range of 12–15°C, symptom development can take up to 3 months

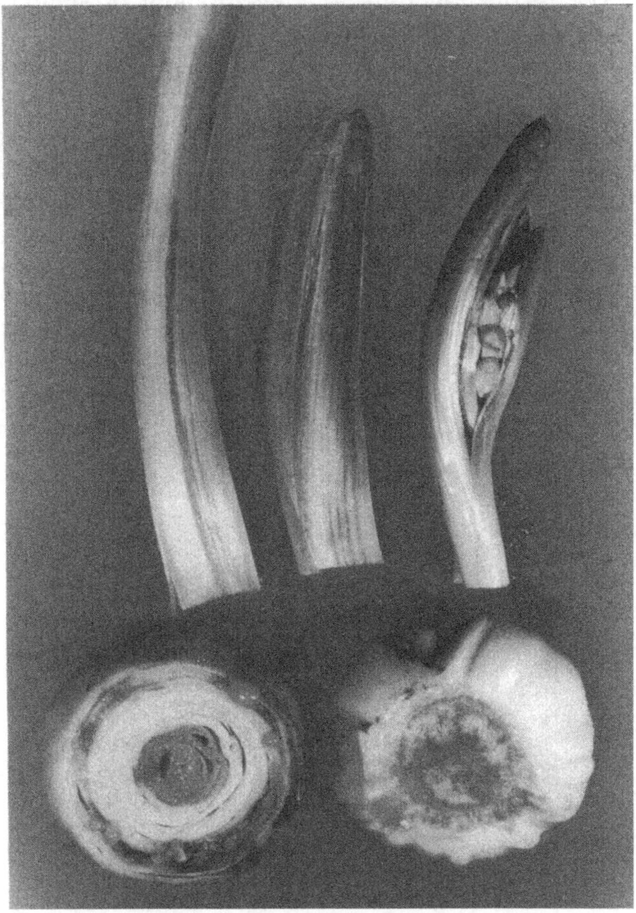

Figure 1.19 *X. c.* pv. *hyacinthi*: bulb halves and sprout from a hyacinth with yellow disease. Bacterial slime is seen between the bulb scales at a cross-sectioned half and near the basal plate. In the corresponding flower stem and leaves, infected xylem beams upward.

(Beyer, 1972). When *Xch* is multiplying in the intercellular spaces in the leaves, the downward movement of the disease towards the bulb nose is slow. When the bulb nose is reached, especially in between the bulb scales, the dispersion of the bacteria is slowed down considerably because of the small size of the intercellular cavities.

The characteristic bulb symptoms of yellow disease are clearly seen when a segment is cut with a sharp knife from the ultimate tip of the bulb nose or lower part of the basal plate; the so-called visitation of the bulb. The incubation time for development of symptoms in the hyacinth bulb stored at 25°C is about 4–6 weeks.

Often *Erwinia carotovora* pv. *atroseptica*, familiar to the hyacinth growers as white snot ('witsnot'), is found in hyacinth bulbs with *Xch* infection. This may hamper diagnosis by obscuring the characteristic yellow disease symptom with its glassy, slimy and resinous growth. Sometimes the infection stops near the top of the bulb to give nose-yellow ('neusgeel'). Once the basal plate is reached, explosive multiplication of *Xch* occurs to give bottom-yellow ('bodemgeel') (Figure 1.19). From the basal plate spread at 20°C can occur in a few days upwards into the scales via the xylem, which may be affected to give scale-yellow

Figure 1.20 *X. c.* pv. *hyacinthi*: leaf with a characteristic secondary infection at the leaf top.

('rokgeel') (Figure 1.19). Secondary symptoms are seen during the next growing season. If the basal plate has been heavily infected the sprout will emerge blind ('blinden'). Bulbs with less severely infected basal plates will sprout more slowly than healthy hyacinths and will wilt suddenly to give drop-outs ('zakkers'). From the infected basal plate *Xch* can infect the growing hyacinth leaves via the xylem, thus giving rise to streamers ('stralers') consisting of glassy dark green stripes in the leaf lamina and yellow stripes in the white underground parts (Figure 1.19).

In early spring hyacinths can be found with one or more wilted leaf tips. The leaf tops hang down, dry out quickly and turn black to give flags ('vlaggers'). These symptoms arise from bulbs with nose-yellow when the top of the emerging sprout, formed for the greater part by the exterior leaf, is infected (Figure 1.20). New infections will occur when bacteria are released from 'stralers', 'spetters' or 'vlaggers'.

Sub-epidermal growth of *Xch* is hygroscopic because of the production of xanthan; leaf tissue will burst at these sites after rainfall. With this phenomenon the life cycle of yellow disease is complete and inoculum is available for wind dispersal.

1.17.3 Isolation and disease diagnosis

Morphologically, *Xch* cannot be distinguished from the other pathovars. The bacterial cells have an average length of about 1.5 µm, possess one polar flagellum, are yellow pigmented, form shining ochre colonies on agar plates, and show an optimum growth at temperatures between 28 and 30°C (Van den Mooter and Swings, 1990).

To assess whether certain symptoms in hyacinth plants or bulbs are caused by the contagious yellow disease, a rapid, specific and sensitive detection method is needed. Until recently, methods such as dilution plating of suspected hyacinth tissue and immunofluorescence studies with polyclonal antisera were used for this purpose, each with their obvious drawbacks of long incubation period, and occurrence of cross-reactions, respectively (Kamerman, 1975; Obata, 1986).

In 1990 a monoclonal antibody was raised that appeared to be useful in testing hyacinth plant extracts by indirect ELISA (Van Doorn *et al.*, 1990). This antibody of the IgG class recognizes a soluble, membrane-associated protein with an apparent molecular mass of about 44 000.

When this monoclonal antibody was tested with some other *X. campestris* pathovars, the pathovars *graminis*, *campestris* and *pruni* showed cross-reaction. However, it is not clear whether these pathovars can be present on hyacinths as epiphytes, and if so, in sufficient numbers to give false positive results. Only *X. campestris* pv. *graminis* has been isolated once or twice from hyacinth plants.

New and complementary monoclonal antibodies will be developed to improve this immunological detection method. The ELISA test based on this monoclonal antibody was used in 1991 in a large-scale test for yellow disease, in collaboration with Dutch hyacinth growers (Van Doorn *et al.*, 1992).

It was found that the test was suitable to discriminate yellow disease from other diseases in hyacinths but failed to detect the first leaf symptoms ('spetters') in all cases. To improve the sensitivity of detection of yellow disease a method, based on the polymerase chain reaction, is now being developed.

1.17.4 Sources of inoculum and disease management

In addition to wind dispersal, spread of yellow disease can also be mediated by infected tools; by inexperienced people walking through the (wet) field; by animals such as rabbits or foxes; by birds; by direct contact of infected leaves with adjacent plants; cutting of flowers with contaminated knives or hands; and wounding of bulbs by digging machines. Also wounding of bulbs during mechanical sorting may easily disperse bacteria (Beyer, 1972; Kamerman, 1975). No data are available on the occurrence of *Xch* on other plant species as an epiphyte, and the bacterium apparently cannot survive in the soil.

In the field, growers try to prevent spread of the disease by eliminating foci of infection as early as possible. Infected plants are removed and destroyed by spraying with hydrochloric acid or formalin; as a precaution

Table 1.4 Heat treatment ('heetstook') of hyacinth bulbs

Period of treatment	*Month*	*Temperature (°C)*	*Purpose*
0–4 weeks after harvest	June–July	25–30	
4–6 weeks	July–August	30	Pretreatment; reduction of flower-setting and bulb metabolism.
2 weeks	August–September	38	Hampering of bulb metabolism will minimize heat damage.
3 days	September	44	Killing of the xanthomonads in the bulb.
Time until planting	October	25–30	Reduction of sprouting.

the adjacent hyacinths without visible symptoms are destroyed as well. During storage, eradication of bacteria in bulbs can be accomplished by a hot-air treatment ('heetstook') (Table 1.4). This method, developed in the early 1920s and perfected with time, is not harmful to the sprout, but severe enough to achieve disinfection with respect to yellow disease. This treatment is performed in large, ventilated heating rooms several weeks after harvesting the hyacinth bulbs. During this heat treatment samples are checked regularly for the presence of heat-damage symptoms by cutting the bulbs in halves. The full treatment as described in Table 1.4 is advised for suspected lots of hyacinth bulbs. Sometimes a treatment of 2 days at 44°C, or even 4 weeks at 38°C is applied without raising the temperature to 44°C.

These treatments are considered to be sufficient for presumed healthy lots or planting stock; sometimes this risk is taken when treating bulbs of heat-sensitive hyacinth cultivars. Slightly diseased bulbs can be cured by this treatment and will produce normal plants in the following season. In contrast, heavily infected bulbs will die before planting.

Temperatures above 44°C will cause desiccation of the bulb scales, necrosis of bulb tissue or severe infection by other microorganisms, e.g. *Aspergillus niger* ('roetbollen'). Furthermore, besides this method to kill putative *Xch* in the bulbs, growers nowadays have available hyacinth cultivars which are less susceptible to yellow disease (Van Tuyl, 1982; Boeder and Van Tuyl, 1989).

1.18 *XANTHOMONAS CAMPESTRIS* PV. *MANGIFERAEINDICAE*: CAUSE OF BACTERIAL BLACK SPOT OF MANGOES

O. PRUVOST
Laboratoire de Phytopathologie, CIRAD/FLHOR

B.Q. MANICOM
ITSC, Nelspruit

1.18.1 Introduction

A bacterial disease of mango with symptoms similar to what is now called bacterial black spot was described in South Africa at the beginning of the twentieth century. The causal agent was reported to be *Bacillus mangiferae* n. sp. (Doidge, 1915). It is possible that this was based on use of mixed cultures. The first description of the causal agent, then designated *Pseudomonas mangiferae-indicae*, was recorded in India by Patel *et al.* (1948a). The causal organism is now named *Xanthomonas campestris* pv. *mangiferaeindicae* (Patel, Moniz and Kulkarni 1948) Robbs, Ribeiro and Kimura 1974. The disease has now been reported in most tropical and subtropical countries where mangoes are cultivated (Patel *et al.*, 1948b; Robbs, 1954; Tarr, 1957; Liao, 1972; Steyn *et al.*, 1974; Daniel *et al.*, 1975; Moffett *et al.*, 1979; Afzal, 1987; El-Goorani, 1987; Buruchara *et al.*, 1990; Fukuda *et al.*, 1990; Lim *et al.*, 1991). Disease incidence is high when rainfall and high temperatures occur at the same time, and bacterial black spot is regarded as one of the most serious diseases in countries where such environmental conditions occur (e.g. the Republic of South Africa).

X. c. pv. *mangiferaeindicae* is pathogenic on mango, but also to other plant species in the family of *Anacardiaceae*, e.g. cashew (*Anacardium occidentale* L.; Patel *et al.*, 1948a,b), and pepper tree (*Schinus terebenthifolius* Radii; Pruvost *et al.*, 1992). Strains isolated from ambarella (*Spondias cytherea* Sonn.) in Martinique Island (Rott and Frossard, 1986) can induce lesions after inoculation on the above-mentioned plant species (Pruvost and Luisetti, 1989), but do not seem able to infect mango under natural conditions.

1.18.2 Disease symptoms

X. c. pv. *mangiferaeindicae* can infect all the aerial parts of mango, with leaf and fruit lesions being the most common. Lesions on the leaves start as small (1–3 mm in diameter) watersoaked lesions. Lesions are

Figure 1.21 *X. c.* pv. *mangiferaeindicae*: infected mango leaf.

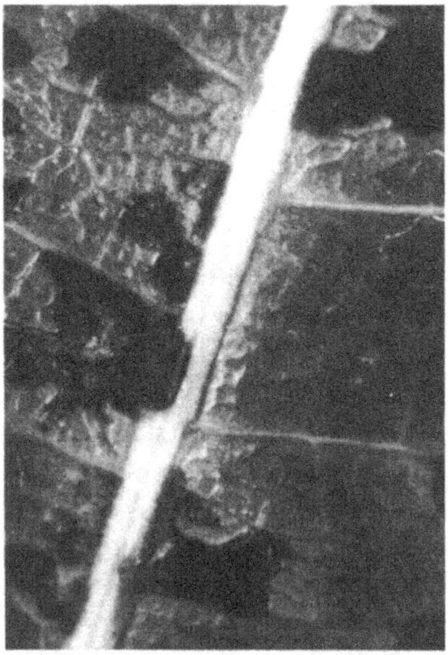

Figure 1.22 *X. c.* pv. *mangiferaeindicae*: infected mango leaf lesions.

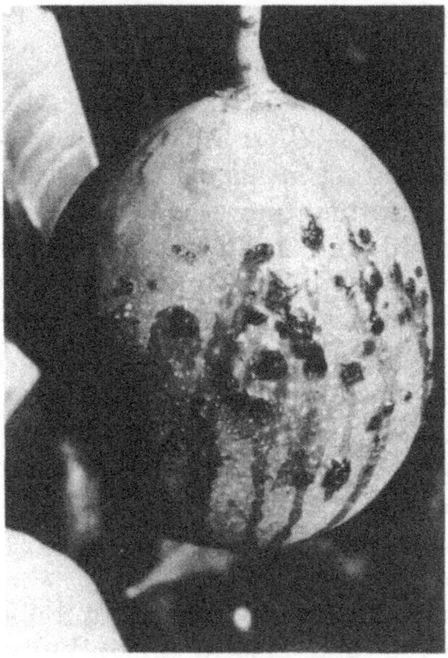

Figure 1.23 *X. c.* pv. *mangiferaeindicae*: infected mango fruit lesions.

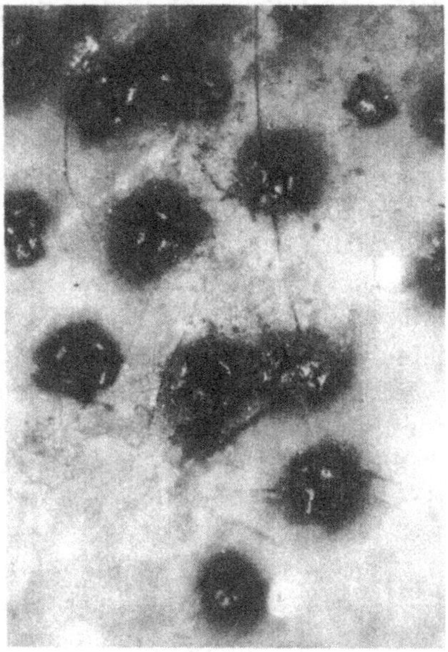

Figure 1.24 *X. c.* pv. *mangiferaeindicae*: infected mango fruit lenticels.

angular and are delimitated by the veins (Figures 1.21 and 1.22). Older lesions turn black, and are raised slightly. A yellowish halo usually forms around each lesion. Some exudates may occur under very humid conditions. They do not increase noticeably in size, but a number of these may coalesce. Very old lesions (1 year or more) become light grey, dry, and crack.

Fruit lesions start as watersoaked halos around lenticels or wounds (Figures 1.23 and 1.24). As lesions enlarge, they become raised, blacken, and crack open. Lesions generally enlarge irregularly up to 15 mm in diameter and extend to a depth of 8–15 mm into the flesh. They usually produce a gummy and very infectious exudate. Fruit can drop, especially when infections start on young fruit or when the fruit stalk becomes infected, causing economically important crop loss.

Stem lesions are black cankers, which generally form longitudinal cracks. Gummy exudates may also occur. Stem cankers occur only on very susceptible cultivars.

During the summer season, bud necrosis induced by *X. c.* pv. *mangiferaeindicae* can be observed only after occurrence of a cyclone. Wounds are necessary for the pathogen to induce such lesions (Pruvost *et al.*, 1993).

1.18.3 Isolation and disease diagnosis

X. c. pv. *mangiferaeindicae*, a non-pigmented xanthomonad, can be isolated easily from diseased plant samples using standard methods of isolation. On media containing glucose (e.g. NGA or YPDA) or sucrose (e.g. SPA or Wilbrink's), colonies appear after incubation for 2–3 days at 28–30°C and are initially light grey, turning into a white to cream colour with age. They are mucoid, round, and shallowly convex. *X. c.* pv. *mangiferaeindicae* has biochemical and physiological properties of *X. campestris* (Manicom and Wallis, 1984). Polyclonal antisera, which may sometimes react with antigens of a few other xanthomonads (e.g. *phaseoli* or *corylina*), are available and are useful for confirmation of identity, but certain identification must be based on pathogenicity. Pathogenicity can be tested by depositing droplets of bacterial inoculum on wounds performed with a sterile needle on attached or detached mango leaves.

Very occasionally, yellow-pigmented colonies, typical of *Xanthomonas*, may be isolated. Pigmented strains have been isolated in Brazil (Robbs *et al.*, 1978), in Reunion Island (Pruvost and Luisetti, 1989), and in South Africa. Yellow-pigmented strains from Brazil and Reunion Island had reduced virulence on several hosts, e.g. mango, cashew and pepper tree (Pruvost and Luisetti, 1989). Lesions induced by yellow-pigmented strains are indistinguishable from those induced by non-pigmented strains. Since no studies have been performed on the taxonomy of this group, it is not known whether these yellow-pigmented strains are variants of pv. *mangiferaeindicae* or belong to another taxon. At the moment, yellow-pigmented strains and non-pigmented strains isolated from *ambarella* are provisionally named pv. *mangiferaeindicae*, but their taxonomic position should be evaluated carefully.

1.18.4 Sources of inoculum and disease management

X. c. pv. *mangiferaeindicae* can overwinter on mango trees. Leaf and stem lesions are the most epidemiologically significant sources of inoculum. There is a direct correlation between the disease incidence on leaves during winter time and fruit infections (Manicom, 1986; Pruvost *et al.*, 1990). The causal agent survives year-round as an epiphyte on various aerial parts of mango trees (Manicom, 1986; Pruvost *et al.*, 1990, 1993; Pruvost and Luisetti, 1991). Its ability to survive as an epiphyte on weeds in not known.

Young leaves are resistant to infection. On mature leaves, *X. c.* pv. *mangiferaeindicae* is essentially a wound pathogen (Manicom, 1986; Pruvost *et al.*, 1990).

Susceptibility of the fruit of cvs Early Gold and José increases with fruit age. The pathogen infects fruit through wounds and lenticels.

There is a positive correlation between epiphytic populations on young mangoes and subsequent infections during the rainy season (Pruvost and Luisetti, 1991).

The ability of the causal agent to survive in the soil is not known precisely, but may be limited to only a few days. The pathogen survived for 2–3 months on plant debris in Reunion Island. However, environmental and edaphic factors that affect this survival are not known.

Young trees tend to be more susceptible to bacterial black spot than older trees.

Wind-driven rains are essential for the spread of *X. c.* pv. *mangiferaeindicae* (Manicom, 1986; Pruvost *et al.*, 1990). Tools used for orchard maintenance can also disseminate the pathogen. Although some reports are available (Kishun, 1986; Van Zyl *et al.*, 1988), the role of insects in the dispersal of the disease is not known precisely. Long-distance movement of the pathogen is essentially with diseased propagating material, such as budwood, rootstock seedlings, and budded plants. There are no conclusive examples of the spread of *X. c.* pv. *mangiferaeindicae* through diseased fruit. Seed transmission does not seem to occur.

Although very drastic eradication campaigns could eliminate the disease, there are no concrete examples available. In areas where bacterial black spot is endemic, integrated disease management is recommended. Mango cultivars vary in susceptibility to bacterial black spot and it is advised that planting very susceptible cvs (e.g. Haden, Kent, Keitt, Irwin, Lippens, Brooks, Zill, Ruby, Eldon) be avoided. Planting of wind-breaks around the orchards decreases greatly the incidence of the disease. Orchard maintenance should not be performed while trees are wet. Copper compounds must be sprayed during the summer season (hot and humid). The number of treatments is dependent on rainfall, susceptibility of the cultivar used, and quantity of inoculum available in the orchard (Manicom and Wood, 1981; Pruvost *et al.*, 1989). One or two additional winter sprays, together with pruning of infected twigs, will reduce the inoculum available in the rainy season and, thus, will improve control significantly.

REFERENCES

Abdel-Fatakh, A.A. and Safiyazov, Z. (1976) Study of the hypersensitive reaction of cotton infected with *Xanthomonas malvacearum*. *Uzbekistan Biologija*, **6**, 13–17 (in Russian).

Afzal, M.A. (1987) Outbreaks and new records. Pakistan. Black spot of mango. *FAO Plant Protection Bulletin*, **35**, 165.

Aggour, A.R., Coyne, D.P., Vidaver, A.K. and Eskridge, K.M. (1989) Transmission of the common blight pathogen in bean seed. *Journal of the American Society for Horticultural Science*, **114**, 1002–8.

Aldrick, S.J., Buddenhagen, I.W. and Reddy, A.P.K. (1973) The occurrence of bacterial leaf blight in wild and cultivated rice in Northern Australia. *Australian Journal of Agricultural Research*, **24**, 219–27.

Allen, S.J. (1986) Bacterial blight of cotton 1984–1985. *Plant Disease Survey New South Wales 1984–1985*, 31–2.

Allen, S.J. and West, K.-L.D. (1987) Diseases of cotton in New South Wales during the 1985–1986 season. *Plant Disease Survey New South Wales 1985–1986*, 27–30.

Allen, S.J. and West, K.-L.D. (1991) Predominance of race 18 of *Xanthomonas campestris* pv. *malvacearum* on cotton in Australia. *Plant Disease*, **75**, 43–4.

Alvarez, A.M., Benedict, A.A. and Mizumoto, C.Y. (1985) Identification of *Xanthomonas* and groupings of *Xanthomonas campestris* pv. *campestris* strains with monoclonal antibodies. *Phytopathology*, **75**, 722–8.

Alvarez, A.M., Benedict, A.A., Mizumoto, C.Y., Pollard, L.W. and Civerolo, E.L. (1991) Analysis of *Xanthomonas campestris* pv. *citri* and *X. campestris* pv. *citrumelo* with monoclonal antibodies. *Phytopathology*, **81**, 857–65.

Alvarez, A.M., Buddenhagen, E.S. and Trujillo, E.E. (1974) Bacterial leaf spot of onion caused by *Xanthomonas* sp. *Proceedings of the American Phytopathological Society*, **1**, 120 (abstr.).

Alvarez, A.M., Cho, J.J. and Hori, T.M. (1987) Black rot of cabbage in Hawaii. Hawaii Agricultural Experiment Station Research Series 051-08.87.

Alvarez, A.M. and Lou, K. (1985) Rapid identification of *Xanthomonas campestris* pv. *campestris* by enzyme-linked immunosorbent assay (ELISA). *Plant Disease*, **69**, 1082–6.

Alves, M.L.B. and Takatsu, A. (1984) Variabilidade em *Xanthomonas campestris* pv. *manihotis*. *Fitopatologia Brasileira*, **9**, 485–94.

Anderson, M.J. and Nameth, S.T. (1990) Development of a polyclonal, antibody-based serodiagnostic assay for the detection of *Xanthomonas campestris* pv. *pelargonii* in geranium plants. *Phytopathology*, **80**, 357–60.

Angeles-Ramos, R., Vidaver, A.K. and Flynn, P. (1991) Characterization of epiphytic *Xanthomonas campestris* pv. *phaseoli* and pectolytic xanthomonads recovered from symptomless weeds in the Dominican Republic. *Phytopathology*, **81**, 677–81.

Anonymous (1978) EPPO Data Sheets on Quarantine Organisms: *Xanthomonas pruni* (E.F. Smith) Dowson, EPPO, Paris, France.

Arnold, M.H. and Brown, S.J. (1968) Variation in the host–parasite relationship of a crop disease. *Journal of Agricultural Science*, **71**, 19–36.

Arunakumari A. and Vidaver, A.K. (1986) Identification of *Xanthomonas campestris* pv. *phaseoli* mutants with altered pathogenicity using a detached bean leaflet bioassay. *Phytopathology*, **76**, 1136 (abstr.).

Atkinson, G.F. (1891) Black 'rot' of cotton. *Alabama Agricultural Experiment Station Bulletin*, **27**, 16 pp.

Atkinson, G.F. (1892) Some diseases of cotton. *Alabama Agricultural Experiment Station Bulletin*, **41**, 19–29.

Autrey, L.J.C., Dookun, A. and Saumtally, S. (1990) Improved serological methods for diagnosis of the leaf scald bacterium *Xanthomonas albilineans*. *Proceedings of the International Society of Sugar Cane Technologists*, **20**, 704–13.

Azad, H. and Schaad, N.W. (1988) The relationship of *Xanthomonas campestris* pv. *translucens* to frost and the effect of frost on black chaff development in wheat. *Phytopathology*, **78**, 95–100.

Bach, E.E., Alba, A.P.C., Pereira, A.L.G., Zagatto, A.G. and Rossetti, V. (1980) Estudos serologicos de isolados do *Xanthomonas citri* (Hasse) Dow. que ocurrem em algunas paises. *Resumos Summa Phytopathologica*, **6**, 12–13, 51.

Ben-David, A., Bashan, Y. and Okon, Y. (1986) Ethylene production in pepper (*Capsicum annuum*) leaves infected with *Xanthomonas campestris* pv. *vesicatoria*. *Physiological and Molecular Plant Pathology*, **29**, 305–16.

Beyer, J.J. (1966) Dr. J.H. Wakker (1859–1927). *Netherlands Journal of Plant Pathology*, **72**, 38–45.

Beyer, J.J. (1972) Course of the infection by *Xanthomonas hyacinthi* (Wakker) Dowson ('yellow disease') in the leaf and bulb of the hyacinth. *Mededeling Landbouwhogeschool Wageningen*, 72–30, 3–52.

Bhatt, V.V. and Patel, M.K. (1954) Comparative study of species of *Xanthomonas* parasitising leguminous plants in India. *Indian Phytopathology*, **7**, 160–80.

Bird, L.S. (1986) Half a century of dynamics and control of cotton disease: bacterial blight. *Proceedings of the Beltwide Cotton Producers Research Conference*, National Cotton Council of America, Memphis, TN, pp. 41–8.

Bird, L.S., Brinkerhoff, L.A. and Davis, R.G. (1981) Bacterial blight, in *Compendium of Cotton Diseases* (ed. G.M. Watkins), American Phytopathological Society, pp. 25–8.

Bishop, A.L. and Basarich, L. (1989) Leaf blight of *Alnus rhombifolia* and *A. cordata* caused by *Xanthomonas campestris*. *Plant Disease*, **73**, 610.

Bitancourt, A.A. (1957) O cancro citrico. *Biologico*, **23**, 101–11.

Blake, J.H., Chase, A.R. and Simone, G.W. (1989) A foliar disease of *Fittonia verschaffeltii* caused by a pathovar of *Xanthomonas campestris*. *Plant Disease*, **73**, 269–72.

Boeder, M.R. and Van Tuyl, J.M. (1989) Onderzoek geelziekresistentie maakt vorderingen. *Bloembollencultuur*, **6**, 26–9.

Bondar, G. (1912) Uma nova molestia bacteriana das hastes da mandioca. *Chacaras e Quintaes*, **5**, 15–18.

Bradbury, J.F. (1984) *Xanthomonas* Dowson 1939, in *Bergey's Manual of Systematic Bacteriology*, Vol. 1 (eds N.R. Krieg and J.G. Holt), Williams & Wilkins, Baltimore, pp. 199–210.

Bradbury, J.F. (1986) *Guide to Plant Pathogenic Bacteria*, CAB International, Farnham House, Farnham Royal, UK.

Brinkerhoff, L.A. (1963) Variability of *Xanthomonas malvacearum*: the cotton bacterial blight pathogen. *Oklahoma Agricultural Experiment Station Technical Bulletin*, T-98, 95 pp.

Brinkerhoff, L.A. (1970) Variation in *Xanthomonas malvacearum* and its relation to control. *Annual Review of Phytopathology*, **8**, 85–110.

Brinkerhoff, L.A. and Hunter, R.E. (1963) Internally infected seeds as a source of inoculum for the primary cycle of bacterial blight of cotton. *Phytopathology*, **53**, 1397–401.

Brown, N.A. (1923) Bacterial leaf spot of geranium in the Eastern United States. *Journal of Agricultural Research*, **23**, 361–72.

Brown, S.J. (1976) Plant pathology, in *Agricultural Research for Development* (ed. M.H. Arnold), pp. 157–73.

Burdekin, D.A. (1972) Bacterial canker of poplar. *Annals of Applied Biology*, **72**, 295–9.

Burkholder, W.H. and Starr, M.P. (1948) The generic and specific characters of pathogenic species of *Pseudomonas* and *Xanthomonas*. *Phytopathology*, **38**, 494–502.

Buruchara, R.A., Mukunya, D.M. and Gathuru, E.M. (1990) Bacterial black spot of mangoes in Kenya, in *Proceedings of the Seventh International Conference on Plant Pathogenic Bacteria*, Budapest, Hungary, 11–16 June, 1989 (ed. Z. Klement), Akademiai Kiado, Budapest, pp. 577–82.

Butare, I. and Banyangabose, F. (1982) Effect of soil fertility on cassava bacterial blight in Rwanda, in *Root Crops in Eastern Africa. Proceedings of a Workshop held in Kigali, Rwanda, 23–27 November 1980*, International Development Research Centre, Ottawa, Canada, pp. 53–5.

Canteros, B.I. (1990) Diversity of plasmids and plasmid-encoded phenotypic traits in *Xanthomonas campestris* pv. *vesicatoria*, PhD Dissertation, University of Florida, Gainesville.

Channon, A.G. and Hissett, R. (1984) The incidence of bacterial wilt caused by *Xanthomonas campestris* pv. *graminis* in pasture grasses in the West of Scotland. *Plant Pathology*, **33**, 113–21.

Charudattan, R., Stall, R.E. and Batchelor, D.L. (1973) Serotypes of *Xanthomonas vesicatoria* unrelated to its pathotypes. *Phytopathology*, **63**, 1260–5.

Chase, A.R. and Jones, J.B. (1989) Leaf spot and blight of *Strelitzia reginae* (Bird-of-Paradise) caused by *Xanthomonas campestris*. *Plant Disease*, **71**, 845–7.

Chen, C.T., Liang, Y.G. and Jiang, P.H. (1990) Leaf scald of sugarcane in Taiwan. *Taiwan Sugar*, **37**, 12.

Choi, S.H., Lee, K.G., Choi, Y.S., Kim, Y.T. and Lee, E.J. (1988) Pathogenicity and bacteriological characterization of *Xanthomonas campestris* from mulberry. *Korean Journal of Plant Pathology*, **4**, 143–8.

Chupp, C. and Sherf, A.F. (1960) Crucifer diseases, in *Vegetable Diseases and their Control* (eds C. Chupp and A.F. Sherf), The Ronald Press Co., New York, pp. 237–40.

Cia, E., Balmer, E., Ferraz, C.A.M., Gridi-Papp, I.L. and Paradela, F.O. (1973) Variability of *Xanthomonas malvacearum* (E.F. Smith) Dowson in Sao Paulo State. *Anais da Escola Superior de Agricultura 'Luiz de Querioz'*, **30**, 457–63 (in Portugese).

Civerolo, E.L. (1974) Temperature effects on the relationship between *Xanthomonas pruni* and its virulent phages. *Phytopathology*, **64**, 1248–55.

Civerolo, E.L. (1975) Quantitative aspects of pathogenesis of *Xanthomonas pruni* in peach leaves. *Phytopathology*, **65**, 258–64.

Civerolo, E.L. (1976) Phage sensitivity and virulence relationships in *Xanthomonas pruni*. *Physiological Plant Pathology*, **9**, 67–75.

Civerolo, E.L. (1984) Bacterial canker disease of citrus. *Journal of the Rio Grande Valley Horticultural Society*, **37**, 127–46.

Civerolo, E.L. (1991) CABI-DC-EPPO Data Sheet on International Organisms: *Xanthomonas campestris* pv. *citri*, CAB International, Wallingford, England (in press).

Civerolo, E.L., Sasser, M., Helkie, C. and Burbage, D. (1982) Selective medium for *Xanthomonas campestris* pv. *pruni*. *Plant Disease*, **66**, 39–43.

Commonwealth Mycological Institute (1987) *Distribution Maps of Plant Diseases*, No. 340, Edition 4.

Cook, A.A. and Stall, R.E. (1982) Distribution of races of *Xanthomonas vesicatoria* pathogenic to pepper. *Plant Disease*, **66**, 388–9.

Coyne, D.P. and Schuster, M.L. (1983) Genetics of and breeding for resistance to bacterial pathogens in vegetable crops. *Hortscience*, **18**, 30–6.

Cunfer, B.M. (1988) Bacterial diseases of wheat and their potential importance in tropical regions, in *Wheat Production Constraints in Tropical Environments* (ed. A.R. Klatt), Mexico D.F., CIMMYT, pp. 263–73.

Currey, D.W. (1989) Eradication of citrus canker from Thursday Island. *Queensland Agricultural Journal*, **115**, 78–9.

Daniel, J.F. and Boher, B. (1981) Fluorescent antibody technique for detection of *Xanthomonas campestris* pv. *manihotis* on cassava leaves, in *Proceedings of the Fifth International Conference on Plant Pathogenic Bacteria, Cali, Colombia, 1981* (ed. J.C. Lozano), Centro Internacional de Agricultura Tropical, Cali, Colombia, pp. 176–80.

Daniel, J.F. and Boher, B. (1985) Epiphytic phase of *Xanthomonas campestris* pathovar *manihotis* on aerial parts of cassava. *Agronomie*, **5**, 111–15.
Daniel, J.F., Gardan, L., Luisetti, J. and Prunier, J.P. (1975) Identification de l'agent responsable de la maladie des taches noires de la mangue (*Mangifera indicae*). *Fruits*, **30**, 625–30.
De Cleene, M. (1989) Scanning electron microscopy of the establishment of compatible and incompatible *Xanthomonas campestris* pathovars on the leaf surface of Italian ryegrass and maize. *Bulletin OEPP/EPPO Bulletin*, **19**, 81–8.
De Cleene, M., Leyns, F., Van den Mooter, M., Swings, J. and De Ley, J. (1981) Reaction of grass varieties grown in Belgium to *Xanthomonas campestris* pv. *graminis*. *Parasitica*, **37**, 29–34.
Dedal, O.I., Palomar, M.K. and Napiere, C.M. (1980) Host range of *Xanthomonas manihotis* Starr. *Annals of Tropical Research*, **2**, 149–55.
De Kam, M. (1981) The identification of the two subspecies of *Xanthomonas populi in vitro*. *European Journal of Forest Pathology*, **11**, 25–9.
De Kam, M. (1984) *Xanthomonas campestris* pv. *populi*, the causal agent of bark necrosis in poplar. *Netherlands Journal of Plant Pathology*, **90**, 13–22.
De Peuter, V. (1989) Verwelking by *Pelargonium* veroorzaakt door *Xanthomonas campestris* pv. *pelargonii*. Thesis, Hoger Instituut der Kempen, Belgium.
Dianese, J.C. (1985) Atividade pectolitica das patovares de *Xanthomonas campestris* que afetam a mandioca. *Revista de Microbiologica*, **16**, 195–202.
Diaz, P.C. (1974) La pústula bacteriana de la soya en Venezuela. *Agronomia Tropical*, **24**, 279–87.
Digat, B. (1978) Mise en evidence de la latence epiphylle du *Xanthomonas pelargonii* (Brown) Starr and Burkholder chez le *Pelargonium*. *Annales de Phytopathologie*, **10**, 61–6.
Doidge, E.M. (1915) A bacterial disease of the mango. *Bacillus mangiferae* n. sp. *Annals of Applied Biology*, **2**, 1–44.
Doidge, E.M. (1929) Further studies on citrus canker. *South African Agriculture Department Bulletin*, **51**, 1–31.
Dopson, R.N. (1964) The eradication of citrus canker. *Plant Disease Reporter*, **48**, 30–1.
Dougherty, D.E., Powell, C.C. and Larsen, P.O. (1974) Epidemiology and control of bacterial leaf spot and stem rot of *Pelargonium hortorum*. *Phytopathology*, **64**, 1081–2.
Du Plessis, H.J. (1983) Chemical control of bacterial spot disease on plums: preliminary evaluation of bactericides. *Deciduous Fruit Grower*, **33**, 413–18.
Du Plessis, H.J. (1984) Scanning electron microscopy of *Xanthomonas campestris* pv. *pruni* in plum petioles and buds. *Phytopathologische Zeitschrift*, **109**, 277–84.
Du Plessis, H.J. (1986) Systemic migration and establishment of *Xanthomonas campestris* pv. *pruni* in plum leaves and twigs. *Journal of Phytopathology*, **116**, 221–7.
Du Plessis, H.J. (1987a) Control of bacterial spot on plums by sprays of calcium oxytetracycline and copper and infusion with N-pyrrolidinomethyl-tetracycline. *Phytophylactica*, **19**, 227–9.
Du Plessis, H.J. (1987b) Resistance of Laetitia plum to *Xanthomonas campestris* pv. *pruni*. *Phytophylactica*, **19**, 249–50.
Du Plessis, H.J. (1987c) Canker development on plum shoots following systemic movement of *Xanthomonas campestris* pv. *pruni* from inoculated leaves. *Plant Disease*, **71**, 1078–9.
Du Plessis, H.J. (1988a) Bacterial spot disease of stone fruits: overview of findings. *Deciduous Fruit Grower*, **38**, 128–32.

Du Plessis, H.J. (1988b) Differential virulence of *Xanthomonas campestris* pv. *pruni* to peach, plum, and apricot cultivars. *Phytopathology*, **78**, 1312–15.

Du Plessis, H.J., Loos, M.A. and Matthee, F.N. (1981) Bacteriophage typing of *Xanthomonas campestris* pv. *pruni* isolates from different stone fruit species. *Phytophylactica*, **13**, 57–61.

Duveiller, E. (1989) Research on '*Xanthomonas translucens*' of wheat and triticale at CIMMYT. *Bulletin OEPP/EPPO Bulletin*, **19**, 97–103.

Duveiller, E. (1990) Screening criteria for bacterial leaf streak in bread wheat, durum wheat and triticale. *Proceedings of the 7th International Conference on Plant Pathogenic Bacteria*, Budapest, Hungary, 1989, 1011–15.

Duveiller, E. and Bragard, C. (1992) Comparison of immunofluorescence and two assays for detection of *Xanthomonas campestris* pv. *undulosa* in seeds of small grains. *Plant Disease*, **76**, 999–1003.

Duveiller, E., Bragard, C. and Maraite, H. (1991) Bacterial diseases of wheat in the warmer areas – Reality or myth?, in *Proceedings of the Wheat for Nontraditional Warm Areas International Conference, Iguazu-Falls, Brazil, July 29–August 3, 1990* (ed. D. Saunders), UNDP/CIMMYT.

Dye, D.W. (1969) Eradicating citrus canker from New Zealand. *New Zealand Journal of Agriculture*, **118**, 20–1.

Dye, D.W., Bradbury, J.F., Goto, M., Hayward, A.C., Lelliott, R.A. and Schroth, M.N. (1980) International standards for naming pathovars of phytopathogenic bacteria and a list of pathovar names and pathotype strains. *Review of Plant Pathology*, **59**, 153–68.

Dye, D.W., Starr, M.P. and Stolp, H. (1964) Taxonomic clarification of *Xanthomonas vesicatoria* based on host specificity, bacteriophage sensitivity, and cultural characteristics. *Phytopathologische Zeitschrift*, **51**, 394–407.

Ebbels, D.L. (1976) Diseases of Upland cotton in Africa. *Review of Plant Pathology*, **55**, 747–63.

Egan, B.T. and Sturgess, O.W. (1980) Commercial control of leaf scald disease by thermotherapy and a clean seed programme. *Proceedings of the International Society of Sugar Cane Technologists*, **17**, 1602–6.

Egli, T., Goto, M. and Schmidt, D. (1975) Bacterial wilt, a new forage grass disease. *Journal of Phytopathology*, **82**, 111–21.

Egli, T. and Schmidt, D. (1982) Pathogenic variation among the causal agents of bacterial wilt of forage grasses. *Journal of Phytopathology*, **104**, 138–50.

Eisenstark, A. and Bernstein, L.B. (1955) Specificity of bacteriophages of *Xanthomonas pruni*. *Phytopathology*, **45**, 596–7.

Eisenstark, A., Goldberg, S.S. and Bernstein, L.B. (1955) Lysogenicity in *Xanthomonas pruni*. *Journal of General Microbiology*, **12**, 402–5.

Elango, F. and Lozano, J.C. (1980) Transmission of *Xanthomonas manihotis* in seed of cassava (*Manihot esculenta*). *Plant Disease*, **64**, 784–6.

Elango, F.N., Lozano, J.C. and Peterson, J.F. (1981) Relationships between *Xanthomonas campestris* pv. *manihotis*, *X. c.* pv. *cassavae* and Colombian yellowish isolates, in *Proceedings of the Fifth International Conference on Plant Pathogenic Bacteria, Cali, Colombia, 1981* (ed. J.C. Lozano), Centro Internacional de Agricultura Tropical, Cali, Colombia, pp. 96–104.

El-Banoby, F.E. and Rudolph, K. (1989) Multiplication of *Xanthomonas campestris* pvs. *secalis* and *translucens* in host and nonhost plants (rye and barley) and development of watersoaking. *Bulletin OEPP/EPPO Bulletin*, **19**, 104–11.

El-Goorani, M.A. (1987) Plant diseases caused by bacteria in United Arab Emirates, in *Proceedings of the Sixth International Conference on Plant Pathogenic Bacteria*, College Park, Maryland, USA, 2–7 June, 1985 (eds E.L. Civerolo, A. Collmer, R.E. Davies and A.G. Gillapsie), Martinus Nijhoff Publishers, Dordrecht, pp. 653–7.

Elliott, C. (1951) *Manual of Bacterial Plant Pathogens*, Chronica Botanica Co., Waltham, MA, 186 pp.

El-Nur, E. (1970) Bacterial blight of cotton, in *Cotton Growth in the Gezira Environment* (eds M.A. Siddiqh and L.C. Hughes), Sudan Agricultural Research Corporation, pp. 179–88.

El-Zik, K.M. (1986) Half a century of dynamics and control of cotton diseases: dynamics of cotton diseases and their control, in *Proceedings of the Beltwide Cotton Producers Research Conference*, National Cotton Council of America, Memphis, TN, pp. 29–33.

Ewbank, E. (1992) Étude du catabolisme d'acides aminés conduisant à la formation de phytotoxines par *Xanthomonas campestris* pv. *manihotis* (agent pathogène de la bactériose du manioc). PhD Thesis, Université Catholique de Louvain, Louvain-la-Neuve, Belgium, 191 pp.

Ewbank, E. and Maraite, H. (1990) Conversion of methionine to phytotoxic 3-methylthiopropionic acid by *Xanthomonas campestris* pv. *manihotis. Journal of General Microbiology*, **136**, 1185–90.

Ezelio, W.N.O. (1977) Control of cassava bacterial blight in Nigeria, in *Report on a Workshop on Cassava Bacterial Blight, Nigeria, 1976* (eds G. Persley, R.E. Terry and R. MacIntyre), International Development Research Centre, Ottawa, Canada, pp. 15–17.

Fahy, P.C. and Persley, G.J. (1983) *Plant Bacterial Diseases: a diagnostic guide*, Academic Press, Sydney, Australia.

Falloon, R. and Hume, D. (1988) Productivity and persistence of prairie grass *Bromus willdenowii* Kunth 1. Effects of head smut fungus *Ustilago bullata* Berk. *Grass and Forage Science*, **43**, 179–84.

Fang, C.T. and Hsu, C.K. (1978) Some problems on the study of bacterial leaf blight of rice. *Scientiae Agricultura Sinica*, **2**, 73–7.

Fang, C.T., Lin, C.F. and Chu, L.L. (1956) A preliminary study on the disease cycle of the bacterial leaf blight of rice. *Acta Phytopathologica Sinica*, **2**, 173–85.

Fawcett, H.S. and Jenkins, A.E. (1933) Records of citrus canker from herbarium specimens of the genus *Citrus* in England and the United States. *Phytopathology*, **23**, 820–4.

Fett, W.F. (1979) Survival of *Pseudomonas glycinea* and *Xanthomonas phaseoli* var. *sojense* in leaf debris and soybean seed in Brazil. *Plant Disease Reporter*, **63**, 79–83.

Fett, W.F. (1984) Accumulation of isoflavonoids and isoflavone glucosides after inoculation of soybean leaves with *Xanthomonas campestris* pv. *glycines* and pv. *campestris* and a study of their role in resistance. *Physiological Plant Pathology*, **24**, 303–20.

Fett, W.F. and Dunn, M.F. (1987) Auxin production by plant-pathogenic pseudomonads and xanthomonads. *Applied and Environmental Microbiology*, **53**, 1839–45.

Fett, W.F., Dunn, M.F., Maher, G.T. and Maleeff, B. (1987) Bacteriocins and temperate phage of *Xanthomonas campestris* pv. *glycines. Current Microbiology*, **16**, 137–44.

Follin, J.C., Girardot, B., Mangano, V. and Benitez, R. (1988) New results on inheritance of immunity to bacterial blight (*Xanthomonas campestris* pv. *malvacearum* (Smith) Dye, race 18 and 20 in the cotton plant (*Gossypium hirsutum* L.). *Coton et Fibres Tropicales*, **43**, 167–74.

Fukuda, T., Uehara, K., Azegami, K., Tabei, H. and Nishiyama, K. (1990) Bacterial canker of Mango in Japan caused by *Xanthomonas campestris* pv. *mangiferaeindicae. Annals of the Phytopathological Society of Japan*, **56**, 474–80.

Gabriel, D.W., Hunter, J.E., Kingsley, M.T., Miller, J.W. and Lazo, G.R. (1988)

Clonal population structure of *Xanthomonas campestris* and genetic diversity among citrus canker strains. *Molecular Plant–Microbe Interactions*, **1**, 59–65.
Gabriel, D.W., Kingsley, M.T., Hunter, J.E. and Gottwald, T. (1989) Reinstatement of *Xanthomonas citri* (ex Hasse) and *X. campestris* (ex. Smith) to species and reclassification of all *X. campestris* pv. *citri* strains. *International Journal of Systematic Bacteriology*, **39**, 14–22.
Gaitskell, A. (1959) *Gezira: a story of development in the Sudan*, Faber & Faber, London.
Gilbertson, R.L., Maxwell, D.P., Hagedorn, D.J. and Leong, S.A. (1989) Development and application of a plasmid DNA probe for detection of bacteria causing common bacterial blight of bean. *Phytopathology*, **79**, 518–25.
Gilbertson, R.L., Rand, R.E. and Hagedorn, D.J. (1990) Survival of *Xanthomonas campestris* pv. *phaseoli* and pectolytic strains of *X. campestris* in bean debris. *Plant Disease*, **74**, 322–7.
Gitaitis, R.D. (1991) Field transplant production, in *Compendium of Tomato Diseases* (eds J.B. Jones *et al.*), APS Press, The American Phytopathological Society, St. Paul, MN.
Gitaitis, R.D., Hamm, J. and Bertrand, P. (1987) Refractive quality of bacterial colonies as a means to differentiate *Xanthomonas campestris* pv. *pruni* from other yellow-pigmented bacteria. *Phytopathology*, **77**, 641 (abstr.).
Gondran, J. and Betin, M. (1989) Susceptibility of Italian ryegrass to *Xanthomonas campestris* pv. *graminis*. *Bulletin OEPP/EPPO Bulletin*, **19**, 149–56.
Goode, M.J. and Sasser, M.J. (1980) Prevention is the key to controlling bacterial spot and bacterial speck of tomato. *Plant Disease*, **64**, 831–4.
Goodman, C.A. and Hattingh, M.J. (1986) Transmission of *Xanthomonas campestris* pv. *pruni* in plum and apricot nursery trees by budding. *Hortscience*, **21**, 995–6.
Goodman, C.A. and Hattingh, M.J. (1988a) Mechanical transmission of *Xanthomonas campestris* pv. *pruni* in plum nursery trees. *Plant Disease*, **72**, 643.
Goodman, C.A. and Hattingh, M.J. (1988b) Effect of plum and apricot bud origin, time cut, and time budded on development of bacterial spot of stone fruit. *Phytophylactica*, **20**, 133–4.
Goto, M. (1962) Studies on Citrus canker. I. *Bulletin of the Faculty of Agriculture Shizuoka University*, **12**, 3–72.
Goto, M., Ohta, K. and Okabe, N. (1975) Studies on saprophytic survival of *Xanthomonas citri* (Hasse) Dowson. 2. Longevity and survival density of the bacterium on artificially infested weeds, plant residues and soils. *Annals of the Phytopathological Society of Japan*, **41**, 141–7.
Goto, M., Serizawa, S. and Morita, M. (1970) Studies on citrus canker disease. II. Leaf infiltration technique for the detection of *Xanthomonas citri* (Hasse), with special reference to the comparison with the phage method. *Bulletin of the Faculty of Agriculture Shizuoka University*, **20**, 1–19.
Goto, M. and Starr, M.P. (1972) Phage–host relationships of *Xanthomonas citri* compared with those of other *Xanthomonas*. *Annals of the Phytopathological Society of Japan*, **38**, 226–48.
Goto, M., Yaguchi, Y. and Hyodo, H. (1980) Ethylene production in citrus leaves infested with *Xanthomonas citri* and its relation to defoliation. *Physiological Plant Pathology*, **16**, 343–50.
Graham, J.H. (1953) Overwintering of three bacterial pathogens of soybean. *Phytopathology*, **43**, 189–92.
Gremmen, J. and Koster, R. (1972) Research on poplar canker (*Aplanobacter populi*) in the Netherlands. *European Journal of Forest Pathology*, **2**, 116–24.
Groth, D.E. and Braun, E.J. (1986) Growth kinetics and histopathology of

Xanthomonas campestris pv. *glycines* in leaves of resistant and susceptible soybeans. *Phytopathology*, **76**, 959–65.

Groth, D.E. and Braun, E.J. (1989) Survival, seed transmission, and epiphytic development of *Xanthomonas campestris* pv. *glycines* in the north-central United States. *Plant Disease*, **73**, 326–30.

Hahn, S.K. (1979) Breeding of cassava for resistance to cassava mosaic disease (CMD) and bacterial blight (CBB) in Africa, in *Diseases of Tropical Food Crops. Proceedings of an International Symposium, U.C.L., Louvain-la-Neuve, Belgium 1978* (eds H. Maraite and J.A. Meyer), Université Catholique de Louvain, Louvain-la-Neuve, Belgium, pp. 211–19.

Halfon-Meiri, A. and Volcani, Z. (1977) A combined method for detecting *Colletotrichum gossypii* and *Xanthomonas malvacearum* in cotton seed. *Seed Science and Technology*, **5**, 129–39.

Hansford, C.G. (1937) Annual report of the plant pathologist, 1936. *Report of the Department of Agriculture, Uganda 1936–7, Part 2*, pp. 47–8.

Hartung, J.S. and Civerolo, E.L. (1987) Genomic fingerprints of *Xanthomonas campestris* pv. *citri* strains from Asia, South America, and Florida. *Phytopathology*, **77**, 282–5.

Hartwig, E.E. and Johnson, H.W. (1953) Effect of the bacterial pustule disease on yield and chemical composition of soybeans. *Agronomy Journal*, **45**, 22–3.

Hartwig, E.E. and Lehman, S.G. (1951) Inheritance of resistance to the bacterial pustule disease in soybean. *Agronomy Journal*, **43**, 226–9.

Hasse, C.H. (1915) *Pseudomonas citri*, the cause of citrus canker. *Journal of Agricultural Research*, **4**, 97–100.

Hayward, A.C. (1964) Characteristics of *Pseudomonas solanacearum*. *Journal of Applied Bacteriology*, **27**, 265–77.

Hayward, A.C. (1974) Latent infection by bacteria. *Annual Review of Phytopathology*, **12**, 87–97.

Hedges, F. (1922) Bacterial pustule of soybean. *Science*, **56**, 111–12.

Hedges, F. (1924) A study of bacterial pustule of soybean and comparison of *Bacterium phaseoli* var. *sojensis* Hedges with *Bac. phaseoli* (E.F. Smith). *Journal of Agricultural Research*, **29**, 229–51.

Heise, P.J. and Vidaver, A.K. (1988) Occurrence and characterization of a bacterial pathogen on milkweed (*Asclepias* spp.) in Nebraska. *Phytopathology*, **78**, 1668 (abstr.).

Hellmers, E. (1952) Bacterial leaf spot of *Pelargonium* (*Xanthomonas pelargonii* Brown) Starr & Burkholder in Denmark. *Transactions of the Danish Academy of Science*, No. 4, 1–35.

Hildebrand, D.C., Palleroni, N.J. and Schroth, M.N. (1990) Deoxyribonucleic acid relatedness of 24 xanthomonad strains representing 23 *Xanthomonas campestris* pathovars and *Xanthomonas fragariae*. *Journal of Applied Bacteriology*, **68**, 263–9.

Hildebrand, D.C., Schroth, M.N. and Wilhelm, S. (1967) Systemic invasion of strawberry by *Xanthomonas fragariae* causing vascular collapse. *Phytopathology*, **57**, 1260–1.

Hill, G.F. (1918) History of citrus canker in the Northern Territory. *Northern Territory of Australia Bulletin*, **18**, 11 pp.

Hokawat, S. (1978) Soybean pustule bacteria in Thailand, M.S. Thesis, Kasetsart University, Bangkok, Thailand (in Thai with abstract in English).

Hokawat, S. (1988) Pathophysiological characterization of the susceptible and resistant reaction of soybeans against the pathogen of pustule disease (*Xanthomonas campestris* pv. *glycines* Nakano, Dye). PhD Thesis, University of Göttingen, Germany, 101 pp.

Hokawat, S. and Rudolph, K. (1988) Effect of extracellular polysaccharides of *Xanthomonas campestris* pv. *glycines* on multiplication and survival of the pathogen. *Proceedings of the 26th National Conference*, 3–5 January 1988, Bangkok, Thailand, pp. 23–30.

Hokawat, S. and Rudolph, K. (1991) Variation in pathogenicity and virulence of strains of *Xanthomonas campestris* pv. *glycines*, the incitant of bacterial pustule of soybean. *Journal of Phytopathology*, **131**, 73–83.

Huang, J.-S. (1986) Ultrastructure of bacterial penetration in plants. *Annual Review of Phytopathology*, **24**, 141–57.

Huang, T.C. and Lee, H.L. (1988) Hot acidified zinc sulfate as seed soaking agent for the control of crucifer black rot. *Plant Protection Bulletin, Taiwan*, **30**, 245–58.

Humaydan, H.S., Harman, G.E., Nedrow, B.L. and Dinitto, L.V. (1980) Eradication of *Xanthomonas campestris*, the causal agent of black rot, from Brassica seeds with antibiotics and sodium hypochlorite. *Phytopathology*, **70**, 127–31.

Hunter, J.E., Abawi, G.S. and Becker, R.F. (1975) Observation on the source and spread of *Xanthomonas campestris* in an epidemic of black rot in New York. *Plant Disease Reporter*, **59**, 384–7.

Hunter, R.S., Brinkerhoff, L.A. and Bird, L.S. (1968) The development of a set of Upland cotton lines for differentiating races of *Xanthomonas malvacearum*. *Phytopathology*, **58**, 830–2.

Hussain, T. and Brinkerhoff, L.A. (1978) Race 18 of the cotton bacterial blight pathogen, *Xanthomonas malvacearum*, identified in Pakistan in 1977. *Plant Disease Reporter*, **62**, 1085–7.

Ikotun, T. (1978) Histopathology of cassava tissues infected by *Xanthomonas manihotis*. *Acta Phytopathologica Academiae Scientiarum Hungaricae*, **13**, 383–90.

Ikotun, T. (1981a) Some characteristics that distinguish *Xanthomonas cassavae* from *Xanthomonas manihotis*. *Fitopatologia Brasileira*, **6**, 1–14.

Ikotun, T. (1981b) Studies on the host range of *Xanthomonas manihotis*. *Fitopatologia Brasileira*, **6**, 15–21.

Ikotun, T. (1984) The nature and function of the extracellular polysaccharide produced by *Xanthomonas campestris* pathovar *manihotis*. *Fitopatologia Brasileira*, **9**, 467–73.

Innes, N.L. (1983) Bacterial blight of cotton. *Biological Reviews*, **58**, 157–76.

Irey, M.S. and Stall, R.E. (1982) The value of xanthomonadins for identification of pigmented *Xanthomonas campestris* pathovars, in *Proceedings of the Fifth International Conference on Plant Pathogenic Bacteria* (ed. J.C. Lozano), CIAT, Cali, Colombia, pp. 85–9.

Ishimaru, C., Eskridge, K.M. and Vidaver, A.K. (1991) Distribution analysis of naturally occurring epiphytic populations of *Xanthomonas campestris* pv. *phaseoli* on dry beans. *Phytopathology*, **81**, 262–8.

Jacquet, C. (1990) Adaptation d'un test ELISA sur membrane (Dot-Blot) pour la detection des agents pathogènes du pelargonium en production certifiée. Thèse, CNRA de Versailles, France.

James, W.C. (1971) An illustrated series of assessment keys for plant diseases, their preparation and usage. *Canadian Plant Disease Survey*, **51**, 39–65.

Janse, J.D. and Defrancq, M. (1988) Characterization of bacterial strains isolated from *Manihot esculenta* and of strains of *Xanthomonas campestris* pv. *oryzae* and *X. campestris* pv. *ricini* from Niger. *Phytopathologia Mediterranea*, **27**, 182–5.

Janse, J.D. and Miller, H.J. (1983) Yellow disease in *Scilla tubergeniana* and related bulbs caused by *Xanthomonas campestris* pv. *hyacinthi*. *Netherlands Journal of Plant Pathology*, **89**, 203–6.

Jindal, J.K., Patel, P.N. and Khan, A.M. (1981) Variability in xanthomonads of grain legumes. II. Pathogenic variability in *Xanthomonas phaseoli* mungbean strain, *X. vignicola* and *X. phaseoli* var. *sojense*. *Phytopathologisches Zeitschrift*, **100**, 1–9.

Jones, J.B., Pohronezny, K.L., Stall, R.E. and Jones, J.P. (1986) Survival of *Xanthomonas campestris* pv. *vesicatoria* in Florida on tomato crop residues, weeds, seeds and volunteer tomato plants. *Phytopathology*, **76**, 430–4.

Jones, J.B. and Van Vuurde, J.W.L. (1989) Magnetic immunoisolation of *Xanthomonas campestris* pv. *pelargonii*. *Proceedings of the Seventh International Conference on Plant Pathogenic Bacteria*, Budapest, 1989, p. 247 (abstr.).

Jones, J.P. (1961) A weed host of *Xanthomonas phaseoli* var. *sojense*. *Phytopathology*, **51**, 206.

Jones, R.K., Barnes, L.W., Gonzalez, C.F., Leach, J.E., Alvarez, A.M. and Benedict, A.A. (1989) Identification of low-virulence strains of *Xanthomonas campestris* pv. *oryzae* from rice in the United States. *Phytopathology*, **79**, 984–90.

Jones, S.B. and Fett, W.F. (1985) Fate of *Xanthomonas campestris* infiltrated into soybean leaves: an ultrastructural study. *Phytopathology*, **75**, 733–41.

Jones, S.B. and Fett, W.F. (1987) Bacterial pustule disease of soybean: microscopy of pustule development in a susceptible cultivar. *Phytopathology*, **77**, 266–74.

Joseph, J. and Elango, F. (1991) The status of cassava bacterial blight caused by *Xanthomonas campestris* pv. *manihotis* in Trinidad. *Journal of Phytopathology*, **133**, 320–6.

Kado, C.I. and Liu, S.T. (1981) Rapid procedure for detection and isolation of large and small plasmids. *Journal of Bacteriology*, **145**, 1365–73.

Kamerman, W. (1975) Biology and control of *Xanthomonas hyacinthi* in hyacinths. *Acta Horticulturae*, **47**, 99–105.

Kechel, H.G. (1982) Untersuchungen über die Resistenz von Pappeln gegenüber dem Erreger des Pappelkrebses *Xanthomonas populi* (RIDÉ) RIDÉ and RIDÉ. Dissertation, Ludwig-Maximilians Universität zu München, 155 pp.

Keil, H.L. (1979) Control of bacterial spot (caused by *Xanthomonas pruni*) in apricot trees by trunk infusion with oxytetracycline. *Plant Disease Reporter*, **63**, 407–9.

Keil, H.L. and Carroll, V.J. (1967) Control of peach bacterial spot with terramycin in Maryland. *Plant Disease Reporter*, **51**, 345–8.

Keil, H.L. and Civerolo, E.L. (1979) Effects of trunk injections of oxytetracycline on bacterial spot disease in peach trees. *Plant Disease Reporter*, **63**, 1–5.

Keil, H.L. and Fogle, H.W. (1974) Orchard susceptibility of some apricot, peach, and plum cultivars and selections to *Xanthomonas pruni*. *Fruit Varieties Journal*, **28**, 16–19.

Kennedy, B.W. and King, T.H. (1962a) Angular leaf spot of strawberry caused by *Xanthomonas fragariae* sp. nov. *Phytopathology*, **52**, 873–5.

Kennedy, B.W. and King, T.H. (1962b) Studies on epidemiology of bacterial angular leafspot on strawberry. *Plant Disease Reporter*, **46**, 360–3.

Kennedy, B.W. and Pfleger, F.L. (1986) Outbreak of bacterial stem rot of geranium in Minnesota. *Plant Disease*, **70**, 78.

Kersters, K., Pot, B., Hoste, B., Gillis, M. and De Ley, J. (1989) Protein electrophoresis and DNA:DNA hybridization of xanthomonads from grasses and cereals. *Bulletin OEPP/EPPO Bulletin*, **19**, 51–5.

Khundet, K. (1989) Dissemination of *Xanthomonas campestris* pv. *glycines* causing bacterial pustule through soybean seed. M.S. thesis, Kasetsart University, Bangkok, Thailand (in Thai with abstract in English).

Kim, B.S. (1986) Testing for detection of *Xanthomonas campestris* pv. *campestris* in crucifer seed and seed disinfection. *Korean Journal of Plant Pathology*, **2**, 96–101.

Kim, H.K., Orser, C., Lindow, S.E. and Sands, D.C. (1987) *Xanthomonas campestris* pv. *translucens* strains active in ice nucleation. *Plant Disease*, **17**, 994–7.

Kimura, O. and Dianese, J.C. (1983) Caracterizacao protéica e isoenzimatica das patovares de *Xanthomonas campestris* que atacam a mandioca. *Pesquisa Agropecuaria Brasileira*, **18**, 1215–28.

Kishun, R. (1986) Role of insects in transmission and survival of *Xanthomonas campestris* pv. *mangiferaeindicae*. *Indian Phytopathology*, **39**, 509–11.

Knauss, J.F. and Tammen, J. (1967) Resistance of *Pelargonium* to *Xanthomonas pelargonii*. *Phytopathology*, **57**, 1178–81.

Knight, R.L. (1949) The genetic control of blackarm disease (*Xanthomonas malvacearum*) in cotton. *5th Congrès International de Pathologie Compareè* (Sudan Government Research Division Memoir No. 19).

Koike, H. (1981) Assessment of sugarcane yield losses due to diseases. *Sugarcane Pathologists Newsletter*, **26**, 1–2.

Köning, H.C. (1938) The bacterial canker of poplars. *Mededelingen Phytopathologisch Laboratorium 'Willie Commelin Scholten' Baarn*, **14**, 3–47.

Kore, S.S. and Birandar, M.V. (1986) Bacterial leaf spot disease of sunflower. *Journal of Maharashtra Agricultural Universities*, **11**, 187–9.

Koucheki, H.K. (1973) The role of seed in transmission of a disease of *Pelargonium hortorum*. *Plant Disease Reporter*, **57**, 909–11.

Kuan, T.-L., Minsavage, G.V. and Schaad, N.W. (1986) Aerial dissemination of *Xanthomonas campestris* pv. *campestris* from crucifer weeds. *Plant Disease*, **70**, 409–13.

Kuhara, S. (1978) Present epidemic status and control of the citrus canker disease (*Xanthomonas citri (Hasse) Dowson) in Japan. Review of Plant Protection Research*, **11**, 132–42.

Kwaje, S.L. (1982) Bacterial blight on cassava. *FAO Plant Protection Bulletin*, **30**, 82–3.

Labruyere, R. (1975) *Xanthomonas graminis* bij grassen. *Jaarverslag Instituut voor Plantenziektenkundig Onderzoek 1975*, 24.

Lagiere, R. (1960) La bactériose du cotonnier (*Xanthomonas malvacearum*) (E.F. Smith) Dowson dans le monde et en République Centrafricaine (Oubangui-Chari), Paris, ICRT, p. 252.

Lambotte, M. and Perreaux, D. (1979) Histopathology of cassava bacterial blight in susceptible and tolerant cultivars, in *Diseases of Tropical Food Crops. Proceedings of an International Symposium, U.C.L., Louvain-la-Neuve, Belgium, 1978* (eds H. Maraite and J.A. Meyer), Université Catholique de Louvain, Louvain-la-Neuve, Belgium, pp. 153–60.

Laviolette, F.A., Athow, K.L., Probst, A.H. and Wilcox, J.R. (1976) Effect of bacterial pustule on yield of soybeans. *Crop Science*, **10**, 150–1.

Leach, J.E., Rhoads, M.L., Vera Cruz, C.M., White, F.F., Mew, T.W. and Leung, H. (1992) Assessment of genetic diversity and population structure of *Xanthomonas oryzae* pv. *oryzae* with a repetitive DNA element. *Applied and Environmental Microbiology*, **58**, 2188–95.

Lee, H.A. (1918) Further data on the susceptibility of rutaceous plants to citrus canker. *Journal of Agricultural Research*, **15**, 661–5.

Leksomboon, C., Thaveechai, N., Kositratana, W. and Paradornuwat, A. (1989) Detection of cassava blight bacteria from leaf and soil. *Kasetsart Journal, 23*, 228–38.

Lelliott, R.A. and Stead, D.E. (1987) *Methods for the Diagnosis of Bacterial Diseases of Plants*, Blackwell Scientific Publications, Oxford.

Lemattre, M. (1966) Étude histologique et histochimique des réactions déclenchées par *Xanthomonas pelargonii* dans les tissus de *Pelargonium zonale*.

Annales des Epiphyties, **17**, 215–28.
Léoville, F. and Coléno, A. (1976) Détection de *Xanthomonas albilineans* (Ashby) Dowson, agent de l'échaudure de la canne à sucre, dans les boutures contaminées. *Annales de Phytopathologe*, **8**, 233–6.
Leyendecker, P.J. (1950) 1949 Plant disease survey for New Mexico. *Plant Disease Reporter*, **34**, 39–40.
Leyns, F., De Cleene, M., Swings, J. and De Ley, J. (1984) The host range of the genus *Xanthomonas*. *The Botanical Review*, **50**, 308–55.
Leyns, F., De Cleene, M., Swings, J., Van den Mooter, M. and De Ley, J. (1981a) Host range studies on some isolates of *Xanthomonas campestris* pv. *graminis* from Belgium. *Mededelingen van de Faculteit Landbouwwetenschappen, Rijksuniversiteit Gent*, **46**, 793–8.
Leyns, F., De Cleene, M., Van Bogaert, G., Van de Wiele, A. and De Ley, J. (1988) Preliminary investigations about the mode of transmission and spread of *Xanthomonas campestris* pv. *graminis* on forage grasses. *Journal of Phytopathology*, **122**, 76–88.
Leyns, F., Van den Mooter, M., Swings, J., De Cleene, M. and De Ley, J. (1981b) Distribution of *Xanthomonas campestris* pv. *graminis* in fields of forage grasses in northern Belgium. *Parasitica*, **37**, 131–3.
Liao, C.H. (1972) Studies on mango fruit spot. I. Symptoms and causal organism. *Journal of Taiwan Agricultural Research*, **21**, 146–52.
Lim, T.K., Sijam, K. and Ong, E.T. (1991) Bacterial black spot of Mango. A new disease in Malaysia, in *Proceedings of the Third International Mango Symposium*, Darwin, Australia, *Acta Horticulturae*, **291**, 317–23.
Lourido, L.C. (1974) Una metodologia para estimar los beneficios y los costos esperados en un programa de investigacion agricola aplicada: el anublo bacterial en la yuca, Thesis Econ., Universidad de los Andes, Bogota, Colombia.
Lovrekovitch, L. and Klement, Z. (1965) Serological and bacteriophage studies on *Xanthomonas vesicatoria* strains isolated from tomato and pepper. *Phytopathologisches Zeitschrift*, **53**, 222–8.
Lozano, J.C. (1977) Identification of bacterial leaf blight in rice caused by *Xanthomonas oryzae* in America. *Plant Disease Reporter*, **61**, 644–8.
Lozano, J.C. (1986) Cassava bacterial blight: a manageable disease. *Plant Disease*, **70**, 1089–93.
Lozano, J.C. and Nolt, B.L. (1989) Pests and pathogens of cassava, in *Plant Protection and Quarantine*, Vol. II, *Selected Pests and Pathogens of Quarantine Significance* (ed. R.P. Kahn), CRC Press, Boca Raton, Florida, pp. 169–82.
Lozano, J.C. and Sequeira, L. (1974a) Bacterial blight of cassava in Colombia; etiology. *Phytopathology*, **64**, 74–82.
Maas, J.L., Finney, M.M., Civerolo, E.L. and Sasser, M. (1985) Association of an unusual strain of *Xanthomonas campestris* with apple. *Phytopathology*, **75**, 438–45.
Manicom, B.Q. (1986) Factors affecting bacterial black spot of mangoes caused by *Xanthomonas campestris* pv. *mangiferaeindicae*. *Annals of Applied Biology*, **109**, 129–35.
Manicom, B.Q. and Wallis, F.M. (1984) Further characterization of *Xanthomonas campestris* pv. *mangiferaeindicae*. *International Journal of Systematic Bacteriology*, **34**, 77–9.
Manicom, B.Q. and Wood, R. (1981) Attempts to control bacterial black spot of mangoes. *Subtropica*, **2**, 12–14.
Maraite, H. and Meyer, J.A. (1975) *Xanthomonas manihotis* (Arthaud-Berthet) Starr, causal agent of bacterial wilt, blight and leaf spots of cassava in Zaïre. *PANS*, **21**, 27–37.

Maraite, H. and Perreaux, D. (1979) Comparative symptom development in cassava after infection by *Xanthomonas manihotis* or *X. cassavae* under controlled conditions, in *Cassava Bacterial Blight in Africa: past, present and future. Report of an Interdisciplinary workshop, IITA, Ibadan, Nigeria, 1978* (eds E.R. Terry, G.J. Persley and S.C.A. Cook), Centre for Overseas Pest Research, London, pp. 17–24.

Maraite, H. and Weyns, J. (1979) Distinctive physiological, biochemical and pathogenic characteristics of *Xanthomonas manihotis* and *X. cassavae*, in *Diseases of Tropical Food Crops. Proceedings of an International Symposium, U.C.L., Louvain-la-Neuve, Belgium, 1978* (eds H. Maraite and J.A. Meyer), Université Catholique de Louvain, Louvain-la-Neuve, Belgium, pp. 103–17.

Maraite, H., Weyns, J., Yimkwan, O., Lipemba, P. and Perreaux, D. (1981) Physiological and pathogenic variations in *Xanthomonas campestris* pv. *manihotis*, in *Proceedings of the Fifth International Conference on Plant Pathogenic Bacteria, Cali, Colombia, 1981* (ed. J.C. Lozano), Centro International de Agricultura Tropical, Cali, Colombia, pp. 358–68.

Marcano, M. and Trujillo, G. (1982) Bacteriosis de la yuca; identificados algunos aspectos epidemiologicos del pathogeno que pueden utilizarse para disminuir los danos ocasionados par la enfermedad. *Fonaiap Divulga (Venezuela)*, **1**, 12–13.

Marcano, M. and Trujillo, G. (1984) Papel de las malezas en relacion a la perpetuacion del anublo bacteriano de la yuca. *Revista de la Facultad de Agronomia (Venezuela)*, **13**, 167–81.

Marco, G.M. and Stall, R.E. (1983) Control of bacterial spot of pepper initiated by strains of *Xanthomonas campestris* pv. *vesicatoria* that differ in sensitivity to copper. *Plant Disease*, **67**, 779–81.

Martin, J.P. and Robinson, P.E. (1961) Leaf scald, in *Sugarcane Diseases of the World*, Vol. 1 (eds J.P. Martin, E.V. Abbott and G. Hughes), Elsevier, Amsterdam, pp. 79–107.

McGuire, R.G., Jones, J.B. and Sasser, M.J. (1986) Tween media for semiselective isolation of *Xanthomonas campestris* pv. *vesicatoria*. *Plant Disease*, **70**, 887–91.

Mew, T.W. (1987) Current status and future prospects of research on bacterial blight of rice. *Annual Review of Phytopathology*, **25**, 359–82.

Mew, T.W. (1989) An overview of the world bacterial blight situation, in *Bacterial Blight of Rice*, International Rice Research Institute, Manila, Philippines, pp. 7–12.

Mew, T.W. and Vera Cruz, C.M. (1979) Variability of *Xanthomonas oryzae*: specificity in infection of rice differentials. *Phytopathology*, **69**, 152–5.

Mew, T.W., Vera Cruz, C.M. and Medalla, E.S. (1992) Monitoring the population shift of *Xanthomonas oryzae* pv. *oryzae* in response to the planting of rice cultivars in the Philippines. *Plant Disease*, **76**, 1029–32.

Millar, R.L. (1955) Studies on the nature of pathogenicity of *X. phaseoli* (E.F. Smith) Dowson and of *X. phaseoli* var. *sojensis* (Hedges) Starr and Burk. PhD thesis, Cornell University, NY, USA.

Miller, J.W. and Chase, A.R. (1986) New disease of *Pellonia* and *Pilea* spp. caused by *Xanthomonas campestris*. *Plant Disease*, **70**, 346–8.

Minsavage, G.V., Canteros, B.I. and Stall, R.E. (1990a) Plasmid-mediated resistance to streptomycin in *Xanthomonas campestris* pv. *vesicatoria*. *Phytopathology*, **80**, 719–23.

Minsavage, G.V., Dahlbeck, D., Whalen, M.C., *et al.* (1990b) Gene-for-gene relationships specifying disease resistance in *Xanthomonas campestris* pv. *vesicatoria*–pepper interactions. *Molecular Plant–Microbe Interactions*, **3**, 41–7.

Mizukami, T. and Wakimoto, S. (1969) Epidemiology and control of bacterial leaf blight of rice. *Annual Review of Phytopathology*, **7**, 51–72.

Moffett, M.L. (1973) Bacterial spot of stone fruit in Queensland. *Australian Journal of Biological Science*, **26**, 171–9.

Moffett, M.L. (1983) Bacterial plant pathogens recorded in Australia, in *Plant Bacterial Disease: a diagnostic guide* (eds P.C. Fahy and G.J. Persley), Academic Press, New York, p. 333.

Moffett, M.L. and Croft, B.J. (1983) *Xanthomonas*, in *Plant Bacterial Diseases: a diagnostic guide* (eds P.C. Fahy and G.J. Persley), Academic Press, New York, pp. 189–228, 192–200.

Moffett, M.L., Peterson, R.A. and Wood, B.A. (1979) Bacterial black spot of mango. *Australasian Plant Pathology*, **8**, 54–6.

Morales, G.A.Y. and Gutierrez, G.H. (1986) Estudo de la pústula bacteriana (*Xanthomonas phaseoli* var. *sojense* (Hedges) Starr y Burk) en el cultivo de la soja (*Glycine max.* (L.) Merrill). *Centro Agrícola*, **13**(1), 33–41.

Mostade, J.M. and Butare, I. (1979) Symptomatologie et épidémiologie de la nécrose bactérienne du manioc causée par *Xanthomonas cassavae* au Rwanda, in *Disease of Tropical Food Crops. Proceedings of an International Symposium, U.C.L., Louvain-la-Neuve, Belgium, 1978* (eds H. Maraite and J.A. Meyer), Université Catholique de Louvain, Louvain-la-Neuve, Belgium, pp. 95–117.

Muhle-Larsen, C. (1966) Le problème du chancre bactérien et d'amélioration des peupliers par voie génétique. FAO/CIP Versailles-Grammont, 18 pp.

Mulrean, E.N. and Schroth, M.N. (1981) A semiselective medium for the isolation of *Xanthomonas campestris* pv. *juglandis* from walnut buds and catkins. *Phytopathology*, **71**, 336–9.

Munnecke, D.E. (1954) Bacterial stem rot and leaf spot of Pelargonium. *Phytopathology*, **44**, 627–32.

Nakano, K. (1919) Bacterial leaf-blight of soybean. *Journal of Plant Protection, Tokyo*, **6**, 217–21.

Namekata, T. and Oliveira, A.R. de (1972) Comparative serological studies between *Xanthomonas citri* and a bacterium causing canker on Mexican lime, in *Proceedings of the Third International Conference on Plant Pathogenic Bacteria* (ed. H.P. Maas Geesteranus), Centre for Agricultural Publication and Documentation, Wageningen, The Netherlands, pp. 151–2.

Nayudu, M.V. (1973) Problem of variation in *Xanthomonas malvacearum*. *Acta Phytopathologica Academiae Scientiarum Hungaricae*, **8**, 77–9.

Nayudu, M.V. and Walker, J.C. (1960) Bacterial spot of tomato as influenced by temperature and by age and nutrition of the host. *Phytopathology*, **50**, 360–4.

Obata, T. (1986) Studies on the hyacinth yellows and the use of antiserum as an aid to its diagnosis in the field. *Research Bulletin of the Plant Protection Service, Japan*, **5**, 15–17.

Odurukwe, S.O. and Arene, O.B. (1980) Effect of N, P, K, fertilizers on cassava bacteria blight and root yield of cassava. *Tropical Pest Management*, **26**, 391–5.

Onsando, J.M. (1987) Management of black rot of cabbage (*Xanthomonas campestris* pv. *campestris*) in Kenya. *Tropical Pest Management*, **33**, 5–6.

Onyango, D.M. and Ramos, A.H. (1979) Bacterial blight of cassava in Kenya, in *Cassava Bacterial Blight in Africa: past, present and future. Report of an Interdisciplinary workshop, IITA, Ibadan, Nigeria, 1978* (eds E.R. Terry, G.J. Persley and S.C.A. Cook), Centre for Overseas Pest Research, London, pp. 25–8.

Otim-Nape, G.W. (1980) Cassava bacterial blight in Uganda. *Tropical Pest Management*, **26**, 274–7.

Otim-Nape, G.W. (1983) The effects of bacterial blight on germination, tuber yield and quality of cassava, in *Proceedings of the 10th International Congress of*

Plant Protection, Brighton 1983, British Crop Protection Council, London, p. 1205.
Ou, S.H. (1985a) Bacterial leaf streak, in *Rice Diseases*, 2nd edn, CAB, Kew, England, pp. 96–103.
Ou, S.H. (1985b) *Rice Diseases*, 2nd edn, CAB, Kew, England.
Pandian, S.K., Somasundaram, T., Jajaraman, J. and Lakshmanan, L. (1982) A new toxin isolated from *Xanthomonas malvacearum* which inhibits mitochondrial ATP-ADP translocase. *Archives of Biochemistry and Biophysics*, **216**, 486–7.
Park, S.J. and Dhanvantari, B.N. (1987) Transfer of common blight (*Xanthomonas campestris* pv. *phaseoli*) resistance from *Phaseolus coccineus* Lam. to *P. vulgaris* L. through interspecific hybridization. *Canadian Journal of Plant Science*, **67**, 685–95.
Patel, M.K., Abhyankar, S.G. and Kulkarni, Y.S. (1949) Black rot of cabbage. *Indian Phytopathology*, **2**, 58–61.
Patel, M.K., Kulkarni, Y.S. and Moniz, L. (1948a) *Pseudomonas mangiferae-indicae*, pathogenic on mango. *Indian Phytopathology*, **1**, 147–52.
Patel, M.K., Moniz, L. and Kulkarni, Y.S. (1948b) A new bacterial disease of *Mangifera indica* L. *Current Science*, **6**, 189–90.
Patel, P.N. and Shekhawat, G.S. (1971) Occurrence of *Xanthomonas translucens* f.sp. *hordei* in India and its pathogenicity to rice. *Plant Disease Reporter*, **55**, 365–8.
Paul, V. and Birkenstaedt, E. (1989) Investigation of methods for testing the susceptibility of forage grasses to *Xanthomonas campestris* pv. *graminis*. *Bulletin OEPP/EPPO Bulletin*, **19**, 137–42.
Paul, V. and Freudenstein, K. (1989) Susceptibility of cultivars of Italian rye-grass to *Xanthomonas campestris* pv. *graminis*. *Bulletin OEPP/EPPO Bulletin*, **19**, 165–72.
Paxton, J.D. (1972) Toxin production by *Xanthomonas phaseoli* var. *sojensis*, in *Phytotoxins in Plant Diseases, Proceedings of the NATO Advanced Study Institute, Pugnochiuso, Italy, June 1970* (eds R.K.S. Wood, A. Ballio and A. Graniti), Academic Press, London, p. 377.
Peltier, G.L. and Frederich, W.J. (1926) Effects of weather on the world distribution and prevalence of citrus canker and citrus scab. *Journal of Agricultural Research*, **32**, 147–64.
Perreaux, D., Maraite, H. and Meyer, J.A. (1978) Histopathological study by fluorescent microscopy of cassava stems infected by *Xanthomonas manihotis*, in *Proceedings of the Fourth International Conference on Plant Pathogenic Bacteria*, INRA, Angers, France, pp. 935–41.
Perreaux, D., Maraite, H. and Meyer, J.A. (1986) Detection of 3(methylthio)propionic acid in cassava leaves infected by *Xanthomonas campestris* pv. *manihotis*. *Physiological and Molecular Plant Pathology*, **28**, 323–8.
Persley, G.J. (1978) Epiphytic survival of *Xanthomonas manihotis* in relation to the disease cycle of cassava bacterial blight, in *Proceedings of the Fourth International Conference on Plant Pathogenic Bacteria*, INRA, Angers, France, pp. 773–7.
Persley, G.J. (1979a) Studies on the survival and transmission of *Xanthomonas manihotis* on cassava seed. *Annals of Applied Biology*, **93**, 159–66.
Persley, G.J. (1979b) Studies on the epidemiology and ecology of cassava bacterial blight, in *Cassava Bacterial Blight in Africa: past, present and future. Report of an Interdisciplinary Workshop, IITA, Ibadan, Nigeria, 1978* (eds E.R. Terry, G.J. Persley and S.C.A. Cook), Centre for Overseas Pest Research, London, pp. 5–7.

Persley, G.J. (1980) Studies on bacterial blight of cassava in Africa, PhD Thesis, University of Queensland, Brisbane, Australia, 182 pp.

Pillai, S.K. and Hildebrandt, A.C. (1968) *In vitro* differentiation of geranium – *Pelargonium hortorum* (Bailey) – plants from apical meristems. *Phyton*, **25**, 81–7.

Pohronezny, K.L., Moss, M.A., Dankers, W. and Schenk, J. (1990) Dispersal and management of *Xanthomonas campestris* pv. *vesicatoria* during thinning of direct-seeded tomato. *Plant Disease*, **74**, 800–5.

Pohronezny, K. and Volin, R.B. (1983) The effect of bacterial spot on yield and quality of fresh market tomatoes. *HortScience*, **18**, 69–70.

Poswal, M.A.T. (1988) Races of *Xanthomonas campestris* pv. *malvacearum* (Smith) Dye, the causal organism of bacterial blight of cotton, in Nigeria. *Journal of Phytopathology*, **123**, 6–11.

Prathuangwong, S. (1985) Soybean bacterial pustule research in Thailand, in *Tropical Legume Improvement. Proceedings of Thai/ACIAR Planning and Coordination Workshop*, Bangkok, Thailand, 10–12 October 1983 (ed. G.J. Persley), ACIAR Proceedings Series No. 8, pp. 40–1.

Prathuangwong, S., Khandet, K. and Thaveechai, N. (1989) Dissemination and yield loss assessment of *Xanthomonas campestris* pv. *glycines* causing bacterial pustule through soybean seed in Thailand. *Proceedings of the 27th National Conference*, 30 January–2 February 1989, Bangkok, Thailand, pp. 115–22.

Pruvost, O., Couteau, A. and Luisetti, J. (1989) Efficacité de différentes formulations chimiques pour lutter contre la maladie des taches noires de la mangue (*Xanthomanas campestris* pv. *mangiferaeindicae*). *Fruits*, **44**, 343–50.

Pruvost, O., Couteau, A. and Luisetti, J. (1990) Development of bacterial black spot of mangoes and epiphytic populations of the pathogen (*Xanthomonas campestris* pv. *mangiferaeindicae*) under natural conditions in Reunion Island. *Fruits*, **45**, 125–40.

Pruvost, O., Couteau, A. and Luisetti, J. (1992) Pepper tree (*Schinus terebenthifolius* Raddi), a new host plant for *Xanthomonas campestris* pv. *mangiferaeindicae*. *Journal of Phytopathology*, **137**, 289–98.

Pruvost, O., Couteau, A., Verniere, C. and Luisetti, J. (1993) Epiphytic survival of *Xanthomonas campestris* pv. *mangiferaeindicae* on Mango buds, in *Proceedings of the Fourth International Mango Symposium*, Miami, Florida, USA, *Acta Horticulturae*, in press.

Pruvost, O. and Luisetti, J. (1989) Strains of *Xanthomonas campestris* isolated from Ambarella (*Spondias cytherea* Sonn.) in the French West Indies are probably a new pathogenic form of *Xanthomonas campestris* pv. *mangiferaeindicae*. *Fruits*, **44**, 539–42.

Pruvost, O. and Luisetti, J. (1991) Effect of time of inoculation with *Xanthomonas campestris* pv. *mangiferaeindicae* on mango fruits susceptibility. Epiphytic survival of *X. c.* pv. *mangiferaeindicae* on mango fruits in relation to disease development. *Journal of Phytopathology*, **133**, 139–51.

Qhobela, M., Claflin, L.E. and Nowell, D.C. (1990) Evidence that *Xanthomonas campestris* pv. *zeae* can be distinguished from other pathovars capable of infecting maize by restriction fragment length polymorphism of genomic DNA. *Canadian Journal of Plant Pathology*, **12**, 183–6.

Rahimian, H. (1989a) Angular leaf spot of *Alnus subcordata* caused by a pathovar of *Xanthomonas campestris*. *Iranian Journal of Plant Pathology*, **25**, 27–8.

Rahimian, H. (1989b) Leaf spot of *Bellis perennis* incited by a pathovar of *Xanthomonas campestris*. *Proceedings of the Ninth Iranian Congress of Plant Protection*, p. 153 (abstr.).

Ramos, L.J. and Volin, R.B. (1987) Role of stomatal opening and frequency on infection of *Lycopersicon* spp. by *Xanthomonas campestris* pv. *vesicatoria*. *Phytopathology*, **77**, 1311–17.

Randhawa, P.S. and Civerolo, E.L. (1985) A detached-leaf bioassay for *Xanthomonas campestris* pv. *pruni*. *Phytopathology*, **75**, 1060–3.

Randhawa, P.S. and Civerolo, E.L. (1987) Indigenous plasmids in *Xanthomonas campestris* pv. *pruni*, in *Plant Pathogenic Bacteria. Proceedings of the Sixth International Conference on Plant Pathogenic Bacteria*, College Park, Maryland (eds E.L. Civerolo, A. Collmer, R.E. Davis and A.G. Gillaspie), Martinus Nijhoff Publishers, Dordrecht, pp. 465–9.

Randhawa, P.S. and Schaad, N.W. (1984) Selective isolation of *Xanthomonas campestris* from crucifer seeds. *Phytopathology*, **74**, 268–72.

Reckhaus, P.M. (1983) Occurrence of bacterial blight of rice in Niger, West Africa. *Plant Disease*, **67**, 1039.

Reddy, A.P.K., Mackenzie, D.R., Rouse, D.I. and Rao, A.V. (1979) Relationship of bacterial leaf blight severity to grain yield of rice. *Phytopathology*, **69**, 967–9.

Reuther, G. (1975) Untersuchungen zur Vermehrung von Pelargonienvarietäten durch Gewebekultur. *Gartenbauwissenschaft*, **4**, 181–7.

Reuther, G. and Sonneborn, H.H. (1983) Untersuchung zur Frühdiagnose von Xanthomonasinfektionen. *Gartenbau und Gärtnerwelt*, **33**, 854–7.

Ricaud, C. and Autrey, L.J.C. (1989) Gumming disease, in *Diseases of Sugarcane. Major Diseases* (eds C. Ricaud, B.T. Egan, A.G. Gillaspie Jr and C.G. Hughes), Elsevier, Amsterdam, pp. 21–38.

Ricaud, C. and Ryan, C.C. (1989) Leaf scald, in *Diseases of Sugarcane. Major Diseases* (eds C. Ricaud, B.T. Egan, A.G. Gillaspie Jr and C.G. Hughes), Elsevier, Amsterdam, pp. 39–58.

Ridé, M. (1958) Sur l'étiologie du chancre suintant du peuplier. *Comptes Rendus de l'Académie des Sciences, Paris*, **246**, 2795–8.

Ridé, M. (1963) Our present knowledge of bacterial canker caused by *Aplanobacter populi*. FAO/CIP Casale Monferrato (translated by RATKO).

Ridé, M., Digat, B. and Ridé, S. (1975) (Abstr.) Detection and epidemiology of *Aplanobacter populi* by immunofluorescence techniques. ISPP Phytobacteriology section. Working group on 'The ecology of plant pathogenic bacteria on aerial structures of plants'. Wye College.

Ridé, M. and Ridé, S. (1978a) *Xanthomonas populi* (Ridé) comb. nov. (Syn. *Aplanobacter populi* (Ridé) spécificité, variabilité et absence de relation avec *Erwinia cancerogena* Ur. *European Journal of Forest Pathology*, **8**, 310–33.

Ridé M. and Ridé, S. (1978b) Factors affecting inoculation success in woody plants. *Proceedings of the Fourth International Conference on Plant Pathogenic Bacteria*, Angers, France, **2**, 957–68.

Ridé, M., Ridé, S. and Poutier, J.C. (1978) Factors affecting the survival of *Xanthomonas populi* on aerial structures of poplar. *Proceedings of the Fourth International Conference on Plant Pathogenic Bacteria*, Angers, France, **2**, 803–14.

Ridé, M., Ridé, S., Steenackers, M. and Steenackers, V. (1985) Artificial infection of different poplar clones with different geographical isolates of *Xanthomonas populi*. *Proceedings of Conference on Wood as a Renewable Raw Material*, pp. 29–44.

Rivera, N., Rodriguez, I. and Ezavin, M. (1985) Medio selectivo para aislar *Xanthomonas albilineans*, agente causal de la escaldadura foliar de la cana de azucar. *Ciencia y Tecnica en la Agricultura, Proteccion de Plantas*, **8**, 75–88.

Robbs, C.F. (1954) Bacterias fitopatogenicas do Brasil. *Agronomia*, **13**, 265–82.

Robbs, C.F., Da Ponte, J.J. and Da Gloria Sales, M. (1978) Nota sobre *Xanthomonas mangiferaeindicae* no nordeste doBrasil. *Fitopatologia Brasileira*, **3**, 215–17.

Robbs, C.F., Ribeiro, R. de L.D. and Kimura, O. (1974) Sobre a posicao taxonomica de *Pseudomonas mangiferaeindicae* Patel *et al.* 1948, agente causal da 'mancha bacteriana' das folhas da mangueira (*Mangifera indica* L.). *Arquivas Universidad Federal Rural Rio de Janeiro*, **4**, 11–14.

Robbs, C.F., Ribeiro, R.D., Kimura, O. and Lakiba, F. (1972) Variaçoes em *Xanthomonas manihotis* (Arthaud Berthet) Starr. *Revista da Sociedade Brasileira de Fitopatologia*, **5**, 67–75.

Roberts, D.L. and Vargas, J.M. (1984) Antigenic relatedness of the North American Toronto bentgrass bacterium to *Xanthomonas campestris* pv. *graminis* from Europe. *Phytopathology*, **74**, 813.

Rodriguez, G.S., Garza-Lopez, J.G., Stapleton, J.J. and Civerolo, E.L. (1985) Citrus bacteriosis in Mexico. *Plant Disease*, **69**, 808–10.

Rott, P., Arnaud, M. and Baudin, P. (1986) Serological and lysotypical variability of *Xanthomonas albilineans* (Ashby) Dowson, causal agent of sugarcane leaf scald disease. *Journal of Phytopathology*, **116**, 201–11.

Rott, P., Chatenet, M. and Baudin, P. (1988a) L'Echaudure des feuilles de canne à sucre provoquée par *Xanthomonas albilineans* (Ashby) Dowson. I. Synthèse bibliographique. *Agronomie Tropicale*, **43**, 236–43.

Rott, P., Chatenet, M., Granier, M. and Baudin, P. (1988b) L'Echaudure des feuilles de canne à sucre provoquée par *Xanthomonas albilineans* (Ashby) Dowson. II. Diagnostic et spectres d'hôtes de l'agent pathogène en Afrique tropicale. *Agronomie Tropicale*, **43**, 244–51.

Rott, P. and Frossard, P. (1986) Un chancre bactérien du Prunier de Cythère (*Spondias cytherea*, Sonn.) en Martinique. *Fruits*, **41**, 605–13.

Roy, M.A. (1988) Use of fatty acids for the identification of phytopathogenic bacteria. *Plant Disease*, **72**, 460.

Ruano, O. and Mohan, S.K. (1982) A new race of *Xanthomonas campestris* pv. *malvacearum* (Smith) Dye, in Paranà State. *Fitopatologia Brasileira*, **7**, 439–42.

Rudolph, K.W.E., Gross, M., Neugebauer, M., *et al.* (1989) Extracellular polysaccharides as determinants of leaf spot diseases caused by pseudomonads and xanthomonads, in *Phytotoxins and Plant Pathogenesis* (eds A. Graniti, R.D. Durbin and A. Ballio), NATO ASI Series, Vol. H27, Springer-Verlag, Berlin, pp. 177–218.

Saaltink, G.J. (1966) The possibility of entrance through hydathodes by *Xanthomonas hyacinthi*. *Mededelingen van de Faculteit Landbouwwetenschappen, Rijksuniversiteit Gent*, **31**, 941–6.

Sabet, K.A., Ishag, F. and Khalil, O. (1969) Studies on the bacterial diseases of Sudan crops. VII. New records. *Annals of Applied Biology*, **63**, 357–69.

Saettler, A.W. (1989) Common bacterial blight, in *Bean Production Problems in the Tropics*, 2nd edn (eds H.F. Schwartz and M.A. Pastor-Corrales), Centro Internacional de Agricultura Tropical, Cali, Colombia, pp. 261–83.

Saettler, A.W. (1991) Diseases caused by bacteria, in *Compendium of Bean Diseases* (ed. R. Hall), APS Press, St Paul, MN, pp. 29–32.

Samson, R., Afri, A. and Carvil, N. (1989) Criteria for identification of *Xanthomonas campestris* causing wilt of graminiae. *Bulletin OEPP/EPPO Bulletin*, **19**, 43–50.

Santos, R.M., Dos, D.B. and Dianese, J.C. (1985) Comparative membrane characterization of *Xanthomonas campestris* pv. *cassavae* and *X. campestris* pv. *manihotis*. *Phytopathology*, **75**, 581–7.

Schaad, N.W. (1976a) Control of black rot, *University of Georgia Research Bulletin 187*, Athens, GA, 19 pp.
Schaad, N.W. (1976b) Immunological comparison and characterization of ribosomes of *Xanthomonas vesicatoria*. *Phytopathology*, **66**, 770–6.
Schaad, N.W. (ed.) (1988) *Laboratory Guide for Identification of Plant Pathogenic Bacteria*, 2nd edn, APS Press, St Paul, MN.
Schaad, N.W. (1989) Detection of *Xanthomonas campestris* pv. *campestris* in crucifers, in *Detection of Plant Pathogenic Bacteria in Seed and Other Planting Material* (eds A.W. Saettler, N.W. Schaad and D.A. Roth), APS Press, St Paul, MN, pp. 68–75.
Schaad, N.W. and Dianese, J.C. (1981) Cruciferous weeds as a source of inoculum of *Xanthomonas campestris* in black rot of crucifers. *Phytopathology*, **71**, 1215–20.
Schaad, N.W. and Donaldson, R.C. (1980) Comparison of two methods of detection of *Xanthomonas campestris* in infected crucifer seeds. *Seed Science and Technology*, **8**, 383–91.
Schaad, N.W. and Forster, R.L. (1985) A semi-selective agar medium for isolating *Xanthomonas campestris* pv. *translucens* from wheat seeds. *Phytopathology*, **75**, 260–3.
Schaad, N.W., Gabrielson, R.L. and Mulanax, M.W. (1980) Hot acidified cupric acetate soaks for eradication of *Xanthomonas campestris* from crucifer seeds. *Applied and Environmental Microbiology*, **39**, 803–7.
Schaad, N.W. and Stall, R.E. (1988) *Xanthomonas*, in *Laboratory Guide for Identification of Plant Pathogenic Bacteria*, 2nd edn (ed. N.W. Schaad), APS Press, St Paul, MN, pp. 81–94.
Schaad, N.W. and Thaveechai, N. (1983) Black rot of crucifers in Thailand. *Plant Disease*, **67**, 1231–4.
Schaad, N.W. and White, W.C. (1974) Survival of *Xanthomonas campestris* in soil. *Phytopathology*, **64**, 1518–20.
Schmidt, D. (1988) Le flétrissement bactérien des graminées fourragères: essais pour limiter la dispersion de la maladie lors du fauchage. *Revue Suisse d'Agriculture*, **20**, 351–7.
Schmidt, D. (1989) Epidemiological studies of bacterial wilt of fodder grasses. *Bulletin OEPP/EPPO Bulletin*, **19**, 89–96.
Schneider, I.U.I. and Iliukhina, M.K. (1983) Bacterial diseases of winter wheat. *Zaschita Rastenii*, **10**, 40–1.
Schoulties, C.L., Civerolo, E.L., Miller, J.W., *et al.* (1987) Citrus canker in Florida. *Plant Disease*, **71**, 388–95.
Schultz, T. and Gabrielson, R.L. (1986) *Xanthomonas campestris* pv. *campestris* in Western Washington seed fields: occurrence and survival. *Phytopathology*, **76**, 1306–9.
Schuster, M.L. and Coyne, D.P. (1981) Biology, epidemiology, genetics and breeding for resistance to bacterial pathogens of *Phaseolus vulgaris* L. *Horticultural Reviews*, **3**, 28–58.
Scott, J.W. and Jones, J.B. (1989) Inheritance of resistance to foliar bacterial spot of tomato incited by *Xanthomonas campestris* pv. *vesicatoria*. *Journal of the American Society for Horticultural Science*, **114**, 111–14.
Serizawa, S., Inoue, K. and Goto, M. (1969) Studies on citrus canker I. Dispersal of the citrus canker organisms. *Bulletin of the Shizuoka Prefecture Citrus Experiment Station*, **8**, 81–5.
Shane, W.W., Baumer, J.S. and Teng, P.S. (1987) Crop losses caused by *Xanthomonas* streak on spring wheat and barley. *Plant Disease*, **71**, 927–30.

Sheppard, J.W., Roth, D.A. and Saettler, A.W. (1989) Detection of *Xanthomonas campestris* pv. *phaseoli* in bean, in *Detection of Bacteria in Seed and Other Planting Material* (eds A.W. Saettler, N.W. Schaad and D.A. Roth), APS Press, St Paul, MN, pp. 17–29.

Simpson, D.M. (1956) Strains of cotton resistant to bacterial blight. *Plant Disease Reporter*, **40**, 549–55.

Sinclair, J.B. (ed.) (1984) *Compendium of Soybean Diseases*, The American Phytopathological Society, Inc., MN, pp. 4–5.

Sinclair, J.B. and Dhingra, O.D. (1975) An annotated bibliography of soybean diseases, 1882–1974. Urbana, International Soybean Program, University of Illinois.

Sletten, A. (1989) Detection methods for *Xanthomonas campestris* pv. *graminis* on forage grasses. *Bulletin OEPP/EPPO Bulletin*, **19**, 69–72.

Smith, H.E. (1960) Report of the disease loss committee. *Proceedings of the 20th Meeting of the Cotton Disease Council*, Memphis, TN, pp. 2–3.

Sowell, G. and Dempsey, A.H. (1977) Additional sources of resistance to bacterial spot of pepper. *Plant Disease Reporter*, **61**, 684–6.

Spence, J.A. (1957) The incidence of leaf scald disease of sugarcane in the West Indies. *Proceedings of the British West Indies Sugar Technologists*, **57**, 140–6.

Srivastava, S.S.L. and Bais, B.S. (1986) Involvement of pectolytic and cellulolytic enzymes in bacterial leaf pustule of soybean. *Farm Science Journal*, **1**, 82–4.

Stall, R.E. and Civerolo, E.L. (1991) Research relating to the recent outbreak of citrus canker in Florida. *Annual Review of Phytopathology*, **29**, 399–420.

Stall, R.E., Civerolo, E.L., Ducharme, E.P., *et al.* (1986) Management of citrus canker by eradication of *Xanthomonas campestris* pv. *citri*, in *Proceedings of the Sixth International Conference on Plant Pathogenic Bacteria* (eds E.L. Civerolo *et al.*), Martinus Nijhoff Publishers, Dordrecht, pp. 900–5.

Stall, R.E. and Hall, C.B. (1984) Chlorosis and ethylene production in pepper leaves infected by *Xanthomonas campestris* pv. *vesicatoria*. *Phytopathology*, **74**, 373–5.

Stall, R.E. and Hodge, N.C. (1989) Use of fatty acid profiles to identify strains of *Xanthomonas campestris* pv. *citri* from the citrus canker epidemic in Florida. *Phytopathology*, **79**, 376 (abstr.).

Stall, R.E., Loschke, D.C. and Jones, J.B. (1986) Linkage of copper resistance and avirulence loci on a self-transmissible plasmid in *Xanthomonas campestris* pv. *vesicatoria*. *Phytopathology*, **76**, 240–3.

Stall, R.E., Miller, J.W., Marco, G.M. and Canteros de Echenique, B.I. (1980) Population dynamics of *Xanthomonas citri* causing cancrosis of citrus in Argentina. *Proceedings of the Florida State Horticultural Society*, **93**, 10–14.

Stall, R.E. and Thayer, P.L. (1962) Streptomycin resistance of the bacterial spot pathogen and control with streptomycin. *Plant Disease Reporter*, **46**, 389–92.

Stapp, C. (1966) *Xanthomonas malvacearum*, der Erreger der Baumwollbakteriose. *Zentralblatt für Bakteriologie, Parasitenkunde, Infektions-Krankheiten und Hygiene, II*, **210**, 509–42.

Starr, M.P. and Nasuno, S. (1967) Pectolytic activity of phytopathogenic xanthomonads. *Journal of General Microbiology*, **46**, 425–33.

Stead, D.E. (1989) Groupings of *Xanthomonas campestris* pathovars of cereals and reassessment by fatty acid profiling. *Bulletin OEPP/EPPO Bulletin*, **19**, 57–68.

Steenackers, V. (1966) La sélection et la création des peupliers résistants aux diverses maladies des peupliers et plus particulièrement à *Aplanobacter populi* (Ridé). FAO/CIP Groupe de travail des maladies, Versailles-Grammont.

Steenackers, V. (1976) Progrès récents dans la sélection des clones de peupliers résistants à *Aplanobacter populi*. FAO/CIP Groupe de travail des maladies, Angers.

Stephens, C.T. and Tuinier, J. (1989) Disease symptomatology and variation in susceptibility of seed-propagated hybrid geranium varieties to *Xanthomonas campestris* pv. *pelargonii*. *Plant Disease*, **73**, 559–62.

Steyn, P.L., Viljoen, N.M. and Kotze, J.M. (1974) The causal organism of bacterial black spot of mangoes. *Phytopathology*, **64**, 1400–4.

Stolp, H. and Starr, M.P. (1964) Bacteriophage reactions and separation of phytopathogenic xanthomonads. *Phytopathologische Zeitschrift*, **51**, 442–78.

Swings, J., De Vos, P., Van den Mooter, M. and De Ley, J. (1983) Transfer of *Pseudomonas maltophilia* Hugh 1981 to the genus *Xanthomonas* as *Xanthomonas maltophilia* (Hugh 1981) comb. nov. *International Journal of Systematic Bacteriology*, **33**, 409–13.

Swings, J., Van den Mooter, M., Vauterin, L. *et al.* (1990) Reclassification of the causal agents of bacterial blight (*Xanthomonas campestris* pv. *oryzae*) and bacterial leaf streak (*Xanthomonas campestris* pv. *oryzicola*) of rice as pathovars of *Xanthomonas oryzae* (ex-Ishiyama 1922) sp. nov., nom. rev. *International Journal of Systematic Bacteriology*, **40**, 309–11.

Takatsu, A., Fukuda, S. and Perin, S. (1979) Epidemiological aspects of bacterial blight of cassava in Brazil, in *Diseases of Tropical Food Crops. Proceedings of an International Symposium, U.C.L., Louvain-la-Neuve, Belgium, 1978* (eds H. Maraite and J.A. Meyer), Université Catholique de Louvain, Louvain-la-Neuve, Belgium, pp. 141–50.

Takhtajan, A.L. (1980) Outline of the classification of flowering plants (*Magnoliophyta*). *The Botanical Review*, **46**, 225–359.

Tarr, S.A.J. (1957) Observations récentes sur les maladies des plantes au Soudan. *Bulletin Phytosanitaire de la F.A.O.*, **5**, 195–8.

Teng, P.S., Mew, T.W., Bonman, J.M. and Leung, H. (1989) Methods for estimating yield loss in tropical rice diseases, in *Tropical Agricultural Research Series* No. 22, TARC, Japan, pp. 141–52.

Timmer, L.W., Gottwald, T.R. and Zitko, S.E. (1991) Bacterial exudation from lesions of Asiatic citrus canker and citrus bacterial spot. *Plant Disease*, **75**, 192–5.

Tomlinson, D.L. (1987) A bacterial leaf disease of taro (*Colocasia esculenta*) caused by *Xanthomonas campestris* in Papua New Guinea. *Tropical Pest Management*, **33**, 353–5.

Topp, B.L., Heaton, J.B., Russell, D.M. and Mayer, R. (1989) Field susceptibility of Japanese-type plums to *Xanthomonas campestris* pv. *pruni*. *Australian Journal of Experimental Agriculture*, **29**, 905–9.

Topp, B.L. and Sherman, W.B. (1990) Sources of bacterial spot resistance in Japanese-type plum cultivars. *Fruit Varieties Journal*, **44**, 32–5.

Tuinier, J.E. and Stephens, C.T. (1989) Use of serology to detect *Xanthomonas campestris* pv. *pelargonii* in aqueous extracts of geranium plants. *Plant Disease*, **73**, 875–8.

Umemura, Y. and Kawano, K. (1983) Field assessment and inheritance of resistance to cassava bacterial blight. *Crop Science*, **23**, 1127–32.

Vakili, N.G. (1967) Importance of wounds in bacterial spot (*Xanthomonas vesicatoria*) of tomatoes in the field. *Phytopathology*, **57**, 1099–103.

Van den Mooter, M., Maraite, H., Meiresonne, L., *et al.* (1987) Comparison between *Xanthomonas campestris* pv. *manihotis* (ISPP List 1980) and *X. campestris* pv. *cassavae* (ISPP List 1980) by means of phenotypic, protein electrophoretic,

DNA hybridization and phytopathological techniques. *Journal of General Microbiology*, **133**, 57–71.

Van den Mooter, M. and Swings, J. (1990) Numerical analysis of 295 features of 266 *Xanthomonas* strains and related strains and an improved taxonomy of the genus. *International Journal of Systematic Bacteriology*, **40**, 348–69.

Van den Mooter, M., Swings, J., De Cleene, M., Leyns, F. and De Ley, J. (1981) Isolation and identification of *Xanthomonas campestris* pv. *graminis* from forage grasses in Belgium. *Parasitica*, **37**, 23–8.

Van den Mooter, M., Swings, J., Gosselé, F., Kersters, K. and De Ley, J. (1987) The taxonomy of *Xanthomonas* Dowson 1939, in *Plant Pathogenic Bacteria* (eds E.L. Civerolo, A. Collmer, R.E. Davis and A.G. Gillaspie), Martinus Nijhoff Publishers, Dordrecht, pp. 795–6.

Van Doorn, J., Boonekamp, P.M. and Roossien, F.F. (1990) Toets op geelziek komt in zicht. *Bloembollencultuur*, **26**, 40–3.

Van Doorn, J., Hollinger, T.C., Boonekamp, P.M. and Van Schadewijk, A.R. (1992) Geelziektoets ter discussie; methode bruikbaar voor herkennen pseudozwart. *Bloembollencultuur*, **28**, 46–7.

Van Heerden, H.G. (1976) Differentiation of local races of *Xanthomonas malvacearum* (E.F. Smith) Dowson with the aid of *Gossypium hirsutum* and *G. barbadense* lines. *Phytophylactica*, **8**, 19–20 (in Africaans).

Van Tuyl, J.M. (1982) Breeding for resistance to yellow disease of hyacinths II. Influence of flowering time, leaf characteristics, stomata and chromosome number on the degree of resistance. *Euphytica*, **31**, 621–8.

Van Vuurde, J.W.L. (1987) New approach in detecting phytopathogenic bacteria by combined immunoisolation and immunoidentification assays. *Bulletin OEPP/EPPO Bulletin*, **17**, 139–48.

Van Zyl, E., Kotze, J.M. and Steyn, P.L. (1988) Isolation of *Xanthomonas campestris* pv. *mangiferaeindicae* from gall fly-induced lesions on mango leaves. *Phytophylactica*, **20**, 89–90.

Vauterin, L., Swings, J., Kersters, K., *et al.* (1990) Towards an improved taxonomy of *Xanthomonas*. *International Journal of Systematic Bacteriology*, **40**, 312–16.

Vauterin, L., Swings, J. and Kersters, K. (1991a) Grouping of *Xanthomonas campestris* pathovars by SDS–PAGE of proteins. *Journal of General Microbiology*, **137**, 1677–87.

Vauterin, L., Yang, P., Hoste, B., *et al.* (1991b) Differentiation of *Xanthomonas campestris* pv. *citri* strains by sodium dodecyl sulphate–polyacrylamide gel electrophoresis of proteins, fatty acid analysis, and DNA–DNA hybridization. *International Journal of Systematic Bacteriology*, **41**, 535–42.

Vauterin, L., Yang, P., Hoste, B., Pot, B., Swings, J. and Kersters, K. (1992) Taxonomy of xanthomonads from cereals and grasses based on SDS–PAGE of proteins, fatty acid analysis and DNA hybridization. *Journal of General Microbiology*, **738**, 1467–77.

Vera Cruz, C.M., Gosselé, F., Kersters, K., *et al.* (1984) Differentiation between *Xanthomonas campestris* pv. *oryzae*, *Xanthomonas campestris* pv. *oryzicola* and the bacterial 'brown blotch' pathogen on rice by numerical analysis of phenotypic features and protein gel electrophoregrams. *Journal of General Microbiology*, **130**, 2983–99.

Verma, J.P. (1986) *Bacterial Blight of Cotton*, CRC Press, Inc., Boca Raton, Florida.

Verma, J.P. and Singh, R.P. (1970) Two new races of *Xanthomonas malvacearum*, the cause of blackarm of cotton. *Cotton Growing Review*, **47**, 203–5.

Verma, J.P. and Singh, R.P. (1975) Studies on the distribution of races of *Xanthomonas malvacearum* in India. *Indian Phytopathology*, **28**, 459–63.

Verma, J.P., Singh, R.P., Borkar, S.G., Sinha, P.P. and Prasad, R. (1980) Management of bacterial blight of cotton with particular reference to variation in the pathogen, *Xanthomonas malvacearum* (Smith) Dowson. *Annals of Agricultural Research*, **1**, 98–107.

Volcani, Z. (1969) The effect of mode of irrigation and wind direction on disease severity caused by *Xanthomonas vesicatoria* on tomato in Israel. *Plant Disease Reporter*, **53**, 459–61.

Voronkevich, I.V. and Shmanenkova, T.N. (1968) Biological role of extracellular polysaccharides of *Xanthomonas phaseoli* var. *sojense* (Hedges) Starr and Burk., the causal agent of soybean bacteriosis (in Russian), cited in Sinclair, J.B. and Dhingra, O.D. (1975) An annotated bibliography of soybean diseases 1882–1974. International Soybean Program, University of Illinois.

Wainwright, S.H. and Nelson, P.E. (1972) Histopathology of *Pelargonium* species infected with *Xanthomonas pelargonii*. *Phytopathology*, **62**, 1337–47.

Wakker, J.H. (1887) Onderzoek der ziekten van hyacinten en andere bol-en knolgewassen gedurende de jaren 1883, 1884 en 1885. *Algemeene Vereeniging voor Bloembollencultuur*, pp. 4–13.

Walker, J.C. (1952) Diseases of crucifers, in *Diseases of Vegetable Crops*, McGraw-Hill, New York, pp. 128–31.

Wang, F.M., Chen, Y.K. and Bai, K.J. (1964) Preliminary observation of the transmission of the bacterial blight pathogen through rice straw. *Plant Protection*, **2**, 9–10.

Weber, C.R., Dunleavy, J.M. and Fehr, W.R. (1966) Effect of bacterial pustule on closely related soybean lines. *Agronomy Journal*, **58**, 544–5.

Whittle, A.M. (1980) Leaf scald. *Proceedings of the First Inter-American Sugar Cane Seminar*, pp. 99–106.

Wickens, G.M. (1953) Bacterial blight of cotton. A survey of present knowledge, with particular reference to possibilities of control of the disease in African rain-grown countries. *Empire Cotton Growing Review*, **30**, 81–103.

Wickens, G.M. (1956) Vascular infection of cotton by *Xanthomonas malvacearum* (E.F. Smith) Dowson. *Annals of Applied Biology*, **44**, 129–37.

Wiehe, P.O. and Dowson, W.J. (1953) A bacterial disease of cassava (*Manihot utilissima*) in Nyasaland. *Empire Journal of Experimental Agriculture*, **21**, 141–3.

Wilkins, P. and Exley, J. (1977) Bacterial wilt of ryegrass in Britain. *Plant Pathology*, **26**, 99.

Williams, P.H. (1980) Black rot: a continuing threat to world crucifers. *Plant Disease*, **64**, 736–42.

Wong, W.C. (1990) Production of monoclonal antibodies to *Pseudomonas syringae* pv. *phaseolicola* and *Xanthomonas campestris* pv. *phaseoli*. *Letters in Applied Microbiology*, **10**, 241–4.

Xie Guanlin, Wang Hanrong, Wang Gongjin and Wang Meili (1991a) Rice seed inspection of *Xanthomonas campestris* pv. *oryzicola*. *Seed*, **51**, 17–19.

Xie Guanlin, Wang Hanrong, Chen, G.A. and Ye, Y.W. (1991b) A study of the ingress of *Xanthomonas oryzae* pv. *oryzicola* to rice plant. *Zhiwu Jianyi*, **5**, 1–4.

Young, J.M. (1987) *Catalogue of the International Collection of Micro-organisms from Plants (ICMP)*, Plant Diseases Division, DSIR, Auckland, New Zealand.

Young, J.M., Bradbury, J.F., Gardan, L., *et al.* (1991a) Comment on the reinstatement of *Xanthomonas citri* (ex Hasse 1915) Gabriel *et al.* 1989 and *X. phaseoli* (ex Smith 1897) Gabriel *et al.* 1989: indication of the need for minimal standards for the genus *Xanthomonas*. *International Journal of Systematic Bacteriology*, **41**, 172–7.

Young, J.M., Bradbury, J.F., Davis, R.E., *et al.* (1991b) Nomenclatural revisions of plant pathogenic bacteria and list of names 1980–1988. *Review of Plant Pathology*, **70**, 211–21.

Yuen, G.Y., Alvarez, A.M., Benedict, A.A. and Trotter, K.J. (1987) Use of monoclonal antibodies to monitor the dissemination of *Xanthomonas campestris* pv. *campestris*. *Phytopathology*, **77**, 366–70.

Zachowski, M.A. (1989) Zur Pathophysiologie von *Xanthomonas campestris* pv. *malvacearum* (Smith) Dye, dem Erreger der Eckigen Blattfleckenkrankheit, und zur Resistenz verschiedener Baumwollsorten (*Gossypium hirsutum* L.) [Studies on the pathophysiology of *Xanthomonas campestris* pv. *malvacearum* (Smith) Dye, and on the resistance of different cotton cultivars (*Gossypium hirsutum* L.)]. PhD thesis, University of Göttingen, 113 pp.

Zachowski, M.A., Cinar, Ö. and Rudolph, K. (1989a) Characterization of *Xanthomonas campestris* pv. *malvacearum* (angular leaf spot of cotton) isolates from Turkey and resistance tests of Turkish cotton cultivars. *Phytopathologia Mediterranea*, **28**, 10–15.

Zachowski, M.A., Ebrahim-Nesbat, F. and Rudolph, K. (1989b) Effect of continuous darkness on production of exopolysaccharides by *Xanthomonas campestris* pv. *malvacearum* in susceptible cotton cotyledons, in *Plant Pathogenic Bacteria. Proceedings of the Seventh International Conference on Plant Pathogenic Bacteria*, Budapest, 11–16 June 1989 (ed. Z. Klement), Akadémiai Kiado, Budapest, pp. 173–8.

Zachowski, A. and Rudolph, K. (1988) Characterization of isolates of bacterial blight of cotton (*Xanthomonas campestris* pv. *malvacearum*) from Nicaragua. *Journal of Phytopathology*, **123**, 344–52.

Zaumeyer, W.J. and Thomas, H.R. (1957) A monographic study of bean diseases and methods for their control. *United States Department of Agriculture Technical Bulletin* 868, 255 pp.

Zurini, I.E., Docea, E. and Carpinisan, T. (1981) Status of the races of *Xanthomonas malvacearum* (Smith) Dowson in Romania. *Analele Institutului de Cercetari de Pentru Plantelor*, **16**, 93–8 (in Romanian).

2

The bacterium *Xanthomonas*

J. Swings, L. Vauterin and K. Kersters
Laboratorium voor Mikrobiologie, Universiteit Gent

2.1 INTRODUCTION

Since the reviews on the biology of *Xanthomonas* by Starr (1981) and Bradbury (1984, 1986), many new data have appeared. In this chapter we will emphasize the common bacteriological features and the physiological potential of *Xanthomonas*. It is our aim to stress the wealth of physiological activities displayed by this bacterium. We will not dwell on the use of the different features in taxonomy. Some topics which have been reviewed by others will only be referred to and will not be treated in detail here.

Although we recognize the excellent studies of many workers on *Xanthomonas* in the past, it is difficult to compare previous data due to the different strains and methodologies used. For this chapter, we have adapted the observations of Van den Mooter (1984) as basic material and supplemented them with other data. We have not tried to reach completeness.

Throughout this chapter we use the subdivision of the genus *Xanthomonas* in seven basic taxa: *X. albilineans* (phenon 4), *X. axonopodis* (phenon 5), the *X. campestris* complex (phenons 8 and 9), *X. fragariae* (phenon 3), *X. maltophilia* (phenon 1), *X. oryzae* (phenon 7), and *X. populi* (phenon 6). These seven taxa and their corresponding phenon numbers have been based on a numerical analysis of phenotypic features (Van den Mooter, 1984; Van den Mooter and Swings, 1990). The species *X. ampelina*, which has been reclassified as *Xylophilus ampelinus* (Willems *et al.*, 1987) is not considered here. Taxonomically the situation is more

Xanthomonas.
Edited by J.G. Swings and E.L. Civerolo.
Published in 1993 by Chapman & Hall, London.
ISBN 0 412 434202

complicated, as the *X. campestris* complex seems to be composed of several species. This point will be discussed in Chapter 3. In the following sections, when no particular reference is mentioned, we explicitly refer to Van den Mooter (1984) and Van den Mooter and Swings (1990).

2.2 THE CELL

2.2.1 Cell morphology

All *Xanthomonas* cells are Gram-negative rods, sometimes slightly curved with rounded ends, measuring 0.2–0.6 μm by 0.8–2.9 μm. They occur mostly alone or in pairs but chains and filaments also occur. Exceptionally, cells up to 1.5 μm wide and 4 μm long were found, e.g. in the strains *X. campestris* A 674 (LMG 944) and in *X. campestris* A 676 (LMG 945). Cell length is variable, even within the same strain. The occurrence of pairs of cells is quite typical for *X. maltophilia* and *X. campestris*, but it is not always clear whether it is a pair of cells or an elongated single cell. Most cells have intracellular granules. Polyhydroxyalkanoates may accumulate in low quantities. No spores or other resting stages occur.

No pili or fimbriae have been reported. Cells are surrounded by xanthan gum (extracellular polysaccharide, EPS), not recognizable as capsules by light or electron microscopy.

2.2.2 The cell envelope

Based on electronmicrographs, *Xanthomonas* has the typical inner and outer membrane structure of Gram-negative bacteria (dos Santos and Dianese, 1985). The outer membrane (OM) is a unique structure of Gram-negative bacteria, composed of a mosaic of six to eight major proteins and 50 or more minor ones, one or two heterogeneous lipopolysaccharide (LPS) molecules and several lipids. The OM fulfils a variety of functions, e.g. protection from the environment, selective permeability and export, receptor and interaction functions and an anchoring function of external structures (Hancock, 1991). Most of our present knowledge on the OM of Gram-negative bacteria stems from the work on Enterobacteriaceae and *Pseudomonas aeruginosa*, but a limited amount of specific data is available on *Xanthomonas*.

The LPS from 17 pathovars of *X. campestris* are composed of uronic acid, phosphate, glucose, mannose, rhamnose and 2-keto-3-deoxyoctonate, but without heptose (Volk, 1966, 1968). Using HPLC, Sutherland and Kennedy (1986) were not able to detect mannose, but they found small quantities of heptose. Neither xylose nor fucose are necessary components of any 'minimal common structure' (Volk, 1968, 1972). The cell wall of *X. maltophilia* is characterized by a great diversity of monosaccharide components (up to 11) and the content of rhamnose is so high that it was thought to occur as a rhamnan (Wilkinson, 1968). Neal and Wilkinson (1982) have further elucidated the structure of the LPS of *X. maltophilia*. They confirmed a tetrasaccharide repeating unit for the O-specific polymer, containing three rhamnopyranosyl molecules and xylose or 3-O-methylxylose. The core oligosaccharide contains D-glucose (two molecules), D-mannose (one molecule), D-galactosamine (one molecule), D-galacturonic acid (one molecule) and 3-deoxyoctulosonic acid, phosphate (two to three molecules), but no aldoheptose. The lipid A backbone of the LPS from *X. campestris* pv. *citri* contains β-1,6-linked glucosamine disaccharides, a structure which seems to be common to many, if not all, Gram-negative bacteria (Hase and Rietschel, 1976).

In most bacteria, the OM contains usually a limited number (three to eight) of major proteins present in high copy number, i.e. 50 000 to 200 000 per cell, making them the predominant proteins in the bacterial cell (Hancock, 1991). It is now generally assumed that OM proteins and LPS are involved in plant–pathogen interactions. dos Santos and Dianese (1985) have compared the proteins from the OM and the inner membrane (IM) fractions from *X. campestris* pv. *manihotis* and *X. campestris* pv. *cassavae*. The protein patterns of the IM fraction contained a larger number of bands than the OM. Total membrane protein patterns from *X. c.* pv. *manihotis* strains were very similar whereas those from *X. c.* pv. *cassavae* strains showed major differences.

A *Xanthomonas* strain (previously thought to be a *Pseudomonas*-like bacterium) was shown to possess a surface layer (Austin *et al.*, 1990), resembling crystalline surface layers in other bacteria (Sleytr and Messner, 1983).

Another structural OM protein is associated with the ice nucleating activity (Ina). At least five bacterial species, *Erwinia ananas*, *Erwinia herbicola*, *Pseudomonas fluorescens*, *Pseudomonas syringae* and *X. c.* pv. *translucens*, are known to nucleate the crystallization of ice. They cause the frost-sensitive plants on which they occur to become more susceptible to frost damage by initiating the formation of ice. In bacteria, the activity is conferred by a protein with periodicities of repetition, aligning large numbers of water molecules (for a review see Lindow, 1983). In the five bacteria cited, a single gene confers the Ina phenotype, which is thought to have been spread through horizontal transmission (Warren and Wolber, 1991). Fourteen *X. campestris* pv. *translucens* strains expressed Ina at temperatures from −2 to −8°C whereas strains from other pathovars and *Xanthomonas* species did not exhibit ice nucleation above −8°C (Kim *et al.*, 1987).

The barrier function of the OM of *X. maltophilia* and its role in the unusual resistance against antibiotics of this species was examined by Yamazaki *et al.* (1989). Solute transport in the OM of *X. maltophilia* is less than 5% of that of the OM of *E. coli*, while the OM pores are only a little smaller than those of *E. coli*. This suggested that *X. maltophilia* has a low copy number of porins.

Debette and Prensier (1989) have shown the presence of an esterase hydrolysing β-naphthyl acetate on the OM of *X. maltophilia*.

A protein (or lipoprotein) of the OM of *X. maltophilia* acts as a binding site of the human chorionic gonadotropin with a high affinity. This binding was not observed with other Gram-negative bacteria (Richert and Ryan, 1977).

A high-molecular-weight immunogen composed of a sugar–protein complex seems to be localized in the IM of *X. c.* pv. *campestris* (Thaveechai and Schaad, 1986).

2.2.3 Motility

Cells are motile, sometimes sluggishly. *X. maltophilia* (phenon 1) and *X. albilineans* (phenon 4), which do not form xanthan gum, both have cells that move rapidly. *X. campestris* pv. *graminis* (phenon 8) is non-motile. *X. populi* (phenon 6) contains motile and non-motile strains. Flagella may be difficult to stain by the method described by Rhodes (1958) as in most cases they are easily detached from the cell. A limited number of cells have polar monotrichous flagellation. *X. maltophilia* is multiflagellated.

Kamoun and Kado (1990) described a switch from non-motile, non-chemotactic, EPS producing cultures (wild type) to motile, chemotactic and non-EPS producing cultures (swarmers) in five different *X. campestris* pathovars. Reversions of the swarmer cells to the wild type cells never occurred *in planta* but were observed on bacteriological culture media.

2.2.4 Parasitism by *Bdellovibrio*

Bdellovibrio strains parasitic towards *X. oryzae* have been isolated (Uematsu, 1980). They present the usual vibrio and spiral forms and penetrate into the periplasmic space of the host cell. A disorganization in the cytoplasm and nuclear area of the host cell occurs immediately upon attachment of *Bdellovibrio*. Finally, the parasite splits into several cells before the host cell wall is disrupted (Uematsu *et al.*, 1978). The parasite may play a role in maintaining populations of plant pathogens at a low level.

2.2.5 Phages

Many attempts have been made (a) to use phage typing in fine characterization schemes for *Xanthomonas* pathovars, mainly intended for epidemiological studies, (b) to use phages as biocontrol agents, (c) to use phages for the detection of the host bacteria in seed, debris, soil, and (d) to use phage counts to forecast disease outbreaks. The literature on phages of *Xanthomonas* is quite extensive. The phages of *Xanthomonas* are mostly tadpole-shaped with a polyhedral head and tail, or filamentous. Although most *Xanthomonas* phages have a narrow host range, phages with a broad host range are also known, e.g. the *Xanthomonas* phage Pg 125, attacking strains from different *X. campestris* pathovars. The Pg 125 phage heads are about 60 nm in diameter and the tails are 180 nm in length and 10–15 nm in width (Sutton *et al.*, 1958).

(a) X. campestris *pv.* campestris *phages*

Specific virulent phages have been studied by Liew and Alvarez (1981a,b), who proposed to differentiate *X. c.* pv. *campestris* into two major lysotypes. These phages have polyhedral heads of 55–65 nm diameter and, based on tail structure, they constitute three morphological groups. One group (OH, OK, HP and HT phages) has contractile tail sheaths averaging 18 × 115 nm, a second group (P phages) has non-contractile flexuous tails averaging 14 × 120 nm, whereas a third group has short wedge-shaped tails 15 nm long.

(b) X. campestris *pv.* citri *phages*

Two similar filamentous phages have been studied, Cf and Cflt (Kuo *et al.*, 1987a). They behave as symbionts: while the host continues to grow and to divide, the phage chromosome replicates as an episome. Filamentous phages are continuously extruded from the growing host cell and normally do not cause lysis or lysogeny of the bacterium. Phage Cflt forms turbid plaques, does not affect the growth of the infected cells and has a low phage yield, whereas phage Cf forms clear plaques, retards the growth of the infected cells and has a high yield (Kuo *et al.*, 1987b). Both phages are approximately 1157 nm long and have a diameter of 6 nm. Phage Cflt is able to incorporate its genome into that of its host and can engage in a lysogenic cycle. A closely related filamentous phage, Cf16, has been described (Dai *et al.*, 1987). The 'neolysogenic' insertion of Cf16-v1 into the host chromosome was accomplished by a site-specific recombination mechanism similar to integration of phage λ (Dai *et al.*, 1988a). The stability of the carrier state varied considerably even for very related phages. The spontaneous curing frequency varied from 1 to 100% (Dai *et al.*, 1988b).

(c) X. campestris *pv.* glycines *phages*

Fett *et al.* (1987) have described a temperate phage characterized by a polygonal head, 50–70 nm in diameter, and a long non-contractile tail

(143–11 nm). Under the electron microscope full and empty phage particles were observed.

(d) X. campestris *pv.* malvacearum *phages*

Virulent phages have been described (Hayward, 1964).

(e) X. campestris *pv.* mangiferaeindicae *phages*

White *et al.* (1985) described four bacteriophages. Phage Xm1 has a non-contractile tail, 250 nm long, and a polyhedral head, 50 nm in diameter. The filamentous phage Xm2 is 9 nm in diameter and 250–1470 nm long. Phages Xm3 and Xm4 have long contractile tails, measuring, respectively, 80 and 72 nm and heads with a diameter of 56 and 48 nm. Mixed infections are only seen rarely. The phages are able to infect a few other pathovars.

(f) X. campestris *pv.* phaseoli *phages*

Sutton and Wallen (1967) used four phages to type strains of *X. c.* pv. *phaseoli*. The phage types seemed to correlate with virulence and geographic distribution. The phages have heads with a diameter from 60 to 70 nm and tails measuring 160–190 nm by 10–15 nm (Katznelson *et al.*, 1954; Sutton and Wallen, 1967).

(g) X. campestris *pv.* pruni *phages*

Virulent and temperate phages have been described (Civerolo, 1970, 1973; Randhawa and Civerolo, 1986).

(h) X. campestris *pv.* vesicatoria *phages*

From the 88 bacterial cultures examined, 77 (90%) were shown to be susceptible to broad host range phages isolated from soils (Dye *et al.*, 1964).

(i) X. oryzae *phages*

OP and SP phages have been isolated from *X. oryzae* pv. *oryzae* and *X. oryzae* pv. *oryzicola*, respectively (Goto, 1965). OP phages were used to distinguish phage types within *X. o.* pv. *oryzae* (Wakimoto, 1960; Goto, 1965). They are tadpole-shaped with a head of 70 nm in diameter and tails measuring 150 nm in length and 15 nm in width for OP1 and 25 nm for OP2. Their host range is restricted to *X. o.* pv. *oryzae*. SP3 and SP6 phages have a broader host range as they attacked strains from different *X. campestris* pathovars (Goto, 1965). Goto *et al.* (1971) found SP phage particle densities up to 7000 per ml in drainage water of paddy fields. They also noticed that the host range of the SP phages in nature was very diverse and extremely unstable, a phenomenon associated with colony-type mutations. Xf and Xf2 are flexible filamentous phages which have been studied extensively (Kuo *et al.*, 1969; Wakimoto and Kamiunten, 1978; Kamiunten and Wakimoto, 1979, 1980). They are 8–10 nm thick and 900–1000 nm and 1450–1650 nm long, respectively. They do not attack other *Xanthomonas* species. One *X. oryzae* strain released Xf and Xf2 phages. Another tadpole-shaped phage, Xp10, has been isolated and used in studies on the regulation of the transcription of its genome. The early Xp10 genes are transcribed by the host polymerase, whereas the late Xp10 genes are transcribed by the Xp10 RNA polymerase (Liao and Kuo, 1986; Liao *et al.*, 1987). The transcription of the Xp10 genome is turned off 15 min after phage infection by modification of the host RNA polymerase through loss of the σ factor. This phenomenon is different from other phage systems. The transcription of the host genome is stopped and the cells are lysed 1 h after infection. In the RNA polymerase from *X. oryzae* a new protein subunit, κ, was found (Chen *et al.*, 1989).

Lesion development due to *X. oryzae* was prevented when rice seedlings were immersed in phage suspensions before inoculation (Reddy *et al.*, 1977).

2.2.6 Pigments

Most *Xanthomonas* strains form yellow, water-insoluble pigments (xanthomonadins) when cultivated on the common growth media such as GYCA, LPGA and NA. The tones vary from very pale yellow to deep yellow. The colour intensity changes upon ageing. The yellow pigment is probably bound to the cellular membrane (Stephens and Starr, 1963; Dianese and Schaad, 1982). The exact chemical structure of the pigments was determined by Andrewes *et al.* (1973, 1976) to be mono- or dibromo-aryl polyenes (xanthomonadins). Besides *Xanthomonas*, only marine organisms are able to concentrate and to incorporate bromine in organic compounds. The xanthomonadin pigments occur esterified to glycerophosphoryl sorbitol or to glycerophosphate (Aririatu and Kester, 1985). Xanthomonadins only occur in the genus *Xanthomonas* and can be used as a taxonomic marker, particularly to differentiate the genus *Xanthomonas* from the other yellow-pigmented plant-associated bacteria, e.g. *Pseudomonas, Erwinia,* and *Flavobacterium* (Starr *et al.*, 1977). One example is given by Jenkins and Starr (1982a), who have shown that *X. populi* contains xanthomonadins, supporting incorporation of this species in the genus *Xanthomonas*. Another example concerns *X. maltophilia*, in which non-pigmented as well as definitely yellow-coloured strains occur. The xanthomonadins of yellow-pigmented *X. maltophilia* strains differ from those of other *Xanthomonas* strains: the former are hexaenes with chlorine, whereas the latter are mostly octaenes with bromine as the halogen (Jenkins and Starr, 1985). The xanthomonadins are thought to function in the protection of the bacteria against photobiological damage (Jenkins and Starr, 1982b).

Several *Xanthomonas* strains do not produce a yellow pigment on GYCA or LPGA media, e.g. *X. c.* pv. *manihotis* (all strains), *X. c.* pv. *populi* (seven strains), *X. c.* pv. *ricini* (five strains), *X. c.* pv. *viticola* (three strains), *X. c.* pv. *phyllanthi* (two strains) and one strain each of *X. c.* pv. *centellae*, *X. c.* pv. *pedalii*, *X. c.* pv. *azadirachtae*, *X. c.* pv. *mangiferaeindicae*, *X. c.* pv. *vesicatoria*, and a number of *X. maltophilia*. In spite of the lack of yellow pigments in 12% of the strains examined, they were all typical *Xanthomonas* (Van den Mooter, 1984). In the past, several of the albino *Xanthomonas* strains have been misnamed *Pseudomonas gardneri*, *Ps. vitiswoodrowii* and *Ps. mangiferaeindicae*.

Strain-specific dark brown to black soluble pigments may be formed by cultures grown on GYCA, e.g. by the strains *X. c.* pv. *bauhiniae* NCPPB 1335, *X. c.* pv. *laureliae* NCPPB 1155, *X. c.* pv. *punicae* NCPPB 466, *X. c.* pv. *ricini* NCCPB 1063, all *X. c.* pv. *phaseoli* var. *fuscans* strains and *X. maltophilia* ATCC 17806 and LMG 960.

2.2.7 Polyamines

Polyamines occur both in prokaryotic and eukaryotic cells and they have been used as chemotaxonomic markers. Auling *et al.* (1991) determined the polyamine pattern of 19 *Xanthomonas* cultures. The common feature was the presence of spermidine as the major polyamine in all strains. Significant quantities of cadaverine occur in some strains. When present, spermine, putrescine and 1,3-diaminopropane occur only in low amounts. Polyamine patterns allow a differentiation of *Xanthomonas* from authentic *Pseudomonas*, the latter being characterized by major amounts of putrescine.

2.2.8 Ubiquinones

Xanthomonas strains are characterized by the ubiquinone Q-8 (Ikemoto *et al.*, 1980).

2.2.9 Fatty acids

Lipids may account for up to 20% of the weight of bacteria (Lechevalier and Lechevalier, 1988). *Xanthomonas* is rich in fatty acids; over 65 fatty acids have been found, which include straight chain fatty acids, iso- and anteiso-branched chain fatty acids, 2- and 3-hydroxy fatty acids, monounsaturated fatty acids and cyclopropane fatty acids. The most common fatty acids are: 10:0, 11:0 iso, 11:0 iso 3OH, 12:0 3OH, 13:0 iso 3OH, 14:0, 15:0 iso, 15:0 anteiso, 16:0 iso, 16:1 *cis* 9, 16:0, 17:1 iso F, 17:0 iso, each appearing in at least 90% of the 966 strains tested (Yang *et al.*, 1993). The following set of nine fatty acids occurs in 99% of over 1000 genuine *Xanthomonas* strains: 11:0 iso, 11:0 iso 3OH, 12:0 3OH, 13:0 iso 3OH, 15:0 iso, 16:1 *cis* 9, 16:0, 17:1 iso F and 17:0. The following fatty acids are 'characteristic' for the genus *Xanthomonas* because they have not been detected so far in the taxa of other plant pathogenic bacteria: 11:0 iso, 11:0 iso 3OH, 13:0 iso 3OH. The most abundant fatty acids are 15:0 iso (on the average 26% of total fatty acids), followed by 16:1 *cis* 9 (17%), 15:0 anteiso (9%), 17:1 iso F (7%), 17:0 iso (7%) and 16:0 (6%) (Yang *et al.*, 1993).

2.2.10 Protein content

A few authors have studied the IM and OM proteins whereas others have analysed the total solubilized cellular proteins. We refer the reader to section 2.2.2 and Chapter 3.

2.2.11 DNA

The average genome size of *Xanthomonas* is 2.5×10^9 dalton. The %G+C content of the DNA ranges from 63 to 71%. For the application of DNA–rRNA and DNA–DNA hybridization and restriction endonuclease analysis in the taxonomy of *Xanthomonas*, refer to Chapter 3.

(a) Indigenous plasmids

Less than half of the investigated *Xanthomonas* strains possess indigenous plasmids. Lazo and Gabriel (1987) screened 117 strains from different *X. campestris* pathovars for plasmids and found that strains from nine pathovars harbour plasmids (pv. *campestris*, pv. *citri*, pv. *cyamopsidis*, pv. *glycines*, pv. *hederae*, pv. *malvacearum*, pv. *phaseoli*, pv. *vesicatoria* and pv. *vignicola*), whereas strains from 13 other pathovars did not contain plasmids. Also, half of the *X. populi* strains harbour plasmids and five of the 12 *X. oryzae* strains tested contain cryptic plasmids (Chen and Tseng, 1988). Fourteen out of 16 pathogenic *X. c.* pv. *glycines* strains contain a plasmid of 18 MDa, whereas non-pathogenic strains do not (Haas and Fett, 1987). Three different plasmids of 18, 40 and 72 MDa were characterized in *X. c.* pv. *manihotis* and shown to have the same %G+C content as the host chromosomal DNA, indicating that the plasmids are indigenous to the bacterium. Other strains from *X. oryzae* and *X. c.* pv. *pruni* did not harbour any plasmids (Lin *et al.*, 1979). The existence of plasmids in the latter pathovar is confirmed (Randhawa and Civerolo, 1987). Within the same pathovar, there seems to be considerable conservation of plasmid DNA sequences.

Lai *et al.* (1977a) transferred the wide host range plasmids RP4 and RK2 to *X. c.* pv. *vesicatoria*. The R plasmids seem to be more stable in bacterial cells on tomato leaves than in bacterial cells on GYCA slants (Lai *et al.*, 1977b).

Many authors have suggested a relationship between the occurrence of plasmids and virulence or ecological fitness.

(b) Transposable elements

Two transposable elements, ISXC4 and ISXC5, have been isolated from *X. c.* pv. *citri*. They originate from the plasmids pXW45N and pXW45J, respectively (Tu *et al.*, 1989; Dai and Chiang, 1990).

(c) Repetitive DNA sequences

Leach *et al.* (1990) found two repetitive DNA sequences in the *X. oryzae* pv. *oryzae* genome. One highly repetitive sequence was approximately 1 kb and was estimated to occur in 81 copies in the *X. oryzae* pv. *oryzae* genome. This suggested that it could be an insertion sequence, but there is no evidence of transposition. The repetitive sequence was found in all strains of the pathovars *oryzae* and *oryzicola* and was distributed throughout the genome. The differences in hybridization patterns and copy number between the pathovars *oryzae* and *oryzicola* suggest a clear divergence of the two populations. Related repetitive elements also occur in the pathovars *holcicola, vasculorum* and *phleipratensis*.

2.3 GROWTH CHARACTERISTICS

2.3.1 Culturing *Xanthomonas*

Xanthomonas cultures, except *X. populi* (phenon 6), can be grown on GYCA medium, composed of D-glucose (1% w/v), yeast extract (0.5%), $CaCO_3$ (3%) and agar (2%) in screw-capped tubes. They are incubated for 2 days at 28°C. The working laboratory cultures can be kept at 4°C and must be transferred monthly.

X. populi can be best cultivated on GPPYA medium, composed of D-glucose (1%), proteose peptone (0.5%), yeast extract (0.5%) and agar

(2%) in distilled water. They are incubated for 5 days at 18°C. The working laboratory cultures are kept at 4°C and transferred after a few weeks.

2.3.2 Colony morphology

Usually, *Xanthomonas* strains form separate colonies on GYCA medium after 3 days of incubation at 28°C. Strains from *X. albilineans* (phenon 4), *X. oryzae* pv. *oryzae* (phenon 7) and *X. c.* pv. *hyacinthi* and *X. c.* pv. *theicola* (phenon 9) form separate colonies after 5–6 days. *X. populi* (phenon 6) and *X. fragariae* (phenon 3) do not form separate colonies, even after 10 days.

Xanthomonas strains have convex, round, yellow, mucoid colonies with an even colony margin. The colonies formed by some *X. maltophilia* (phenon 1) strains are umbonate, do not have an even colony margin and do not form slime.

'Mature' *Xanthomonas* colonies have a diameter of more than 2 mm, sometimes up to 10 mm. Exceptions are *X. albilineans, X. oryzae* pv. *oryzae, X. c.* pv. *hyacinthi* and *X. c.* pv. *theicola,* which have a colony diameter of less than 2 mm. The use of the colony diameter as a differentiating character cannot be advocated as many strains of *X. campestris* (phenon 9) are able to form small colony mutants. Colony variants have been described for *X. c.* pv. *phaseoli* (Corey and Starr, 1957), *X. c.* pv. *malvacearum* (Brinkerhoff, 1963), *X. oryzae* (Goto, 1972; Reddy and Kauffmann, 1977) and *X. c.* pv. *campestris* (Cadmus *et al.*, 1976), but they seem to occur in all other pathovars as well. In most cases, a careful examination of several colonial variants from one strain leads to the conclusion that there is almost no associated variation in their phenotypic features or cellular protein fingerprints. There have been a few attempts to correlate colony characteristics with other features. The results presented by Peros (1988) seem to indicate a correlation between colony characteristics and the aggressiveness of *X. c.* pv. *vasculorum* on sugarcane. Choi *et al.* (1981) have tried to correlate the colony variants of *X. oryzae* to their serological and auto-agglutination properties.

Not all *Xanthomonas* strains are abundant slime-formers: the species *X. maltophilia* and *X. albilineans,* together with some pathovars and colony mutants of *X. campestris,* do not form slime or only very small amounts.

2.3.3 Selective media

There is no general enrichment procedure or a selective medium that can be used for all members of *Xanthomonas* (Starr, 1981). Semi-selective media have been proposed for the isolation of different pathovars of *X. campestris* (Schaad and Stall, 1990).

A selective medium has been developed for the isolation of *X. maltophilia* from soils and rhizospheres (Juhnke and Jardin, 1989).

2.3.4 Conservation of *Xanthomonas* cultures

Three conservation methods have been extensively tested and are very satisfactory. In the first method, cultures are grown in screw-capped tubes for 3 days at 28°C on GYCA medium. After cultivation, they are closed in order to prevent desiccation during storage at 4°C. This can also be achieved by covering the culture with sterile mineral oil. Most strains are still viable after several months. The second method, lyophilization, is the most widely used method and guarantees survival for many years. In the third method, cultures grown on GYCA medium can be conserved for at least 10 years at −79°C with 20% glycerol as cryoprotectant by the method described by Feltham *et al.* (1978). The preservation on silica gel is sometimes preferred because of its low cost and its possibility of repeated sampling, particularly by field laboratories with limited equipment (Sleesman and Leben, 1978).

2.3.5 Growth temperature

X. populi is unable to grow at temperatures above 23°C, whereas all other species grow optimally at 28°C. The temperature sensitivity of *X. populi* was noted by Ridé and Ridé (1978). *Xanthomonas* is unable to grow at 4°C. At 37°C, only a few *X. maltophilia* and *X. campestris* strains are able to grow. According to Gilardi (1978), *X. maltophilia* can grow at 42°C.

2.3.6 pH sensitivity

Xanthomonas is not able to grow at an initial pH of 4.5 or lower. At pH 6.5, all *Xanthomonas* strains grow well, except those of *X. fragariae*, which require a pH value of 7.5.

2.3.7 Osmo-tolerance

X. campestris and *X. maltophilia* are the only species which tolerate 10% glucose, a few strains even tolerate 20%. *Xanthomonas* cannot grow in 30% glucose.

Xanthomonas is not very tolerant of NaCl. *X. fragariae* does not tolerate 0.5% NaCl, the lowest concentration tested. Within each individual species, NaCl tolerance may vary between 0.5 and 5%.

2.3.8 Oxygen requirements

Xanthomonas is an obligate aerobe and uses oxygen as a terminal electron acceptor. Oxidase activity occurs only in *X. maltophilia* and *X. campestris* strains, but not in the other taxa. In the oxidation–fermentation test of glucose (O/F test, Hugh and Leifson test), acid is formed in open tubes by all strains of the *X. campestris* complex, *X. oryzae*, *X. populi* and *X. fragariae*. No acid is produced, either in open or in sealed tubes, by *X. axonopodis*, *X. albilineans* and *X. maltophilia*.

Difference spectra of whole cells of *X. campestris* indicated the presence of *b*-, *c*-, aa_3- and *o*-type cytochromes and possibly also cytochrome a_1 (Rye *et al.*, 1988).

2.3.9 Resistance to heavy metal ions and to miscellaneous inhibitors

All *Xanthomonas* strains are able to grow in the presence of 0.01% sodium salicylate, 0.01% KSCN, 0.01% $(NH_4)_6Mo_7O_{24}$, 0.001% antimony potassium tartrate, 0.01% $Ba(NO_3)_2$, and 0.01% Congo Red, but not in the presence of 0.01% $CuCl_2$, 0.01% $Cu(CH_3COO)_2$, 0.01% $Cu(NO_3)_2$, 0.01% $CuSO_4$, 0.01% $Cd(CH_3COO)_2$, 0.01% $CdSO_4$, 0.01% $CoSO_4$, 0.01% $CoCl_2$, or 0.01% $ZnCl_2$ (except for some strains of *X. campestris* and *X. maltophilia*). *X. maltophilia* is the most resistant species to the compounds tested, followed by the *X. campestris* complex. All other *Xanthomonas* species are more sensitive, particularly *X. fragariae*.

X. maltophilia is not only very resistant to different salts but has also the ability to accumulate metallic compounds. It forms one or a few intracellular particles, mainly composed of zinc and phosphorus in an amorphous state (Sakurai *et al.*, 1990). Charley and Bull (1979) described bioaccumulation of silver by a stable community in a chemostat enrichment culture composed of *X. maltophilia*, *Staphylococcus aureus* and a coryneform. Silver is normally a very toxic metal to heterotrophs. Nevertheless, maximum accumulations of over 300 mg silver per gram dry weight of biomass were recorded. *X. maltophilia* was responsible for the silver uptake.

The resistance to copper bactericides by *X. c.* pv. *vesicatoria* is a problem in vegetable crops, e.g. tomato and pepper. Self-transmissible copper resistance plasmids (pXvCu and pXV10A) have been described (Stall *et al.*, 1986; Bender *et al.*, 1990). Plasmid pXV10A was readily transferable to other pathovars.

According to Wilkinson's (1967) procedure to determine the sensitivity towards EDTA, *X. maltophilia* was classified as resistant. Of 10 quaternary ammonium compounds tested, only decyl- and hexadecyltrimethylammonium bromides were decomposed by *Pseudomonas* and *Xanthomonas* strains isolated from sewage and soil. The *Xanthomonas* strain metabolized decyltrimethylammonium bromide producing 9-carboxynonyl- and 7-carboxyheptylammonium (Dean-Raymond and Alexander, 1977).

2.3.10 Susceptibility towards antibiotics

X. maltophilia is the most resistant species to all antibiotics tested, whereas *X. fragariae* is the most sensitive species. Because *X. maltophilia* is an important nosocomial pathogen in humans, its great resistance to most of the commonly used antibiotics has made it a test organism for new antibacterial compounds (Tofte and Crossley, 1980; O'Donnell *et al.*, 1982; Fukasawa *et al.*, 1983; Matsuda *et al.*, 1985; Barry *et al.*, 1986; Jones *et al.*, 1986; Hirose *et al.*, 1987; Noble and Parekh, 1987; Strandberg *et al.*, 1983; Khardori *et al.*, 1990a,b). The resistance of *X. maltophilia* to all β-lactams is thought to be due to a combination of low OM permeability and broad-spectrum β-lactamases which are induced synchronously (Mett *et al.*, 1988).

Almost all strains from all phenons other than *X. maltophilia* are sensitive to tetracycline (30 mg per Oxoid disk), chloramphenicol (30 mg per disk) and streptomycin (10 mg per disk).

Over 55% of the 548 *X. c.* pv. *vesicatoria* strains were resistant to 100 mg streptomycin sulphate per millilitre. The locus for resistance towards this antibiotic is maintained on a 68 kb plasmid (Minsavage *et al.*, 1990).

2.4 PHYSIOLOGY AND METABOLISM

2.4.1 Carbon utilization

Compared to *Pseudomonas*, *Xanthomonas* strains are much more restricted with respect to growth on different carbon sources. In *Xanthomonas*, carbohydrates are oxidized, never fermented. In the O/F test (Hugh and Leifson, 1953), the species *X. fragariae*, *X. populi*, *X. oryzae* and *X. campestris* oxidize glucose only in open tubes. Traditionally, the assimilation of different carbohydrates is checked either by acidification or by growth tests (auxanography). The phenons that acidify the highest numbers of carbohydrates are, in decreasing order, *X. campestris*, *X. oryzae* and *X. populi*. The other species are more limited in their capacity to acidify carbohydrates. *Xanthomonas* does not acidify sorbitol, adonitol, L-sorbose or *meso*-inositol.

The acidification of cellobiose, trehalose and maltose is a useful criterium for the differentiation of *Xanthomonas* taxa. The acidification data which are published agree well with each other and with those from Van den Mooter and Swings (1990).

Xanthomonas never grows on the following carbon sources: L-rhamnose, adonitol, sodium 2-ketogluconate, inulin, glycolic acid, L-norleucine, L-leucine, L-valine, L-arginine, L-isoleucine, L-threonine, glycine, L-tryptophan, L-lysine, L-serine, L-asparagine, L-cysteine, ethanolamine, sodium hippurate, acetamide, *n*-butyric acid, DL-β-hydroxybutyric acid, pelargonic acid, naphthalene, methanol, iso-butanol, *meso*-erythritol, diethylene glycol, sodium benzoate, potassium sodium tartrate and sodium oxalate.

The growth patterns on different compounds are useful for differentiation of the taxa. Phena 9 and 8 are differentiated by their ability (respectively + and −) to grow on L-alanine, trisodium citrate, sodium acetate and sodium propionate.

Over 90% of the strains from *X. campestris*, *X. oryzae* and *X. populi* are able to grow on glucose, D-mannose, sucrose, trehalose, D-galactose, sodium succinate, sodium fumarate and sodium L-malate, whereas *X. albilineans* shows no or very weak growth on these carbohydrates. The phena *X. campestris*, *X. oryzae* and *X. populi* behave similarly in all growth tests. Low growth responses in carbon utilization tests for *X. albilineans* may be due to a lack of L-methionine or L-cysteine in the basal medium. Negative growth tests with *X. maltophilia* should also be interpreted cautiously as some strains require methionine (Iizuka and Komagata, 1964; Stanier *et al.*, 1966; Ikemoto *et al.*, 1980).

X. maltophilia grows very well on sodium succinate, sodium fumarate, sodium L-malate, trisodium citrate, sodium acetate, calcium gluconate, calcium lactate and sodium propionate. All *X. axonopodis* strains grow on sucrose and calcium lactate, some also on glucose, trehalose and cellobiose. All *X. fragariae* strains grow on glucose and trehalose, sodium succinate, sodium fumarate and sodium L-malate.

2.4.2 Nitrogen and growth factor requirements

To our knowledge, nitrogen fixation does not occur in *Xanthomonas*. Starr (1946) determined that most *X. campestris* strains can grow on a simple ammonium–glucose–salt medium. In a complex medium containing yeast extract, the growth of *X. campestris* is faster than in a defined minimal medium, but molar growth yields are the same. The maximum molar growth yield of 53.6 obtained for *X. campestris* is low compared with the other aerobic heterotrophic bacteria which are characterized by values of over 100 (Rye *et al.*, 1988). Van den Mooter (1984) confirmed that all *X. campestris* strains from phenon 9 are able to grow on an ammonium–glucose–salt medium whereas strains from phenon 8 are not. All other phena are unable to utilize inorganic nitrogen. *X. campestris* phenon 9 strains do not require growth factors. *X. c.* pv. *pruni* ATCC 10017 is able to synthesize NAD from L-tryptophan

(Starr, 1946; Wilson and Henderson, 1963; Brown and Wagner, 1970). Tryptophan and a number of tryptophan analogues can induce the three first enzymes from the kynureninase pathway (Brown and Wagner, 1970). This pathway is absent in most bacteria, but present in several Eukaryotes. Shetty and Gaertner (1978) stressed the dual physiological function of the kynureninase-type enzymes of *X. c.* pv. *pruni*, e.g. in L-tryptophan degradation as well as in NAD synthesis. Whereas during evolution most organisms have adapted their kynureninase-type activities to either catabolic or biosynthetic purposes, *X. c.* pv. *pruni* seems to be an evolutionary branching-point organism, still possessing both functions.

The addition of a vitamin mixture to the ammonium–glucose–salt medium stimulates the growth of a few additional strains from *X. campestris* (phenon 8), *X. oryzae* (phenon 7), *X. axonopodis* (phenon 5) and *X. fragariae* (phenon 3). When 0.3% vitamin-free casamino acids are added to the ammonium–glucose–salt–vitamins medium, a luxurious growth is obtained for all phena, comparable to the growth on GYCA medium. The casamino acid mixture may be replaced by individual amino acids at 0.2%. *X. campestris* (phenon 9) is clearly the most versatile group as it is able to use inorganic and organic nitrogen. Some *X. campestris* strains are even able to use either 0.1% L-alanine, 0.1% glutamic acid, L-hydroxyproline or L-proline as a sole source of carbon and nitrogen. Mori *et al.* (1988) reported that *X. c.* pv. *translucens* IFO 13558 could grow on L-carnitine or DL-carnitine as carbon and nitrogen source.

X. campestris (phenon 8) strains are able to grow on inorganic nitrogen, but may require growth factors, e.g. nicotinic acid. *X. c.* pv. *pruni* strains require nicotinic acid for growth (Starr, 1946). Some authors have studied this NAD requirement. In a medium containing glucose, salts and growth factors, strains from phenon 8 are able to use organic nitrogen in the form of vitamin-free casamino acids or some individual amino acids, e.g. 0.02% of L-aspartic acid, L-cysteine, L-glutamic acid, L-glutamine, L-histidine, L-isoleucine, L-methionine, L-valine or L-ornithine. Growth is inhibited when the casamino acids are replaced by 0.02% L-phenylalanine, L-serine, L-tyrosine or L-proline.

Only strains from the *X. campestris* complex (but not all strains) are able to utilize amino acids as the sole source of nitrogen.

X. oryzae (phenon 7) requires growth factors. Although a medium containing inorganic nitrogen, glucose, salts and growth factors sustains growth, the addition of 0.3% casamino acids or 0.02% of the individual amino acids L-arginine, L-asparagine, L-aspartic acid, L-cysteine, L-glutamic acid, L-glutamine, L-histidine, L-methionine, L-phenylalanine, L-proline or L-tryptophan will increase growth considerably.

X. populi (phenon 6) strains require growth factors and organic nitrogen for growth. The latter may be satisfied by 0.3% casamino acids or by 0.02% L-cysteine, L-glutamic acid or L-methionine.

X. axonopodis (phenon 5) strains require growth factors and organic nitrogen for growth. The latter are required as mixtures of L-amino acids; individual amino acids cannot satisfy the nitrogen requirements.

X. albilineans (phenon 4) requires growth factors and organic nitrogen for growth. The latter can be satisfied by 0.3% casamino acids, 0.02% L-cysteine or 0.02% L-methionine.

Some, but not all, *X. fragariae* (phenon 3) strains may require growth factors. In order to obtain good growth, organic nitrogen should be supplied in the growth medium, e.g. as 0.3% casamino acids or 0.02% L-methionine.

X. maltophilia (phenon 1) requires organic nitrogen, which can be supplied as 0.3% casamino acids or 0.02% L-cysteine or 0.02% L-methionine. According to Stanier *et al.* (1966), *X. maltophilia* requires only L-methionine as a growth factor. Some strains do not depend on L-methionine for growth (Ikemoto *et al.*, 1980).

2.4.3 Metabolic pathways

Knowledge about the metabolic pathways in *Xanthomonas* is more limited than in *Pseudomonas*. Most of the biochemical work has been performed on *X. campestris* pathovars. Metabolically, *Xanthomonas* is related to *Pseudomonas fluorescens*.

Glucose, but not gluconate, is readily oxidized by intact *Xanthomonas* cells (Katznelson, 1955, 1958). The glucose catabolism by *Xanthomonas* has been examined by radiorespirometry (Zagallo and Wang, 1967). In *Xanthomonas*, the Entner–Doudoroff pathway in conjunction with the tricarboxylic acid cycle pathway is the predominant mechanism for glucose catabolism, using 81–93% of the glucose. A small portion (7–19%) of the glucose is routed via the pentose phosphate pathway. The hexose pathway does not play any significant role.

For glucose uptake, two discrete systems seem to exist: one periplasmic oxidative system requiring glucose dehydrogenase activity and a second one, intracellular and phosphorylative (Whitfield *et al.*, 1982).

A strict aerobe such as *Xanthomonas* is dependent on the citric acid cycle (TCA) for its energy production. The TCA cycle also allows metabolism of a wide variety of compounds. Its citrate synthase is inhibited by NADH, but AMP may reactivate the enzyme. These properties seem to be characteristic for the 'metabolic life-style' of the strict aerobes. As they depend on the citric acid cycle for energy production, it seems logical that the first enzyme of the cycle should be sensitive to AMP, a low-energy metabolic signal (Weitzman, 1980).

Comparative enzymological studies have revealed some unique characteristics in the metabolism of aromatic compounds of *Xanthomonas*. Byng *et al.* (1980) performed a comparative enzymological study of tyrosine biosynthesis via the arogenate and 4-hydroxyphenylpyruvate routes. All *Xanthomonas* strains (*X. maltophilia* included) fell into the same group, characterized by NAD-specific prephenate and arogenate dehydrogenases, the former enzyme being very sensitive towards feedback inhibition by tyrosine. Enzymological patterns reflect phylogeny (Byng *et al.*, 1983).

Whitaker *et al.* (1981) studied the allostery of the enzyme 3-deoxy-D-arabino-heptulosonate-7-phosphate synthase (DAHP), the first enzyme in the shikimate pathway leading to aromatic amino acid biosynthesis. *Xanthomonas* strains have a unique DAHP synthase sensitive to feedback inhibition by chorismate.

The metabolism of different aromatic substances by *X. oryzae* and *X. campestris* pathovars (10 strains) typically seems to follow the meta-cleavage scheme (William and Mahadevan, 1980). The ortho-cleavage was shown to be also operative in a *Xanthomonas* strain (Kern *et al.*, 1984).

2.4.4 Enzymatic activities

Based on an examination of the enzymatic activities of over 200 *Xanthomonas* strains from different species (and different pathovars from *X. campestris*), using API ZYM (bioMérieux; 38390 La Balme-les-Grottes, France), the following enzymes were always present in all cultures: acid and alkaline phosphatase, esterase lipase (caprylate), leucine arylamidase, and naphthol-AS-BI-phosphohydrolase. The following enzymes were never present: cystine arylamidase, α-galactosidase, β-glucuronidase, mannosidase and α-fucosidase. Both α- and β-mannosidases seem to be inducible in *X. campestris*.

(a) β-Glucosidase

This activity occurs in most strains from the *X. campestris* complex, *X. oryzae* and *X. albilineans*. This constitutive enzyme is mainly extracellular, although activity is also present in the periplasm.

(b) D-Xylose catabolizing enzymes

D-Xylose isomerase, D-xylose permease and xylulose kinase genesare localized on a DNA fragment of *Xanthomonas* XA1-1 of approximately 15 kb (Liu *et al.*, 1987).

(c) β-Galactosidase

There is great interest in the production of xanthan starting from industrial wastes, e.g. whey, containing 4–5% lactose but also proteins, minerals and vitamins. Although many *Xanthomonas* strains produce β-galactosidase (Hayward, 1977), xanthan gum is not readily produced in lactose-containing media. Frank and Somkuti (1979) determined the general enzymatic properties of β-galactosidase from *X. campestris* VB 1459 and concluded that a negative feedback control together with a low substrate affinity determine the inability of *X. campestris* to synthesize large amounts of extracellular polysaccharide in lactose-based media. Chemical hydrolysis of lactose to glucose and galactose allows a better xanthan production but a direct enhanced production of xanthan by a *Xanthomonas* strain capable of using the lactose more efficiently is preferred. Attempts have been made to introduce the β-galactosidase genes into *Xanthomonas* (Walsh *et al.*, 1984). A mobilizable β-galactosidase expression plasmid, pKMFLT, was constructed and introduced into *X. campestris*, which was able to produce similar amounts of xanthan from whey as from glucose (Fu and Tseng, 1990).

(d) β-Mannanase

X. campestris strains produce extracellular, as well as cell-bound, β-mannanases constitutively. Carob galactomannan is the most easily degradable polysaccharide (Dekker and Candy, 1979).

(e) Esculin hydrolysis

All *Xanthomonas* species, except *X. fragariae*, are able to hydrolyse esculin.

(f) Hydrolysis of starch

This feature is not uniformly present in all species of the genus *Xanthomonas*, but it is present in at least 75% of the strains from the *X. campestris* complex. Negative results should be interpreted with care, as the incubation time may influence the results.

(g) Lignin breakdown

It has been known for years that white rot fungi are able to simultaneously decompose lignin and cellulose from wood. Actinomycetes and other bacteria also seem to play a role in the biodegradation of lignin (Trojanowski *et al.*, 1977; Crawford, 1978; Odier and Monties, 1978; Odier *et al.*, 1981). Ligninolytic *Xanthomonas* strains grow aerobically in a mineral medium on lignin as the sole carbon and energy source.

Odier and Monties (1978) described a *Xanthomonas* strain which was able to degrade wheat lignin. The ligninolytic activity also occurred in the absence of oxygen when nitrates and glucose were present. The results obtained by Kern and Kirk (1987) suggest that *Xanthomonas* is unable to depolymerize the lignin backbone, but degrades low-molecular-weight portions of lignin (mol. wt 600–1000) to carbon dioxide.

(h) Cellulolytic activity

Cellulolytic activity has been demonstrated in several *Xanthomonas* cultures (Knösel and Garber, 1968). Patel and Vaughn (1973) isolated cellulolytic cultures associated with sloughing spoilage of ripe olives. Two *Xanthomonas* strains were able to disintegrate olive tissue and liquefy carboxymethylcellulose, but not to attack filter paper.

Endoglucanase genes from *Xanthomonas albilineans* XA1-1 (Lucas *et al.*, 1987) and from *X. c.* pv. *campestris* (Gough *et al.*, 1988, 1990) have been cloned. Endoglucanase ENGXCA seems to play only a minor role in pathogenesis. Nevertheless, the enzyme is highly active towards carboxymethylcellulose and represents more than 90% of the protein found in the culture filtrate of *X. c.* pv. *campestris* (Gough *et al.*, 1988).

(i) Pectinolytic activity

The symptoms caused in plants by *Xanthomonas* (blight, canker, necrosis) do not suggest extensive production of pectolytic enzymes. Nevertheless, many *Xanthomonas* strains are able to degrade pectin gels (Burkholder and Starr, 1948; Smith, 1958; Hildebrand, 1971). Also, some virulent strains do not have any pectinolytic activity, neither *in vitro* nor *in planta* (Reddy *et al.*, 1969).

Pectinolytic enzymes are induced and/or produced constitutively (Abo-El-Dahab, 1964; Starr and Nasuno, 1967; Pandey and Prasad, 1983; Ikotun, 1984a; Prasad and Bhusan, 1987). Pectin methyl esterase (PME) is inducible in *X. c.* pv. *manihotis* but constitutive in *X. c.* pv. *cassavae*, another pathogen of cassava (Ikotun, 1984b).

Pectinolytic *X. campestris* strains have also been clearly associated with soft rots in fruits and vegetables (Liao and Wells, 1987; Gitaitis *et al.*, 1987). The soft-rotting *Xanthomonas* strains behave like opportunists, infecting plants which are completely unrelated to the original host plant.

(j) Chitinase

X. maltophilia has been described as a bacterial symbiont of the sugar-beet maggot (*Tetanops myopaeformis*) together with *Serratia liquefaciens* and *Serratia marcescens*. The bacteria appear to penetrate the inner surface of the exterior puparial wall. *X. maltophilia* was also associated with the larval gut. Chitinase production, which was demonstrated for the three organisms cited, is thought to play a role in the weakening of the puparial wall, thus facilitating imago emergence (Iverson *et al.*, 1984).

The majority of the *X. maltophilia* isolates from the rhizosphere are chitinolytic, but the presence of this enzyme is not correlated to their antifungal activity.

(k) Alginate hydrolysis

Half of the *X. maltophilia* strains are able to hydrolyse alginate (von Riesen, 1980). It is not clear why a terrestrial organism such as *X. maltophilia* developed a need to degrade alginate, a structural polysaccharide of seaweeds.

(l) Hyaluronidase, chondroitin sulphatase and mucinase

These enzymatic activities were shown in some *X. maltophilia* strains isolated from fleece rot in sheep (MacDiarmid and Burrell, 1986).

(m) Proteolytic activity

Almost all strains from the *X. campestris* complex, *X. oryzae*, *X. albilineans*, *X. fragariae* and *X. maltophilia* hydrolyse gelatin, whereas *X. populi* and *X. axonopodis* do not.

Knösel (1975) reported that several plant pathogenic *Xanthomonas* strains were able to decompose proteins. It is not known whether the proteinases play a role in the nutrition or whether they also help to overcome host defense barriers. By cloning the protease genes and Tn5 mutagenesis it was shown that the proteases are not critically important for the plant pathogenicity (Tang *et al.*, 1987).

The protease and elastase activities produced by the opportunist *X. maltophilia* seem to be involved in severe cutaneous manifestations of bacteraemia in humans (Bottone *et al.*, 1986). *X. maltophilia* from fleece rot in sheep also produced collagenase and elastase (MacDiarmid and Burrell, 1986).

Kobayashi *et al.* (1985) described an alkaline protease from a *X. maltophilia* strain isolated from an alkaline detergent. The enzyme was clearly different from other microbial alkaline proteinases.

Oda *et al.* (1987) described the properties of a carboxyl proteinase (i.e. an aspartic proteinase) isolated from a *Xanthomonas* strain.

Two other major proteases have been isolated and fractionated:

PRT1, a serine protease requiring Ca^{2+} ions and PRT2, a zinc-requiring metalloprotease. PRT1 and PRT2 each belong to different protease classes, showing a different pattern of peptide bond cleavage (Dow *et al.*, 1990).

(n) Aminopeptidases

The aminopeptidase galleries AP1 to AP6 from API (bioMérieux), applied to 32 strains from different *Xanthomonas* species, have revealed a broad spectrum of very active aminopeptidases. No activity occurred on the following substrates: L-pyrrolidonyl-NA, N-benzoyl-L-leucyl-NA, β-alanine-NA, N-benzoyl-L-alanine-4-methoxy-NA.

(o) Peptidyl dipeptidases

A peptidyl dipeptidase-4 (PDP-4) that hydrolyses the carboxy peptide bond, constitutes a major exopeptidase from *X. maltophilia*. It is a promiscuous enzyme capable of hydrolysing a spectrum of peptides including the human hormones angiotensin 1, bradykinin, enkephalins and atriopeptin 2, releasing dipeptide units. Its catalytic site is quite unique (Dasarathy *et al.*, 1989). The enzyme was present in all five *X. maltophilia* strains examined (Stevens *et al.*, 1990).

(p) DNase

This activity is not very pronounced in *Xanthomonas* although over 80% of the strains display some DNase activity. Positive strains occur in *X. maltophilia*, *X. axonopodis* and the *X. campestris* complex.

(q) RNase

Arella and Sylvestre (1979) have described a *X. maltophilia* strain capable of producing an extracellular RNase with optimal activity over the wide pH range 7–10.

(r) Nucleotide and nucleoside breakdown

Hoshino *et al.* (1989) screened microorganisms for nucleoside oxidase activity and found that *X. maltophilia* isolates from the environment were the strongest producers of this enzyme. Nucleotide breakdown occurred in *X. c.* pv *phaseoli* (Hochster and Madsen, 1959) and in *X. oryzae* (Huang *et al.*, 1975; Yang *et al.*, 1975).

(s) ADP-ribosylation

This protein modification process occurs widely in nature and may have diverse biological functions. Many bacterial toxins are ADP-ribosyltransferases (ADPRTs) and exert their toxic activity by mono-ADP-ribosylating host-cell proteins (Gaal and Pearson, 1986). ADP-ribosylation of two proteins has been demonstrated in *X. maltophilia* (Edmonds *et al.*, 1989).

(t) Restriction enzymes

Several restriction enzymes have been purified from *Xanthomonas* strains and are currently applied in molecular genetics; they are indicated by the following symbols: PmaCI, Pma I, Xam I, Xba I, Xcy I, Xho I, Xma I, Xma II, Xma III, Xmn I, Xni I, Xpa I, Xor I and Xor II. Specificities, references and other details can be found in specialized lists (e.g. Kessler and Manta, 1990) and commercial catalogues (e.g. New England Biolabs Catalog).

(u) Hydrolysis of Tweens

The Tweens 80 and 60 are hydrolysed by most strains of *X. oryzae*, the *X. campestris* phenon 9, *X. axonopodis*, *X. albilineans* and *X. maltophilia*. *X. fragariae* and the *X. campestris* phenon 8 are negative.

(v) Lecithinase

Lecithinase occurs in *X. campestris* phenon 9, *X. axonopodis*, *X. albilineans* and *X. maltophilia*.

(w) β-Lactamases

X. maltophilia, which is resistant to β-lactam antibiotics, produces several β-lactamases (Cullmann and Dick, 1990). Two inducible β-lactamases have been well studied: a zinc containing penicillin β-lactamase (L1) (Saino *et al.*, 1982) and a cephalosporin β-lactamase (L2) (Saino *et al.*, 1984). Taken together, the two enzymes hydrolyse the entire range of β-lactams. Both β-lactamases are produced simultaneously under variable conditions (Rosta and Mett, 1989). The β-lactamase L1 has been further characterized biochemically (Bicknell *et al.*, 1985) and its gene has been cloned (Dufresne *et al.*, 1988).

(x) Urease

This activity is always absent in *Xanthomonas*.

(y) Catalase

All *Xanthomonas* strains display catalase activity.

(z) Litmus milk reaction

Litmus milk is not acidified by *Xanthomonas*. Alkalinization and peptonization occur in most strains from the *X. campestris* complex, *X. oryzae*, *X. populi* and *X. maltophilia*.

(aa) Nitrate reduction

Xanthomonas cannot grow aerobically or anaerobically on nitrate as a nitrogen source. The reduction of nitrate occurs only within some strains of *X. maltophilia*. Hugh and Gilardi (1980) found that nitrate reduction occurred in 41% of the 390 *X. maltophilia* strains studied. Nitrate-reducing strains contain a respiratory nitrate reductase. It might enable cells to use nitrate as an electron sink under semi-aerobic conditions (Woodard *et al.*, 1990). Since the inclusion of *X. maltophilia* in the genus *Xanthomonas*, the absence of nitrate reduction cannot be retained as a generic feature of *Xanthomonas*.

2.4.5 Miscellaneous products and activities of *Xanthomonas*

Whereas a wide variety of products from *Pseudomonas* and other related bacteria are known and characterized, our knowledge is more limited for *Xanthomonas*. Two important exceptions should be mentioned: xanthan, which is a very important industrial polysaccharide (see Chapter 8); and the different restriction enzymes which have become important laboratory tools in molecular biology.

(a) H_2S production

Over 90% of all *Xanthomonas* strains are able to produce H_2S in L-cysteine-containing media. Exceptions are a few strains of *X. albilineans* and *X. fragariae*.

(b) Indole formation

This feature is never found in *Xanthomonas*.

(c) Acetoin formation (Voges Proskauer reaction)

Acetoin is generally not formed by *Xanthomonas*, the only exception being *X. oryzae* pv. *oryzicola*.

(d) Volatile metabolites

X. maltophilia produces significant amounts of isopentanol, 2-butanone and 1-butanol in trypticase soy broth (Labows *et al.*, 1980; Zechman and Labows, 1985).

(e) Bacteriocins and other antagonistic effects

Bacteriocins are responsible for antagonisms against highly related bacterial strains. Bacteriocin production by *Xanthomonas* is scarcely documented. Hamon *et al.* (1961) described six *Xanthomonas* strains which produce bacteriocins. *X. c.* pv. *glycines* strains produce bacteriocins (glycinecins) inhibitory towards *X. c.* pv. *glycines* strains but also against the pathovars *phaseoli* and *vesicatoria*. Some strains produce more than one bacteriocin. Inhibition zones extended from 1 to 4 mm from the edge of the producer strain (Fett *et al.*, 1987).

Xanthomonas is also involved in antagonisms against fungi. *X. oryzae* inhibited the growth and the conidial germination of a number of phytopathogenic fungi (Narain and Mohanty, 1983). Among the *X. maltophilia* strains isolated from the rhizosphere, 85% were antagonistic *in vitro* against *Pythium ultimum*, 80% against *Gaeumannomyces graminis* var. *tritici* and 36% against *Rhizoctonia solani* (Leyns *et al.*, 1990). To our knowledge, antifungal compounds in *X. maltophilia* have not been determined.

The antagonistic relationship between the fungus *Karlingia rosea* and *X. maltophilia* has been described by Willoughby (1983). The bacterial colonization appeared as an investment of the sporangia, apparently bounded by an outer wall.

The antibiotic albicin, which is characterized by several aromatic rings and produced by *X. albilineans*, inhibits DNA synthesis. Affected bacterial cells do not lyse but accumulate large polyphosphate granules (Birch and Patil, 1985).

(f) Transformation of α-pinene

X. maltophilia S 201–1 was able to utilize α-pinene (a constituent of essential oils from plants) as a sole carbon and energy source. This compound has a commercial utility as a raw material in several industries. The strain mentioned above accumulated limonene, borneol, camphor, penicillic acid and 2-(4-methyl-3-cyclohexenylidene) propionic acid from α-pinene in culture broth (Narushima *et al.*, 1982).

(g) Formation of optically active α-hydroxyketones

X. oryzae IAM 1657 can reduce 1,2-diketones to optically active α-hydroxyketones. Optically active α-hydroxyketones are a useful source for the synthesis of physiologically active chiral compounds. The reactivity of the dialkyl diketones was lower than that of the aromatic diketones (Ohta *et al.*, 1987).

(h) α-Amino acid ester hydrolase

Kato *et al.* (1980) described an α-amino acid ester hydrolase in *X. c.* pv. *citri* IFO 3835, capable of N-acylation of 7-aminocephem compounds with α-amino acid esters to synthesize the corresponding semi-synthetic cephalosporins. The enzyme may carry out the synthesizing and hydrolytic functions.

(i) Assimilation of polyethers

The synthetic polymers polyethylene glycol (PEG), polypropylene glycol (PPG) and polytetramethylene glycol (PTMG) are polyethers. The latter is exclusively used for the synthesis of polyurethane. *X. maltophilia* is able to assimilate PEG and PTMG as a carbon and energy source, but not PPG (Kawai and Moriya, 1991).

REFERENCES

Abo-El-Dahab, M.K. (1964) Production of pectic and cellulolytic enzymes by *Xanthomonas malvacearum*. *Phytopathology*, **54**, 597–601.

Andrewes, A.G., Hertzberg, S., Liaaen-Jensen, S. and Starr, M.P. (1973) The *Xanthomonas* 'carotenoids' – non-carotenoid, brominated, aryl-polyene esters. *Acta Chemica Scandinavica*, **27**, 2382–95.

Andrewes, A.G., Jenkins, C.L., Starr, M.P., Shepherd, J. and Hope, H. (1976) Structure of xanthomonadin I, a novel dibrominated arylpolyene pigment produced by the bacterium *Xanthomonas juglandis*. *Tetrahedron Letters*, **45**, 4023–4.

Arella, M. and Sylvestre, M. (1979) Production of an extracellular ribonuclease by *Pseudomonas maltophilia*. *Canadian Journal of Microbiology*, **25**, 321–8.

Aririatu, L.E. and Kester, A.S. (1985) Isolation and characterization of the pigment esters of *Xanthomonas juglandis (campestris)*. *Journal of General Microbiology*, **131**, 2047–52.

Auling, C., Busse, H.-J., Pilz, F., Webb, L., Kneifel, H. and Claus, D. (1991) Rapid differentiation, by polyamine analysis, of *Xanthomonas* strains from phytopathogenic pseudomonads and other members of the class *Proteobacteria* interacting with plants. *International Journal of Systemic Bacteriology*, **41**, 223–8.

Austin, J.W., Stewart, M. and Murray, R.G.E. (1990) Structural and chemical characterization of the S layer of a *Pseudomonas*-like bacterium. *Journal of Bacteriology*, **172**, 808–17.

Barry, A.L., Thornsberry, C. and Jones, R.N. (1986) *In vitro* evaluation of A-56619 and A-56620, two new quinolones. *Antimicrobial Agents in Chemotherapy*, **29**, 40–3.

Bender, C.L., Malvick, D.K., Conway, K.E., George, S. and Pratt, P. (1990) Characterization of pXV10A, a copper resistance plasmid in *Xanthomonas campestris* pv. *vesicatoria*. *Applied and Environmental Microbiology*, **56**, 170–5.

Bicknell, R., Emanuel, E.L., Gagnon, J. and Waley, S.G. (1985) The production and molecular properties of the zinc β-lactamase of *Pseudomonas maltophilia* IID 1275. *Biochemical Journal*, **229**, 791–7.

Birch, R.G. and Patil, S.S. (1985) Preliminary characterization of an antibiotic produced by *Xanthomonas albilineans* which inhibits DNA synthesis in *Escherichia coli*. *Journal of General Microbiology*, **131**, 1069–75.

Bottone, E.J., Reitano, M., Janda, J.M., Troy, K. and Cuttner, J. (1986) *Pseudomonas maltophilia* exoenzyme activity as correlate in pathogenesis of ecthyma gangrenosum. *Journal of Clinical Microbiology*, **24**, 995–7.

Bradbury, J.F. (1984) Genus II. *Xanthomonas* Dowson 1939, 187^{AL}, in *Bergey's Manual of Systematic Bacteriology*, Vol. 1 (ed. N.R. Krieg), Williams & Wilkins, Baltimore, pp. 199–210.

Bradbury, J.F. (1986) *Xanthomonas* Dowson 1939, 187, in *Guide to Plant Pathogenic Bacteria*, CAB International Mycological Institute, Slough, pp. 198–260.

Brinkerhoff, L.A. (1963) Variability of *Xanthomonas malvacearum*. The cotton bacterial blight pathogen. Technical Bulletin T-98. Agricultural Experiment Station, Oklahoma State University, Stillwater.

Brown, A.T. and Wagner, C. (1970) Regulation of enzymes involved in the conversion of tryptophan to nicotinamide adenine dinucleotide in a colorless strain of *Xanthomonas pruni*. *Journal of Bacteriology*, **101**, 456–63.

Burkholder, W.H. and Starr, M.P. (1948) The generic and specific characters of phytopathogenic species of *Pseudomonas* and *Xanthomonas*. *Phytopathology*, **38**, 494–502.

Byng, G.S., Johnson, J.L., Whitaker, R.J., Gherna, R.L. and Jensen, R.A. (1983) The evolutionary pattern of aromatic amino acid biosynthesis and the emerging phylogeny of pseudomonad bacteria. *Molecular Evolution*, **19**, 272–82.

Byng, G.S., Whitaker, R.J., Gherna, R.L. and Jensen, R.A. (1980) Variable enzymological patterning in tyrosine biosynthesis as a means of determining natural relatedness among the *Pseudomonadaceae*. *Journal of Bacteriology*, **144**, 247–57.

Cadmus, M.C., Rogovin, S.P., Burton, K.A., Pittsley, J.E., Knutson, C.A. and Jeanes, A. (1976) Colonial variation in *Xanthomonas campestris* NRRL B-1459 and characterization of the polysaccharide from a variant strain. *Canadian Journal of Microbiology*, **22**, 942–8.

Charley, R.C. and Bull, A.T. (1979) Bioaccumulation of silver by a multispecies community of bacteria. *Archives of Microbiology*, **123**, 239–44.

Chen, C.-K., Tu, J. and Kuo, T.-T. (1989) A new protein subunit k for RNA polymerase from *Xanthomonas campestris* pv. *oryzae*. *Journal of Biological Chemistry*, **264**, 4362–6.

Chen, L.-J. and Tseng, Y.-H. (1988) Cryptic plasmids of *Xanthomonas campestris* pv. *oryzae*. *Plant Protein Bulletin*, **30**, 78–85.

Choi, J.E., Matsuyama, N. and Wakimoto, S. (1981) Colony type variants of *Xanthomonas campestris* pv. *oryzae* and their serological properties. *Annals of the Phytopathological Society of Japan*, **47**, 244–51.

Civerolo, E.L. (1970) Comparative relationships between two *Xanthomonas pruni* bacteriophages and their bacterial host. *Phytopathology*, **60**, 1385–8.

Civerolo, E.L. (1973) Relationships of *Xanthomonas pruni* bacteriophages to bacterial spot disease in Prunus. *Phytopathology*, **63**, 1279–84.

Corey, R.R. and Starr, M.P. (1957) Colony types of *X. phaseoli*. *Journal of Bacteriology*, **74**, 137–40.

Crawford, D.L. (1978) Lignocellulose decomposition by selected *Streptomyces* strains. *Applied and Environmental Microbiology*, **35**, 1041–5.

Cullmann, W. and Dick, W. (1990) Heterogeneity of beta-lactamase production in *Pseudomonas maltophilia*, a nosocomial pathogen. *Chemotherapy*, **36**, 117–26.

Dai, H. and Chiang, K.-S. (1990) Sequence alternation of indigenous plasmid in *Xanthomonas campestris* pv. *citri* in response to filamentous phage infection. *Microbios*, **63**, 21–8.

Dai, H., Chow, T.-Y., Liao, H.-J., Chen, Z.-Y. and Chiang, K.-S. (1988a) Nucleotide sequences involved in the neolysogenic insertion of filamentous phage Cf16-v1 into the *Xanthomonas campestris* pv. *citri* chromosome. *Virology*, **167**, 613–20.

Dai, H., Liao, H.-J. and Chiang, K.-S. (1988b) Differential stability of filamentous phage genomes in *Xanthomonas campestris* pv. *citri*. *Microbios*, **56**, 157–67.

Dai, H., Tsay, S.-H., Kuo, T.-T., Lin, Y.-H. and Wu, W.-C. (1987) Neolysogenization of *Xanthomonas campestris* pv. *citri* infected with filamentous phage Cf16. *Virology*, **156**, 313–20.

Dasarathy, Y., Stevens, J., Fanburg, B.L. and Lanzillo, J.J. (1989) A peptidyl dipeptidase-4 from *Pseudomonas maltophilia*: purification and properties. *Archives of Biochemistry and Biophysics*, **270**, 255–66.

Dean-Raymond, D. and Alexander, M. (1977) Bacterial metabolism of quaternary ammonium compounds. *Applied and Environmental Microbiology*, **33**, 1037–41.

Debette, J. and Prensier, G. (1989) Immunoelectron microscopic demonstration of an esterase on the outer membrane of *Xanthomonas maltophilia*. *Applied and Environmental Microbiology*, **55**, 233–9.

Dekker, R.F.H. and Candy, G.P. (1979) The β-mannanases elaborated by the phytopathogen *Xanthomonas campestris*. *Archives of Microbiology*, **122**, 297–9.

Dianese, J.C. and Schaad, N.W. (1982) Isolation and characterization of inner and outer membranes of *Xanthomonas campestris* pv. *campestris*. *Phytopathology*, **72**, 1284–9.

dos Santos, R.M.D.B. and Dianese, J.C. (1985) Comparative membrane characterization of *Xanthomonas campestris* pv. *cassavae* and *X. campestris* pv. *manihotis*. *Phytopathology*, **75**, 581–7.

Dow, J.M., Clarke, B.R., Milligan, D.E., Tang, J.-L. and Daniels, M.J. (1990) Extracellular proteases from *Xanthomonas campestris* pv. *campestris*, the black rot pathogen. *Applied and Environmental Microbiology*, **56**, 2994–8.

Dufresne, J., Vézina, G. and Levesque, R.C. (1988) Molecular cloning and expression of the imipenem-hydrolyzing β-lactamase gene from *Pseudomonas maltophilia* in *Escherichia coli*. *Review of Infectious Diseases*, **10**, 806–17.

Dye, D.W., Starr, M.P. and Stolp, H. (1964) Taxonomic clarification of *Xanthomonas vesicatoria* based upon host specificity, bacteriophage sensitivity, and cultural characteristics. *Phytopathologische Zeitschrift*, **51**, 394–407.

Edmonds, C., Griffin, G.E. and Johnstone, A.P. (1989) Demonstration and partial characterization of ADP-ribosylation in *Pseudomonas maltophilia*. *Biochemical Journal*, **261**, 113–18.

Feltham, R.K.A., Power, A.K., Pell, P.A. and Sneath, P.H.A. (1978) A simple method for storage of bacteria at −76°C. *Journal of Applied Bacteriology*, **44**, 313–16.

Fett, W.F., Dunn, M.F., Maher, G.T. and Maleeff, B.E. (1987) Bacteriocins and temperate phage of *Xanthomonas campestris* pv. *glycines*. *Current Microbiology*, **16**, 137–44.

Frank, J.F. and Somkuti, G.A. (1979) General properties of beta-galactosidase of *Xanthomonas campestris*. *Applied and Environmental Microbiology*, **38**, 554–6.

Fu, J.-F. and Tseng, Y.-H. (1990) Construction of lactose-utilizing *Xanthomonas campestris* and production of xantham gum from whey. *Applied and Environmental Microbiology*, **56**, 919–23.

Fukasawa, M., Noguchi, H., Okuda, T., Komatsu, T. and Yano, K. (1983) *In vitro* antibacterial activity of SM-1652, a new broad-spectrum cephalosporin with antipseudomonal activity. *Antimicrobial Agents in Chemotherapy*, **23**, 195–200.

Gaal, J.C. and Pearson, C.K. (1986) Covalent modification of proteins by ADP-ribosylation. *TIBS*, **11**, 171–5.

Gilardi, G.L. (1978) Identification of *Pseudomonas* and related bacteria, in *Glucose Nonfermenting Gram Negative Bacteria in Clinical Microbiology* (ed. G.L. Gilardi), CRC Press, Inc., Boca Raton, FL, pp. 25–38.

Gitaitis, R.D., Sasser, M.J., Beaver, R.W., McInnes, T.B. and Stall, R.E. (1987) Pectolytic xanthomonads in mixed infections with *Pseudomonas syringae* pv. *syringae*, *P. syringae* pv. *tomato*, and *Xanthomonas campestris* pv. *vesicatoria* in tomato and pepper transplants. *Phytopathology*, **77**, 611–15.

Goto, M. (1965) Phage-typing of the causal bacteria of bacterial leaf blight (*Xanthomonas oryzae*) and bacterial leaf streak (*X. translucens* f. sp. *oryzae*) of rice in the tropics. *Annals of the Phytopathological Society of Japan*, **30**, 253–7.

Goto, M. (1972) Interrelationship between colony type, phage susceptibility and virulence in *Xanthomonas oryzae*. *Journal of Applied Bacteriology*, **35**, 505–15.

Goto, M., Frank, P.G. and Ou, S.H. (1971) A study on the phage–bacteria relationship of *Xanthomonas translucens* f. sp. *oryzicola* (Fang *et al.*) Bradbury. *Annals of the Phytopathological Society of Japan*, **37**, 249–58.

Gough, C.L., Dow, J.M., Barber, C.E. and Daniels, M.J. (1988) Cloning of two endoglucanase genes of *Xanthomonas campestris* pv. *campestris*: analysis of the role of the major endoglucanase in pathogenesis. *Molecular Plant–Microbe Interactions*, **1**, 275–81.

Gough, C.L., Dow, J.M., Keen, J., Henrissat, B. and Daniels, M.J. (1990) Nucleotide sequence of the *eng*XCA gene encoding the major endoglucanase of *Xanthomonas campestris* pv. *campestris*. *Gene*, **89**, 53–9.

Haas, M.J. and Fett, W.F. (1987) Detection of plasmids in the plant pathogenic bacterium *Xanthomonas campestris* pathovar *glycines*, in *Biotechnology in Agricultural Chemistry* (eds H.M. LeBaron, R.O. Mumma, R.C. Honeycutt, J.H. Duesing, J.F. Philips and M.J. Haas), American Chemical Society, Washington DC, pp. 148–54.

Hamon, Y., Véron, M. and Péron, Y. (1961) Contribution à l'étude des propriétés lysogènes et bactériocinogènes dans le genre *Pseudomonas*. *Annales de l'Institute Pasteur*, **101**, 738–53.

Hancock, R.E.W. (1991) Bacterial outer membranes: evolving concepts. *ASM News*, **57**, 175–82.

Hase, S. and Rietschel, E.T. (1976) Isolation and analysis of the lipid A backbone. Lipid A structure of lipopolysaccharides from various bacterial groups. *European Journal of Biochemistry*, **63**, 101–7.

Hayward, A.C. (1964) Bacteriophage sensitivity and biochemical group in *Xanthomonas malvacearum*. *Journal of General Microbiology*, **35**, 287–98.

Hayward, A.C. (1977) Occurrence of glycoside hydrolases in plant pathogenic and related bacteria. *Journal of Applied Bacteriology*, **43**, 407–11.

Hildebrand, D.C. (1971) Pectate and pectin gels for differentiation of *Pseudomonas* sp. and other bacterial plant pathogens. *Phytopathology*, **61**, 1430–6.

Hirose, T., Okezaki, E., Kato, H., Ito, Y., Inoue, M. and Mitsuhashi, S. (1987) *In vitro* and *in vivo* activity of NY-198, a new difluorinated quinolone. *Antimicrobial Agents in Chemotherapy*, **31**, 854–9.

Hochster, R.M. and Madsen, N.B. (1959) The breakdown of adenosine phosphates in extracts of *Xanthomonas phaseoli*. *Canadian Journal of Biochemistry and Physiology*, **37**, 639–49.

Hoshino, M., Isono, Y. and Sudo, T. (1989) Production of a new enzyme, nucleoside oxidase, by *Pseudomonas maltophilia* LB-86. *Agricultural and Biological Chemistry*, **53**, 399–403.

Huang, T.-C., Lin, F.-H. and Kuo, T.-T. (1975) Properties of membrane-bound adenosine triphosphatase from *Xanthomonas oryzae*. *Botanical Bulletin Academia Sinica*, **16**, 36–44.

Hugh, R. and Gilardi, G.L. (1980) *Pseudomonas*, in *Manual of Clinical Microbiology*, 3rd edn (eds A. Balows, W.J. Hausler, Jr and J.P. Truant), American Society for Microbiology, Washington DC, pp. 288–317.

Hugh, R. and Leifson, E. (1953) The taxonomic significance of fermentative versus oxidative metabolism of carbohydrates by various Gram-negative bacteria. *Journal of Bacteriology*, **66**, 24–6.

Iizuka, H. and Komagata, K. (1964) Microbiological studies on petroleum and natural gas. I. Determination of hydrocarbon-utilizing bacteria. *Journal of General and Applied Microbiology*, **10**, 207–21.

Ikemoto, S., Suzuki, K., Kaneko, T. and Komagata, K. (1980) Characterization of strains of *Pseudomonas maltophilia* which do not require methionine. *International Journal of Systemic Bacteriology*, **30**, 437–47.

Ikotun, T. (1984a) Pectolytic enzymes produced by *Xanthomonas campestris* pathovar *cassavae* and their involvement in pathogenesis. *Zeitschrift für Allgemeine Mikrobiologie*, **24**, 363–8.

Ikotun, T. (1984b) Inducible production of pectin-methylesterase by *Xanthomonas campestris* pathovar *manihotis*. *Zeitschrift für Allgemeine Mikrobiologie*, **24**, 139–42.

Iverson, K.L., Bromel, M.C., Anderson, A.W. and Freeman, T.P. (1984) Bacterial symbionts in the sugar beet root maggot, *Tetanops myopaeformis* (von Röder). *Applied and Environmental Microbiology*, **47**, 22–7.

Jenkins, C.L. and Starr, M.P. (1982a) The pigment of *Xanthomonas populi* is a nonbrominated aryl-heptaene belonging to xanthomonadin pigment group 11. *Current Microbiology*, **7**, 195–8.

Jenkins, C.L. and Starr, M.P. (1982b) The brominated aryl-polyene (xanthomonadin) pigments of *Xanthomonas juglandis* protect against photobiological damage. *Current Microbiology*, **7**, 323–6.

Jenkins, C.L. and Starr, M.P. (1985) Formation of halogenated aryl-polyene (xanthomonadin) pigments by the type and other yellow-pigmented strains of *Xanthomonas maltophilia*. *Annales de l'Institut Pasteur Microbiologie*, **136B**, 257–64.

Jones, R.N., Fuchs, P.C., Barry, A.L., Ayers, L.W., Gerlach, E.H. and Gavan, T.L. (1986) Antimicrobial activity of Ro 15-8074, active metabolite of a new oral cephalosporin (Ro 15-8075), against 7,775 recent clinical isolates. *Antimicrobial Agents in Chemotherapy*, **30**, 961–3.

Juhnke, M.E. and Jardin, E.D. (1989) Selective medium for isolation of *Xanthomonas maltophilia* from soil and rhizosphere environments. *Applied and Environmental Microbiology*, **55**, 747–50.

Kamiunten, H. and Wakimoto, S. (1979) Biological properties of the filamentous phages released from *Xanthomonas oryzae* isolates. *Annals of the Phytopathological Society of Japan*, **45**, 174–81.

Kamiunten, H. and Wakimoto, S. (1980) Physico-chemical properties of filamentous phage Xf2 of *Xanthomonas campestris* pv. *oryzae*. *Annals of the Phytopathological Society of Japan*, **46**, 315–21.

Kamoun, S. and Kado, C.I. (1990) Phenotypic switching affecting chemotaxis, xanthan production, and virulence in *Xanthomonas campestris*. *Applied and Environmental Microbiology*, **56**, 3855–60.

Kato, K., Kawahara, K., Takahashi, T. and Kakinuma, A. (1980) Purification of α-amino acid ester hydrolase from *Xanthomonas citri*. *Agricultural and Biological Chemistry*, **44**, 1069–74.

Katznelson, H. (1955) The metabolism of phytopathogenic bacteria. I. Comparative studies on the metabolism of representative species. *Journal of Bacteriology*, **70**, 469–75.

Katznelson, H. (1958) Metabolism of phytopathogenic bacteria. II. Metabolism of carbohydrates by cell-free extracts. *Journal of Bacteriology*, **75**, 540–3.

Katznelson, H., Sutton, M.D. and Bayley, S.T. (1954) The use of bacteriophage of *Xanthomonas phaseoli* in detecting infection in beans, with observations on its growth and morphology. *Canadian Journal of Microbiology*, **1**, 22–9.

Kawai, F. and Moriya, F. (1991) Bacterial assimilation of polytetramethylene glycol. *Journal of Fermentation and Bioengineering*, **71**, 1–5.

Kearney, B. and Staskawicz, B.J. (1990) Characterization of IS476 and its role in bacterial spot disease of tomato and pepper. *Journal of Bacteriology*, **172**, 143–8.

Kern, H.W. and Kirk, T.K. (1987) Influence of molecular size and ligninase pretreatment on degradation of lignins by *Xanthomonas* sp. strain 99. *Applied and Environmental Microbiology*, **53**, 2242–6.

Kern, H.W., Webb, L.E. and Eggeling, L. (1984) Characterization of a ligninolytic bacterial isolate: taxonomic relatedness and oxidation of some lignin related compounds. *Systematic Applied Microbiology*, **5**, 433–47.

Kessler, C. and Manta, V. (1990) Specificity of restriction endonucleases and DNA modification methyltransferases – a review (Edition 3). *Gene*, 1–248.

Khardori, N., Elting, L., Wong, E., Schable, B. and Bodey, G.P. (1990a) Nosocomial infections due to *Xanthomonas maltophilia* (*Pseudomonas maltophilia*) in patients with cancer. *Review of Infectious Diseases*, **12**, 997–1003.

Khardori, N., Reuben, A., Rosenbaum, B., Rolston, K. and Bodey, G.P. (1990b) *In vitro* susceptibility of *Xanthomonas* (*Pseudomonas*) *maltophilia* to newer antimicrobial agents. *Antimicrobial Agents in Chemotherapy*, **34**, 1609–10.

Kim, H.K., Orser, C., Lindow, S.E. and Sands, D.C. (1987) *Xanthomonas campestris* pv. *translucens* strains active in ice nucleation. *Plant Disease*, **71**, 994–7.

Knösel, D. (1975) Extracelluläre proteolytische Aktivität bei phytopathogenen Bakterien. *Phytopathologische Zeitschrift*, **82**, 100–6.

Knösel, D. and Garber, E.D. (1968) Separation of pectolytic and cellulolytic enzymes in culture filtrates of phytopathogenic bacterial species by starch gel zone electrophoresis. *Phytopathologische Zeitschrift*, **61**, 292–8.

Kobayashi, T., Ogasawara, A., Ito, S. and Saitoh, M. (1985) Purification and some properties of alkaline proteinase produced by *Pseudomonas maltophilia*. *Agricultural and Biological Chemistry*, **49**, 693–8.

Kuo, T.-T., Chao, Y.-S., Lin, Y.-H., Lin, B.-Y., Liu, L.-F. and Feng, T.-Y. (1987a) Integration of the DNA of filamentous bacteriophage Cflt into the chromosomal DNA of its host. *Journal of Virology*, **61**, 60–5.

Kuo, T.-T., Huang, T.-C. and Chow, T.-Y. (1969) A filamentous bacteriophage from *Xanthomonas oryzae*. *Virology*, **39**, 548–55.

Kuo, T.-T., Lin, Y.-H., Huang, C.-M., Chang, S.-F., Dai, H. and Feng, T.-Y. (1987b) The lysogenic cycle of the filamentous phage Cflt from *Xanthomonas campestris* pv. *citri*. *Virology*, **156**, 305–12.

Labows, J.N., McGinley, K.J., Webster, G.F. and Leyden, J.J. (1980) Headspace analysis of volatile metabolites of *Pseudomonas aeruginosa* and related species by gas chromatography–mass spectrometry. *Journal of Clinical Microbiology*, **12**, 521–6.

Lai, M., Panopoulos, N.J. and Shaffer, S. (1977a) Transmission of R plasmids among *Xanthomonas* spp. and other plant pathogenic bacteria. *Phytopathology*, **67**, 1044–50.

Lai, M., Shaffer, S. and Panopoulos, N.J. (1977b) Stability of plasmid-borne antibiotic resistance in *Xanthomonas vesicatoria* in infected tomato leaves. *Phytopathology*, **67**, 1527–30.

Lazo, G.R. and Gabriel, D.W. (1987) Conservation of plasmid DNA sequences and pathovar identification of strains of *Xanthomonas campestris*. *Phytopathology*, **77**, 448–53.

Leach, J.E., White, F.F., Rhoads, M.L. and Leung, H. (1990) A repetitive DNA sequence differentiates *Xanthomonas campestris* pv. *oryzae* from other pathovars of *X. campestris*. *Molecular Plant–Microbe Interactions*, **3**, 238–46.

Lechevalier, H. and Lechevalier, M.P. (1988) Chemotaxonomic use of lipids – an overview, in *Microbial Lipids* (eds C. Ratledge and S.G. Wilkinson), Academic Press, London, pp. 869–902.

Leyns, F., Lambert, B., Joos, H. and Swings, J. (1990) Antifungal bacteria from different crops, in *Biological Control of Soil-borne Plant Pathogens* (ed. D. Hornby), CAB International, Wallingford, pp. 437–44.

Liao, C.H. and Wells, J.M. (1987) Association of pectolytic strains of *Xanthomonas campestris* with soft rots of fruits and vegetables at retail markets. *Phytopathology*, **77**, 418–22.

Liao, Y.-D. and Kuo, T.-T. (1986) Loss of s-factor of RNA polymerase of *Xanthomonas campestris* pv. *oryzae* during phage Xp10 infection. *Journal of Biological Chemistry*, **261**, 13714–19.

Liao, Y.-D., Tu, J. and Kuo, T.-T. (1987) Regulation of transcription of the Xp10 genome in bacteriophage-infected *Xanthomonas campestris* pv. *oryzae*. *Journal of Virology*, **61**, 1695–9.

Liew, K.W. and Alvarez, A.M. (1981a) Phage typing and lysotype distribution of *Xanthomonas campestris*. *Phytopathology*, **71**, 274–6.

Liew, K.W. and Alvarez, A.M. (1981b) Biological and morphological characterization of *Xanthomonas campestris* bacteriophages. *Phytopathology*, **71**, 269–73.

Lin, B.-C., Day, H.-J., Chen, S.-J. and Chien, M.-C. (1979) Isolation and characterization of plasmids in *Xanthomonas manihotis*. *Botanical Bulletin Academia Sinica*, **20**, 157–71.

Lindow, S.E. (1983) The role of bacterial ice nucleation in frost injury to plants. *Annual Review of Phytopathology*, **21**, 363–84.

Liu, C.-Q., Lucas, R.J., Daly, J.M., Rickard, P.A.D. and Dunn, N.W. (1987) Cloning of *Xanthomonas* DNA that expresses D-xylose catabolic enzymes. *Journal of Biotechnology*, **6**, 159–65.

Lucas, R.J., Austen, R.A. and Dunn, N.W. (1987) Cloning of genes encoding endoglucanases from a cellulolytic xanthomonad. *Journal of Biotechnology*, **6**, 83–90.

MacDiarmid, J.A. and Burrell, D.H. (1986) Characterization of *Pseudomonas maltophilia* isolates from fleece rot. *Applied and Environmental Microbiology*, **51**, 346–8.

Matsuda, K., Sasaki, K., Inoue, K., Kondo, H., Inoue, M. and Mitsuhashi, S. (1985) *In vitro* antibacterial activity of Sch 34343 and its stability to β-lactamases and renal dehydropeptidase 1. *Antimicrobial Agents in Chemotherapy*, **28**, 684–8.

Mett, H., Rosta, S., Schacher, B. and Frei, R. (1988) Outer membrane permeability and β-lactamase content in *Pseudomonas maltophilia* clinical isolates and laboratory mutants. *Review of Infectious Diseases*, **10**, 765–9.

Minsavage, G.V., Canteros, B.I. and Stall, R.E. (1990) Plasmid-mediated resistance to streptomycin in *Xanthomonas campestris* pv. *vesicatoria*. *Phytopathology*, **80**, 719–23.

Mori, N., Kasugai, T., Kitamoto, Y. and Ichikawa, Y. (1988) Purification and some properties of carnitine dehydrogenase from *Xanthomonas translucens*. *Agricultural and Biological Chemistry*, **52**, 249–50.

Narain, A. and Mohanty, A.P. (1983) Bacterial antagonists of some phytopathogenic fungi. *Indian Journal of Mycology and Plant Pathology*, **13**, 28–31.

Narushima, H., Omori, T. and Minoda, Y. (1982) Microbial transformation of α-pinene. *European Journal of Applied Microbiology and Biotechnology*, **16**, 174–8.

Neal, D.J. and Wilkinson, S.G. (1982) Lipopolysaccharides from *Pseudomonas maltophilia*. *European Journal of Biochemistry*, **128**, 143–9.

Noble, R.C. and Parekh, M.C. (1987) *In vitro* comparison of ceftazidime, aztreonam, cefpirome, gentamicin, imipenem, enoxacin, and ticarcillin plus clavulanic acid against 108 strains of *Pseudomonas maltophilia*. *Current Microbiology*, **15**, 177–9.

Oda, K., Nakazima, T., Terashita, T., Suzuki, K.-i. and Murao, S. (1987) Purification and properties of an S-PI(Pepstatin Ac)-insensitive carboxyl proteinase from a *Xanthomonas* sp. bacterium. *Agricultural and Biological Chemistry*, **51**, 3073–80.

Odier, E., Janin, G. and Monties, B. (1981) Poplar lignin decomposition by Gram-negative aerobic bacteria. *Applied and Environmental Microbiology*, **41**, 337–41.

Odier, E. and Monties, B. (1978) Biodégradation de la lignine de blé par *Xanthomonas* 23. *Annals of Microbiology*, **129A**, 361–77.

O'Donnell, E.D., Freimer, E.H., Gilardi, G.L. and Raeder, R. (1982) Comparative *in vitro* activities of *N*-formimidoyl thienamycin and moxalactam against nonfermentative aerobic Gram-negative rods. *Antimicrobial Agents in Chemotherapy*, **21**, 673–5.

Ohta, H., Konishi, J., Kato, Y. and Tsuchihashi, G.-I. (1987) Microbial reduction of 1,2-diketones to optically active α-hydroxyketones. *Agricultural and Biological Chemistry*, **51**, 2421–7.

Pandey, P.K. and Prasad, M. (1983) *In vitro* secretion of pectolytic enzymes by *Xanthomonas campestris*. *Zentralblatt für Mikrobiologie*, **138**, 71–7.

Patel, I.B. and Vaughn, R.H. (1973) Cellulolytic bacteria associated with sloughing spoilage of California ripe olives. *Applied Microbiology*, **25**, 62–9.

Peros, J.P. (1988) Variability in colony type and pathogenicity of the causal agent of sugarcane gumming *Xanthomonas campestris* pv. *vasculorum* (Cobb) Dye. *Journal of Plant Disease Protection*, **95**, 591–8.

Prasad, M. and Bhushan, A. (1987) The influence of carbon nutrition on activation and activities of pectolytic enzyme in cultures of *Xanthomonas campestris* pv. *vignicola*. *Zentralblatt für Mikrobiologie*, **142**, 631–8.

Randhawa, P.S. and Civerolo, E.L. (1986) Interaction of *Xanthomonas campestris* pv. *pruni* with pruniphage and epiphytic bacteria on detached peach leaves. *Phytopathology*, **76**, 549–53.

Randhawa, P.S. and Civerolo, E.L. (1987) Indigenous plasmids in *Xanthomonas campestris* pv. *pruni*, in *Plant Pathogenic Bacteria. Proceedings of the Sixth International Conference on Plant Pathogenic Bacteria*, College Park, Maryland (eds E.L. Civerolo, A. Collmer, R.E. Davis and A.G. Gillaspie), Martinus Nijhoff Publishers, Dordrecht, pp. 465–9.

Reddy, A.P.K. and Kauffman, H.E. (1977) Loss of virulence associated with aging of *Xanthomonas oryzae* cultures. *Indian Phytopathology*, **30**, 106–11.

Reddy, M.N., Stuteville, D.L. and Sorensen, E.L. (1969) Pectolytic activity by *Xanthomonas alfalfae in vitro* and in infected alfalfa plants. *Phytopathology*, **59**, 887–8.

Reddy, P.R., Raychaudhuri, S.P. and Rao, Y.P. (1977) Effect of bacteriophages on the infectivity of *Xanthomonas oryzae*, the incitant of bacterial leaf blight of rice. *Journal of Plant Disease Protection*, **84**, 592–6.

Rhodes, M.E. (1958) The cytology of *Pseudomonas* spp. as revealed by a silver-plating staining method. *Journal of General Microbiology*, **18**, 639–48.

Richert, N.D. and Ryan, R.J. (1977) Specific gonadotropin binding to *Pseudomonas maltophilia*. *Proceedings of the National Academy of Sciences, U.S.A.*, **74**, 878–82.

Ridé, M. and Ridé, S. (1978) *Xanthomonas populi* (Ridé) comb. nov. (syn. *Aplanobacter populi* Ridé) spécifié, variabilité, et absence de relation avec *Erwinia cancerogena*. *European Journal of Forest Pathology*, **8**, 310–33.

Rosta, S. and Mett, H. (1989) Physiological studies of the regulation of β-lactamase expression in *Pseudomonas maltophilia*. *Journal of Bacteriology*, **171**, 483–7.

Rye, A.J., Drozd, J.W., Jones, C.W. and Linton, J.D. (1988) Growth efficiency of *Xanthomonas campestris* in continuous culture. *Journal of General Microbiology*, **134**, 1055–61.

Saino, Y., Inoue, M. and Mitsuhashi, S. (1984) Purification and properties of an inducible cephalosporinase from *Pseudomonas maltophilia* GN12873. *Antimicrobial Agents in Chemotherapy*, **25**, 362–5.

Saino, Y., Kobayashi, F., Inoue, M. and Mitsuhashi, S. (1982) Purification and properties of inducible penicillin β-lactamase isolated from *Pseudomonas maltophilia*. *Antimicrobial Agents in Chemotherapy*, **22**, 564–70.

Sakurai, I., Kawamura, Y., Koike, H., *et al.* (1990) Bacterial accumulation of metallic compounds. *Applied and Environmental Microbiology*, **56**, 2580–3.

Schaad, N.W. and Stall, R.E. (1990). In *Laboratory Guide for Identification of Plant Pathogenic Bacteria*, 2nd edn (ed. N.W. Schaad), APS Press, St Paul, Minnesota, pp. 81–94.

Shetty, A.S. and Gaertner, F.H. (1978) Kynureninase-type enzymes from two strains of *Xanthomonas pruni*. *FEMS Microbiology Letters*, **3**, 259–63.

Sleesman, J.P. and Leben, C. (1978) Preserving phytopathogenic bacteria at −70°C or with silica gel. *Plant Disease Reporter*, **62**, 910–13.

Sleytr, U.B. and Messner, P. (1983) Crystalline surface layers on bacteria. *Annual Review of Microbiology*, **37**, 311–39.

Smith, W.K. (1958) A survey of the production of pectic enzymes by plant pathogenic and other bacteria. *Journal of General Microbiology*, **18**, 33–41.

Stall, R.E., Loschke, D.C. and Jones, J.B. (1986) Linkage of copper resistance and avirulence loci on a self-transmissible plasmid in *Xanthomonas campestris* pv. *vesicatoria*. *Phytopathology*, **76**, 240–3.

Stanier, R.Y., Palleroni, N.J. and Doudoroff, M. (1966) The aerobic pseudomonads: a taxonomic study. *Journal of General Microbiology*, **43**, 159–271.

Starr, M.P. (1946) The nutrition of phytopathogenic bacteria. I. Minimal nutritive requirements of the genus *Xanthomonas*. *Journal of Bacteriology*, **51**, 131–43.

Starr, M.P. (1981) The genus *Xanthomonas*, in *The Prokaryotes*, Vol. 1 (eds M.P. Starr, H. Stolp, H.G. Trüper, A. Balows and H.G. Schlegel), Springer Verlag, Berlin, pp. 742–63.

Starr, M.P., Jenkins, C.L., Bussey, L.B. and Andrewes, A.G. (1977) Chemotaxonomic significance of the xanthomonadins, novel brominated arylpolyene pigments produced by bacteria of the genus *Xanthomonas*. *Archives of Microbiology*, **113**, 1–9.

Starr, M.P. and Nasuno, S. (1967) Pectolytic activity of phytopathogenic xanthomonads. *Journal of General Microbiology*, **46**, 425–33.

Stephens, W.L. and Starr, M.P. (1963) Localization of carotenoid pigment in the cytoplasmic membrane of *Xanthomonas juglandis*. *Journal of Bacteriology*, **86**, 1070–4.

Stevens, J., Fanburg, B.L. and Lanzillo, J.J. (1990) Determination of peptidyl dipeptidase activity in 24 bacterial species. *Canadian Journal of Microbiology*, **36**, 56–9.

Strandberg, D.A., Jorgensen, J.H. and Drutz, D.J. (1983) Activities of aztreonam and new cephalosporins against infrequently isolated Gram-negative bacilli. *Antimicrobial Agents in Chemotherapy*, **24**, 282–6.

Sutherland, I.W. and Kennedy, F.D. (1986) Comparison of bacterial lipopolysaccharides by high-performance liquid chromatography. *Applied and Environmental Microbiology*, **52**, 948–50.

Sutton, M.D., Katznelson, H. and Quadling, C. (1958) A bacteriophage that attacks numerous phytopathogenic *Xanthomonas* species. *Canadian Journal of Microbiology*, **4**, 493–7.

Sutton, M.D. and Wallen, V.R. (1967) Phage types of *Xanthomonas phaseoli* isolated from beans. *Canadian Journal of Botany*, **45**, 267–80.

Tang, J.L., Gough, C.L., Barber, C.E., Dow, J.M. and Daniels, M.J. (1987) Molecular cloning of protease genes(s) from *Xanthomonas campestris* pv. *campestris*: expression in *Escherichia coli* and role in pathogenicity. *Molecular and General Genetics*, **210**, 443–8.

Thaveechai, N. and Schaad, N.W. (1986) Immunochemical characterization of a subspecies-specific antigenic determinant of a membrane protein extract of *Xanthomonas campestris* pv. *campestris*. *Phytopathology*, **76**, 148–53.

Tofte, R.W. and Crossley, K. (1980) Antimicrobial susceptibility of *Acinetobacter calcoaceticus*, *Serratia marcescens*, *Pseudomonas fluorescens* and *Pseudomonas maltophilia* to three novel cephalosporin-like antibiotics. *Journal of Antimicrobial Chemotherapy*, **6**, 799–800.

Trojanowski, J., Haider, K. and Sundman, V. (1977) Decomposition of ^{14}C-labelled lignin and phenols by a *Nocardia* sp. *Archives of Microbiology*, **114**, 149–53.

Tu, J., Wang, H.-R., Chang, S.-F. *et al.* (1989) Transposable elements of *Xanthomonas campestris* pv. *citri* originating from indigenous plasmids. *Molecular and General Genetics*, **217**, 505–10.

Uematsu, T. (1980) Ecology of *Bdellovibrio* parasitic to rice bacterial leaf blight pathogen, *Xanthomonas oryzae*. *Review of Plant Protection Research*, **13**, 12–26.

Uematsu, T., Nagao, N. and Wakimoto, S. (1978) Ultrastructural aspects of host–parasite relationships between *Xanthomonas oryzae* and *Bdellovibrio* isolates. *Annals of the Phytopathological Society of Japan*, **44**, 561–9.

Van den Mooter, M. (1984) De Taxonomie van het Plantpathogene Bakteriën-Geslacht *Xanthomonas* Dowson 1939, PhD Thesis, Rijksuniversiteit, Gent.

Van den Mooter, M. and Swings, J. (1990) Numerical analysis of 295 phenotypic features of 266 *Xanthomonas* strains and related strains and an improved taxonomy of the genus. *International Journal of Systemic Bacteriology*, **40**, 348–69.

Volk, W.A. (1966) Cell wall lipopolysaccharides from *Xanthomonas* species. *Journal of Bacteriology*, **91**, 39–42.

Volk, W.A. (1968) Quantitative assay of polysaccharide components obtained from cell wall lipopolysaccharides of *Xanthomonas* species. *Journal of Bacteriology*, **95**, 980–2.

Volk, W.A., Salomonsky, N.L. and Hunt, D. (1972) *Xanthomonas sinensis* cell wall lipopolysaccharide. *Journal of Biological Chemistry*, **247**, 3881–7.

von Riesen, V.L. (1980) Digestion of algin by *Pseudomonas maltophilia* and *Pseudomonas putida*. *Applied and Environmental Microbiology*, **39**, 92–6.

Wakimoto, S. (1960) Classification of strains of *Xanthomonas oryzae* on the basis of their susceptibility against bacteriophages. *Annals of the Phytopathological Society of Japan*, **25**, 193–8.

Wakimoto, S. and Kamiunten, H. (1978) Filamentous bacteriophages of *Xanthomonas oryzae*, in *Proceedings of the Fourth International Conference on Plant Pathogenic Bacteria*, Angers, France, pp. 375–85.

Walsh, P.M., Hass, M.J. and Somkuti, G.A. (1984) Genetic construction of lactose-utilizing *Xanthomonas campestris. Applied and Environmental Microbiology*, **47**, 253–7.

Warren, G. and Wolber, P. (1991) Molecular aspects of microbial ice nucleation. *Molecular Microbiology*, **5**, 239–43.

Weitzman, P.D.J. (1980) Citrate synthase and succinate thiokinase, in *Classification and Identification*, Academic Press, London, pp. 107–25.

Whitaker, R.J., Byng, G.S., Gherna, R.L. and Jensen, R.A. (1981) Comparative allostery of 3-deoxy-D-*arabino*-heptulosonate 7-phosphate synthetase as an indicator of taxonomic relatedness in pseudomonad genera. *Journal of Bacteriology*, **145**, 752–9.

White, C.L., Melkjorsen, G.B. and Da Graça, J.V. (1985) Isolation of four bacteriophages from the mango bacterial black spot pathogen *Xanthomonas campestris* pv. *mangiferaeindicae. Phytophylactica*, **17**, 11–14.

Whitfield, C., Sutherland, I.W. and Cripps, R.E. (1982) Glucose metabolism in *Xanthomonas campestris. Journal of General Microbiology*, **128**, 981–5.

Wilkinson, S.G. (1967) The sensitivity of pseudomonads to ethylene-diaminetetra-acetic acid. *Journal of General Microbiology*, **47**, 67–76.

Wilkinson, S.G. (1968) Studies on the cell walls of *Pseudomonas* species resistant to ethylenediaminetetra-acetic acid. *Journal of General Microbiology*, **54**, 195–213.

Willems, A., Gillis, M., Kersters, K., Van den Broecke, L. and De Ley, J. (1987) Transfer of *Xanthomonas ampelina* Panagopoulos 1969 to a new genus, *Xylophilus* gen. nov., as *Xylophilus ampelinus* (Panagopoulos 1969) comb. nov. *International Journal of Systemic Bacteriology*, **37**, 422–30.

William, F. and Mahadevan, A. (1980) Degradaton of aromatic compounds by *Xanthomonas* species. *Journal of Plant Disease Protection*, **87**, 738–44.

Willoughby, L.G. (1983) A new kind of antagonistic association, between bacteria and aquatic fungi. *Transactions of the British Mycology Society*, **80**, 91–7.

Wilson, R.G. and Henderson, L.M. (1963) Tryptophan–niacin relationship in *Xanthomonas pruni. Journal of Bacteriology*, **85**, 221–9.

Woodard, L.M., Bielkie, A.R., Eisses, J.F. and Ketchum, P.A. (1990) Occurrence of nitrate reductase and molybdopterin in *Xanthomonas maltophilia. Applied and Environmental Microbiology*, **56**, 3766–71.

Yamazaki, E., Ishii, J., Sato, K. and Nakae, T. (1989) The barrier function of the outer membrane of *Pseudomonas maltophilia* in the diffusion of saccharides and β-lactam antibiotics. *FEMS Microbiology Letters*, **60**, 85–8.

Yang, P., Vauterin, L., Vancanneyt, M., Swings, J. and Kersters, K. (1993) Application of fatty acid methyl enters for the toxonomic analysis of the genus *Xanthomonas. Systematic and Applied Microbiology*, **16**, 47–71.

Yang, S.-E., Lin, F.-H. and Kuo, T.-T. (1975) The utilization of exogenously supplied nucleotide by *Xanthomonas oryzae. Botanical Bulletin Academia Sinica*, **16**, 61–5.

Zagallo, A.C. and Wang, C.H. (1967) Comparative glucose catabolism of *Xanthomonas* species. *Journal of Bacteriology*, **93**, 970–5.

Zechman, J.M. and Labows Jr, J.N. (1985) Volatiles of *Pseudomonas aeruginosa* and related species by automated headspace concentration–gas chromatography. *Canadian Journal of Microbiology*, **31**, 232–7.

3

Taxonomy of the genus *Xanthomonas*

[1] *Laboratorium voor Microbiologie, Faculteit Wetenschappen, Universiteit Gent*
[2] *Department of Plant Pathology, College of Tropical Agriculture and Human Resources, University of Hawaii at Manoa*

3.1 INTRODUCTION

From the origin of bacteriology until the 1960s, the taxonomy of bacteria has been dominated by a phenetic approach, based mainly on morphological, biochemical and physiological features for description and differentiation of taxa. Taxonomic systems were based on subjective interpretations of 'important' phenotypic features. In many cases this has led to artificial classifications, which are no longer tenable in the light of modern phylogenetic taxonomy.

Since the 1960s, three important events have changed bacterial taxonomy considerably. (a) Computer-assisted numerical analysis of phenotypic data (Sokal and Sneath, 1963) has enabled bacterial taxonomists to interpret results in a more rational and unbiased way by giving equal weight to each feature tested. In addition, it has allowed comparison of large numbers of features and strains. (b) DNA homology

Xanthomonas.
Edited by J.G. Swings and E.L. Civerolo.
Published in 1993 by Chapman & Hall, London.
ISBN 0 412 434202

has become the ultimate criterion for the definition of bacterial species. DNA–DNA hybridization techniques are currently used to determine the extent of DNA homology between closely related bacteria, whereas DNA–rRNA hybridization and rRNA sequencing techniques are used to establish relationships between distantly related taxa. (c) It is nowadays generally accepted that only polyphasic approaches, based on phenotypic, chemotaxonomic and genotypic data may lead to stable general purpose classifications. Earlier taxonomic systems were quite commonly based on monophasic approaches.

Among the latest developments in bacterial taxonomy are fingerprinting techniques and other systems yielding a large number of discriminating data from one experiment. Most of these systems are directed towards practical diagnosis. They are characterized by rapidity, automation, ease of standardization, and possess the possibility to generate databases for identification.

The taxonomy of *Xanthomonas* has followed the general tendencies and developments of bacterial taxonomy. The first report of a xanthomonad dates from the end of the last century (Wakker, 1883). Since then, the list of new xanthomonads has increased steadily. Although we will not discuss the history of the systematics of the genus, it is worth mentioning that a single phenotypic feature, i.e. host-specificity, has played a crucial role in the classification. Almost every new species of *Xanthomonas* was described solely on the basis of its specific pathogenicity towards a particular host plant. The application of this practice in taxonomy, which has been called the 'new host–new species' concept (Starr, 1981) is not restricted to *Xanthomonas*, although it has unquestionably culminated in this genus. Up to 1974, over 100 *Xanthomonas* species were recognized. Dye and Lelliott (1974) reduced the number of *Xanthomonas* species to five, the vast majority of the former nomenspecies being condensed into *X. campestris*. This drastic reduction was justified by the impossibility of reliably differentiating the species from one another on the basis of phenotypic tests (Colwell and Liston, 1961; Dye, 1962; Dye and Lelliott, 1974). Later, in order to meet the practical need of plant pathologists to name pathogens that are pathogenic for particular hosts, a special-purpose nomenclature was established (Dye *et al.*, 1980), in which nearly all former *Xanthomonas* species were renamed as pathovars of *X. campestris*. Up to now, over 140 *X. campestris* pathovars have been described.

3.2 CURRENT STATUS OF THE TAXONOMY OF *XANTHOMONAS*

Based on DNA–rRNA hybridization studies (Palleroni *et al.*, 1973; De Vos and De Ley, 1983) the genus *Xanthomonas* constitutes a distinct branch in the gamma group of the *Proteobacteria* (Stackebrandt *et al.*, 1988). Based on the latest comprehensive synopsis of classification of plant pathogenic xanthomonads (Bradbury, 1986), and the recent literature, the following species are now recognized within the genus *Xanthomonas*:

X. *albilineans* (Ashby) Dowson 1943, the causal agent of leaf scald and white stripe of sugarcane;
X. *axonopodis* Starr & Garces 1950, which causes gummosis of *Axonopus* spp.;
X. *campestris* (Pammel) Dowson 1939, which consists of more than 140 phenotypically indistinguishable pathovars causing diverse diseases on a vast range of host plants;
X. *fragariae* Kennedy & King 1962, causing leafspot disease of *Fragaria* spp.;
X. *populi* (Ridé and Ridé 1978) is responsible for bacterial canker of *Populus* spp.;
X. *maltophilia* (Hugh) Swings *et al.* 1983, which is not a plant pathogenic species but a heterogeneous collection of opportunistic bacteria, isolated from diverse environments;
X. *oryzae* (Ishiyama) Swings *et al.* 1991 is an important pathogen of rice, consisting of pathovars X. *oryzae* pv. *oryzae*, causing bacterial blight, and X. *oryzae* pv. *oryzicola*, causing leaf streak of rice.

'X. *ampelina*', causing a bacterial blight of grapes, has been transferred to a new genus as *Xylophilus ampelinus* on the basis of polyphasic examination (Willems *et al.*, 1987).

The original plant pathogenic members of *Xanthomonas* share several characteristic phenotypic features (see Chapter 2). X. *maltophilia* differs in a number of features from the other xanthomonads, and therefore, the transfer of this species from *Pseudomonas* to *Xanthomonas* (Swings *et al.*, 1983) seems to be controversial. DNA-rRNA hybridization has revealed that this species is closely related to *Xanthomonas*, although binding degrees were significantly lower than among the other species of the genus (De Vos and De Ley, 1983). In addition, the habitats and phenotypic behaviour of X. *maltophilia* are very diverse, and it is questionable whether all strains belong in one single species.

A list of all currently known X. *campestris* pathovars, which updates the latest survey by Bradbury (1986), is given in Chapter 1. The current classification of all phenotypically indistinguishable nomenspecies as pathovars of X. *campestris* is generally considered to be a compromise between the viewpoint of plant pathologists on the one hand, who need names for bacteria that are pathogenic for specific hosts, and the viewpoint of bacterial taxonomists on the other hand, who need differentiating phenotypic features for the description of species. However, creation of special-purpose classifications for particular taxa does not contribute to the development of a general and sound taxonomic system, and can only be supported as a provisional solution until a more natural classification is established. An important consequence of the pathovar system in *Xanthomonas* is that non-pathogenic xanthomonads, which are becoming frequently isolated from healthy or diseased plants, cannot be classified properly. As the taxonomy of *Xanthomonas* progresses, it is obvious that the current pathovar classification is not a reflection of the real relationships within the genus (Vauterin *et al.*, 1990b, and further in this chapter).

3.3 PHENETIC BASIS OF THE TAXONOMY OF *XANTHOMONAS*

3.3.1 Description of the genus *Xanthomonas* Dowson 1939

The present description is based on the phenotypic description by Bradbury (1984) and revised according to Van den Mooter and Swings (1990), and accomplished with results of Auling *et al.* (1991) and Yang *et al.* (1993).

Most xanthomonads form smooth yellow mucoid colonies. The yellow pigments are halogenated arylpolyenes and are characteristic for the genus. The cells are Gram-negative rods, measuring 0.4–0.6 by 1.0–2.9 μm. They occur mostly alone or in pairs, but chains and filaments occur as well. Cells are usually motile by one to several polar flagella, sometimes sluggishly. Catalase is present, but urease and indole are not produced. Nitrate is not reduced to nitrite, except by some strains of *X. maltophilia*. Oxidase is absent or only weakly reactive. Acetoin is not formed. Litmus milk is not acidified. There is no growth at pH 4.5, at 4°C or in the presence of 6% NaCl or 30% glucose. Small amounts of acid are produced from many carbohydrates but not from L-rhamnose, adonitol, sorbose, D-sorbitol, *meso*-erythritol or *meso*-inositol. Glycine, L-glutamine and L-asparagine are not utilized as sole carbon and nitrogen sources. *Xanthomonas* strains are chemo-organotrophic. One or more growth factors may be required. More than 99% of the *Xanthomonas* strains contain the following nine fatty acids: 11:0 iso, 11:0 iso 3OH, 12:0 3OH, 13:0 iso 3OH, 15:0 iso, 16:1 *cis* 9, 16:0, 17:1 iso F and 17:0 iso. Among these fatty acids, 11:0 iso, 11:0 iso 3OH and 13:0 iso 3OH are characteristic for the genus *Xanthomonas* and serve as a useful criterion to differentiate *Xanthomonas* from other bacteria. *Xanthomonas* strains contain spermidine as the major polyamine, with the exception of *X. maltophilia,* which also has cadaverine as the major polyamine in addition to spermidine. Most xanthomonads are plant pathogens or associated with plants. *X. maltophilia* strains also occur in many non-plant habitats in nature, e.g. as opportunistic human pathogens.

3.3.2 Numerical analysis of phenotypic features

A large number of plant pathogenic xanthomonads are phenotypically very uniform. This accounts for the inclusion of most of the former species into *X. campestris*. In the most extensive phenotypic study performed on *Xanthomonas* (Van den Mooter and Swings, 1990), a collection of 266 *Xanthomonas* and related strains were examined by 295 morphological, physiological and biochemical features. After numerical analysis, the following nine phena were distinguished.

Phenon 1 consisted of five *X. maltophilia* strains, including the type strain ATCC 13637, and the type strains of *Pseudomonas betle* NCPPB 323 and *Pseudomonas hibiscicola* NCPPB 1683. The restricted number of strains did not allow valid conclusions about the phenotypic homogeneity within the species *X. maltophilia*. The following features did not fit the generic description of *Xanthomonas* by Bradbury (1984): reduction of nitrates, the lack of slime production, growth in the presence of 0.1% 2,3,5-triphenyltetrazolium chloride, and the multitrichous flagellation. However, tolerance of 0.1% 2,3,5,-triphenyltetrazolium chloride was observed for 19% of the *X. campestris* strains studied by Van den Mooter and Swings (1990), and production of slime is not a stable characteristic of *Xanthomonas*.

Phenon 2 consisted of nine strains of *Xylophilus ampelinus*. The previous inclusion of this species in *Xanthomonas* was based on phenotypic similarity only. Willems *et al.* (1987) reported that these strains are unrelated to *Xanthomonas*.

Phenon 3: four *X. fragariae* strains.

Phenon 4: seven *X. albilineans* strains.

Phenon 5: three *X. axonopodis* strains.

Phenon 6: eight strains of *X. populi*.

Phenon 7: 10 strains of *X. oryzae*.

Phenon 8: eight strains of *X. c.* pv. *graminis* Egli and Schmidt 1982.

Phenon 9: 189 *X. campestris* strains, including the type strain NCPPB 528.

The differentiation of phena 1–9 was based on 29 features. The misnamed *'Pseudomonas' vitiswoodrowii* NCPPB 1014, *'Pseudomonas' mangiferaeindicae* NCPPB 490 and *'Pseudomonas' gardneri* NCPPB 881 belonged to phenon 9 of *X. campestris*, confirming previous DNA–rRNA hybridization data (De Vos *et al.*, 1985). Within this phenon, it was impossible to correlate the phytopathogenic specialization of the strains to one or more phenotypic features, confirming the previous statements by several authors in the past. In the study of Van den Mooter and Swings, however, only a very limited number of strains were examined per pathovar.

For a few pathovars, the phenotypic study has been extended to a larger number of strains, i.e. for the *X. oryzae* pathovars, for *X. campestris* pvs *manihotis* and *cassavae* and the pathovars isolated from grasses.

Thirty-five *X. oryzae* pv. *oryzae* strains and 14 *X. oryzae* pv. *oryzicola* strains were examined for 133 conventional phenotypic features (Vera Cruz *et al.*, 1984). Strains of both pathovars clustered in separate phena and could be differentiated by four phenotypic tests.

In another study, 267 phenotypic features of 23 *X. c.* pv. *manihotis* and six *X. c.* pv. *cassavae* strains were analysed numerically (Van den Mooter *et al.*, 1987a). The two pathovars could be differentiated by five features.

Numerical analysis of 45 strains from grasses, examined by 257 phenotypic features (Van den Mooter *et al.*, 1987b) revealed three phena, corresponding respectively to *X. c.* pv. *graminis*, *X. c.* pv. *phleipratensis*, and *X. c.* pvs *arrhenatheri* and *poae*. These phena could be differentiated by 10 phenotypic features.

These three examples illustrate that, to a certain extent, phenotypic features can differentiate pathovars, provided that large numbers of strains are screened with large numbers of phenotypic features. However, other examples, such as the phenotypic relatedness of *Xylophilus ampelinus* to *Xanthomonas*, clearly indicate that the application of phenotypic analysis, even based on objective numerical analysis of numerous features and strains, does not necessarily reflect phylogenetic relationships. Therefore, phenotypic relatedness must always be confirmed by genomic techniques.

Nowadays, a number of miniaturized systems that permit testing of a palette of phenotypic features are available commercially. Such systems seem to be promising in solving the problem of the phenotypic uniformity of the *X. campestris* pathovar complex. The enzymatic API ZYM gallery (API Identification Systems, bioMérieux, France) and the Biolog Identification System (Biolog Inc., Hayward, CA, USA) have been evaluated to discriminate about 30 *X. campestris* pathovars that have been studied extensively by electrophoresis of proteins. The enzymatic gallery of API Systems (API ZYM) was used successfully to differentiate *X. c.* pvs *begoniae* and *pelargonii* (Vauterin *et al.*, 1990a). These pathovars could be distinguished from each other by three enzymatic activities: chymotrypsin and α-D-galactosidase (positive in *X. c.* pv. *begoniae* and negative in *X. c.* pv. *pelargonii*) and α-D-glucosidase (positive in *X. c.* pv. *pelargonii* and negative in *X. c.* pv. *begoniae*). The application of API ZYM has been further extended to 220 strains of 27 other pathovars of *X. campestris* and *X. oryzae* and the other *Xanthomonas* species, but was not discriminative at pathovar level when larger numbers of pathovars were included.

The Biolog method is a microplate system based on the ability of an organism to differentially assimilate 95 different carbon sources.

Changes in redox potential of the suspension are recorded. In *Xanthomonas*, this system has been used to analyse 280 strains belonging to all *Xanthomonas* species and 40 pathovars of *X. campestris* and *X. oryzae*. Each of the *Xanthomonas* species could be separated, whereas only some *X. campestris* pathovars (i.e. *X. c.* pvs *hyacinthi, theicola, pelargonii, juglandis, cucurbitae,* and a group of *arrhenatheri, poae,* and *phlei*) could be delineated as homogeneous subgroups by numerical analysis.

3.3.3 Serology

(a) Polyclonal antibodies

In a multiphasic approach to taxonomy, serological relationships among bacteria have been used as indicators of phenotype. In the pioneering work of Elrod and Braun (1947a–c) some *Xanthomonas* species were separated by agglutinin and precipitin tests provided that heat-killed cells were used to avoid cross-reactive flagellar antigens. Nevertheless, taxonomic groupings did not adequately correlate with nomenspecies or pathotype.

Numerous attempts were subsequently made to identify nomenspecies of phytopathogenic xanthomonads by the precipitin (either in solution or by immunodiffusion) or the agglutinin reaction (Schaad, 1979; Starr, 1981; Bouzar and Moore, 1990); however, for the most part comparisons were made for only a few strains of each nomenspecies or pathotype, and no comprehensive study was undertaken for the xanthomonads. In a study of 28 strains of *X. albilineans* from different geographical regions, Rott *et al.* (1986) found that strains could be separated into three serotypes (serovars), and that these groupings corresponded well with the lysotypical properties of the same strains. For *X. vesicatoria* (Lovrekovich and Klement, 1965; O'Brien *et al.*, 1967; Charudattan *et al.*, 1973) and *X. oryzae* (Obata and Tsuboi, 1972; Mahanta and Addy, 1977) conflicting data were obtained concerning the relationship between pathotype and serotype, and given the heterogeneous nature of polyclonal antisera, it was not possible to resolve these discrepancies.

In the attempt to achieve greater specificity in polyclonal antisera, partially purified antigen preparations, including ribosomes (Schaad, 1976, 1978; Thaveechai and Schaad, 1984), membrane proteins (Minsavage and Schaad, 1983; Thaveechai and Schaad, 1986; Azad and Schaad, 1988) and enzymes (Debette and Prensier, 1989) have been used as immunogens, and immunoassays more sensitive than the agglutinin and precipitin tests, such as immunofluorescence (IF), enzyme-linked immunosorbent assay (ELISA), radioimmunoassay (RIA), and dot-blot assays have been employed (Schaad, 1978; Richaud *et al.*, 1979; Vruggink, 1979; Alvarez and Lou, 1985; Leach *et al.*, 1987). Regardless of the nature of the immunogen, however, whether it be crude or partially purified, polyclonal antisera are of limited value in *Xanthomonas* taxonomy because they are heterogeneous mixtures of antibodies, and cross-reactions among pathovars of *X. campestris* have not been eliminated. For example, using the less sensitive immunodiffusion assays, 25 strains of *X. c.* pv. *campestris* were typed into four serovars using antiserum prepared to a ribosomal extract from *X. c.* pv. *campestris* (Thaveechai and Schaad, 1984). Nevertheless, the more sensitive IF tests revealed cross-reactions between *X. c.* pv. *campestris* and three other *X. campestris* pathovars (Thaveechai and Schaad, 1984). Antisera to membrane proteins also were used in immunodiffusion to group 26 strains of *X. c.* pv. *translucens* into two serovars, but again cross-reactions occurred between these strains using the same sera with the more sensitive IF tests (Azad and Schaad, 1988).

As long as specificity is determined for each pathogen and cross-reactions with other bacteria are described, polyclonal antisera may be used for tentative identification of certain pathovars (Vruggink, 1979). A polyclonal antiserum to whole cells of *X. c.* pv. *pelargonii* reacted with several other pathovars of *X. campestris*, but these bacteria would not be expected to be found in geranium tissue (Anderson and Nameth, 1990). The presence of avirulent xanthomonads in test materials could present problems, however.

(b) Monoclonal antibodies

In contrast to polyclonal antibodies, monoclonal antibodies (mAbs) react to a single antigenic determinant (epitope), having been produced by a process of cell fusion of β-lymphocytes with mouse myeloma cells to form hybridomas. This is followed by selection and hybridoma culture by which each antibody-producing hybridoma cell line is isolated from a heterogeneous mixture (Kohler and Milstein, 1975). By screening hybridoma cell lines for desired specificity, the reaction spectrum for each mAb may be determined. Finally, through culturing and frozen storage, hybridoma cell lines, each secreting a mAb of defined specificity, can be perpetuated and distributed to attain reproducible results in multiple laboratories (Ball *et al.*, 1990).

Monoclonal antibodies have been developed for several pathovars of *X. campestris* (Alvarez *et al.*, 1985, 1991; Benedict *et al.*, 1989, 1990; Permar and Gottwald, 1989; Alvarez and Benedict, 1990; Franken, 1990; Wong, 1990; Duveiller and Bragard, 1992) as well as the species *X. albilineans* (Tsai *et al.*, 1990) and *X. maltophilia*. Selected mAbs are used for rapid detection of pathogens in plant samples and as indicators of population diversity in epidemiological studies (Yuen *et al.*, 1987; Lipp, 1988; Roberts, 1991; Lipp *et al.*, 1992).

Table 3.1 Taxon-specific monoclonal antibodies produced for xanthomonads

Genus *Species/pathovar*	*mAb(s)*	*No. of target strains tested*	*Reference*
Xanthomonas	X1, X11	436 >2000	Alvarez *et al.* (1985, unpublished)
X. albilineans	12 mAbs	1	Tsai *et al.* (1990)
	8 mAbs	43	Alvarez and Benedict (unpublished)
X. campestris			
pv. *armoraciae*	XcaB35	24	Alvarez *et al.* (1987)
pv. *begoniae*	Xcb-1	26	Benedict *et al.* (1990)
pv. *campestris*	X9, X13, X17, X21, A11	573 329	Alvarez *et al.* (1985, 1987, unpublished)
pv. *citri*	A1, B1, B2, C1, X-4600	48 30	Alvarez *et al.* (1991) Permar and Gottwald (1989)
pv. *citrumelo*	CBS1, CBS1a	225	Alvarez *et al.* (1991)
pv. *dieffenbachiae*	Xcd1, Xcd3, Xcd7, Xcd108	329 92	Lipp *et al.* (1992) Benedict (unpublished)
pv. *pelargonii*	Xcp-1	76	Benedict *et al.* (1990)
pv. *phaseoli*	XP2	18	Wong (1990)
	Xcphas1, Xcphas2	31	Benedict (unpublished)
pv. *undulosa*	AB3-B6	44	Duveiller and Bragard (1992)
pv. *vesicatoria*	Xcv1 to Xcv25	171	Alvarez, Benedict, Bouzar, and Jones (unpublished)
X. campestris from ti (*Cordyline terminalis*)	Xct	10	Alvarez *et al.* (1991)
X. maltophilia	Xm1, Xm2	26	Alvarez *et al.* (1987), Benedict (unpublished)
X. oryzae			
pv. *oryzae*	Xco1, Xco2, Xco5	178	Benedict *et al.* (1989)
pv. *oryzicola*	Xcola	8	Benedict *et al.* (1989)

Monoclonal antibodies generated for various groups of xanthomonads are listed in Table 3.1. To verify the specificity of mAbs to the target organisms, large numbers of strains from various ecosystems and geographical regions were tested in most cases. Genus-specific mAbs X1 and X11, for example, have been tested with over 2000 strains of *Xanthomonas* species and *X. campestris* pathovars in addition to numerous other genera and species. MAb X1 reacts with all xanthomonads tested to date, including most strains of *X. maltophilia*; mAb X11 reacts with many strains of *X. maltophilia* and most (approximately 98%) but not all other xanthomonads tested.

In most of the above studies either heat- or formalin-treated whole cell preparations were used as immunogens, indicating that most of the mAbs generated were specific for outer membrane molecules. Depending on the selections made, the mAbs were capable of identifying xanthomonads at the genus, species, or pathovar level. This raises interesting questions regarding unique molecular properties of the outer envelope of these pathogens.

Analysis of antigens reacting with genus-specific mAbs X1 and X11 provides some insight on common features shared by xanthomonads. The antigen for mAb X1 appears to be a 20–30 kDa protein associated with the outer membrane, whereas mAb X11 is associated with the lipopolysaccharide (LPS) as are many of the pathovar-specific mAbs generated to xanthomonads (Wong, 1990; Alvarez *et al.*, 1991). MAbs prepared against the exopolysaccharide xanthan from *X. c.* pv. *campestris* reacted with trisaccharide side-chains from several different xanthan preparations from this pathovar (Haaheim *et al.*, 1989). A pyruvylated terminal mannose was the immunodominant epitope for one of the mAbs, whereas for three others the inner mannose-glucuronic acid appeared to be immunodominant (Haaheim *et al.*, 1989). Further comparative structural studies of xanthans may provide useful taxonomic data, and if more strains are examined, such mAbs might be used for detection of xanthomonads in the environment.

Yellow-pigmented oxidative bacteria are commonly encountered in and on plant parts as well as on seed. *Xanthomonas*-like bacteria recovered from rice seed on semi-selective media reacted in ELISA with mAbs X1 and X11 but not to mAbs Xco1 and Xco2 specific for *X. oryzae* pv. *oryzae* (Benedict *et al.*, 1989). Through phenotypic tests and fatty acid analysis these bacteria were identified as *Xanthomonas* sp., but in subsequent pathogenicity tests, they were non-pathogenic to rice (Benedict *et al.*, 1989; Jones *et al.*, 1989). In contrast, all strains that reacted positively with the *X. oryzae* pv. *oryzae*-specific mAbs in ELISA were pathogenic.

Yellow-pigmented bacteria that resembled *X. c.* pv. *campestris* also were recovered on semi-selective media from crucifer seeds (Alvarez *et al.*, 1987). These bacteria formed round, mucoid colonies, hydrolysed starch and had other phenotypic characteristics of *X. campestris* including production of pigments that co-migrated with xanthomonadin (identical R_f values in thin-layer chromatography). In inoculation studies on cabbage seedlings they were weakly virulent, and in contrast to moderate and highly virulent strains of *X. c.* pv. *campestris* that reacted with two or more of the *X. c.* pv. *campestris*-specific mAbs (X9, X13, X17, X21, A11), the weakly virulent strains reacted with only mAb X17 or X21. Since both virulent and avirulent strains react with the genus-specific mAbs X1 and X11, these mAbs are useful for separating non-pathogenic xanthomonads from the pathogenic xanthomonads in seed or plant tissues.

In The Netherlands, mAbs have been produced that detect *X. c.* pv. *campestris* in crucifer seed (Franken, 1990). In these studies as with those reported previously (Alvarez *et al.*, 1985), no single mAb was found that reacted with all strains of this pathogen.

The geographical distribution of diverse populations of a pathogen also has been traced with mAbs. For example, *X. c.* pv. *campestris* strains from cabbage in Georgia, Florida and South Carolina had higher proportions of a particular serotype (serotype 5) than strains originally isolated in California, Hawaii and Japan (Alvarez *et al.*, 1985). Nevertheless, serotypes 1 and 2 prevalent in Hawaii and Thailand also were recovered from crop debris in Costa Rica, Panama and Argentina. Although there is great serological diversity of *X. c.* pv. *campestris* in strains from all areas tested, nearly all (>99%) of the pathogenic *X. c.* pv. *campestris* strains reacted with one or more of the mAbs already described (Alvarez *et al.*, 1985).

Serotypes of *X. c.* pv. *citri* were related both to geographical origin of strains and their pathogenicity. Strains causing severe citrus bacterial canker (CBC), originally termed 'pathotype A' (now, CBC-A) and found throughout the world on many *Citrus* sp., reacted with mAb A1, whereas the milder 'pathotype B' (CBC-B) strains localized in several groves of lemon and grapefruit in Northern Argentina and Uruguay reacted with mAbs B1 and B2 but not with mAb A1 (Alvarez *et al.*, 1991). Recently, a unique group of CBC-A strains from India and Saudi Arabia that are negative for mAb A1 but react with new strain-specific mAbs has been detected.

Both serological and genetic evidence indicate that except for seven strains from Saudi Arabia and India, the CBC-A strains are quite uniform; moreover, they are distinct from the B/C/D strains associated with milder forms of citrus canker in Latin America (Gabriel *et al.*, 1988; Alvarez *et al.*, 1991). MAb B2, generated to an Argentine B strain, reacted with nearly all CBC-B strains from Argentina, a 'pathotype C' (CBC-C) strain (XC70) recovered from *Citrus aurantifolia* (Mexican lime) in Brazil, and the 'pathotype D' (XC90) strain recovered from *Citrus aurantifolia* in Colima, Mexico. Although the CBC-C strain reacted with mAb B2, it also could be distinguished from the B/D group by a unique epitope that reacted with mAb C1 found on no other xanthomonad tested.

Both serological and restriction fragment length polymorphism (RFLP) analyses indicate that the D strain XC90 was sufficiently similar to the Argentine CBC-B strains to be included in the same group (Gabriel *et al.*, 1988; Alvarez *et al.*, 1991). The D strain was moderately virulent in whole plant and detached leaf assays, whereas nine other strains of xanthomonads also isolated from *Citrus aurantifolia* in Mexico were weakly virulent or avirulent (Rodriguez *et al.*, 1985) and did not react with mAb B2 (Alvarez *et al.*, 1991).

In contrast to citrus canker, the outbreak of citrus bacterial spot (CBS) on Florida nursery trees in 1984 (Schoulties *et al.*, 1987) was associated with a serologically and genetically diverse group of xanthomonads. The CBS strains were first considered 'pathotype E' of *X. c.* pv. *citri* but were later termed simply CBS strains or *X. c.* pv. *citrumelo* (Gabriel *et al.*, 1989). Clear distinctions between the CBC-A, CBC-B/C/D and the CBS strains were shown by serological reactions and other forms of analysis (RFLP and detached leaf assay (Gabriel) *et al.*, 1988; Hartung and Civerolo, 1989; Graham and Gottwald, 1990; Alvarez *et al.*, 1991; Gottwald *et al.*, 1991). SDS–PAGE of proteins and fatty acid analysis also grouped the CBS strains and differentiated them from the CBC-A strains of *X. c.* pv. *citri* (Vauterin *et al.*, 1991b) but did not show the heterogeneity of CBS strains observed in RFLP, detached leaf assays and serology (Gottwald *et al.*, 1991).

Using mAbs the CBS strains were separated into eight serological groups (Alvarez *et al.*, 1991). Despite this heterogeneity, however, a common epitope on a LPS antigen was shared by many of the most aggressive and some of the moderately aggressive CBS strains (Alvarez *et al.*, 1991; Gottwald *et al.*, 1991). All but four of 98 highly aggressive CBS strains reacted with mAb CBS1. Some moderately and weakly aggressive CBS strains reacted with mAb CBS1, but unlike the aggressive strains, the latter also reacted with another mAb, Xct, and were thereby differentiated from the aggressive strains (Alvarez *et al.*, 1991; Gottwald *et al.*, 1991). The close correspondence between the mAb reaction patterns, which measure antigen variability, RFLP analysis, which measures variability in the DNA organization, and detached leaf assays, which measure levels of aggressiveness (Gottwald *et al.*, 1991), is evidence that carefully selected mAbs detect specific surface antigens that are markers of biologically and taxonomically significant characteristics. Because of their simplicity in application, mAbs can be used effectively for rapid identification, particularly when large sample numbers are required.

The serological heterogeneity or homogeneity of a bacterial pathogen has implications in taxonomy. As noted previously, uniformity with respect to surface antigens often corresponds to pathogens with a relatively narrow host range (Alvarez and Benedict, 1990). For example, *X. c.* pvs *begoniae, pelargonii,* and *X. o.* pvs *oryzae* and *oryzicola* all have a narrow host range, affecting only one or a few host genera within a plant family. These pathogens are characterized by a single common antigen possessed by most strains of the pathogen (Table 3.1). In contrast, *X. c.* pvs *campestris, citri,* and *dieffenbachiae* that affect several genera of hosts within a plant family, appear to share no common antigens unique to the pathovar, but collectively must be identified with a panel of mAbs.

The tomato and pepper leaf blight pathogen, *X. c.* pv. *vesicatoria,* represents the most complex and serologically diverse group of xanthomonads studied to date. Pathovar-specific mAbs for *X. c.* pv. *vesicatoria* are rare; conversely, this pathogen is antigenically similar to numerous strains of phenotypically unrelated pathovars. Selected mAbs generated to surface antigens of representative strains delineate subpopulations of this pathovar and indicate their serological similarities with other pathovars of *X. campestris.*

The nature of pathovar-specific surface antigens is of interest because it appears that certain surface antigens may be related to host recognition phenomena. Except for the *X. oryzae* pv. *oryzae*-specific mAb, Xco1, and mAb C1, specific for the pathotype C strain of *X. c.* pv. *citri,* all of the pathovar-specific or nearly pathovar-specific mAbs are associated with LPS. Furthermore, as indicated by silver staining and Western blot analysis using mAbs, particular LPS banding patterns are associated with different *X. campestris* pathovars, including *citri, begoniae* and *pelargonii* and *X. oryzae* (Benedict *et al.*, 1989, 1990; Alvarez *et al.*, 1991). The chain lengths responsible for ladder-like patterns of LPS in polyacrylamide gels appear to be highly conserved phenotypic characteristics. For example, LPS banding patterns of all CBC-A strains of *X. c.* pv. *citri* were similar to each other and different from LPS banding patterns of all CBS strains. Furthermore, despite the heterogeneity of CBS strains in terms of aggressiveness and RFLP analysis, the CBS strains all shared the same LPS banding patterns (Alvarez *et al.*, 1991). Such findings reflect groupings found by analysis of SDS-protein profiles and fatty acid analysis (Vauterin *et al.*, 1991b).

In summary, mAbs have been generated and selected for identification of xanthomonads at different levels of taxonomic differentiation. Depending on the distinctions required, mAbs can be used to identify the genus, species, pathovar, or strains within the pathovar. When specificity is clearly defined such mAbs have value in efficient diagnostic procedures or for epidemiological studies.

3.3.4 Electrophoresis of proteins

The usefulness of protein electrophoresis in the taxonomy of *Xanthomonas* was first demonstrated by El-Sharkawy and Huisingh (1971). Although their study was performed on a small number of strains, esterase patterns varied considerably among strains of different nomenspecies. The development of methods to determine similarity between patterns using computer-aided numerical analysis (Kersters and De Ley, 1975) made it possible to objectively compare and group larger numbers of protein patterns. In the studies by Vera Cruz *et al.* (1984) and Van den Mooter *et al.* (1987a,b), computer-aided cluster analysis of native protein patterns was performed and compared with phenotypic data. In the three cases, the results of both methods correlated well, but in general, protein electrophoresis revealed finer differences. Minsavage and Schaad (1983) characterized *X. c.* pv. *campestris* strains from diverse origin by electrophoresis of total cell envelope proteins and distinguished them from other *X. campestris* pathovars. Similarly, dos Santos and Dianese (1985) reported differences in total membrane electrophoresis patterns of *X. c.* pvs *manihotis* and *cassavae*.

Vauterin *et al.* (1990a) compared 155 *X. c.* pv. *begoniae* and *X. c.* pv. *pelargonii* strains isolated world-wide by numerical analysis of SDS–protein electrophoresis patterns and concluded that both pathovars are homogeneous taxa, distinct from *X. c.* pv. *campestris*.

These examples illustrate clearly that protein electrophoresis may provide a valuable tool for taxonomy and identification of *Xanthomonas*. It has been stressed that the very complex taxonomic structure of the pathovar system of *X. campestris* will only be resolved by the examination of large sets of representative strains (Vauterin *et al.*, 1990b). Therefore, screening techniques are required that are rapid, easy to standardize, and at the same time accurate enough to discriminate at a fine taxonomic level. SDS-electrophoresis according to Laemmli (1970) of whole-cell solubilized proteins combined with computer-assisted analysis of patterns is one of the most suitable techniques in this respect.

In an extensive SDS-electrophoresis analysis of whole-cell proteins of 307 *Xanthomonas* strains comprising all species and 27 *X. campestris* pathovars (Vauterin *et al.*, 1991a), 19 significant protein clusters were delineated. The species *X. campestris*, *X. fragariae*, *X. populi* and *X. axonopodis* showed a common basic SDS–PAGE protein pattern, whereas the species *X. albilineans*, and particularly *X. maltophilia*, had very different profiles. Within *X. campestris*, only limited correspondence was found with the existing pathovar subdivision. Six pathovars could be defined clearly on the basis of protein electrophoresis, namely *X. c.* pvs *campestris*, *graminis*, *hyacinthi*, *pelargonii*, *pruni* and *theicola*. Remarkably, five pathovars from leguminous hosts (*Fabaceae*), i.e. *X. c.* pvs *phaseoli* (including var. *fuscans*), *cajani*, *vignicola*, *alfalfae* and *glycines* displayed very similar protein profiles. The majority of the pathovars, however, consisted each of a homogeneous core from which one or more strains were aberrant. Even the pathovar reference strain was aberrant in *X. c.* pvs *ricini* and *vitians*. Clear examples of pathovars consisting of two protein electrotypes were *X. c.* pvs *vesicatoria, dieffenbachiae* and *juglandis*. Other heterogeneous pathovars included *X. c.* pvs *alfalfae, vignicola, melonis* and particularly *X. c.* pv. *poinsettiicola*, which comprised three very different protein types.

SDS–PAGE of proteins has been applied further for analysing *X. campestris* strains from citrus. *X. campestris* strains from citrus are subdivided in to five different pathological groups (see section 3.3.3). In a comparative study of 61 strains representing all four disease forms of citrus bacterial canker (CBC groups A, B, C and D) and citrus bacterial spot (CBS or group E) (Vauterin *et al.*, 1991b), SDS–PAGE of proteins, fatty acid analysis and DNA hybridization were used to reveal the taxonomic relationships among the different groups. The results of the three methods revealed two distinct entities: group A of CBC and group E (CBS), respectively. Both SDS–PAGE of proteins and fatty acid analysis suggested that group E (CBS) was less heterogeneous than was concluded from restriction fragment length polymorphism analysis (Gabriel *et al.*, 1988; Hartung and Civerolo, 1989), although protein patterns of CBS clearly showed more variation than CBC-A strains. The relationships of CBC groups B, C and D were not certain, due to the limited number of strains available of these groups. According to protein electrophoresis, the B and D strains were highly similar and characteristic whereas the C strains constituted a separate group, which is similar to serological findings based on mAbs.

In another polyphasic study, 144 strains belonging to all known *Xanthomonas* species and pathovars pathogenic for cereals, grasses and other *Poaceae* were compared by the same three methods (Vauterin *et al.*, 1992). The results of the three methods correlated well, although it was found that SDS–PAGE of proteins was more reliable than fatty acid analysis in detecting quantitative relationships, when both techniques are compared to DNA homology data. Interestingly, a number of pathovars from related cereal crops were indistinguishable from or very similar to one another. These included *X. c.* pvs *translucens, hordei, secalis, cerealis* and *undulosa* from cereals, *poae, arrhenatheri* and *phlei* from grasses (Egli and Schmidt, 1982), and *X. c.* pv. *phleipratensis*. *X. c.* pv. *graminis* had a distinct protein profile, although DNA hybridization showed that genomically this pathovar also belongs to the aforementioned group of pathovars of cereal crops. The species *X. axonopodis* was shown to be heterogeneous and constituted two distinct groups (type A and B). Type B strains of *X. axonopodis* had atypical protein and fatty acid profiles and were concluded not to be xanthomonads. Subgroup A of *X. axonopodis* fell together with a group of *X. c.* pv. *vasculorum* strains. *X. c.* pv. *vasculorum* is another example of a pathovar which consists of two distinct biological entities that are pathogenic for the same host. As demonstrated by protein electrophoresis, type A includes the pathovar reference strain and has a protein profile similar to *X. axonopodis* type A, whereas type B protein patterns are indistinguishable from *X. c.* pv. *holcicola*, a pathovar of *Sorghum*, another member of the *Poaceae*.

At present the protein electrophoresis analysis has been extended to all known pathovars of the species *X. campestris*. A number of other pathovars, e.g. *X. c.* pv. *corylina*, *X. c.* pv. *incanae* and *X. c.* pv. *carotae*, appeared to be heterogeneous on the basis of protein patterns. The pathovar reference strain of *X. c.* pv. *carotae* had a strongly aberrant protein profile from all other xanthomonads, indicating that this strain is misnamed.

In conclusion, electrophoresis of SDS-solubilized whole-cell proteins has contributed to a large extent to the elucidation of the natural relationships between the currently defined *X. campestris* pathovars. Many unexpected data from other phenotypic assays which conflict with the pathovar system can be explained by comparing the results to groupings obtained by clustering of protein patterns. This technique has illustrated that pathovars are clearly not always biological entities. Many fictive biological groups have been created for strains that have one phenotypic feature in common, i.e. pathogenicity towards the same host.

Based on grouping of protein patterns of large numbers of strains, a restricted number of representative strains from the defined clusters have been selected for genomic analysis of the genus *Xanthomonas* by DNA–DNA hybridization. As discussed further in this chapter, the protein electrophoretic groupings are largely confirmed by DNA homology data.

3.3.5 Fatty acid composition

Since Abel *et al.* (1963) first demonstrated the feasibility of using fatty acid contents for the classification of bacteria, the application of fatty acid analysis by gas chromatography has become increasingly used in bacterial classification and identification. This technique is characterized as sensitive, accurate, highly reproducible and relatively simple to perform. With the development of computer software, automatic quantitative and qualitative analysis of fatty acid profiles became possible. Establishment of fatty acid databases of well-characterized bacteria taxa allowed rapid identification of unknown isolates (Sasser *et al.*, 1984). Currently this technique is routinely used for bacterial identification in many laboratories.

Within the genus *Xanthomonas*, the fatty acid profiles of *X. maltophilia* have been investigated (Moss *et al.*, 1973; Ikemoto *et al.*, 1978, 1980; Neal and Wilkinson, 1982). Long before its transfer to the genus *Xan-*

thomonas (Swings *et al.*, 1983), fatty acid analysis results showed that this bacterium was distinct from other *Pseudomonas* species as it contained large amounts of iso-branched fatty acids (Moss *et al.*, 1973). Rietschel *et al.* (1975) found that the fatty acids derived from the lipopolysaccharides of *Xanthomonas* isolates were very similar to those of *X. maltophilia* but distinct from those of other bacteria. The similarity in fatty acid composition between *Xanthomonas* strains and *X. maltophilia* was further confirmed by Ikemoto *et al.* (1980).

Stead (1989) investigated the fatty acid profiles of xanthomonads isolated from cereals and grasses. He found that the strains isolated from cereals, i.e. *X. c.* pv. *translucens*, *X. c.* pv. *hordei*, *X. c.* pv. *undulosa*, *X. c.* pv. *secalis* and *X. c.* pv. *cerealis* shared similar fatty acid profiles, while *Xanthomonas* strains isolated from grasses, i.e. *X. c.* pv. *graminis*, *X. c.* pv. *poae* and *X. c.* pv. *phleipratensis* grouped together. *X. oryzae* pv. *oryzae*, *X. o.* pv. *oryzicola* and *X. c.* pv. *holcicola* each had distinct fatty acid profiles.

In another comparative study (Vauterin *et al.*, 1992), 144 *Xanthomonas* strains including isolates from cereals and grasses as well as other *Xanthomonas* species were investigated by fatty acid profiles, SDS–PAGE and DNA–DNA hybridization. As mentioned previously (see section 3.3.4), there was excellent agreement among the three mentioned techniques.

Stall and Hodge (1989) and Graham *et al.* (1990) investigated the fatty acid profiles of *X. c.* pv. *citri* group E (CBS) strains (see also sections 3.3.3 and 3.3.4). In the latter study, the results of fatty acid analysis, phytopathogenicity tests and RFLP analysis were compared. There was good correlation among the results of the different techniques. However, in the above studies only *X. c.* pv. *citri* group E strains were analysed and the number of the strains was restricted. In a polyphasic study (Vauterin *et al.*, 1991b), 61 *X. c.* pv. *citri* strains including CBC groups A, B, C and D and CBS (group E) were compared. On the basis of fatty acid profiles, these strains could be divided into three FAME types: CBC groups A + B, CBC groups C + D and CBS (group E).

In an extensive study on the FAME composition of the genus *Xanthomonas* (Yang *et al.*, 1993), 966 *Xanthomonas* strains which included all seven *Xanthomonas* species and more than 130 *X. campestris* pathovars were investigated. The results showed that the genus *Xanthomonas* contains a great diversity of fatty acids and the profiles are very complex. At least 65 different fatty acids were found. Typical for the fatty acid profiles of the genus *Xanthomonas* is the occurrence of many branched and hydroxy-branched fatty acids. The following nine fatty acids were present in over 99% of the *Xanthomonas* strains tested: 11:0, 11:0 iso 3OH, 12:0 3OH, 13:0 iso 3OH, 15:0 iso, 16:1 *cis* 9, 16:0, 17:1 iso F and 17:0 iso. Among these, 11:0 iso, 11:0 iso 3OH and 13:0 iso

3OH allow differentiation of the genus *Xanthomonas* from other phytopathogenic bacteria.

Cluster analysis based on the fatty acid profiles of all *Xanthomonas* species and *X. campestris* pathovars revealed 31 FAME clusters. Five of the clusters corresponded to the *Xanthomonas* species *X. albilineans*, *X. axonopodis*, *X. fragariae*, *X. maltophilia* and *X. populi*. The two pathovars of *X. oryzae*, *oryzae* and *oryzicola*, were clearly different from each other and from other *Xanthomonas* as well. The fatty acid profiles of strains of *X. campestris* were very complex and comprised 24 clusters. The following *X. campestris* pathovars possessed distinct FAME profiles: *cassavae*, *musacearum*, *spermacoces*, *cyamopsidis*, *patelii*, *hyacinthi* and *theicola*, while the other *X. campestris* pathovars clustered within 16 different groups.

As observed also by protein electrophoresis, some of the *X. campestris* pathovars were grouped in accordance with the host plants from which they were isolated. For example, all the *X. campestris* pathovars pathogenic for crucifers, i.e. *campestris*, *incanae*, *raphani*, *aberrans*, *barbareae* and *armoraciae* grouped together on the basis of FAME composition and shared similar protein electrophoretic patterns. Most of the *X. campestris* pathovars isolated from leguminous host plants also grouped together both in fatty acid analysis and protein electrophoresis. Another example was found for the *X. campestris* pathovars pathogenic for cereals, i.e. *translucens*, *secalis*, *undulosa*, *cerealis* and *hordei*.

Most *Xanthomonas* species and *X. campestris* pathovars were homogeneous on the basis of fatty acid composition, although atypical strains were frequently encountered. However, some *X. campestris* pathovars displayed very heterogeneous fatty acid profiles. A typical example is *X. c.* pv. *vasculorum* which was divided into four different FAME subgroups and a few aberrant strains. Two of the FAME subgroups were confirmed by protein electrophoretic data. Nevertheless, these subdivisions were not correlated with the host plants, with the geographic distribution of the strains or with any other known trait.

3.3.6 Other chemotaxonomic markers

Besides electrophoresis of proteins and fatty acid analysis, few chemotaxonomic markers have contributed towards improvement of the taxonomy of *Xanthomonas*. Methods such as determination of polyamines, isoprenoid quinones, and xanthomonadin pigments have found limited application, except at the generic level.

(a) Polyamines

Both the quantitative and qualitative composition of polyamines are thought to be conserved and characteristic for a given organism, and therefore, polyamine determination has been used to study relationships within the *Proteobacteria* (Busse and Auling, 1988). Auling *et al.* (1991) compared the polyamine contents of 19 *Xanthomonas* strains with phytopathogenic strains of the *Pseudomonas fluorescens* complex. They found spermidine to be the major polyamine compound of *Xanthomonas* whereas putrescine was the main polyamine compound of the pseudomonads studied. Within the genus *Xanthomonas*, the species *X. axonopodis* and *X. fragariae* contained significantly lower amounts of polyamines than the others. *X. campestris* strains exhibited a very similar profile and contained only traces of cadaverine, whereas *X. maltophilia* had high amounts of this compound.

(b) Isoprenoid quinones

The structural variations of isoprenoid quinones among different bacteria are useful chemotaxonomic markers (Collins and Jones, 1981). The major isoprenoid quinone present in *Xanthomonas* (including *X. maltophilia*) is ubiquinone containing eight isoprene units (Q8) (Ikemoto *et al.*, 1980). Most fluorescent pseudomonads have ubiquinone Q9 as the predominant component (Oyaizu and Komagata, 1983).

(c) Xanthomonadins

Xanthomonadins are brominated arylpolyenes (Andrewes *et al.*, 1976; see also Chapter 2) which are hitherto not found in bacteria other than *Xanthomonas*. The chemotaxonomic significance of xanthomonadins in the genus *Xanthomonas* was studied by Starr *et al.* (1977). These authors concluded that xanthomonadins are a useful criterion for identifying bacteria as xanthomonads. Pigments of yellow *X. maltophilia* strains are recognized as chlorinated hexaenes, whereas xanthomonadins are brominated, mostly octaenes (Jenkins and Starr, 1985).

3.4 GENETIC BASIS OF THE CLASSIFICATION

3.4.1 Percentage G+C determination

In earlier taxonomic studies, the percentage guanine plus cytosine of the genome (%G+C) was used as an important genotypic parameter for classification. This parameter is useful in that bacteria having quite different %G+C cannot be closely related. Similar %G+C values, however, do not necessarily indicate that bacteria are closely related.

The %G+C range of the genus *Xanthomonas* is quite well characterized. In Table 3.2 we have compiled our own data supplemented with published results from De Vos and De Ley (1983), Vera Cruz *et al.* (1984), Van den Mooter *et al.* (1987a,b) and Vauterin *et al.* (1990a). More

Table 3.2 Percentage guanine plus cytosine (mol% G+C) of genomic DNA of the genus *Xanthomonas*

Xanthomonas species and pathovar (no. of strains)	*mol% G+C content*
X. *albilineans* (1)	64.5
X. *axonopodis* (1)	65.0
X. *fragariae* (1)	63.3
X. *maltophilia* (3)	65.5–67.5
X. *oryzae* pvs *oryzae* (2), *oryzicola* (1)	64.5–65.0
X. *populi* (9)	63.2–65.2
X. *campestris* pvs *alfalfae* (1), *armoraciae* (2), *barbareae* (1), *begoniae* (1), *cajani* (1), *campestris* (1), *cassavae* (2), *citri* (1), *corylina* (4), *cucurbitae* (2), *dieffenbachiae* (2), *glycines* (1), *juglandis* (1), *malvacearum* (1), *manihotis* (1), *melonis* (1), *pelargonii* (2), *phaseoli* + var. *fuscans* (3), *pisi* (1), *poinsettiicola* (4), *pruni* (1), *ricini* (5), *vesicatoria* (2), *vignicola* (1)	64.6–67.5
X. *campestris* pv. *incanae* (2)	68.0–68.3
X. *campestris* pvs *graminis* (2), *phlei* (2), *poae* (1), *hyacinthi* (3), *theicola* (1)	69.1–69.7

than 70 values of %G+C have been determined for all *Xanthomonas* species and 31 X. *campestris* pathovars. The genus *Xanthomonas* has a high %G+C ranging between 63 and 69.7. Although the majority of the X. *campestris* pathovars have a rather narrow range around 66% G+C, some individual pathovars clearly have a higher G+C content, namely X. *c.* pvs *graminis, poae, phlei, hyacinthi* and *theicola* (more than 69% G+C), suggesting that there are significant differences within the species X. *campestris* at the genomic level.

3.4.2 DNA–rRNA hybridization

Genes coding for ribosomal RNA (rRNA) constitute a very conservative part of the genome. Therefore, estimation of the degree of binding between rRNA and DNA from different organisms is a good parameter to reveal distant relationships among taxa. DNA–rRNA hybridization techniques have been used to construct phylogenetic classification systems (De Ley *et al.*, 1978; De Vos and De Ley, 1983; De Vos *et al.*, 1985, 1989; Stackebrandt *et al.*, 1988).

DNA–rRNA hybridization has been of primary importance in the delineation of the genus *Xanthomonas*, and particularly the separation of the genera *Xanthomonas* and *Pseudomonas*. Palleroni *et al.* (1973) used a competitive rRNA hybridization technique to analyse the relationships between *Pseudomonas* and *Xanthomonas*. They revealed five *Pseudomonas* clusters and one *Xanthomonas* group, to which *Pseudomonas maltophilia* was closely related. These results have been refined and extended in several other phylogenetic studies based on rRNA homology (De Vos and De Ley, 1983; Stackebrandt *et al.*, 1988; De Vos *et al.*, 1989). A reclassification of *P. maltophilia* to *Xanthomonas* (Swings *et al.*, 1983) was based mainly on the genotypic relationship between these taxa, found by DNA–rRNA hybridization.

De Vos and De Ley (1983) hybridized DNA from 27 genuine *Xanthomonas* strains with ^{14}C-radiolabelled 23S rRNA from *X. c.* pv. *campestris* NCPPB 528^{T}. Within the *Xanthomonas* group, $T_{m(e)}$ (thermal elution temperature) values ranged from 80.0 to 81.0°C. *X. populi* was included in this group, which supported the earlier proposal of Ridé and Ridé (1978) to reclassify '*Aplanobacter populi*' as *X. populi*. '*X. ampelina*', however, displayed $T_{m(e)}$ values ranging from 61.0 to 63.0°C, which clearly indicated that this species was not related to *Xanthomonas*. Later, Willems *et al.* (1987) reclassified this species in the new genus *Xylophilus* as *X. ampelinus*. In the study of De Vos and De Ley (1983), four *P. maltophilia* strains exhibited $T_{m(e)}$ values of 76.0–78.0°C with *Xanthomonas*, which was the closest related genus. This species was transferred to *Xanthomonas* by Swings *et al.* (1983).

In another DNA–rRNA study, De Vos *et al.* (1989) showed that a number of species formerly classified as *Pseudomonas* actually belong to *Xanthomonas*, i.e. '*P. betle*', '*P. gardneri*', '*P. hibiscicola*', '*P. mangiferaeindicae*', '*P. viticola*' and '*P. vitiswoodrowii*'. The latter three species have been reclassified as pathovars of *X. campestris*, whereas for the former three, the position within the genus *Xanthomonas* is not clear.

3.4.3 Ribosomal RNA sequence analysis

The 16S and 23S rRNA are universal and conserved molecules. These features together with the high copy number of rRNA molecules in any living cell make 16S and 23S rRNA molecules an excellent tool for sequence analysis in phylogenetic studies (Woese, 1987) and a basis for probes used for identification (Giovannoni *et al.*, 1988). De Parasis and Roth (1990) sequenced 600–800 nucleotides of 16S rRNA of 26 *Xanthomonas* strains belonging to seven *X. campestris* pathovars, *X. oryzae* and *X. fragariae*, together with 16 *Pseudomonas* and 10 *Erwinia* strains. The most variable region of the phytopathogenic bacteria studied was the region between bases 1064 and 1090. Within *Xanthomonas*, however, variation was only found in bases 1064–1068, which is clearly not sufficient to differentiate within the genus. The authors concluded that probes based on the region 1064–1090 are suitable as a genus-specific assay, e.g. to differentiate *Xanthomonas* rapidly from other phytopathogenic bacteria, but that other variable regions of 16S rRNA or 23S rRNA need to be found in order to differentiate pathovars.

3.4.4 DNA restriction analysis

A number of genomic fingerprinting techniques are based on the capacity of restriction endonucleases to recognize and cleave specific sequences (cleavage sites) of double-stranded DNA. In it simplest form, a bacterial genome is digested with a given restriction enzyme, and the pool of restriction fragments is separated by agarose gel electrophoresis. The resulting banding pattern is a genomic fingerprint of the organism.

This technique has been applied for the differentiation of phytopathological groups of *X. c.* pv. *citri* strains (Hartung and Civerolo, 1987) (see also section 3.3.3). The groupings obtained were later confirmed by other genomic and phenotypic methods.

For bacteria with large genomes such as *Xanthomonas*, the banding patterns thus obtained are very complex and difficult to interpret. In addition, patterns are usually not clear due to non-specific fragmentation of the DNA prior to digestion. To overcome these problems, the DNA is transferred from the gel to nitrocellulose membranes (Southern blotting) and hybridized against a cloned and radiolabelled DNA probe. The hybridization patterns are then compared. This technique, known as restriction fragment length polymorphism (RFLP) analysis, has been applied in several studies of *Xanthomonas*.

Lazo *et al.* (1987) compared RFLP patterns of 93 strains belonging to 26 *X. campestris* pathovars. They concluded that RFLP analysis is suitable for differentiating pathovars and might help resolving the taxonomy of *Xanthomonas*.

X. c. pv. *citri* has been subjected to several DNA restriction endonuclease digestion analyses. Besides the genomic restriction fingerprint study of Hartung and Civerolo (1987), RFLP analyses of this pathovar have been performed by Gabriel *et al.* (1988) and Hartung and Civerolo (1989). The results of these different studies were parallel in that the most widespread CBC-A strains constituted a clonal group, whereas the Florida E strains or CBS strains (see also section 3.3.3) were heterogeneous on the basis of RFLP patterns. The CBC-B, C and D strains together formed another clonal group. The relationships between strains of CBC-A and CBC-B, C and D as revealed by the different studies were not concordant.

This discrepancy is probably due to the use of different combinations of probes and restriction enzymes. Indeed, RFLP patterns reflect only congruence of a small fraction of the entire genome, i.e. the fraction corresponding to the probe used. In addition, the polymorphisms revealed are dependent on the cleavage site of the restriction enzyme used. Although the application of molecular techniques in taxonomy should be favoured unconditionally, this example shows that every molecular method has its limitations and drawbacks, and should therefore be compared to other methods.

A third variant of DNA restriction analysis is based on digestion of genomic DNA with rare-cutting restriction endonucleases. This also results in a reduced number of bands and facilitates interpretation of patterns. However, laborious purification techniques are needed to obtain high-molecular-weight DNA, and the resulting high-molecular-weight restriction fragments are separated by pulsed field electrophoresis (Cooksey and Graham, 1989). This method has been applied again for a genomic study of the *X. c.* pv. *citri* pathogenicity groups (Egel *et al.*, 1991). Previous findings based on RFLP analysis were confirmed by this study, but not all CBS strains could be compared, due to excessive shearing of DNA during purification and manipulation.

3.4.5 DNA–DNA hybridization

DNA–DNA hybridization techniques offer the advantage over other genomic methods that they measure levels of total genomic homology. DNA homology as estimated by DNA–DNA hybridization is currently the most reliable parameter for determining taxonomic relationships at the subgeneric rank, and is considered as the conventional method for the delineation of species (Schleifer and Stackebrandt, 1983; Johnson, 1984; Wayne *et al.*, 1987; Murray *et al.*, 1990).

In the genus *Xanthomonas*, DNA–DNA hybridization has revealed groupings that are not consistent with the current classification. The first DNA hybridization experiments of *Xanthomonas* nomenspecies were reported by Murata and Starr (1973), who observed that up to 50% of the DNA was characteristic for the nomenspecies studied. Schroth and Hildebrand (1983) have stressed the need for genotypic data to clarify the taxonomic relationships within the genera *Pseudomonas* and *Xanthomonas*. However, further DNA homology data were scanty until the end of the 1980s.

Van den Mooter *et al.* (1987a) reported DNA homologies below 40% D (degree of binding) between *X. c.* pvs *manihotis, cassavae* and *campestris* using the spectrophotometric method of De Ley *et al.* (1970). Similarly, Vauterin *et al.* (1990a) found homology values ranging from 15 to 39% between *X. c.* pvs *begoniae, pelargonii, campestris, pisi, cassavae, hyacinthi* and *juglandis*, whereas within the pathovars *begoniae, pelargonii* and *campestris*, DNA homology amounted to 85–100% D. Hildebrand *et al.* (1990) determined DNA relatedness of 23 *X. campestris* pathovars and *X. fragariae* using the S1 nuclease method. They found one major cluster consisting of the legume pathovars *X. c.* pvs *phaseoli, glycines, vignicola* and *phaseoli* var. *fuscans* together with four non-legume pathovars. Two minor clusters comprised a group of *X. c.* pvs *pelargonii, taraxaci, carotae* and '*X. gardneri*', and a group of *X. c.* pvs *juglandis* and *celebensis*. The low genotypic relationships between many *X. campestris* pathovars thus far studied are in sharp contrast with the phenotypic uniformity of the whole *X. campestris* complex.

In the context of a global taxonomic investigation of the genus *Xanthomonas*, extensive DNA–DNA hybridization studies have been executed since 1987 (Vauterin *et al.*, 1990a, b, 1991, 1992). All DNA homology data from the aforementioned references together with new data are compiled in Table 3.3. In order to reduce the size of the matrix of %D values, uniform groups were delineated above 60% and the internal mean %D values together with their standard deviations were calculated.

Between the 21 DNA homology groups shown in Table 3.3, DNA binding degrees were generally below 40%. Some of the DNA groups consist of pathovars attacking related host plants. This is clearly the case for the six pathovars from crucifers (Brassicaceae) which constitute a distinct group XII, and for the pathovars from grasses and cereals (Poaceae), recovered in group IX. A third group contained all but one pathovar from legumes (Fabaceae) in DNA homology group VIII. The only exception was *X. c.* pv. *pisi*, constituting a separate DNA homology group XIII, respectively. However, there was no apparent host–pathogen relationship between the pathovars from legumes and a number of other pathovars that also belong to group VIII, e.g. *X. c.* pvs *begoniae, citri, malvacearum, manihotis, ricini*, and even the 'species' *X. axonopodis*. The opposite case, where strains that attack highly related hosts or even the same host, but occur in different DNA homology groups, exists as well, e.g. *X. c.* pvs *cassavae* A and *manihotis*, occurring in groups XIV and VIII, respectively, and *X. c.* pvs *coracanae* and *vasculorum*, occurring outside DNA homology group IX of the Poaceae.

X. c. pv. *citri* is an example where all strains belong to the same DNA homology group, but constitute different subgroups, delineated at 60–70% D (Vauterin *et al.*, 1991b). These subgroups correlate well with the pathogenicity of the strains, i.e. the different pathogenic groups recognized in *X. c.* pv. *citri*. However, other *X. campestris* pathovars are related at very high levels of DNA homology to these groups. *X. c.* pv. *glycines* is related at over 90% with pathogenicity group A of *X. c.* pv. *citri*, whereas *X. c.* pv. *alfalfae* has 85% DNA homology with group E. This again illustrates that pathogenicity is not correlated with genomic relatedness, and hence, is not a reliable taxonomic marker.

Table 3.3 DNA homology groups of the genus *Xanthomonas*

Group[a]	*Xanthomonas species and pathovars*[a]	*Mean %D*[b]	*SD (%)*	*n*[c]
I (2)	*X. albilineans* (2)	97	–	1
II (2)	*X. fragariae* (2)	91	–	1
III (1)	*X. maltophilia* (1)	–	–	–
IV (3)	*X. populi* (3)	96	1	3
V (3)	*Bromus* isolates (3)	88	6	2
VI (6)	*X. campestris* pvs *holcicola* (3), *vasculorum* B (3)	90	4	6
VII (6)	*X. oryzae* pvs *oryzae* (3), *oryzicola* (3)	91	7	6
VIII (82)	*X. axonopodis* (2), *X. campestris* pvs *coracanae* (2), *vasculorum* A (4), *cyamopsidis* (1), *desmodii* (1), *patelii* (1), *desmodiirotundifolii* (1), *desmodiigangetici* (1), *lespedezae* (1), *cassiae* (1), *tamarindi* (2), *erythrinae* (1), *bauhiniae* (1), *clitoriae* (1), *desmodiilaxiflorii* (1), *rhynchosiae* (1), *vignaeradiatae* (1), *sesbaniae* (1), *glycines* (2), *cajani* (2), *citri* (20), *vesicatoria* B (3), *malvacearum* (2), *phaseoli-fuscans* (3), *vignicola* (3), *alfalfae* (2), *cassavae* B (1), *phaseoli* (2), *manihotis* (4), *begoniae* (7), *ricini* (4), *poinsettiicola* A (1), *dieffenbachiae* (2)	78	15	213
IX (23)	*X. campestris* pvs *graminis* (2), *arrhenatheri* (2), *phlei* (2), *phleipratensis* (1), *translucens* (3), *undulosa* (3), *cerealis* (2), *secalis* (2), *hordei* (4), *poae* (2)	78	11	48
X (3)	*X. campestris* pv. *hyacinthi* (3)	98	2	3
XI (4)	*X. campestris* pv. *pelargonii* (4)	98	4	4
XII (17)	*X. campestris* pvs *campestris* (7), *raphani* (3), *armoraciae* (2), *barbareae* (2), *incanae* (2), *aberrans* (1)	87	7	36
XIII (2)	*X. campestris* pv. *pisi* (2)	99	–	1
XIV (4)	*X. campestris* pv. *cassavae* A (4)	94	2	4
XV (10)	*X. campestris* pvs *juglandis* (2), *corylina* (4), *pruni* (2), *poinsettiicola* C (1), *populi* (1)	79	15	16
XVI (2)	*X. campestris* pv. *vesicatoria* A (2)	99	–	1
XVII (2)	*X. campestris* pv. ? (2)	98	–	1
XVIII (1)	*X. campestris* pv. *theicola* (1)	–	–	–
XIX (2)	*X. campestris* pv. *poinsettiicola* B (2)	99	–	1
XX (2)	*X. campestris* pv. *cucurbitae* (2)	88	–	1
XXI (2)	*X. campestris* pv. *melonis* (2)	88	–	1

[a] Number of strains in parentheses.
[b] D = degree of binding.
[c] Total number of hybridization values determined.

Another interesting case is represented by DNA homology group XV, in which *X. c.* pvs *corylina, juglandis* and *pruni* were interrelated at 92% D. These three pathovars attack hazelnut, walnut and prune, respectively, all being trees from the temperate regions. The very high genomic relatedness of these organisms suggests that they probably originated from the same organism, which had the ability to infect and colonize the vascular system of trees in the temperate regions. *X. c.* pv. *populi*, a pathogen of poplar, is related to this group at 63% D.

The present data illustrate also that a number of pathovars are composed of two or more unrelated genotypes. This is the case for *X. c.* pv. *vesicatoria*, consisting of two unrelated subgroups A and B. Subgroup A contains the pathotype strain and constitutes DNA homology group XVI, whereas subgroup B falls in DNA homology group VIII together with a number of leguminous and other pathovars. Similarly, *X. c.* pv. *vasculorum* consists of subgroups A and B. Subgroup A contains the pathotype strain and is represented in DNA homology group VIII, whereas subgroup B is related to *X. c.* pv. *holcicola* in group VI. The most heterogeneous pathovar is *X. c.* pv. *poinsettiicola*, composed of subgroups A, B and C. Subgroup A contains the pathotype strain and occurs in DNA homology group VIII, whereas subgroup C is related to the tree pathogens in group XV and subgroup B constitutes a separate DNA homology group XIX.

Here we are confronted with the major criticism of the pathovar system. Both phenotypic and genomic data have shown that pathovars do not always represent biological populations. In a number of cases, different genomic populations have been given the same pathovar name because they caused disease on the same host plants, whereas in other cases new pathovars have been defined only because they were isolated from a new host. The taxonomic consequences of this one-sided approach are schematically represented in Figure 3.1. Some pathovars are composed of two or more taxonomic groups, whereas many others are only different names for the same organisms.

3.5 PERSPECTIVES FOR AN IMPROVED TAXONOMY OF *XANTHOMONAS*

Taxonomic conclusions based on protein electrophoresis, fatty acid analysis, DNA–DNA hybridization and other characterization techniques such as mAbs and RFLP analysis have drawn a new picture of the internal relationships of the genus *Xanthomonas*. The groupings obtained by the three methods are generally concordant, as illustrated in several comparative analyses (Vauterin *et al.*, 1991b, 1992; and this chapter). It also becomes evident that the *X. campestris* pathovars do not constitute a single species as suggested by the phenotypic similarity,

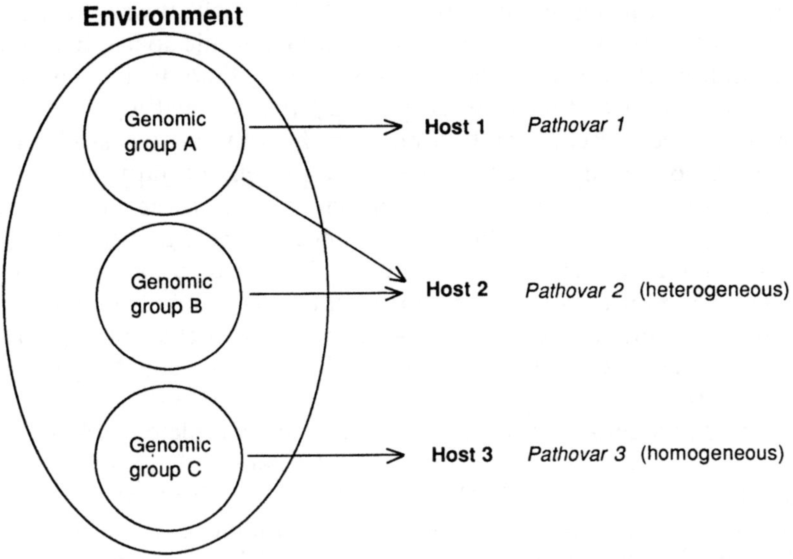

Figure 3.1 Schematic representation of the possible situations within the *Xanthomonas campestris* pathovar system based on phytopathogenicity. Strains of genomic group A are pathogenic for two different hosts (host 1 and host 2). The strains isolated from host 1 are named as pathovar 1, whereas the strains isolated from host 2 are named pathovar 2. Strains of another genomic group (group B) are also pathogenic for host 2, and consequently, are classified in the same pathovar. A clear example of this situation is found in *X. c.* pv. *vasculorum* and *X. c.* pv. *holcicola*. Strains of two different genomic groups isolated from sugarcane (host 2) are named *X. c.* pv. *vasculorum*, whereas those strains of genomic group A that are isolated from *Sorghum* (host 1) are named *X. c.* pv. *holcicola*. Genomic group C represents the case where a single homogeneous genomic group corresponds to one pathovar. An example is *X. c.* pv. *hyacinthi*, which is pathogenic for hyacinths.

but are composed of at least 16 DNA homology groups (Table 3.3). Another important conclusion is that only in a few cases does there exist a clear correlation between the genomic groups and the hosts from which they were isolated. In most cases, however, strains belonging to the same pathovar are genomically highly related.

It is interesting to note that most of the DNA homology groups are clearly delineated. However, only half of the number of pathovars have been studied up to now, and it is likely that when additional pathovars are hybridized, some of the DNA homology groups may become less clearly defined. In this respect, it should be emphasized that the current classification system, or any system which attempts to divide biological populations into discrete taxa, will always be artificial because of its inconsistency with the real taxonomic space, harbouring a continuum of cloudy condensed nodes rather than distinct entities. A clear illustration of this view is given by the non-pathogenic xanthomonads which occur epiphytically on healthy or diseased plants. Attempts to classify these strains in a protein or fatty acid profile database of *Xanthomonas* have shown that many of these strains occupy separate positions within the genus without being significantly related to any known taxon, nor to each other. This observation suggests that the pool of xanthomonads present in the environment is much more complex and diffuse than what we see from the restricted subpopulations of pathogenic strains, which have actually been the sole subject of our attention until now.

In view of a general purpose taxonomy, the complete DNA sequence is advanced as reference to determine phylogenetic relationships (Wayne *et al.*, 1987; Murray *et al.*, 1990). As DNA–DNA hybridization measures levels of total genomic homology, this technique is the most reliable substitute for DNA sequence comparison and has been recommended as the standard criterion for the designation of species. DNA homology of 60–70% is considered to be the minimum for strains within a species (Schleifer and Stackebrandt, 1983; Wayne *et al.*, 1987). Consequently, each of the present DNA homology groups of *Xanthomonas* should constitute a separate species of the genus. At present it is obvious that in contrast to earlier classifications of *Xanthomonas*, the future species classification will not be correlated to phytopathogenic specialization. An immediate advantage of this general classification is that avirulent xanthomonads, to which increasing importance is given from an ecological point of view, can also be accommodated.

Although it is possible to differentiate most of the DNA homology groups by fingerprinting techniques such as SDS–PAGE of proteins and fatty acid analysis, they cannot be distinguished by phenotypic tests. This lack of descriptive classical phenotypic features that help define the genospecies has long been a major problem in the taxonomy of the genus *Xanthomonas*. Methods such as fatty acid analysis, protein electrophoresis and monoclonal antibody typing are used increasingly as diagnostic tools to replace classical phenotypic tests and can serve as alternative sources for descriptive phenotypic features. Moreover, with the availability of modern molecular methods such as DNA restriction fingerprinting and probe techniques, one can question whether it is still desirable to adhere to the principle that distinctive classical phenotypic features are necessary for the definition of species.

In conclusion, the currently recognized species *X. albilineans*, *X. oryzae* and *X. fragariae* and *X. populi* remain distinct and homogeneous taxa and can retain their current species status. *X. axonopodis*, however, falls within DNA homology group VIII and thus can no longer be considered as a separate *Xanthomonas* species. *X. maltophilia* is a complex of diffuse DNA homology clusters which represent different species. More work is needed in order to propose a new classification of this species. The exclusion of *X. maltophilia* from the genus *Pseudomonas* (Swings *et al.*, 1983) was certainly sound from a phylogenetic point of view. The transfer of this species to *Xanthomonas* can be justified by a number of arguments. However, arguments can be found also that support reclassification of *X. maltophilia* as a new genus (Van Zyl and Steyn, 1992). This is an example of relationships at the limit of the genus level, which illustrates again that the division of natural populations into hierarchical branches is artificial. Besides the recognized *Xanthomonas* species, 16 new genospecies can be defined within *X. campestris*. Taxonomically each of them deserves the full species status and should be named accordingly. For reasons of uniformity, it would be desirable to revive the former species names whenever possible. From a practical point of view, it might be justifiable to retain a number of pathovars within the new species. However, the practical usefulness of the new classification will only appear to full advantage when rapid diagnostic tools based on genomic relatedness become available.

REFERENCES

Abel, K., deSchmertzing, H. and Peterson, J.I. (1963) Classification of microorganisms by analysis of chemical composition. *Journal of Bacteriology*, **85**, 1039–44.

Alvarez, A.M. and Benedict, A.A. (1990) Relationships among phytopathogenic bacteria distinguished with monoclonal antibodies, in *Plant Pathogenic Bacteria* (ed. Z. Klement), Akademiai Kiado, Budapest, pp. 859–63.

Alvarez, A.M., Benedict, A.A. and Mizumoto, C.Y. (1985) Identification of xanthomonads and groupings of *Xanthomonas campestris* pv. *campestris* strains with monoclonal antibodies. *Phytopathology*, **75**, 722–8.

Alvarez, A.M., Benedict, A.A., Mizumoto, C.Y., Pollard, L.W. and Civerolo, E.L. (1991) Analysis of *Xanthomonas campestris* pv. *citri* and *X. campestris* pv. *citrumelo* with monoclonal antibodies. *Phytopathology*, **81**, 857–65.

Alvarez, A.M., Benedict, A.A., Or, G. and Mizumoto, C.Y. (1987) Identification of xanthomonads from crucifer seeds with monoclonal antibodies. *Phytopathology*, **77**, 1725 (abstr.).

Alvarez, A.M. and Lou, K. (1985) Rapid identification of *Xanthomonas campestris* pv. *campestris* by ELISA. *Plant Disease*, **69**, 1082–6.

Anderson, M.J. and Nameth, S.T. (1990) Development of a polyclonal antibody-based serodiagnostic assay for the detection of *Xanthomonas campestris* pv. *pelargonii* in geranium plants. *Phytopathology*, **80**, 357–60.

Andrewes, A.G., Jenkins, C.L., Starr, M.P., Shepherd, J. and Hope, H. (1976) Structure of xanthomonadin I, a novel dibrominated arylpolyene pigment produced by the bacterium *Xanthomonas juglandis*. *Tetrahedron Letters*, **45**, 4023–4.

Auling, C., Busse, H.-J., Pilz, F., Webb, L., Kneifel, H. and Claus, D. (1991) Rapid differentiation, by polyamine analysis, of *Xanthomonas* strains from phytopathogenic pseudomonads and other members of the class *Proteobacteria* interacting with plants. *International Journal of Systematic Bacteriology*, **41**, 223–8.

Azad, H. and Schaad, N.W. (1988) Serological relationships among membrane proteins of strains of *Xanthomonas campestris* pv. *translucens*. *Phytopathology*, **78**, 272–7.

Ball, E.M., Hampton, R.O., De Boer, S.H. and Schaad, N.W. (1990) Polyclonal antibodies, in *Serological Methods for Detection and Identification of Viral and Bacterial Plant Pathogens* (eds R. Hampton, E. Ball and S. De Boer), APS Press, St Paul, MN, pp. 33–54.

Benedict, A.A., Alvarez, A.M., Berestecky, J., *et al.* (1989) Monoclonal antibodies specific for *Xanthomonas campestris* pv. *oryzae* and pv. *oryzicola*. *Phytopathology*, **79**, 322–8.

Benedict, A.A., Alvarez, A.M. and Pollard, L.W. (1990) Pathovar-specific antigens of *Xanthomonas campestris* pv. *begoniae* and *X. campestris* pv. *pelargonii* detected with monoclonal antibodies. *Applied and Environmental Microbiology*, **56**, 572–4.

Bouzar, H. and Moore, L.W. (1990) Ouchterlony double diffusion, Plates: bacteria, in *Serological Methods for Detection and Identification of Viral and Bacterial Plant Pathogens* (eds R. Hampton, E. Ball and S. De Boer), APS Press, St Paul, MN, pp. 129–40.

Bradbury, J.F. (1984) Genus II. *Xanthomonas* Dowson 1939, 187^{AL}, in *Bergey's Manual of Systematic Bacteriology*, Vol. 1 (eds N.R. Krieg and J.G. Holt), Williams & Wilkins, Baltimore, pp. 199–210.

Bradbury, J.F. (1986) *Xanthomonas* Dowson 1939, 187, in *Guide to Plant Pathogenic Bacteria*, CAB International Mycological Institute, Slough, pp. 198–260.

Busse, J. and Auling, G. (1988) Polyamine pattern as chemotaxonomic marker within the *Proteobacteria*. *Systematic Applied Microbiology*, **11**, 1–8.

Charudattan, R., Stall, R.E. and Batchelor, D.L. (1973) Serotypes of *Xanthomonas vesicatoria* unrelated to its pathotypes. *Phytopathology*, **63**, 1260–5.

Collins, M.D. and Jones, D. (1981) Distribution of isoprenoid quinone structural types in bacteria and their taxonomic implications. *Microbiological Reviews*, **45**, 316–54.

Colwell, R.R. and Liston, J. (1961) Taxonomic analysis with the electronic computer of some *Xanthomonas* and *Pseudomonas* species. *Journal of Bacteriology*, **82**, 913–19.

Cooksey, D.A. and Graham, J.H. (1989) Genomic fingerprinting of two pathovars of phytopathogenic bacteria by rare-cutting restriction enzymes and field inversion electrophoresis. *Phytopathology*, **79**, 745–50.

Debette, J. and Prensier, G. (1989) Immunoelectron microscopic demonstration of an esterase on the outer membrane of *Xanthomonas maltophilia*. *Applied and Environmental Microbiology*, **55**, 233–9.

De Ley, J., Cattoir, H. and Reynaerts, A. (1970) The quantitative measurement of DNA hybridization from renaturation rates. *European Journal of Biochemistry*, **12**, 133–42.

De Ley, J., Segers, P. and Gillis, M. (1978) Intra- and intergeneric similarities of *Chromobacterium* and *Janthinobacterium* ribosomal ribonucleic acid cistrons. *International Journal of Systematic Bacteriology*, **28**, 154–68.

De Parasis, J. and Roth, D.A. (1990) Nucleic acid probes for identification of phytobacteria: identification of genus-specific 16s rRNA sequences. *Phytopathology*, **80**, 618–21.

De Vos, P. and De Ley, J. (1983) Intra- and intergeneric similarities of *Pseudomonas* and *Xanthomonas* ribosomal ribonucleic acid cistrons. *International Journal of Systematic Bacteriology*, **33**, 487–509.

De Vos, P., Goor, M., Gillis, M. and De Ley, J. (1985) Ribosomal ribonucleic acid cistron similarities of phytopathogenic *Pseudomonas* species. *International Journal of Systematic Bacteriology*, **35**, 169–84.

De Vos, P., Van Landschoot, A., Segers, P., *et al.* (1989) Genotypic relationships and taxonomic localization of unclassified *Pseudomonas* and *Pseudomonas*-like strains by deoxyribonucleic acid:ribosomal ribonucleic acid hybridizations. *International Journal of Systematic Bacteriology*, **39**, 35–49.

dos Santos, R.M.D.B. and Dianese, J.C. (1985) Comparative membrane characterization of *Xanthomonas campestris* pv. *cassavae* and *X. campestris* pv. *manihotis*. *Phytopathology*, **75**, 581–7.

Duveiller, E. and Bragard, C. (1992) Comparison of immunofluorescence and two assays for detection of *Xanthomonas campestris* pv. *undulosa* in seeds of small grains. *Plant Disease*, **76**, 999–1003.

Dye, D.W. (1962) The inadequacy of the usual determinative tests for the identification of *Xanthomonas* spp. *New Zealand Journal of Science*, **5**, 393–496.

Dye, D.W., Bradbury, J.F., Goto, M., Hayward, A.C., Lelliott, R.A. and Schroth, M.N. (1980) International standards for naming pathovars of phytopathogenic bacteria and a list of pathovar names and pathotype strains. *Review of Plant Pathology*, **59**, 153–68.

Dye, D.W. and Lelliott, R.A. (1974) Genus II. *Xanthomonas*, in *Bergey's Manual of Determinative Bacteriology*, 8th edn (eds R.E. Buchanan and N.E. Gibbons), Williams & Wilkins, Baltimore, pp. 243–9.

Egel, D.S., Graham, J.H. and Riley, T.D. (1991) Population dynamics of strains of *Xanthomonas campestris* differing in aggressiveness on Swingle citrumelo and grapefruit. *Phytopathology*, **81**, 666–71.

Egli, T. and Schmidt, D. (1982) Pathogenic variation among the causal agent of bacterial wilt of forage grasses. *Phytopathologische Zeitschrift*, **104**, 138–50.

Elrod, R.P. and Braun, A.C. (1947a) Serological studies of the genus *Xanthomonas* I. Cross-agglutination relationships. *Journal of Bacteriology*, **53**, 509–18.

Elrod, R.P. and Braun, A.C. (1947b) Serological studies of the genus *Xanthomonas* II. *Xanthomonas translucens* groups. *Journal of Bacteriology*, **53**, 519–24.

Elrod, R.P. and Braun, A.C. (1947c) Serological studies of the genus *Xanthomonas*. III. The *Xanthomonas vasculorum* and *Xanthomonas phaseoli* groups. *Journal of Bacteriology*, **54**, 349–57.

El-Sharkawy, T.A. and Huisingh, D. (1971) Electrophoretic analysis of esterases and other soluble proteins from representatives of phytopathogenic bacterial genera. *Journal of General Microbiology*, **68**, 149–54.

Franken, A.A.J.M. (1990) Production, characterization and application of monoclonal antibodies for detecting *Xanthomonas campestris* pv. *campestris* in cabbage seeds, in *Proceedings of the Symposium on Perspectives for Monoclonal Antibodies in Agriculture*, PUDOC, Wageningen, The Netherlands, pp.27–36.

Gabriel, D.W., Hunter, J.E., Kingsley, M.T., Miller, J.W. and Lazo, G.R. (1988) Clonal population structure of *Xanthomonas campestris* and genetic diversity among citrus canker strains. *Molecular Plant–Microbe Interactions*, **1**, 59–65.

Gabriel, D.W., Kingsley, M.T., Hunter, J.E. and Gottwald, T.R. (1989) Reinstatement of *Xanthomonas citri* (ex Hasse) and *X. phaseoli* (ex Smith) to species and reclassification of all *X. campestris* pv. *citri* strains. *International Journal of Systematic Bacteriology*, **39**, 14–22.

Giovannoni, S.J., DeLong, E.F., Olsen, G.J. and Pace, N.R. (1988) Phylogenetic group-specific oligonucleotide probes for identification of single microbial cells. *Journal of Bacteriology*, **170**, 720–6.

Gottwald, T.R., Alvarez, A.M., Hartung, J.S. and Benedict, A.A. (1991) Diversity of *Xanthomonas campestris* pv. *citrumelo* strains associated with epidemics of citrus bacterial spot in Florida citrus nurseries: correlation of detached leaf, monoclonal antibody, and restriction fragment length polymorphism assays. *Phytopathology*, **81**, 749–53.

Graham, J.H. and Gottwald, T.R. (1990) Variation in aggressiveness of *Xanthomonas campestris* pv. *citrumelo* associated with citrus bacterial spot in Florida citrus nurseries. *Phytopathology*, **80**, 190–6.

Graham, J.H., Hartung, J.S., Stall, R.E. and Chase, A.R. (1990) Pathological, restriction-fragment length polymorphism, and fatty acid profile relationship between *Xanthomonas campestris* from citrus and noncitrus hosts. *Phytopathology*, **80**, 829–36.

Haaheim, L.R., Kleppe, G. and Sutherland, I.W. (1989) Monoclonal antibodies reacting with the exopolysaccharide xanthan from *Xanthomonas campestris*. *Journal of General Microbiology*, **135**, 605–12.

Hartung, J.S. and Civerolo, E.L. (1987) Genomic fingerprints of *Xanthomonas campestris* pv. *citri* strains from Asia, South America, and Florida. *Phytopathology*, **77**, 282–5.

Hartung, J.S. and Civerolo, E.L. (1989) Restriction fragment length polymorphisms distinguish *Xanthomonas campestris* strains isolated from Florida citrus nurseries from *X. c.* pv. *citri*. *Phytopathology*, **79**, 793–9.

Hildebrand, D.C., Palleroni, N.J. and Schroth, M.N. (1990) Deoxyribonucleic acid relatedness of 24 xanthomonad strains representing 23 *Xanthomonas campestris* pathovars and *Xanthomonas fragariae*. *Journal of Applied Bacteriology*, **68**, 263–9.

Ikemoto, S., Kuraishi, H. and Komagata, K. (1978) Cellular fatty acid composition in *Pseudomonas* species. *Journal of General and Applied Microbiology*, **24**, 199–213.

Ikemoto, S., Suzuki, K., Kaneko, T. and Komagata, K. (1980) Characterization of strains of *Pseudomonas maltophilia* which do not require methionine. *International Journal of Systematic Bacteriology*, **30**, 437–47.

Jenkins, C.L. and Starr, M.P. (1985) Formation of halogenated aryl-polyene (xanthomonadin) pigments by the type and other yellow-pigmented strains of *Xanthomonas maltophilia*. *Annales de l'Institut Pasteur Microbiologie*, **136**B, 257–64.

Johnson, J.L. (1984) Nucleic acids in bacterial classification, in *Bergey's Manual of Systematic Bacteriology*, Vol. 1 (eds N.R. Krieg and J.G. Holt), Williams & Wilkins, Baltimore, pp. 8–11.

Jones, R.K., Barnes, L.W., Gonzalez, C.F., Leach, J.E., Alvarez, A.M. and Benedict, A.A. (1989) Identification of low-virulence strains of *Xanthomonas campestris* pv. *oryzae* from rice in the United States. *Phytopathology*, **79**, 984–90.

Kersters, K. and De Ley, J. (1975) Identification and grouping of bacteria by numerical analysis of their electrophoretic protein patterns. *Journal of General Microbiology*, **87**, 333–42.

Kohler, G. and Milstein, C. (1975) Continuous cultures of fused cells secreting antibody of predefined specificity. *Nature*, **256**, 495–7.

Laemmli, U.K. (1970) Cleavage of structural proteins during the assembly of the head of bacteriophage T4. *Nature*, **227**, 680–5.

Lazo, G.R., Roffey, R. and Gabriel, D.W. (1987) Pathovars of *Xanthomonas campestris* are distinguishable by restriction fragment-length polymorphism. *International Journal of Systematic Bacteriology*, **37**, 214–21.

Leach, J.E., Ramundo, B.A., Bearson, D.L. and Claflin, L.E. (1987) Dot-immunobinding assay for detecting *Xanthomonas campestris* pv. *holcicola* in sorghum. *Plant Disease*, **712**, 30–3.

Lipp, R.L. (1988) Use of monoclonal antibodies and pathogenicity tests to identify and differentiate strains of *Xanthomonas campestris* pv. *dieffenbachiae*. M.S. Thesis, University of Hawaii.

Lipp, R.L., Alvarez, A.M., Benedict, A.A. and Berestecky, J. (1992) Use of monoclonal antibodies and pathogenicity tests to characterize strains of *Xanthomonas campestris* pv. *dieffenbachiae* from aroids. *Phytopathology*, **82**, 677–82.

Lovrekovich, L. and Klement, Z. (1965) Serological and bacteriophage sensitivity studies on *Xanthomonas vesicatoria* strains isolated from tomato and pepper. *Phytopathologische Zeitschrift*, **39**, 129–33.

Mahanta, I.C. and Addy, S.K. (1977) Serological specificity of *Xanthomonas oryzae* incitant of bacterial blight of rice. *International Journal of Systematic Bacteriology*, **27**, 383–5.

Minsavage, G.V. and Schaad, N.W. (1983) Characterization of membrane proteins of *Xanthomonas campestris* pv. *campestris*. *Phytopathology*, **73**, 747–55.

Moss, C.W., Samuels, S.B., Liddle, J. and Mckinney, R.M. (1973) Occurrence of branched-chained hydroxy fatty acids in *Pseudomonas maltophilia*. *Journal of Bacteriology*, **114**, 1018–24.

Murata, G.R. and Starr, M.P. (1973) A concept of the genus *Xanthomonas* and its species in the light of segmental homology of deoxyribonucleic acids. *Phytopathologische Zeitschrift*, **77**, 285–323.

Murray, R.G.E., Brenner, D.J., Colwell, R.R., *et al.* (1990) Report of the *ad hoc* committee on approaches to taxonomy within the Proteobacteria. *International Journal of Systematic Bacteriology*, **40**, 213–15.

Neal, D.J. and Wilkinson, S.G. (1982) Lipopolysaccharides from *Pseudomonas maltophilia*. *European Journal of Biochemistry*, **128**, 143–9.

Obata, T. and Tsuboi, F. (1972) Studies on the serological diagnosis of bacterial plant diseases. I. Cross-agglutination and gel-diffusion tests with some *Xanthomonas* nomen species. *Res. Bull. Plant Prot. Ser. Jpn.*, **10**, 8–16.

O'Brien, L.M., Morton, D.J., Manning, W.J. and Sheetz, R.W. (1967) Serological differences between apparently typical pepper and tomato isolates of *Xanthomonas vesicatoria*. *Nature*, **215**, 532–3.

Oyaizu, H. and Komagata, K. (1983) Grouping of *Pseudomonas* species on the basis of cellular fatty acid compositions and the quinone system with special reference to the existence of 3-hydroxy-fatty acids. *Journal of General and Applied Microbiology*, **29**, 17–40.

Palleroni, N.J., Kunisawa, R., Contopoulou, R. and Doudoroff R. (1973) Nucleic acid homologies in the genus *Pseudomonas*. *International Journal of Systematic Bacteriology*, **23**, 333–9.

Permar, T.A. and Gottwald, T.R. (1989) Specific recognition of a *Xanthomonas campestris* Florida nursery strain by a monoclonal antibody probe in a microfiltration enzyme immunoassay. *Phytopathology*, **79**, 780–3.

Richaud, C., Felix, S. and Ferre, P. (1979) A simple serological technique for the precise diagnosis of leaf scald disease in sugarcane, in *Proceedings of the Fourth International Conference on Plant Pathogenic Bacteria*, Angers, France, 1978 (ed. Station de Pathologie Végétale et Phytobacteriologie), Institut National de la Recherche Agronomique, Beaucouzé, pp. 337–40.

Ridé, M. and Ridé, S. (1978) *Xanthomonas populi* (Ridé) comb. nov. (syn. *Aplanobacter populi* Ridé) spécificité, variabilité, et absence de relation avec *Erwinia cancerogena. European Journal of Forest Pathology*, **8**, 310–33.

Rietschel, E.T., Lüderitz, O. and Volk, W.A. (1975) Nature, type of linkage, and absolute configuration of (hydroxy) fatty acids in lipopolysaccharides from *Xanthomonas sinensis* and related strains. *Journal of Bacteriology*, **122**, 1180–8.

Roberts, P.D. (1991) Evaluation and application of monoclonal antibodies specific to *Xanthomonas oryzae* pv. *oryzae* as epidemiological tools. M.S. thesis, University of Hawaii.

Rodriguez, G.S., Garza, L.I.G., Stapleton, J.J. and Civerolo, E.L. (1985) Citrus bacteriosis in Mexico. *Plant Disease*, **69**, 808–10.

Rott, P., Arnaud, M. and Baudin, P. (1986) Serological and lysotypical variability of *Xanthomonas albilineans* (Ashby) Dowson, causal agent of sugarcane leaf scald disease. *Journal of Phytopathology*, **116**, 201–11.

Sasser, J.M., Fieldhouse, D.J. and Carter, C.N. (1984) Computer assisted identification of bacteria based on fatty acid analysis. *Phytopathology*, **74**, 882.

Schaad, N.W. (1976) Immunological comparison and characterization of ribosomes of *Xanthomonas vesicatoria. Phytopathology*, **6**, 770–6.

Schaad, N.W. (1978) Use of direct and indirect immunofluorescence tests for identification of *Xanthomonas campestris. Phytopathology*, **68**, 249–52.

Schaad, N.W. (1979) Serological identification of plant pathogenic bacteria. *Annual Review of Phytopathology*, **17**, 123–47.

Schleifer, K.H. and Stackebrandt, E. (1983) Molecular systematics of prokaryotes. *Annual Review of Microbiology*, **37**, 143–87.

Schoulties, C.L., Civerolo, E.L., Miller, J.W., *et al.* (1987) Citrus canker in Florida. *Plant Disease*, **71**, 388–95.

Schroth, M.N. and Hildebrand, O.C. (1983) Towards a sensible taxonomy of bacterial plant pathogens. *Plant Disease*, **67**, 128.

Sokal, R.R. and Sneath, P.H.A. (1963) *Principles of Numerical Taxonomy*, W. H. Freeman, London.

Stackebrandt, E., Murray, R.G.E. and Trüper, H.G. (1988) *Proteobacteria* classis nov., a name for the phylogenetic taxon including the "purple bacteria and their relatives". *International Journal of Systematic Bacteriology*, **38**, 321–5.

Stall, R.E. and Hodge, N.C. (1989) Use of fatty acid profiles to identify strains of *Xanthomonas campestris* pv. *citri* from the citrus canker epidemic in Florida. *Phytopathology*, **79**, 376 (abstr.).

Starr, M.P. (1981) The genus *Xanthomonas*, in *The Prokaryotes*, Vol. 1 (eds M.P. Starr, H. Stolp, H.G. Trüper, A. Balows and H.G. Schlegel), Springer Verlag, Berlin, pp. 742–63.

Starr, M.P., Jenkins, C.L., Bussey, L.B. and Andrewes, A.G. (1977) Chemotaxonomic significance of the xanthomonadins, novel brominated arylpolyene pigments produced by bacteria of the genus *Xanthomonas. Archives of Microbiology*, **113**, 1–8.

Stead, D. (1989) Grouping of *Xanthomonas campestris* pathovars of cereals and grasses by fatty acid profiling. *Bulletin OEPP/EPPO Bulletin*, **19**, 57–68.

Swings, J., De Vos, P., Van den Mooter, M. and De Ley, J. (1983) Transfer of *Pseudomonas maltophilia* Hugh 1981 to the genus *Xanthomonas* as *Xanthomonas maltophilia* (Hugh 1981) comb. nov. *International Journal of Systematic Bacteriology*, **33**, 409–13.

Thaveechai, N. and Schaad, N.W. (1984) Comparison of different immunogen preparations for serological identification of *Xanthomonas campestris* pv. *campestris. Phytopathology*, **74**, 1065–70.

Thaveechai, N. and Schaad, N.W. (1986) Immunological characterization of a subspecies-specific antigenic determinant of a membrane protein extract of *Xanthomonas campestris* pv. *campestris*. *Phytopathology*, **76**, 148–53.

Tsai, C.C., Lin, C.P. and Chen, C.T. (1990) Characterization and monoclonal antibody production of *Xanthomonas albilineans* (Ashby) Dowson, the causal agent of sugarcane leaf scald disease. *Plant Protein Bulletin*, **32**, 125–35.

Van den Mooter, M., Maraite, H., Meiresonne, L., *et al.* (1987a) Comparison between *Xanthomonas campestris* pv. *manihotis* (ISPP List 1980) and *X. campestris* pv. *cassavae* (ISPP List 1980) by means of phenotypic, protein electrophoretic, DNA hybridization and phytopathological techniques. *Journal of General Microbiology*, **133**, 57–71.

Van den Mooter, M., Steenackers, M., Maertens, C., *et al.* (1987b) Differentiation between *Xanthomonas campestris* pv. *graminis* (ISPP List 1980), pv. *phleipratensis* (ISPP List 1980) emend., pv. *poae* Egli and Schmidt 1982 and pv. *arrhenatheri* Egli and Schmidt 1982 by numerical analysis of phenotypic features and protein gel electrophoregrams. *Journal of Phytopathology*, **118**, 135–56.

Van den Mooter, M. and Swings, J. (1990) Numerical analysis of 295 phenotypic features of 266 *Xanthomonas* strains and related strains and an improved taxonomy of the genus. *International Journal of Systematic Bacteriology*, **40**, 348–69.

Van Zyl, E. and Steyn, P.L. (1992) Reinterpretation of the taxonomic position of *Xanthomonas maltophilia* and taxonomic criteria in this genus. Request for an opinion. *International Journal of Systematic Bacteriology*, **42**, 193–8.

Vauterin, L., Vantomme, R., Pot, B., Hoste, B., Swings, J. and Kersters, K. (1990a) Taxonomic analysis of *Xanthomonas campestris* pv. *begoniae* and *X. campestris* pv. *pelargonii* by means of phytopathological, phenotypic, protein electrophoretic and DNA hybridization methods. *Systematic and Applied Microbiology*, **13**, 166–76.

Vauterin, L., Swings, J., Kersters, K., *et al.* (1990b) Towards an improved taxonomy of *Xanthomonas*. *International Journal of Systematic Bacteriology*, **40**, 312–16.

Vauterin, L., Swings, J. and Kersters, K. (1991a) Grouping of *Xanthomonas campestris* pathovars by SDS–PAGE of proteins. *Journal of General Microbiology*, **137**, 1677–87.

Vauterin, L., Yang, P., Hoste, B., *et al.* (1991b) Differentiation of *Xanthomonas campestris* pv. *citri* strains by sodium dodecyl sulfate–polyacrylamide gel electrophoresis of proteins, fatty acid analysis, and DNA–DNA hybridization. *International Journal of Systematic Bacteriology*, **41**, 535–42.

Vauterin, L., Yang, P., Hoste, B., Pot, B., Swings, J. and Kersters, K. (1992) Taxonomy of xanthomonads from cereals and grasses based on SDS–PAGE of proteins, fatty acid analysis and DNA–DNA hybridization. *Journal of General Microbiology*, **138**, 1467–77.

Vera Cruz, C.M., Gosselé, F., Kersters, K., *et al.* (1984) Differentiation between *Xanthomonas campestris* pv. *oryzae*, *X. campestris* pv. *oryzicola* and the 'brown blotch' pathogen on rice by numerical analysis of phenotypic features and protein gel electrophoregrams. *Journal of General Microbiology*, **130**, 2983–99.

Vruggink, H. (1979) Enzyme-linked immunosorbent assay (ELISA) in the serodiagnosis of plant pathogenic bacteria, in *Proceedings of the Fourth International Conference on Plant Pathogenic Bacteria*, Angers, France, 1978 (ed. Station de Pathologie Végétale et Phytobacteriologie), Institut National de la Recherche Agronomique, Beaucouzé, pp. 307–10.

Wakker, J.H. (1883) Vorläufige Mitteilungen über Hyacinthenkrankheiten. *Botanisches Centralblatt*, **14**, 315–17.

Wayne, L.G., Brenner, D.J., Colwell, R.R., *et al.* (1987) Report of the *ad hoc* committee on reconciliation of approaches to bacterial systematics. *International Journal of Systematic Bacteriology*, **37**, 463–4.
Willems, A., Gillis, M., Kersters, K., Van den Broecke, L. and De Ley, J. (1987) Transfer of *Xanthomonas ampelina* Panagopoulos 1969 to a new genus, *Xylophilus* gen. nov., as *Xylophilus ampelinus* (Panagopoulos 1969) comb. nov. *International Journal of Systematic Bacteriology*, **37**, 422–30.
Woese, C.R. (1987) Bacterial evolution. *Microbiological Reviews*, **51**, 221–71.
Wong, W.C. (1990) Production of monoclonal antibodies of *Pseudomonas syringae* pv. *phaseolicola* and *Xanthomonas campestris* pv. *phaseoli*. *Letters in Applied Microbiology*, **10**, 241–4.
Yang, P., Vauterin, L., Vancanneyt, M., Swings, J. and Kersters, K. (1993) Application of fatty acid methyl esters for the taxonomic analysis of the genus *Xanthomonas*. *Systematic Applied Microbiology*, **16**, 47–71.
Yuen, G.Y., Alvarez, A.M., Benedict, A.A. and Trotter, K.J. (1987) Use of monoclonal antibodies to monitor the dissemination of *Xanthomonas campestris* pv. *campestris*. *Phytopathology*, **77**, 366–71.

4

Infection of the plant by *Xanthomonas*

K. Rudolph
Institut für Pflanzenpathologie und Pflanzenschutz, Universität Göttingen

4.1 EPIPHYTIC GROWTH

Cells of xanthomonads which are deposited on plant surfaces by water splash, aerosols or other means need to invade the plant tissue in order to build up high populations and to cause disease. Only when favourable conditions exist may the bacteria temporarily multiply on plant surfaces by a process similar to epiphytes. Thus, for *X. oryzae* and several pathovars of *X. campestris*, e.g. *juglandis, manihotis, malvacearum, phaseoli, phaseoli* var. *fuscans, pruni* and *vesicatoria*, epiphytic growth has been reported (Hirano and Upper, 1983). Azad and Schaad (1988) showed that *X. c.* pv. *translucens* is capable of epiphytic growth on 2-week-old wheat plants, in controlled conditions at 25°C.

However, in general, phytopathogenic xanthomonads are not capable of progressively colonizing plant surfaces. Especially on leaves, epiphytic populations of xanthomonads fluctuate widely. In most cases the limiting factor is availability of free water (Leben and Daft, 1967). In addition, sunlight and particularly u.v. light or extreme temperatures (>50°C) can damage bacterial cells. Occasionally, xanthomonads of unknown affiliation have been isolated which seem to be well adapted to the phyllosphere but are not pathogenic to known plant species (Angeles-Ramos *et al.*, 1991). Phytopathogenic xanthomonads are

Xanthomonas.
Edited by J.G. Swings and E.L. Civerolo.
Published in 1993 by Chapman & Hall, London.
ISBN 0 412 434202

highly specialized to multiply in the plant tissue and may only casually multiply externally. To understand the nature of this epiphytic phase it was necessary to determine preferred sites of multiplication, methods of bacterial attachment to plant surfaces, and the effects of humidity and nutrients on the final level of epiphytic populations. Depending on the disease, a certain epiphytic growth may precede the infection process regularily or very seldom, if at all. The resident epiphytic phase may also play a decisive role for survival of phytopathogenic xanthomonads, not only on the susceptible plant but also on resistant and non-host plants.

4.1.1 The nature of epiphytic growth

Epiphytic populations of xanthomonads do not necessarily indicate epiphytic growth because they can also be created by **bacterial exudation** from lesions (see section 4.4.1 and Figure 4.5). When lesions of citrus canker were placed in water, 10^4–10^5 cells/ml of *X. c.* pv. *citri* exuded immediately, and within 30 min the bacterial concentration in the bathing solution increased 10-fold (Timmer *et al.*, 1991). These studies indicated that the bacterial population determined in the bathing solution was the result of exudation of bacteria already existing in the lesions. Little or no multiplication appeared to occur on the surface of the leaves in water. Often, the development of external populations parallels internal populations, although on a lower level when expressed as bacterial cells per cm^2 leaf.

This dynamic exchange between internal (e.g. substomatal cavities and intercellular spaces) and external bacterial populations probably depends on the prevailing conditions in and on the leaf. Thus, Egel *et al.* (1991) determined a positive relationship between internal and external populations of *X. c.* pv. *citri* and bacteria associated with citrus bacterial spot (CBS) disease and concluded that external populations represented bacteria exuded onto the moist leaf surface through the ruptured epidermis of the lesions. Also, the population trends of *X. c.* pv. *phaseoli* on the surface of bean leaves were similar to those of total populations throughout the assay period of 1–32 days after inoculation (Cafati and Saettler, 1980).

Little is known about **preferred sites** of epiphytic bacteria. Bacterial cells have been observed in depressions between epidermal cells and on the basal parts of leaf hairs, rarely near the guard cells or in stomatal openings (Hirano and Upper, 1983). Scanning electron microscopy revealed that cells of *X. c.* pv. *corylina* first spread at random over the leaf and were then located in the hollows, on veins and sometimes on the lamina (Pruvost and Gardan, 1988), whereas masses of cells of *X. c.* pv. *oryzae* on rice leaves were located around or in the water pore openings of hydathodes 24 h after inoculation (Mew *et al.*, 1984).

Attachment mechanisms can increase and stabilize the resident population. Thus, it was shown by DeCleene (1989) that *X. o.* pvs *oryzae* and *oryzicola* attached to trichomes on maize leaves. On tomato leaf surfaces xanthomonads appeared to survive in small aggregates in an extracellular matrix (Timmer *et al.*, 1987), and cells of *X. c.* pv. *corylina* on hazelnut leaves agglutinated in a mucus (Pruvost and Gardan, 1988). Bacterial cells of *X. c.* pv. *vesicatoria* on pepper were bound to the fruit surface by fibrillar material (Bashan and Okon, 1986). On the fruit surface, bacteria multiplied in small aggregates, submerged in slime, which consisted mainly of sucrose units. Recently, it was reported by Romantschuk (1993) that cells of *X. c.* pvs *hyacinthi* and *vesicatoria* produced fimbriae or pili which attached to plant surfaces.

The **populations** of epiphytic xanthomonads vary considerably during the day and are dependent on environmental conditions. Populations of *X. c.* pv. *citri* on symptomatic leaves varied from 10^4 to 10^6 per leaf in the early morning and declined to 10^2 to 10^5, respectively, by mid-afternoon (Timmer, 1989). In wheat fields in Mexico, 10^6 cfu of *X. c.* pv. *undulosa* per leaf were determined (Duveiller, 1992), and on hazelnut epiphytic populations of *X. c.* pv. *corylina* remained between 10^6 and 10^7 cfu per leaf (Pruvost and Gardan, 1988).

Epiphytic bacterial populations tend to increase at high **humidity** when plants are wet, e.g. *X. c.* pv. *juglandis* (Mulrean and Schroth, 1982). Also, leaf surface populations of *X. c.* pv. *vesicatoria* increased 10- to 100-fold on tomato leaves at high relative humidity (RH) (>90%), fluctuated erratically at moderate RH (50–65%) and declined to very low levels at low RH (10–25%) (Timmer *et al.*, 1987). In Venezuela, *X. c.* pv. *phaseoli* grew epiphytically during the rainy period, but not during the dry period, on the leaves of several non-host crop and weed species (Trujillo, 1989).

Little is known of the effect of **nutrition**. Early in the season, epiphytic populations of *X. c.* pv. *vesicatoria* were inversely correlated with the foliar concentration of potassium; at harvest, however, populations were inversely related to magnesium content (McGuire *et al.*, 1991).

4.1.2 Epiphytic growth preceding infection

In several instances epiphytic populations of xanthomonads have been determined before disease symptoms appeared. Thus, populations of *X. c.* pv. *pelargonii* remained on the leaf surface for at least 2 months without producing symptoms (Digat, 1978a). Under experimental conditions, *ca.* 10^3 bacteria/cm^2 had to be present for at least 30 days at

20°C and high RH before leaf spots appeared. Also, *X. c.* pv. *vitians* can have a long epiphytic phase not necessarily resulting in disease. However, after prolonged rainy periods, symptoms appeared on mature plants (Zoina and Volpe, 1992). In Mexican wheat fields high populations of *X. c.* pv. *undulosa* per leaf were observed one month before disease appearance (Duveiller, 1992). Since the bacterial populations decreased after heavy rainfall in some cases, epiphytic colonization of wheat leaves was assumed. A presymptomatic multiplication was also observed for *X. c.* pv. *vesicatoria* on the surface of pepper leaves (Sharon *et al.*, 1982). Common blight of bean was always detected under field conditions after a population level of at least 5×10^6 cells per navy bean leaflet of *X. c.* pv. *phaseoli* was exceeded (Weller and Saettler, 1980).

However, in most of these cases it has never been proven that the observed prolonged epiphytic growth was necessary for disease initiation. Thus in the case of CBS, higher surface populations often preceded disease development (Gottwald *et al.*, 1992). Nevertheless, the authors concluded that surface bacteria did not necessarily contribute to the disease but may have existed as casual leaf surface inhabitants. Bacteria associated with CBS disease showed a short-lived epiphytic phase on leaf surfaces, and the populations were rather low ($\leq 10^3$ cfu/ml).

4.1.3 Role of epiphytic phase for survival

Epiphytic populations of xanthomonads which survive from previous seasons are potential inoculum sources. The bacteria can remain dormant during the quiescent period in association with perennial plant hosts or parts, seeds, and plant residues. Thus, epiphytic survival was reported for *X. c.* pv. *manihotis* (Persley, 1979), *X. populi* (Ridé *et al.*, 1979), *X. c.* pv. *juglandis* on buds, twigs, catkins and nuts (Esterio and Latorre, 1982), and *X. c.* pv. *malvacearum* in buds (Wrather *et al.*, 1986). *X. c.* pv. *vesicatoria* was found to overwinter on wheat roots (Diachun and Valleau, 1946). However, it is still questionable whether the rhizoplane bacteria represent a partial and possibly transitory extension of the phylloplane phase.

In most annual plants xanthomonads may survive non-pathogenically on or inside the testa of dry seeds. Thus, *X. c.* pv. *malvacearum* overwinters principally in cotton seed fuzz which remains on the seed coat after delinting. *X. o.* pv. *oryzae* overwinters on diseased grains as well as in rice straw and is normally found in husk tissues (Schuster and Coyne, 1974). *X. c.* pv. *phaseoli* can be isolated from 15-year-old bean seed (Schuster and Sayre, 1967).

Xanthomonads may also survive in plant debris in or on the soil as long as the host tissue is not decomposed. Thus, *X. c.* pv. *campestris* survived in cabbage stalks in soil for as long as 2 years (Schaad and White, 1974). Similar observations have been reported for *X. c.* pv. *malvacearum* in dry plant tissues, *X. c.* pv. *phaseoli* in infested straw, and *X. c.* pv. *vesicatoria* in infested debris in tomato fields (Schuster and Coyne, 1974).

Also, non-host plants, especially weeds, can be colonized by epiphytic populations of xanthomonads and, thus, serve as inoculum source for disease on susceptible hosts. Many examples for survival of xanthomonads on non-host plants have been reported. If inoculated non-host plants were held under conditions of high moisture for 48 h and then placed under adverse conditions, surface populations persisted for several weeks without causing disease symptoms (Timmer *et al.*, 1987). Populations of 10^3 to 10^5 cfu per leaf of the non-host hazelnut were determined for *X. c.* pv. *juglandis* (Pruvost and Gardan, 1988). *X. c.* pv. *phaseoli* placed onto 10 non-host plants could be recovered up to 30 days later during the rainy period (Trujillo, 1989). Nevertheless, Angeles-Ramos *et al.* (1991) concluded for the same pathogen that weeds may not be an important inoculum. Similarly, weeds appeared to be questionable sources of primary inoculum of *X. c.* pv. *vesicatoria* in Florida (Jones *et al.*, 1986). The highest chance for inoculum survival can be expected on botanically related wild plants, as has been reported for *X. o.* pv. *oryzae* (Gonzalez *et al.*, 1991) and *X. c.* pv. *campestris* (Schaad and Dianese, 1981).

4.2 PENETRATION

Xanthomonads lack active mechanisms for penetrating the protective barriers of plants. Therefore, the bacteria can enter the host plant only through wounds and through natural openings. The latter may be stomata, water pores of hydathodes, lenticels or nectaries. Most of the bacteria which are randomly dispersed over the leaf surface by water splash or aerosols soon disappear if they are not capable of epiphytic growth (see section 4.1) or gain entrance into the plant. Mechanisms which facilitate plant penetration are chemotaxis and aerotaxis, a fast multiplication in the substomatal cavity, or a passive ingress with the guttation fluid. Specific pathways of penetration exist for certain plant parts, such as seeds and fruits. Plant penetration is favoured by external conditions, especially availability of water, and by structure and status of plant openings.

4.2.1 Portals of entry

The **stomata** probably serve as the most common portal, since they are distributed universally over the leaves, and also occur on stems and fruits. In 1915 Rolfs reported that the upper surfaces of peach leaves resisted aerosols of *X. c.* pv. *pruni* because they lacked stomata. Many studies by light and scanning electron microscopy confirmed the importance of stomata as infection sites, e.g. for *X. c.* pvs *cajani* (Shekhawat and Patel, 1979), *manihotis* (Verdier *et al.*, 1990) or *pruni* (Miles and Daines, 1977).

Also, **hydathodes** appear to be easy entry ports for xanthomonads. Hydathodes usually occur on the marginal teeth or serrations of leaves or at the leaf tips of *Brassica* and plants belonging to the Gramineae, Crassulaceae and Saxifragaceae. Penetration through hydathodes has been well documented for *X. c.* pv. *campestris* (Huang, 1986; Ruissen and Gielink, 1993). *X. o.* pv. *oryzae* generally enters the rice leaf tissue through hydathodes but not through stomata. Scanning electron microscopy revealed that, 24 h after inoculation, virulent bacteria had multiplied outside the water pores of the hydathodes, and some bacteria had gained entrance through these pores (Mew *et al.*, 1984). The bacteria then multiplied in the epithem and invaded the vessels through the so-called vascular pass (Tabei, 1967).

In many cases, xanthomonads infect plants successfully through **wounds**. Wounds can be caused by breakage of leaf hairs or trichomes, pruning or mowing of plants, insects, leaf fall, or transplanting. Wallis *et al.* (1973) obtained typical disease symptoms of *X. c.* pv. *campestris* at the wounding sites made in tertiary veins. Electron microscopy revealed that upon entering the vessels, bacteria acted on the primary walls at the sites between spiral thickenings and then penetrated into adjacent vessels. Pre-inoculation wounding of pepper leaves by rubbing them with carborundum powder or spraying with diluted wax solvents markedly increased disease severity (Diab *et al.*, 1982a). Similarly, sandblasting of tomato leaves or breakage of leaf hairs by scraping resulted in thousands of lesions per plant caused by *X. c.* pv. *vesicatoria* (Vakili, 1967).

Several examples have been reported for transmission of xanthomonads by insect feeding injuries; for example, *X. c.* pvs *phaseoli* and *vignicola* were transmitted by five feeding insects, bacteria were isolated from fresh insect faeces, and several xanthomonads survived for up to 19 days on the bodies of live and dead insects (Kaiser and Vakili, 1978). Other reports describe the transmission of *X. c.* pv. *malvacearum* by a mirid (Martin *et al.*, 1988), *X. c.* pv. *citri* by leaf miners (Cook, 1988), *X. c.* pv. *manihotis* by a bug (Maraite and Meyer, 1975), and *X. c.* pv. *campestris* by a flea beetle (Shelton and Hunter, 1985).

The main means of transmission of *X. c.* pv. *graminis* in fodder grasses is by mowing tools (Schmidt, 1989). Similarly, pruning transmitted *X. o.* pv. *oryzae* in rice (Dutta and Rafey, 1982) and *X. c.* pv. *pruni* in plum trees (Goodman and Hattingh, 1988). Also, wounds caused by transplanting or leaf fall increased susceptibility to *X. o.* pv. *oryzae* and *X. c.* pv. *pruni*, respectively (Zaragoza and Mew, 1979; Gasperini *et al.*, 1984).

4.2.2 Mechanisms

From the natural openings of plants, especially the hydathodes, water and nutrients may exude onto the plant surface. Motile bacterial cells with **chemotactic** response are attracted to these sites. Feng and Kuo (1975) reported that cells of *X. o.* pv. *oryzae* were attracted to exudates from susceptible rice plants. From non-motile isolates of several *X. campestris* pathovars, strongly chemotactic cells evolved during cultivation in bacteriological medium (Kamoun and Kado, 1990; see section 4.6.9), which reverted to non-motile cells in radish leaves. It can be assumed that chemotaxis plays a role only during the transitional infection phase in the life cycle of xanthomonads.

Whether **aerotaxis** helps xanthomonads to find stomata is unknown. However, motile and aerotactic strains of *Pseudomonas solanacearum* were avirulent (Kelman and Hruschka, 1973).

After the bacteria have reached the **substomatal cavity** actively or passively they multiply rapidly before infection is visible to the unaided eye (Huang, 1986). This initial proliferation appears necessary for penetration into the leaf tissue, so that bacterial masses emerging from the substomatal cavities serve as inocula for secondary infection.

A specific process of penetration occurs in plants with hydathodes. **Guttation water** droplets are secreted by the hydathodes in the morning hours and may be sucked back into the leaf as the stomata open and transpiration accelerates. Bacteria suspended in the water droplets can thus penetrate into the leaf. Host plants with reduced ability to guttate should therefore be less susceptible to bacterial diseases. Indeed, a positive correlation between guttation of cabbage cultivars and lesion development caused by *X. c.* pv. *campestris* was reported (Ruissen and Gielink, 1993).

The capability for **ice nucleation** has been demonstrated in many strains of *X. c.* pv. *translucens* and some not completely characterized strains of *X. campestris* (see section 4.6.7). The wounds caused by frost damage due to these strains may facilitate penetration and infection (Kim *et al.*, 1987; Azad and Schaad, 1988).

4.2.3 Infection of different plant parts

The penetration and infection of leaves and stems through stomata, hydathodes and wounds is described in sections 4.2.1 and 4.2.2.

Seeds may be infected by xanthomonads (e.g. *X. c.* pvs *campestris*, *incanae* and *vignicola*) through the vascular elements. In bean, cowpea, cotton, cucurbits and tomato the vascular tissue (raphe), a continuation of the funiculus, provides favourable sites for internal transmission of bacteria. The pathogens are retained in the raphe and may also invade the embryo (Schuster and Coyne, 1974; Shekhawat *et al.*, 1977). Also, rice seed infection by *X. o.* pv. *oryzae* occurs via the vascular system. Bacterial streaming was detected in the vascular strands at the panicle base and in seed pedicles (Durgapal *et al.*, 1980). Alternatively, xanthomonads may penetrate into seeds through openings, such as hilum, micropyle and threshing injuries, e.g. *X. c.* pvs *phaseoli* (Schuster and Coyne, 1974) and *vignicola* (Shekhawat *et al.*, 1977). In tomato seeds the natural openings through seed coats into the space between the testa and endosperm cuticle are penetrated easily (Schuster and Coyne, 1974).

In pepper **fruits**, possible sites for penetration by *X. c.* pv. *vesicatoria* were through dead flowers and by proliferation of bacteria in the wart (small protuberance) area (Bashan and Okon, 1986).

Dormant **buds** may be infected by entry of bacteria through leaf traces, as was assumed by Goodman (1982) for *X. c.* pvs *pruni* and *juglandis*.

Infection of **roots** has been reported for *X. o.* pv. *oryzae* (Zaragoza and Mew, 1979) and *X. c.* pv. *graminis* (Leyns *et al.*, 1988). Wounding increased disease symptoms, but unwounded roots were also infected. In the latter case it is unknown whether the bacteria penetrated into the roots via small, invisible wounds or other means.

4.2.4 Conditions which favour penetration

It has long been known that **moisture** is necessary for bacteria to penetrate. Thus, Rolfs (1915) reported that heavy dews and rain favoured stomatal penetration of peach leaves by *X. c.* pv. *pruni*. Experiments of Diab *et al.* (1982b) showed that high RH with free moisture on pepper leaves for long periods favoured infection by *X. c.* pv. *vesicatoria*. Even when inoculated plants were exposed to high RH (>85%) for a few hours, during 1 or 2 days, disease symptoms appeared. Short periods of unfavourable low RH after inoculation temporarily prevented disease development, but it continued later under high RH (see section 4.8.1).

In several cases **stomatal aperture** and **density** have been correlated with susceptibility to penetration by xanthomonads. Thus, a peach cultivar with high susceptibility towards *X. c.* pv. *pruni* was characterized by a wide stomatal aperture, promoted by nutritional factors, so that water congestion of the substomatal chamber and neighbouring intercellular space was favoured (Matthee and Daines, 1969). Similar differences in stomatal aperture have been reported in *X. c.* pv. *citri*-susceptible (*Citrus grandis*) and resistant (*Citrus nobilis*) cultivars (Huang, 1986). Also, tetraploid hyacinth cultivars with few, large stomata (0.245–0.258 mm in length) tended to be susceptible to *X. c.* pv. *hyacinthi*, whereas diploid cultivars with many small stomata (0.182–0.192 mm) were more resistant (van Tuyl, 1982). Ramos *et al.* (1992) found a correlation of the stomatal width as well as the stomatal frequency with the number of lesions caused by *X. c.* pv. *vesicatoria* on tomato leaves.

The **openings** in **hydathode** water pores of susceptible rice leaves were three times as wide as those in *Leersia japonica* with resistance to *X. o.* pv. *oryzae* (Horino, 1984).

Delay of **wound healing** can also favour infection by xanthomonads. For instance, infection of wounded citrus leaves by *X. c.* pv. *citri* was completely inhibited after the development of meristem tissue composed of thin dividing cells (Koizumi, 1983). The latter did not appear during incubation in a moist environment, so that the leaves remained susceptible until the cell walls in the outer tissues were lignified. Similarly, only leaf scars lacking complete suberization after premature leaf removal were susceptible to *X. c.* pv. *pruni* (Feliciano and Daines, 1970; Gasperini *et al.*, 1984).

4.3 INITIAL EVENTS IN PATHOGENESIS

4.3.1 Histopathology

The concept has been developed that the colonization of leaf surfaces by plant pathogenic bacteria is dependent on **recognition** mechanisms similar to those that have been shown to operate in the attachment of bacteria to other surfaces, such as those of epithelial cells in animals (Sequeira, 1980). Bacteria that cause disease in animals stick tenaciously and specifically to host surfaces, for instance on the intestinal mucosa, lung tissues or human teeth (Costerton *et al.*, 1978). In these cases adhesion is the first step in the progression of disease, mediated by the bacterial glycocalyx components. Although it is indicated that the xanthomonads colonize plant surfaces by adhering to certain components or structures (see section 4.1.1), the nature of this phenomenon is largely unknown. Cell wall constituents, such as extracellular polysaccharides and lipopolysaccharides, and pili, appear to have governing roles in these processes. Thus, it was shown by De Cleene (1989) that *X. o.* pvs *oryzae* and *oryzicola* attached to trichomes on maize leaves. Such mechanisms may increase and stabilize the resident population. However, once the bacteria have penetrated into the intercellular space, a specific attachment to inner surfaces of the mesophyll tissue obviously does not occur in the compatible interaction.

In contrast to mammalian pathology, phytopathogenic bacteria do not need to protect themselves against washing away by a continuous stream of water.

The first step after penetration into the intercellular space of the leaf is the development of microcolonies on the surfaces of the mesophyll tissue. In this stage, colonization of stomata and their surroundings occurs, as was depicted for *X. c.* pv. *manihotis* (Verdier *et al.*, 1990) or for *X. o.* pv. *oryzicola* with clusters of bacterial cells in the substomatal vesicle (Shekhawat and Srivastava, 1972). A little later high numbers of bacteria can be observed in the substomatal cavities and between palisade and spongy parenchyma cells (e.g. *X. c.* pv. *malvacearum*; Cason *et al.*, 1977). The first microcolonies of *X. c.* pv. *glycines* were observed by Jones and Fett (1985) 26 h post inoculation bridging mesophyll cells, usually at host cell junctions (Figure 4.1). The bacteria are now surrounded by fibrillar material, probably bacterial extracellular

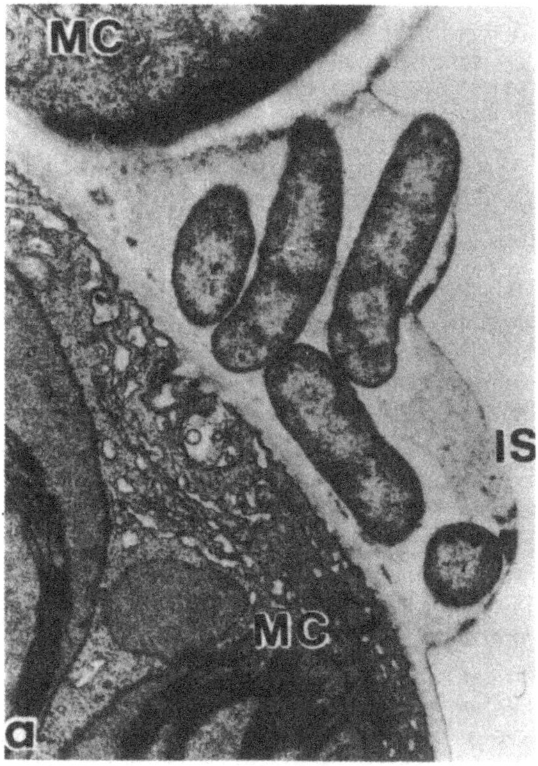

Figure 4.1 Microcolony of *X. c.* pv. *glycines* in the intercellular space (IS) of a susceptible soybean 26 h after inoculation at the junction of two mesophyll cells (MC) (from Jones and Fett, 1985).

polysaccharide (EPS) (see section 4.6.5). By 48 h, the bacteria may be surrounded by abundant EPS that fills the intercellular space. Staining with Ruthenium Red often reveals filamentous particles surrounding bacterial cells in the intercellular leaf space (Figure 4.2). The structures appear to be of bacterial origin by comparison with bacterial-free EPS preparations from *in vitro* cultures of *X. c.* pv. *pelargonii* (Figure 4.3; Rudolph *et al.*, 1987). In later stages the bacteria multiply rapidly in the intercellular spaces, often until the entire parenchyma between two adjacent veins is completely invaded, giving rise to the typical watersoaked angular leaf spots (see section 4.5.1).

The next step in the spread of the pathogen involves, in many cases, entry into the vascular tissue. It is not exactly known how the bacteria enter the vascular tissue from the surrounding parenchyma. However, not in every case are the bacteria able to infect the xylem. Thus, Cason *et al.* (1977) found bacterial cells of *X. c.* pv. *malvacearum* densely distributed through the intercellular spaces of the palisade and spongy parenchyma cells of cotton cotyledons, but no bacteria were detected in the xylem. A typical xylem invader is *X. c.* pv. *pelargonii*. Initially, the

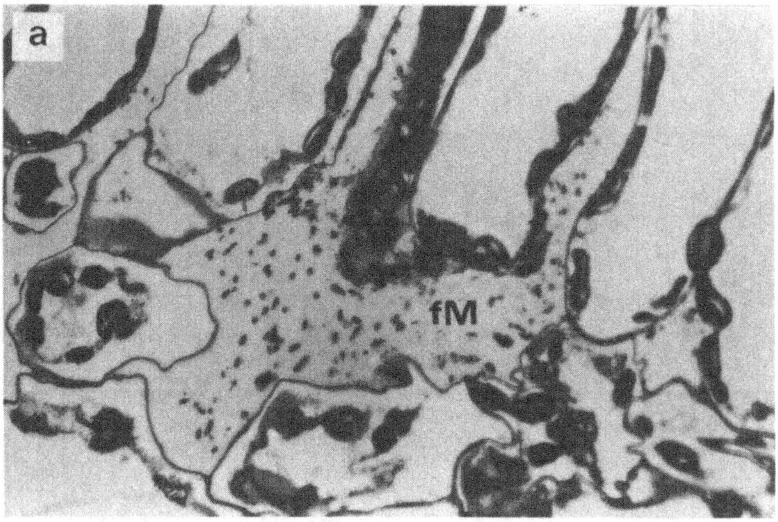

Figure 4.2 Cells of *X. c.* pv. *malvacearum* filling the intercellular spaces of cotton: (a) light microscopy reveals bacterial cells embedded in a Methylene Blue positive material, 24 h post inoculation of cotyledons; (b) electron microscopy and staining with Ruthenium Red show masses of fibrillar material (fM) around bacterial cells; (c) intercellular space of cotton leaf mesophyll filled with fibrillar material (F) 6 days post inoculation, bacterial cells (B) are surrounded by capsules, C = chloroplast, Cw = cell wall. (a) and (b) from Zachowski (1989), (c) from Al-Mousawi *et al.* (1982a).

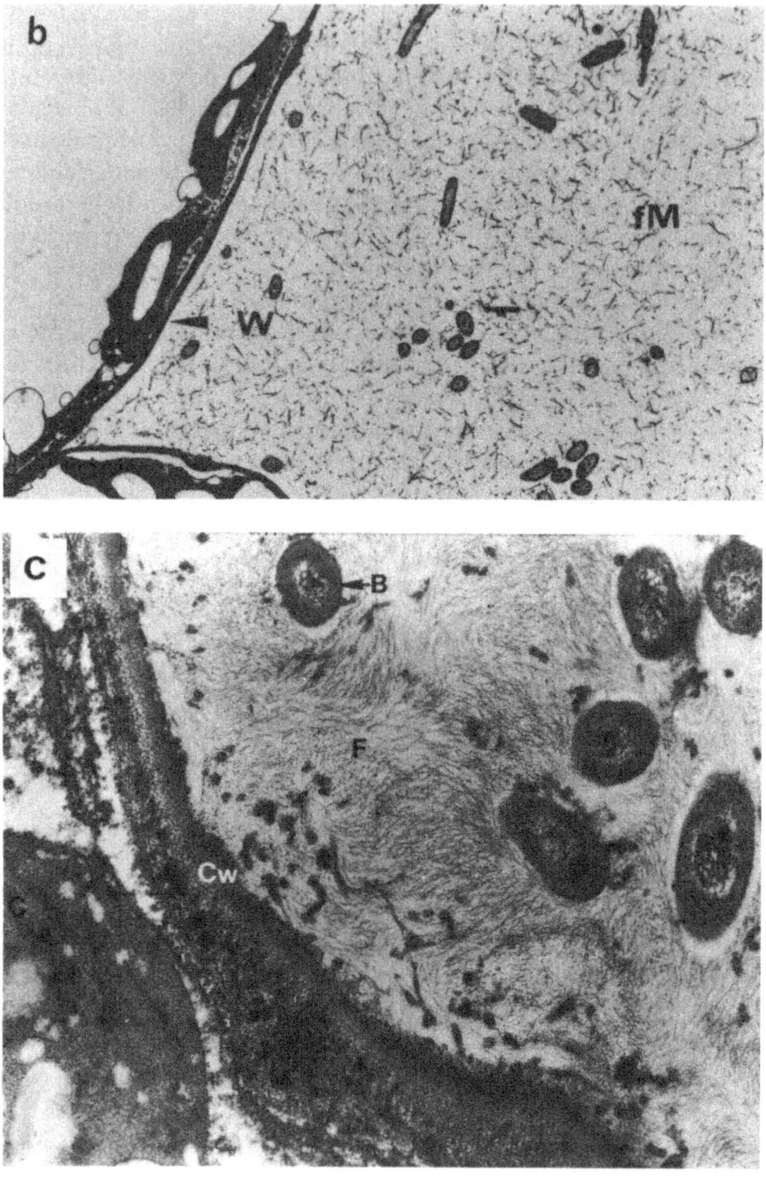

Figure 4.2 *Continued*

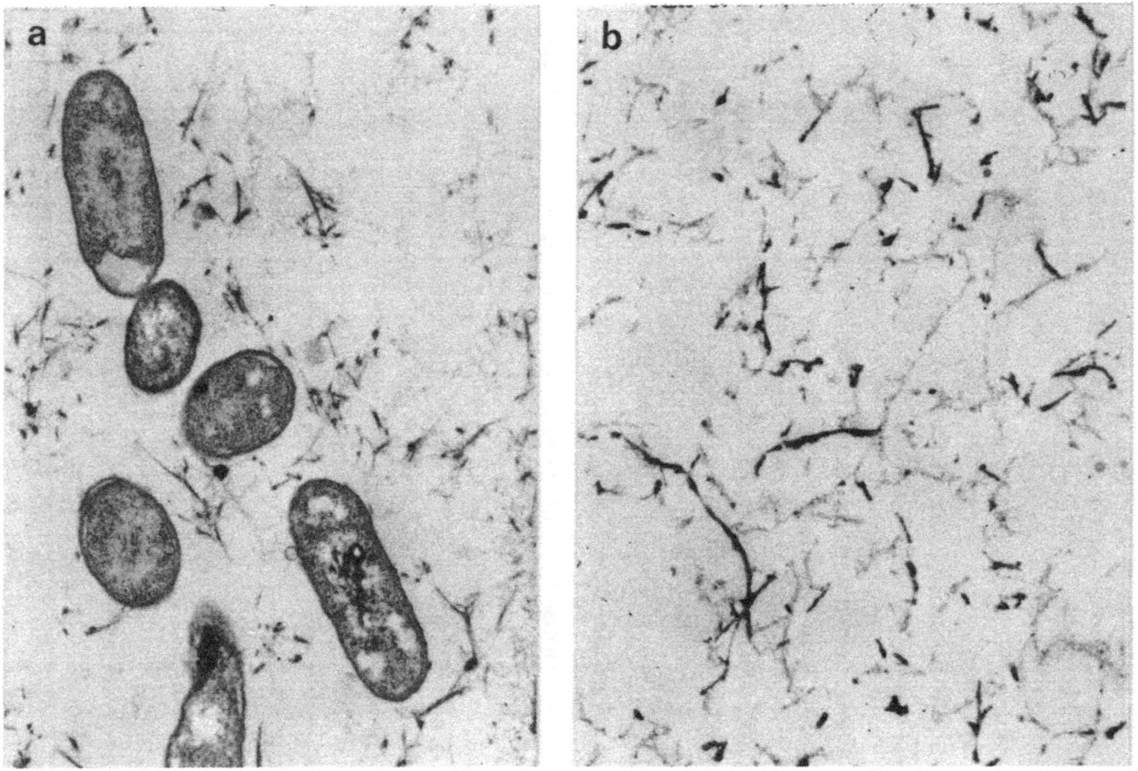

Figure 4.3 Filamentous particles of bacterial slime stained by Ruthenium Red surround cells of *X. c.* pv. *pelargonii* in the intercellular space of pelargonium leaves 3 days post inoculation (a). Similar structures were visible when bacteria-free slime from *in vitro* cultures was mixed with agar (b) (from Rudolph *et al.*, 1987).

bacteria move through the plant in the xylem vessel elements and subsequently move laterally into adjoining parenchyma cells. Bacterial pockets may be formed around affected protoxylem vessel elements, enlarging to encompass all xylem cells in the fascile, and finally portions of the cambium, phloem, cortex and epidermis (Wainwright and Nelson, 1972). It appears unlikely that the bacteria move of their own volition in plants (see section 4.4.2). Instead bacterial cells may be moved by the normal processes occurring in vascular tissue.

Accompanied with bacterial multiplication *in planta* is a gradual **degeneration** of the plant tissue. However, ultrastructural changes of the susceptible host tissue appear rather late. Thus, Morgham *et al.* (1988) observed that in cotton leaves it was at least 3 days after inoculation by *X. c.* pv. *malvacearum* before some chloroplast thylakoid lamellae were separated, vesiculation occurred between the cell wall and plasma membrane and some cells had electron dense cytoplasm.

A typical sequence of degenerative changes of plant tissue has been described by Al-Mousawi *et al.* (1982a) for susceptible cotton **leaves** inoculated by *X. c.* pv. *malvacearum*:

> Early changes included formation of vesicles between the plasmalemma and cell wall, followed by disappearance of the granal and stromal membranes of the chloroplasts. Degeneration of mitochondria occurred after ultrastructural alteration in the chloroplasts. The last cellular organelles to undergo structural degeneration were the nucleus and plasmalemma. Fibrillar material was present at the external cell surfaces near bacteria at 5 days post inoculation. By day 6 many cell walls were broken and fragmented. Bacteria had entered the mesophyll cells, and specific organelles could no longer be distinguished. Large accumulations of electron-dense fibrillar material were present in the intercellular spaces at 6 days post inoculation, when water-soaking symptoms were apparent.

A similar process of tissue degeneration was also described for rice leaves infected by *X. o.* pv. *oryzicola* (Shekhawat and Srivastava, 1972) and for tomato leaves infected by *X. c.* pv. *vesicatoria* (Hseu *et al.*, 1990). Wallis *et al.* (1973) studied the ultrastructure of cabbage leaf **vessels** infected by *X. c.* pv. *campestris* and observed a complete dissolution of the primary wall which enabled the bacteria to pass into adjacent vessels. Separation of the spiral thickening from the primary wall was accompanied by a 'shredding' of the latter, a phenomenon not observed in healthy tissue. Such shredding resulted in swelling of the primary wall residue and release of masses of fibrillar material similar to that which surrounded the bacteria. Fibrillar material was not observed in electron micrographs of cultured *X. campestris* cells. In addition, a reticulate substance which appeared to be composed largely of host material, possibly a mixture of disorganized cellulose and lignin microfibrils, completely or partially plugged some of the vessels in infected veins.

The first symptoms of **water-soaking** are generally observed 3–4 days after inoculation, and the originally small translucent areas enlarge up to 6 or 7 days post inoculation when the maximum size is reached, as has been described in rye and barley leaves inoculated by *X. c.* pvs *secalis* and *hordei*, respectively (El-Banoby and Rudolph, 1989) or in cotton leaves by *X. c.* pv. *malvacearum* (Essenberg *et al.*, 1979).

An exceptional case is the **hypertrophy** of cells, e.g. in the case of soybean leaves infected by *X. c.* pv. *glycines*. The response is first observed and generally develops to its greatest extent in the spongy mesophyll tissue. The most dramatic change noted by Groth and Braun (1986) is the decrease in the amount of endoplasmic reticulum. Within the cytoplasm an increase in the relative volume of the hyaloplasm was accompanied by a decrease in the relative volume of plastids. Paracrystalline arrays of phytoferritin, which were rarely seen in healthy leaves, were quite common in the infected tissue. Jones and Fett (1985) observed formation of many lucent vesicles (possibly Golgi vesicles) within the cytoplasm and invagination of the host plasmalemma in apposition to the bacterial colony.

4.3.2 Physiological and biochemical reactions of the infected plant

One of the earliest changes in infected leaf tissue is an increased **membrane permeability**. In rice leaves infected with *X. o.* pv. *oryzae* an elevated membrane permeability was detected before symptom appearance (18–20 h post inoculation), and later on the difference was very striking from 4 to 18 days post inoculation (Rao and Nayudu, 1979; Figure 4.4). When *X. c.* pv. *citri* was injected into the mesophyll of citrus leaves, nutrients were released from host cells into intercellular spaces after 6–9 h (Goto *et al.*, 1979). Obviously, the permeability changes are essential to provide enough nutrients for bacterial multiplication in the intercellular space. These observations are in accordance with measurements of Novacky and Ullrich-Eberius (1982) of the transmembrane potential (E_m) in *X. c.* pv. *malvacearum*-infected cotton cotyledons. Diseased leaves were characterized by a rapid change of the energy-dependent portion of E_m, since a high E_m was maintained only in the light and not in the dark. Because the ATP level was higher in diseased than in healthy tissue and higher in the dark than in the light, it was concluded that the effect of the bacteria on ATP was at the plasmalemma, probably by uncoupling the H^+ pump from its energy supply, in the dark. These observations suggest a slow decline in the cell's 'viability' in diseased tissue (Novacky, 1980). Whether this effect

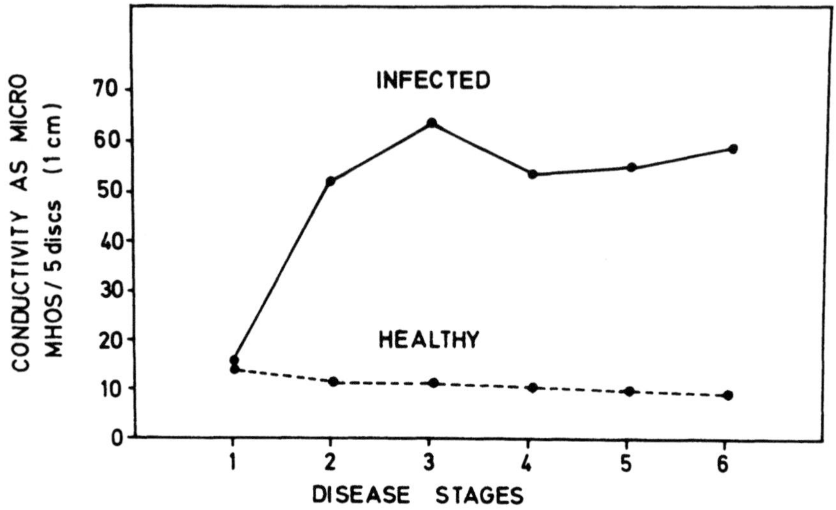

Figure 4.4 Membrane permeability increase in rice leaves infected by *X. o.* pv. *oryzae*, determined as conductivity of bathing solution after 2 h incubation (from Rao and Nayudu, 1979).

is caused by NH_4^+ formation (Novacky and Ullrich-Eberius, 1982), bacterial toxins (see section 4.6.3) or by free radicals (Adam *et al.*, 1989) remains to be clarified.

Similar to many other pathogens, xanthomonads cause an increase in **respiration**. Thus, Watanabe and Asaumi (1975) determined a steady increase in respiration of susceptible rice cultivars infected by *X. o.* pv. *oryzae*. Susceptible pepper fruits showed an increased respiration about 30 h after inoculation with *X. c.* pv. *vesicatoria* (Fischer, 1981). The elevated respiration coincided with a decrease in the ATP level. An increase in related enzymes (peroxidase, phenylalanine ammonia lyase, glutamic dehydrogenase, ascorbic acid oxidase) was reported in rice leaves infected by *X. o.* pv. *oryzae* (Rao and Nayudu, 1979). Phenol oxidation results in development of brown or black pigmentation in plant tissue. This **necrotization** is characteristic of many *Xanthomonas*-caused leaf spots. With progression of pathogenesis, plasmalemma, lysosomes and tonoplast become leaky, and phenols from the vacuole react with cytoplasmic and membrane-bound oxidases and glycosidases. Also, lignin synthesis may be accelerated by infection (Ridé, 1983).

Photosynthesis is primarily unaffected in diseased tissue, for instance in *X. c.* pv. *malvacearum*-infected cotton cotyledons (Novacky and Ullrich-Eberius, 1982). Concomitant with gradual degeneration of infected plant tissue, including the chloroplasts, photosynthetic activity decreases. However, the green regions of *X. o.* pv. *oryzae*-infected susceptible rice leaves had a higher photosynthetic activity than those from tolerant leaves (Philip and Devadath, 1981). Novacky and Ullrich-Eberius (1982) concluded that the photosynthesis apparatus of infected cotton cotyledons may be still intact and that the reduced photosynthesis in later stages is due to water-soaking caused by bacterial extracellular polysaccharides.

The **nitrogen metabolism** of *Xanthomonas*-infected plant tissue is characterized by an increased protein catabolism and NH_4^+ formation in later disease stages (Ullrich and Novacky, 1982). Also, qualitative changes have been reported. Thus, Bennett and Carter (1990) reported the appearance of new proteins in peach leaves infected by *X. c.* pv. *pruni* associated with the water-soaking symptom; and Zapata and Vidaver (1991) found changes in polypeptides during the early stages of pathogenesis in tepary and bush bean leaves after inoculation with *X. c.* pv. *phaseoli*. In bacterial blight (*X. o.* pv. *oryzae*) of infected rice leaves 10 new polypeptides were detected, two of which resembled PR (pathogenicity related) proteins (Kim and Yoo, 1992).

Sugar metabolism is also affected by the gradual degeneration of infected tissue. Misawa and Miyazaki (1972) analysed *X. o.* pv. *oryzae*-infected rice leaves and found that both reducing and non-reducing sugars and crude starch increased in the early stages of infection, but total carbohydrate and hemicellulose carbohydrate content showed the reverse trend.

4.4 MULTIPLICATION AND SPREAD IN THE PLANT

4.4.1 Multiplication *in planta*

After penetration through natural openings or wounds into the intercellular space, phytopathogenic xanthomonads begin to multiply. Even single bacterial cells may initiate infections, as was shown for *X. c.* pv. *malvacearum* (Essenberg *et al.*, 1979). As pointed out by Novacky and Ullrich-Eberius (1982), the intercellular space of higher plants contains all the essential **nutrients** (2–3 mM amino acids, 3–6 mM glucose and 0.1–500 mM sucrose) to enable bacteria to multiply in the first phase of the infection process. Cell walls may be another nutrient source, since most of the xanthomonads excrete cellulases (see section 4.6.2). Thus, *X. c.* pv. *campestris* causes cell wall dissolution early in the infection process (Wallis *et al.*, 1973) and so may be using these polysaccharides as carbon sources. In the intercellular spaces most xanthomonads develop gummy masses (due to secretion of the exopolysaccharide xanthan; see section 4.6.5) that enlarge and push aside and kill the adjacent cells (Figure 4.2a).

Bacterial multiplication in the intercellular space often leads to a

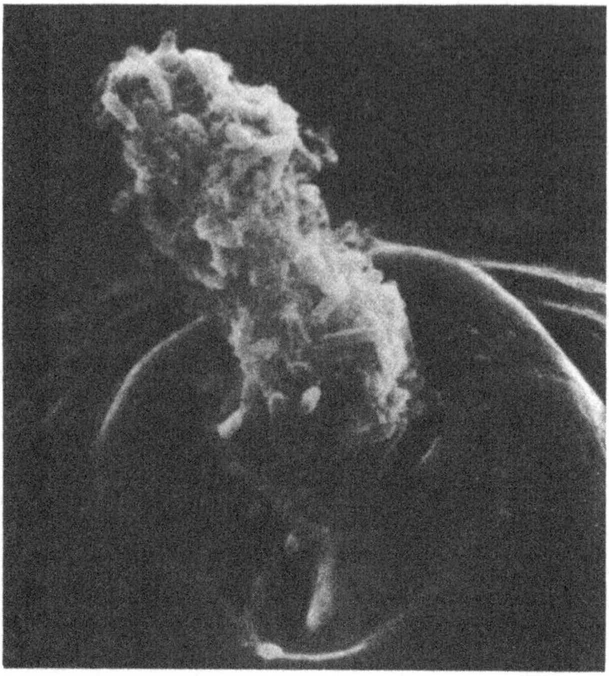

Figure 4.5 Bacterial mass of *X. c.* pv. *pruni* exuding from a peach leaf stoma 9 days after inoculation (from Miles *et al.*, 1977).

physical pressure so that the bacteria are extruded out of the tissue. Thus, it was shown by scanning electron microscopy that masses of cells of *X. c.* pv. *pruni* are pressed out of the stomata 6 days after inoculation, before visible disease symptoms appear (Miles *et al.*, 1977; Figure 4.5).

The **maximum bacterial population** which can be obtained in the plant tissue (Table 4.1) depends on the concentration of inoculum, the inoculation method, the degree of plant susceptibility and the nutrition and growth condition of the plant (Rudolph, 1990b; see section 4.8). Many workers determined bacterial growth in **leaf spot diseases** caused by xanthomonads (Table 4.1). Maximum bacterial concentrations have been estimated as 1×10^8 to 3×10^9 bacterial cells/cm^2 leaf (Figure 4.6 and Table 4.1) which is equal to a concentration of approximately 5×10^{10} to 3×10^{11} bacterial cells/ml. In some instances an initial lag phase of bacterial growth *in planta* was observed (Figure 4.6). During the phase of exponential growth (log phase) the generation (doubling) time is determined as the index for the speed of bacterial multiplication. Under optimal conditions *in vitro* generation times

Table 4.1 Maximum populations and generation times of xanthomonads *in planta* and *in vitro*

X. campestris pv.	*Host*	*Maximum cfu/cm² leaf*	*Generation time (h) during log phase*	*Authors*
phaseoli var. *fuscans*	Bush bean leaf	10^9	6.1	Cinar and Rudolph (1972)
phaseoli	Bush bean leaf in the field	7×10^5	21–25	Cafati and Saettler (1980)
glycines	Soybean leaf	2×10^8	4.1–8.7	Hokawat and Rudolph (1988b)
malvacearum	Cotton leaf	2.5×10^8	5.6	Morgham *et al.* (1988)
malvacearum	Cotton cotyledon	5×10^9	6.5	Zachowski (1989)
secalis	Rye leaf	10^9	2.0–2.1	El-Banoby and Rudolph (1989)
translucens	Barley leaf	10^9	1.7–4.5	
campestris	*In vitro*	–	1.2–1.6	Mason (1935)
phaseoli	*In vitro*	–	2.3–2.7	
pruni	*In vitro*	–	1.5	Young *et al.* (1977)

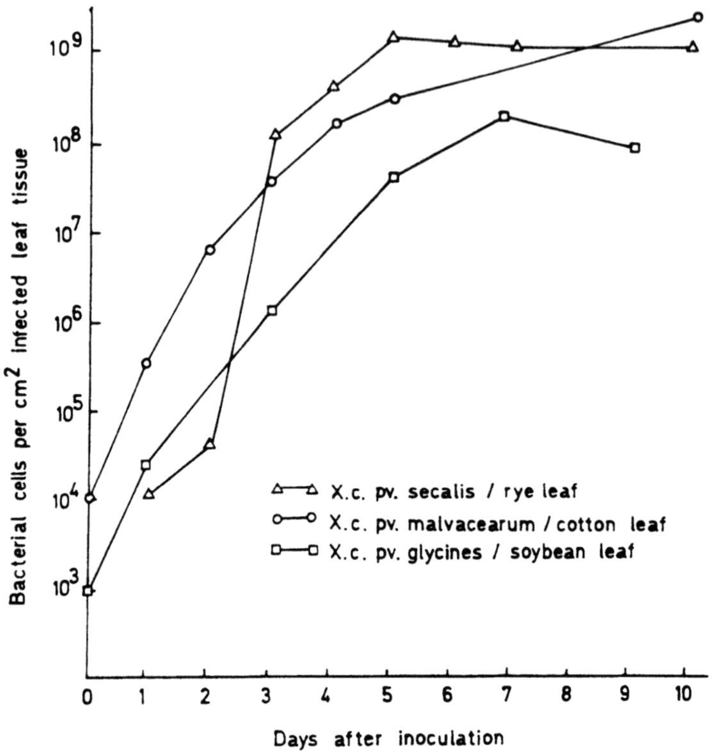

Figure 4.6 Multiplication of *Xanthomonas campestris* pathovars in host plant leaves: pv. *secalis* (from El-Banoby and Rudolph, 1989), pv. *malvacearum* (from Zachowski, 1989), and pv. *glycines* (from Hokawat and Rudolph, 1988b).

between 1.2 and 2.7 h have been determined (Table 4.1). Obviously, the growth *in planta* is slower in most of the cases. Especially in the field, the optimum conditions for growth will never be reached (e.g. cool temperature at night). Bacterial growth can also be characterized by infectivity titration, a procedure for measuring virulence of the pathogen and susceptibility of the host (Ercolani, 1984). Infectivity titrations have been described for *X. c.* pv. *glycines* (Groth and Braun, 1986).

4.4.2 Spread in the plant

Many xanthomonads enter the intact plant leaf through natural openings, mainly stomata and hydathodes, and multiply initially in the substomatal cavity (see section 4.3.1). From there the bacteria progress **intercellularly** in the parenchyma. Leaf veins often act as barriers against lateral spread of the pathogen, giving rise to the typical watersoaked angular leaf spots or streaks parallel to the veins in monocotyledonous plants. A typical example of the latter is leaf streak of rice caused by *X. o.* pv. *oryzicola*. Shekhawat and Srivastava (1972) showed that this bacterium multiplied in large masses in the air spaces near the mid-rib and also in the motor cells, causing loss of turgor of the latter. In advanced stages, the parenchymatous tissues were replaced by the bacterial mass, which oozed out through the stomata and was deposited on the leaf surface in various forms, viz. small compact beads with or without cirrhus-like stalks, large diffused masses of irregular shapes and cirrhi, depending upon the prevailing moisture conditions. Similar exudates appear on plant lesions caused by many different xanthomonads.

Several observations indicate that the xanthomonads do not spread in the plant by **active movement** with the help of flagella. Thus, Du Plessis (1984) never observed flagella on bacterial cells of *X. c.* pv. *pruni* in infected plum tissue. Also, Essenberg *et al.* (1979) concluded that motility is not important in spreading infection of *X. c.* pv. *malvacearum* within cotton leaves during the week following inoculation. Probably, free water in the intercellular spaces is not sufficient to permit bacterial motility.

In many cases xanthomonads spread in the plant by entry into the **vascular tissue**. Even those species and pathovars which cause primarily leaf spot diseases may infect the vascular tissue. Aggour *et al.* (1989) studied the transmission of *X. c.* pv. *phaseoli* in bean seeds. Inoculation of the pedicles of the flower buds and small pods resulted in transmission of the bacteria through the vascular tissue of the pod to the seeds, causing internal infection without any external symptoms shown either by the pods or seeds. Studies of Weller and Saettler (1980) showed that systemic movement of *X. c.* pv. *phaseoli* in bean plants contributed to infection of the leaf canopy, besides rain splash and bud colonization. The bacteria eventually colonized the entire root system after systemically moving downward in the stem. Typical xylem-invading bacteria are *X. albilineans*, the sugarcane leaf scald pathogen, or *X. c.* pv. *campestris* causing black rot of cruciferae. Wallis *et al.* (1973) showed by ultrastructural studies that *X. c.* pv. *campestris* proliferates intensely in the xylem vessels of cabbage leaves following inoculation of tertiary veins. Obviously, the bacteria spread both laterally and distally in the xylem vessels. The bacteria successively cause plasmolysis of adjacent xylem parenchyma cells, initiate primary wall dissolution between some contiguous vessels, and degrade the primary wall where it connects with secondary thickening, which causes detachment of secondary thickenings. The shredding of the primary wall releases masses of fibrillar material that surround the bacteria and appear to associate with bacterial capsules. Protuberances form on the weakened vessel walls, permitting bacteria to enter the intercellular space between adjacent parenchyma cells.

A rapid move of *X. c.* pv. *pelargonii*, the more so downwards, was determined by Preece (1982) in inoculated stem cuttings: 5 cm of the length of a 7 cm cutting contained the bacterium within 15 min. In less than one week bacterial numbers were high throughout the cuttings. Water-soaking symptoms were seen after 5 days (= 10^8 bacteria/g wet weight) and necrosis after 26 days, when the population was somewhat higher. The bacteria proceeded systemically throughout the cuttings, but the typical stem rot symptoms only occurred when the pathogen invaded the parenchymatous tissue. Symptom production was associated with peak numbers of bacteria in the pelargonium tissues.

A **systemic movement** in the vascular bundles was also reported for *X. c.* pv. *pruni*. Scanning electron microscopic studies of Du Plessis (1984) revealed that the main route of ingress of the bacterium into plum twigs occurs systemically through petioles of infected leaves. Summer cankers occurred only if the pathogen had been introduced into petioles and main veins within 5 weeks of shoot emergence (Du Plessis, 1987). Spring cankers during budbreak on trees were caused when leaves had been spray inoculated into the intercellular spaces, indicating a systemic movement of the pathogen from infected leaves. Also the invasion of plum seeds and fruits occurs by systemic movement of *X. c.* pv. *pruni* through stalks (Du Plessis, 1990). Similarly, *X. c.* pv. *malvacearum* enters seeds and survives as shiny masses in and around the micropyle (Preece, 1982) so that chemical control of the pathogen is difficult (Zachowski and Rudolph, 1991).

4.4.3 Latent infection

The colonization or survival of bacteria in plant tissue without symptom expression is defined as latent infection. A short latent period usually follows every bacterial inoculation into plant tissue, in the compatible as well as incompatible combination (Klement, 1982). In tolerant plants high levels of bacteria may be reached without causing marked disease symptoms, as was shown by Bozzano (1993) for certain bush bean cultivars after inoculation with *X. c.* pv. *phaseoli*. Latent infections can be dangerous in programmes to rid vegetative plant material of bacteria. Thus, latent infection of *Iris* with *X. campestris* during *in vitro* culture resulted in plants with typical leaf spots after they were weaned into the soil (Cooke *et al.*, 1993). Low numbers of bacteria are sometimes very difficult to detect and require very sensitive tests, such as immunofluorescence for *X. c.* pvs *pelargonii* and *begoniae* (Digat,

1978b). For the latter bacterium the periods of latency in *Begonia elatior* were determined as 10–14 days during the spring and summer and 20–40 days in the winter (Bech and Paludan, 1990). In infected seeds or other dry plant material xanthomonads can survive for long periods, probably because enclosure of many bacterial cells in dried extracellular polysaccharides prevents a complete desiccation of the bacterial cells. Thus, viable *X. c.* pv. *phaseoli* cells have been recovered from 30-year-old infected bean seed stored at room temperature (Saettler and Mount, 1982). Zvoutete *et al.* (1991) determined that the recovery of cells of *X. c.* pv. *holcicola* from the seed decreased gradually from 8.4×10^5 cfu at harvest to 4.4×10^3 cfu after 24 months in storage.

4.5 PATHOLOGICAL MANIFESTATION

The most common symptoms produced by xanthomonads are leaf, stem and fruit spots and necroses. But wilts (vascular infections), canker and die-back diseases, hypertrophy and rots (tissue maceration) can be also caused by xanthomonads.

4.5.1 Water-soaking

In the early stage of infection watersoaked spots develop on leaves, stems, fruits and pods, as the intercellular spaces become congested with bacteria embedded in slime with a high water-holding capacity. The watersoaked spots of leaves are translucent or greasy when observed in transmitted light. Because the watersoaked spots are often restricted by veins they appear as angular leaf spots on dicotyledons and as streaks or stripes on monocotyledonous plants. Watersoaked spots are typical of bacterial diseases and are rarely caused by fungi. The mechanism for induction of water-soaking is described in section 4.6.5. Typical watersoaked spots on cotton leaves are shown in Figure 1.2.

4.5.2 Bacterial exudates and slimes

In humid or wet weather bacterial slime is extruded from the lesions so that slightly yellow-coloured slimy droplets (called exudates or ooze) appear on the surface of leaves, stems and fruits. Exudates produced by *X. populi* on shoots and by *X. c.* pv. *undulosa* on leaves are shown in Figures 1.8 and 1.14. Ooze formation is characteristic of bacterial diseases and does not appear after infections by fungi or viruses. It should not be confused with gumming of certain trees (e.g. *Prunus* spp.). In later stages exudates may dry to glistering yellowish crusts. Bacterial slime is also a typical symptom in hyacinth bulbs infected by *X. c.* pv. *hyacinthi* (Figure 1.18). In gumming disease of sugarcane (*X. c.*

pv. *vasculorum*), yellow bacterial slime may form below the terminal bud, leading to the death of the top of the plant.

4.5.3 Spots and necroses

In most cases the original watersoaked spots turn into necrotic spots during later stages, probably due to continuous tissue disintegration (see section 4.6.2). These brown spots on leaves, stems, fruits, pods and capsules are the most conspicuous symptoms of many *Xanthomonas*-caused diseases. In some cases, the brown spots are surrounded by small watersoaked areas. In severe infections the localized spots can coalesce so that large areas of the plant surface are destroyed. The resulting disease is often referred to as blight. The increase of lesion radii was found to be a linear function of time, and rates in the greenhouse were approximately 1 mm/day following artificial inoculation of *Anthurium* with *X. c.* pv. *dieffenbachiae* (Yuen, 1989). Abscision layers may develop around the necrotic leaf spots, so that the necrotic zone finally falls out, giving rise to the shot-hole symptom (e.g. *X. c.* pv. *pruni*). On monocotyledonous leaves the spots appear as streaks or stripes (e.g. *X. c.* pvs *translucens*, *graminis* and *hordei*). The leaf blight symptoms of *X. o.* pv. *oryzae* on rice leaves are originally white and later become greyish due to the growth of saprophytic fungi. Some diseases are characterized by a dark-brown necrosis of certain plant parts, such as glumes (blotching caused by *X. c.* pv. *undulosa*), veins (*X. c.* pv. *campestris*), buds (*X. c.* pv. *corylina*) or bark (*X. populi*). Hyacinth leaves show black rim symptoms after infection by *X. c.* pv. *hyacinthi* (Figure 1.17), and *X. albilineans* causes dark-red necrosis of vascular bundles (Figure 1.13).

4.5.4 Chlorosis

The necrotic leaf spots are often surrounded by chlorotic haloes (Figure 1.7), which may be caused by bacterial toxins (see section 4.6.3). Also, a general leaf yellowing (*X. c.* pv. *campestris*) or yellowing and withering (*X. o.* pvs *oryzae* and *oryzicola*) occurs. The gumming disease of sugarcane (*X. c.* pv. *vasculorum*) is characterized by chlorotic leaf stripes which may become necrotic. The typical symptoms of leaf scald of sugarcane (*X. albilineans*) are narrow, white 'pencil-line' stripes on leaf blades and sheaths (Figure 1.12). The responsible toxin is described in section 4.6.3.

4.5.5 Defoliation

When lesions caused by *X. c.* pv. *pruni* on stone fruit leaves are close together, the surrounding tissue yellows and leaf fall frequently occurs.

Rapid leaf fall is also a consequence of bacterial infection of herbaceous plants, such as pepper infected by *X. c.* pv. *vesicatoria*, or cassava infected by *X. c.* pv. *manihotis*.

4.5.6 Plant collapse

Death of plants due to infection by xanthomonads seldom occurs, and if so during the susceptible seedling stage. Stem rotting of *Pelargonium* infected by *X. c.* pv. *pelargonii* results in wilting and plant death. The acute phase of leaf scald of sugarcane (*X. albilineans*) is characterized by the sudden death of affected cane.

Heavy nursery infection of rice plants by *X. o.* pv. *oryzae* ('Kresek') results in the complete death of entire young plants. Similarly, when the growing tips of *Zinnia* seedlings are affected by *X. c.* pv. *zinniae* the plants may collapse. A damping-off also occurs after infection of alfalfa seedlings by *X. c.* pv. *alfalfae*.

4.5.7 Wilting and stunting

Several xanthomonads cause wilting. Mainly due to rotting of the stems, wilting and stunting occur in *Begonia* (*X. c.* pv. *begoniae*) and *Pelargonium* (*X. c.* pv. *pelargonii*; Figure 1.15). Due to systemic invasion wilt is caused in cassava by *X. c.* pv. *manihotis*. The final wilting of bacterial blight (*X. o.* pv. *oryzae*)-infected rice plants ('Kresek') appears to be caused by mechanical occlusion of xylem vessels by bacterial cells and chemically active bacterial slimes (Kaul and Sharma, 1987; see section 4.6.5). A similar symptom is caused by the related *X. c.* pv. *graminis* in grasses.

4.5.8 Twig die back and cankers

In young shoots, infection of the shoot tip or via blossom truss usually results, under favourable conditions, in rapid colonization of parenchyma and collenchyma tissues. The tip then withers and dies as the infection progresses down the shoot. This symptom is referred to as **die back** and is caused by the *X. c.* pvs *pruni, corylina, manihotis* and others.

Die back of twigs and branches can also be caused by cankers. **Cankers** are sunken necrotic lesions on stems or branches arising from disintegration of tissues outside the xylem cylinder. They are sometimes limited in extent by host reactions which can result in more or less massive overgrowth of surrounding tissues. Young branches of *Populus* spp. infected by *X. populi* develop swollen areas that crack and exude bacterial ooze in spring. These may develop into large cankers over several seasons or be infected by other organisms and cause death of the branches above (Figure 1.9). Cankering on branches and shoots is also caused by *X. c.* pvs *citri, pruni, mangiferaeindicae, juglandis* and *corylina*. The cankers may girdle the stem resulting in die back of twigs and branches (Figure 1.10). Cankering may also occur on herbaceous plants, e.g. elongated cracked lesions on tomato stems caused by *X. c.* pv. *vesicatoria* or cankers on *Vigna* spp. caused by *X. c.* pv. *vignicola*.

4.5.9 Hypertrophy and hyperplasia

Very few xanthomonads induce hypertrophic growth in the host plant (see section 4.6.6). The most conspicuous example is bacterial pustule disease of soybean caused by *X. c.* pv. *glycines* (Figures 1.6 and 4.7). According to Jones and Fett (1987) softening of the cell wall and increased wall synthesis may be involved in the enlargement of cells composing the pustules. Lesions of Asiatic citrus canker caused by *X. c.* pv. *citri* consist of hyperplastic mesophyll tissue that erupts through the epidermis (Lawson *et al.*, 1989).

4.5.10 Rotting

Although most xanthomonads produce cellulolytic enzymes, and many strains secrete pectinolytic enzymes (see section 4.6.2), typical soft rots are caused very rarely. Stem rotting occurs after infection by *X. c.* pvs *pelargonii, begoniae* or *malvacearum*. *X. c.* pv. *hyacinthi* causes bulb rotting and *X. c.* pv. *vasculorum* top rotting of sugarcane. A soft rotting of vegetables in storage by incompletely identified *Xanthomonas* species was also reported (Liao and Wells, 1987).

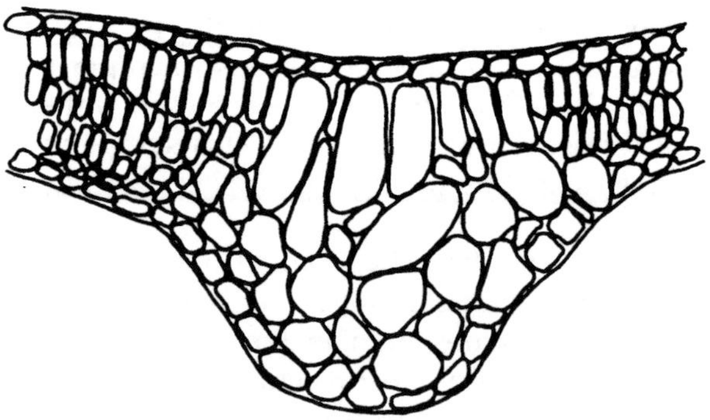

Figure 4.7 Transverse section through young soybean leaf pustule caused by *X. c.* pv. *glycines* showing hypertrophic cells (drawing by Mrs Renate Wolf).

4.6 BACTERIAL DETERMINANTS FOR VIRULENCE AND/OR PATHOGENICITY

4.6.1 Membrane-active substances

All *Xanthomonas* species and pathovars which induce local lesions on their host plants have the capacity to induce the **hypersensitive response (HR)** on non-host plants (Klement, 1982). The HR is characterized by a fast collapse of plant tissue (within 7–18 h) after infiltration of living bacteria in high concentrations ($>10^7$ cells/ml). Before the tissue collapse becomes visible the efflux of ions is markedly increased, indicating damage to the plant membranes. The HR-inducing principle has not been elucidated completely. However, it was suggested that oxygen radicals play a significant role during the HR (Adam *et al.*, 1989; Baker, 1992). The capability to induce the HR is predicative of pathogenicity (Stead, 1990), and it has therefore been used for many years as a diagnostic criterion. HR-negative *Xanthomonas*-like bacteria are usually not phytopathogenic. It can be concluded, therefore, that the capability to induce the HR is needed by the xanthomonads during pathogenesis for induction of cell leakage and release of nutrients in susceptible host plants (see section 4.3.2 and Figure 4.4). It was pointed out earlier by Klement (1982) that during the development of the susceptible as well as the resistant reaction after bacterial inoculation almost identical physiological and biochemical changes occur in the plant, except that they develop more slowly in the compatible interaction.

Therefore, it was not surprising that genetical analysis revealed that the same gene cluster was responsible for pathogenicity on the susceptible plant and for the HR on the resistant plant (Bonas *et al.*, 1991). Schulte *et al.* (1991) reported that DNA sequences homologous to the *hrp* region of *X. c.* pv. *vesicatoria* occurred in the pathovars *alfalfae, armoraciae, begoniae, campestris, citri, dieffenbachiae, glycines, malvacearum, phaseoli, translucens* and *vignicola*, and *X. oryzae*. DNA of a non-pathogenic strain of *X. campestris*, T55, did not hybridize at all. These results indicate a general occurrence of the *hrp* gene cluster in leaf spot-causing xanthomonads. Also, Kamoun *et al.* (1991) concluded that the *hrpXc* gene of *X. c.* pv. *campestris* is a basic pathogenicity gene required for the survival and growth of the bacterium in plant tissue. Transcription of *hrp* genes is not regulated by genes that control the synthesis of extracellular enzymes, which are also required for pathogenicity (Arlat *et al.*, 1991). A common core of *hrp* genes among major representatives of the Gram-negative phytopathogenic bacteria, including *X. campestris*, was characterized by Gough *et al.* (1993). Some of these genes code for proteins as described for the human and animal pathogen

Yersinia, which are components of a machinery secreting pathogenicity-determining proteins.

The so-called **avirulence (*avr*) genes** are responsible for the hypersensitive response of resistant plant cultivars. There is some evidence that the products of some *avr* genes may function as enzymes in secondary metabolic pathways, so that toxin-like substances are produced (Gabriel, 1989). The *avr* genes alone are not sufficient to induce the HR but require the presence of *hrp* genes, as was shown for the avirulence gene *avrBs3* from *X. c.* pv. *vesicatoria* (Bonas *et al.*, 1989) and many others. Another avirulence gene, *avrBs2*, which occurs in all strains of *X. c.* pv. *vesicatoria*, obviously plays a role in virulence (Kearny and Staskawicz, 1990; Bent *et al.*, 1991). Mutants without *avrBs2* were less fit and did not grow to as high a level as the wild-type strain in susceptible pepper plants. Also, De Feyter and Gabriel (1991) reported that spontaneous mutants of *X. c.* pv. *malvacearum* that had lost the avirulence gene *avrb6* had reduced water-soaking ability, suggesting a possible role of that gene in conditioning virulence of the pathogen on a susceptible host. Vice versa, it was shown that the pathogenicity gene *pthA* of *X. c.* pv. *citri* functioned as an avirulence gene in *X. c.* pvs *phaseoli* and *malvacearum* (Swarup *et al.*, 1992). Therefore, the earlier assumption that *avr* genes of phytopathogenic bacteria are entirely dispensable, and without any apparent effect on microbial parasitic capacity (Gabriel, 1989) is not generally valid. The *avr* genes appear to be superimposed on the basic ability to parasitize. The *avr* gene products may induce an increased ionic leakage resulting in enhanced availability of nutrients in the susceptible cultivar and a hypersensitive cell collapse in the resistant cultivar. The highly mutable character of *avr* genes may explain why some races of *Xanthomonas* pathovars are unstable. Thus, we often observed mutation of race 7 of *X. c.* pv. *malvacearum* into the more virulent race 10 (Zachowski *et al.*, 1989).

4.6.2 Enzymes

Several hydrolytic enzymes are secreted by xanthomonads. Often they are cellulolytic, pectolytic, proteolytic, lipolytic, and/or amylolytic. The excretion of enzymes is required for pathogenicity as was shown by Dow *et al.* (1989a). The role of specific enzymes during pathogenesis has not been elucidated completely. Histopathological studies (see section 4.3.1) indicated that plant tissue is degraded during colonization by xanthomonads. Especially, cell wall degrading enzymes have often been described for xanthomonads.

Cellulose and **hemicellulose** degradation can be concluded from micrographs of Wallis *et al.* (1973) that showed separation of spiral thickening from the primary wall of cabbage by *X. c.* pv. *campestris*.

Shredding occurred initially on the exposed wall layer and resulted in swelling and subsequent release of masses of fibrillar material. Goto and Okabe (1959, 1960) examined the degenerative activity of *X. c.* pvs *campestris* and *citri* by measuring the quantity and quality of cellulose in healthy and diseased leaf tissue. They found a marked decrease in total cellulose in the diseased tissue and were also able to distinguish decreases in an alpha-fraction of cellulose in the parasitized tissue and a concomitant increase in beta- and gamma-cellulose fractions. Most, if not all, pathovars of *X. campestris* secrete cellulase. Thus Prasad (1987) reported the exogenous extrusion of pronounced proportions of CMC (carboxymethylcellulose)-adaptive CM (carboxymethyl)-cellulase by the *X. c.* pvs *malvacearum, vesicatoria, campestris* and *vignicola,* and a constitutive production in pv. *phaseoli.* When we determined cellulase activity of 16 different pathovars of *X. campestris,* 14 pathovars showed cellulose degradation, and within the pv. *glycines* all the virulent strains tested produced considerable amounts of cellulase (Hokawat, 1988). A mutant with strongly reduced virulence constructed by reverse genetics from *X. c.* pv. *campestris* (De Crecy-Lagard *et al.,* 1990) was characterized by a very low activity of cellulase.

Pectolytic enzymes are less common within the xanthomonads than cellulases. Thus 10 out of 19 pathovars of *X. campestris* liquefied a nutrient pectate gel in studies of Starr and Nasuno (1967). Seven pathovars excreted polygalacturonate transeliminase which was produced inducible on pectin as was pectinesterase by most of the cultures. Polygalacturonate was degraded in a random manner, the major end-products being unsaturated di- and tri-galacturonic acids, accompanied by lesser amounts of saturated mono-, di- and tri-galacturonic acids. *X. c.* pv. *glycines* produced several different pectolytic enzymes in culture, and it was concluded that endopectin transeliminase, although present in low activity, is involved in pathogenesis (Hokawat, 1988). Also, in *in vitro* cultures of *X. c.* pv. *vignicola* several pectolytic enzymes were secreted, from which polygalacturonate transeliminase was synthesized constitutively, and endopolygalacturonase had preponderance over exopolygalacturonase (Prasad and Bhushan, 1987). Three major isozymes of polygalacturonate lyase were described for *X. c.* pv. *campestris* (Dow *et al.,* 1987). Molecular cloning experiments of Dow *et al.* (1989b) indicated that isozyme I of polygalacturonate lyase is not absolutely necessary for black rot pathogenesis. Beaulieu *et al.* (1991) characterized a pectate lyase from *X. c.* pv. *vesicatoria* by biochemical and genetic analysis. Only 22% of 522 strains were pectolytic. Pectate lyase negative mutants were neither altered in their ability to evoke disease symptoms, nor in their growth *in planta.* Thus, pectate lyase appeared to play no essential role in pathogenesis.

However, soft-rotting *Xanthomonas* strains have been isolated from rotted specimens of tomato, bell pepper, cucumber and papaya (Liao and Wells, 1987). These strains exclusively attack plant materials after harvest and cause only soft rot symptoms. Obviously, excretion of pectolytic enzymes, such as pectate lyase, pectin esterase and polygalacturonase, play a decisive role in pathogenesis of these strains. The pectate lyase of a soft-rotting *X. campestris* was characterized biochemically by Liao (1989).

Other cell wall degrading enzymes are **xylanases** that appear to play a causal role in pathogenesis of *X. c.* pv. *alfalfae* (Reddy *et al.*, 1974), and **β-mannases** that are present in *X. campestris* cultures as cell wall or membrane-bound constituents, but also in the extracellular medium (Dekker and Candy, 1979).

Proteinase activity has been detected in several xanthomonads in *in vitro* cultures as well as *in planta*, e.g. in the *X. c.* pvs *zinniae* (Sun and Ritchie, 1990), *alfalfae* (Reddy *et al.*, 1971) and *phaseoli* (Srivastava and Prasad, 1989). In several cases it was shown that protease-deficient mutants were reduced in pathogenicity and multiplication *in planta*, such as for the *X. c.* pvs *malvacearum* (Gholson *et al.*, 1989), *campestris* (Dow *et al.*, 1990) and *X. oryzae* (Xu and Gonzalez, 1989). Fujimoto *et al.* (1991) identified proteinase genes from the pathovar *phaseoli*.

Other enzymes include α-amylase, invertase and α-glucosidase which increased in tomato fruits infected by *X. c.* pv. *vesicatoria* and caused hydrolysis of starch, sucrose and maltose, leading to a change in the balance of reducing and non-reducing sugars (Kumar and Prasad, 1993).

Some enzymes appear to be pathogenicity related only for a restricted number of *Xanthomonas* pathovars. Examples are **amylase** and **glycoside hydrolases** which are determined for diagnosis of xanthomonads (Hayward, 1977; Bradbury, 1986).

4.6.3 Toxins

Several phytotoxic **carboxylic acids** are produced by different *X. campestris* pathovars, such as phenylacetic, 3-methylthio-propionic, *trans*-3-methylthioacrylic, isovaleric, tiglic, benzoic, *m*-hydroxybenzoic, succinic and fumaric acid by *X. oryzae* (Egawa *et al.*, 1968; Noda *et al.*, 1980) and *X. c.* pv. *campestris* (Robeson and Cook, 1985). The earlier report of Perreaux *et al.* (1982), that 3-methylthio-propionic acid may be the main toxin responsible for the necrotic symptoms produced by *X. c.* pv. *manihotis*, was recently revised by Ewbank (1992). The author characterized two bacterial enzymes which can synthesize several carboxylic acids from amino acids, such as isoleucine, leucine, phenylalanine and tryptophane, by a transamination to the corresponding keto acid followed by a decarboxylation. Since these carboxylic acids also possess phytotoxic properties, it has been suggested that the blight symptom induced by *X. c.* pv. *manihotis* may be caused by a cumulative toxic effect of several carboxylic acids (Ewbank and Maraite, 1990a,b; Ewbank, 1992).

The chlorosis in sugarcane leaf scald disease (*X. albilineans*) appears to be induced by an **albicidin**-like phytotoxin (Birch and Patil, 1987a,b). The authors suggest that the albicidin family of antimicrobial compounds preferentially inhibit plastid DNA replication, resulting in blocked chloroplast differentiation.

Other toxic constituents have not been chemically identified yet, such as a strongly acid polyalcohol of *X. c.* pv. *vasculorum*, detected *in vitro* as well as *in planta* (Ducourneau and Auriol, 1986a,b); a phytotoxin of *X. c.* pv. *malvacearum* possessing an aromatic ring and capable of causing typical blight symptoms (Pandian *et al.*, 1982); phytotoxic constituents from *X. c.* pvs *pelargonii* and *pruni* (Hammerschlag and Bottino, 1978) which affect the transmembrane electrical potential of mesophyll cells (Hammerschlag, 1984); or a glycopeptide toxin from *X. c.* pv. *betlicola* (Tripathi and Vyas, 1990). Tamura *et al.* (1988) suggested that *X. c.* pv. *phormiicola* might produce coronatine, a phytotoxin known from *Pseudomonas syringae* pvs *tomato* and *glycinea* (Rudolph, 1990a). However, studies of Mitchell (1991) revealed that this organism can only produce two analogues of coronatine lacking the cyclopropane ring structure, namely N-coronafacoyl-2-valine and N-coronafacoyl-L-isoleucine.

4.6.4 Lipopolysaccharides

The lipopolysaccharides (LPS) of xanthomonads, like LPS of other Gram-negative bacteria (Lüderitz *et al.*, 1982), consist of a complex polysaccharide portion, to which a lipid component, termed lipid A, is bound covalently. A comparison of lipid-free polysaccharides from xanthomonads by using HPLC of underivatized hydrolysates revealed that, contrary to earlier reports (Volk, 1968), the LPS contained heptose, together with rhamnose and galactose, but no mannose (Sutherland and Kennedy, 1986).

Lipid A represents the endotoxic centre of the LPS molecule. Biological studies on cell wall LPS from *Xanthomonas* species suggested that these LPS are analogous to the endotoxin, as judged by their lethality in mice and their ability to provoke the local Schwartzmann reaction in rabbits (Volk, 1966). Also, the purified LPS endotoxin of *X. sinensis* showed the same toxic effect in guinea pigs as preparations from non-phytopathogenic bacteria, although on a lower level (Helander *et al.*, 1980). It was concluded that fatty acid is an important functional component of the LPS toxophore. Comparable toxic effects of LPS or LPS moieties on plant tissue have never been reported. A long-known biological effect on plants is the induction of resistance by LPS preparations from pseudomonads (Mazzucchi *et al.*, 1979). The fatty acid composition of *Xanthomonas* LPS was analysed by Rietschel *et al.* (1975).

However, it is still unknown to what extent fatty acids of the LPS participate in the total fatty acids which are determined for diagnostic purposes. Analysis of fatty acids from 1400 *Xanthomonas* strains including all seven *Xanthomonas* species and over 130 *X. campestris* pathovars revealed 31 major fatty acid clusters and allowed rapid identification of unknown strains often at pathovar level (Yang *et al.*, 1992). Therefore, lipids appear to play a decisive role in pathovar specificity.

However, the specific composition of the O-specific LPS side-chain may be a determinant of specificity. Monoclonal antibodies specific for LPS antigens of *X. c.* pvs *begoniae* and *pelargonii* reacted with all of their respective pathovar strains and not with 130 strains of other xanthomonads or 89 non-xanthomonads tested (Benedict *et al.*, 1990); and Kamiunten and Fujita (1990) observed strain-specific patterns of LPS from *X. o.* pv. *oryzae* after SDS–PAGE electrophoresis. Since LPS is released under *in vitro* cultured conditions, the same process may occur *in planta*, and the release of LPS, probably complexed with outer membrane proteins or peptides (Dianese and Schaad, 1982), may fulfil a role in pathogenicity.

4.6.5 Extracellular polysaccharides

All phytopathogenic xanthomonads synthesize the heteropolysaccharide **xanthan**, the predominant component of the bacterial slime. Because of their high specific viscosity and other interesting rheological properties xanthan gums are produced in big fermentation plants for many uses in diverse applications (see Chapter 8). The typical xanthan consists essentially of a cellulose backbone, substituted on alternate residues with trisaccharide side-chains (mannose, glucuronic acid, mannose), so that the repeating unit is a pentasaccharide composed of D-glucose, D-mannose and D-glucuronic acid in the ratio of 2:2:1 and of varying amounts of pyruvic and acetic acid (Figures 8.1 and 8.2). The latter substitutes have a role in specific polymer–polymer or polymer–solid interactions. The molecular weight ranges between 1 and 9×10^6, and a double-helix molecular structure is assumed (Miles *et al.*, 1991). It was suggested that the xanthan molecule is less regular in structure than was at first proposed (Sutherland, 1985). A xylose moiety was found in the extracellular polysaccharides (EPS) of *X. c.* pvs *phaseoli* and *juglandis* and rhamnose in *X. c.* pv. *juglandis* (Lawson and Symes, 1977). D-Galactose and KDO were detected in the EPS of unidentified *Xanthomonas* species (Sutherland, 1985).

EPS from pseudomonads as well as xanthomonads are regarded as

determinants of disease due to several effects (Rudolph *et al.*, 1989). It was reported early that enclosure of bacterial cells (*X. c.* pv. *phaseoli*) in EPS resulted in long **survival**, increased the thermal death point and protected the bacteria against UV light (Leach *et al.*, 1957). Some cells survived the storage conditions for almost 10 years (Wilson *et al.*, 1965). Also the addition of 1 mg EPS/ml to cell suspensions of *X. c.* pv. *glycines* in sterile distilled water increased survival (Hokawat and Rudolph, 1988a). Similarly, xanthan gum as a dry powder formulation prolonged survival of plant growth-promoting rhizobacteria in both short- and long-term storage (Kloepper and Schroth, 1981; Suslow and Schroth, 1982).

Reports on **wilt**-induction by exudates from *X. c.* pv. *phaseoli* (Leach *et al.*, 1957) probably do not reflect the situation *in planta*. However, in the case of bacterial blight of rice the wilting induced by several partially purified EPS preparations (Kuo *et al.*, 1970) appears to play a role during pathogenesis. Vidhyasekaran *et al.* (1989) concluded that EPS increase cell membrane leakage so that both blockage and leakage result in wilting of rice plants.

A more general effect of the bacterial EPS may be the induction of **water-soaking** in the intercellular space, which is a prerequisite for bacterial colonization (El-Banoby and Rudolph, 1989). It was pointed out by Young (1974) that most bacteria grow in a range of relative humidities between 0.965 and 0.999. The relative humidity of the intercellular spaces of respiring leaf tissue is greater than 0.96, thus indicating that water may be available for bacterial metabolism. However, the cell walls of leaf parenchyma cells may be partially cutinized or suberized at their outer extremities. Thus, the exposed internal surfaces are hydrophobic to some degree (Häusermann, 1944) and may appear visibly dry rather than moist (Slatyer, 1967). Therefore, it seems obvious that phytopathogenic bacteria need a mechanism by which they can induce and maintain water congestion in the intercellular space. Thus, Verma *et al.* (1983) described how cells of *X. c.* pv. *malvacearum* secreted slime through pores in the surface layers in the form of a jet or a spray, creating a water-soaking of host tissue in advance, thereby helping the rapid spread and multiplication of the bacteria. EPS of a virulent, but not of an avirulent mutant, of *X. c.* pv. *malvacearum* produced persistent water-soaking in leaves of the sensitive cotton cultivars Acala 44 and 1–10B but not in leaves of the resistant cultivar 101–102B (Borkar and Verma, 1989). Water-soaking was also induced by infiltration in non-host plants, such as tobacco, castor and cowpea. Similar results were obtained by Zachowski (1989). Hokawat and Rudolph (1988) reported that EPS of *X. c.* pv. *glycines* added to the inoculum suspension promoted bacterial growth in soybean leaves.

Little is known concerning the **mechanism** of water-soaking induced by *Xanthomonas* EPS. Morris *et al.* (1977) reported from rheological studies that the ordered form of xanthan can bind co-operatively to certain plant cell wall polysaccharides in a way that suggests a role in the colonization of the plant host by the bacterial pathogen. Xanthan is unique in showing a marked preference for interaction with β-1, 4-glucose-containing polysaccharides (including derivatized cellulose) rather than mannan (Dea and Morris, 1977). Experiments of Takahashi and Doke (1984) indicated that cells of *X. c.* pv. *citri* adhered to host tissue through interaction between bacterial EPS and plant agglutinin. The observation that water-soaking is not induced in the resistant cultivar (Borkar and Verma, 1989) may be due to the different composition of plant polymeric substances in resistant cultivars, as was also described in other bacterial diseases (Wydra and Rudolph, 1992). Thus, Morris (1979) suggested that the solution behaviour of certain polysaccharide systems may be determined largely by transient but specific intermolecular interactions. A so-called 'fleeting' tertiary structure appears consistent with the blocking of specific binding sites by segments which are long enough to form one co-operative junction, but not two. Thus, a gel network formation may be competitively inhibited by short segments in the resistant reaction. On the other hand, a synergistic interaction between xanthan and plant polymers (Tako, 1991a,b) may initiate the susceptible reaction, so that the bacteria are embedded in a **gel-like slime** in the intercellular space (Rudolph *et al.*, 1987; see also Figure 4.2). It was suggested earlier (Rudolph *et al.*, 1989) that this gel promotes bacterial colonization of plant tissue by several means, such as:

(a) inhibition of a close morphological contact of the bacteria with plant cell walls so that resistant reactions are not triggered;
(b) inhibition of bacterial agglutination;
(c) retardation of desiccation of bacteria microcolonies in the normally air-filled intercellular spaces;
(d) protection of bacteria from bacteriostatic compounds; and
(e) decrease of oxygen tension by locally hindering gas exchange of the leaf mesophyll cells.

This assumed role in pathogenesis probably explains why the production of large amounts of slime (xanthan) is unique in the *X. campestris* pathovars in contrast to other miscellaneous yellow bacteria. Therefore, Pierce *et al.* (1990) designed a simple diagnostic test for preliminary identification of *X. campestris* strains based on the presence of viscid substance(s) in agar-grown bacterial cultures. Numerous publications which cannot be listed here report a correlation between virulence and amount of EPS produced *in vitro* when different strains of *X. campestris* were compared. Thus, in 1957 Corey and Starr described a series of colonial variants in *X. c.* pv. *phaseoli* with a parallel increase in quantity of polysaccharide slime produced and in virulence for bean in progressing through the series R–S–SM–M. Ramirez *et al.* (1988) exploited this correlation in selecting high-quality xanthan gum producers. Also, swarmer-type derivatives of *X. c.* pv. *campestris* which were EPS-deficient showed reduced virulence (Kamoun and Kado, 1990). Non-mucoid **mutants** of *X. o.* pv. *oryzae* showed attenuated virulence (Tseng *et al.*, 1983), but other pleiotropic effects were also observed. The correlation between decreased virulence and reduced EPS production was recently confirmed by genetic analysis of a host-specific virulence mutant of *X. c.* pv. *citrumelo* (Kingsley and Gabriel, 1991). A mixture with cell-free EPS of the wild type restored the pathogenic response of this mutant. A mutant of *X. c.* pv. *campestris* constructed by reverse genetics showed strongly reduced virulence and produced less xanthan which was characterized by a markedly reduced viscosity (DeCrecy-Lagard *et al.*, 1990).

4.6.6 Hormones and growth substances

In a few instances phytopathogenic xanthomonads induce hypertrophy or hyperplasia of plant tissue, the most conspicuous being soybean leaf pustules caused by *X. c.* pv. *glycines* (Figure 4.7). Other examples are *X. c.* pv. *citri* (Koizumi, 1977), *X. c.* pv. *phaseoli* on tepary bean and several other *Phaseolus* species (Çinar and Rudolph, 1972; Zapata *et al.*, 1988), bacterial pustule of cowpea caused by *X. c.* pv. *vignicola* in Africa (Kaiser and Ramos, 1979), or stem infection of pigeon pea by *X. c.* pv. *cajani* (Shekhawat and Patel, 1979).

Studies of Fett *et al.* (1987) revealed that not only *X. c.* pvs *glycines* and *phaseoli* but also eight other pathovars which do not induce hypertrophy of host tissue produce **indole acetic acid (IAA)**, as well as several additional indoles, from L-tryptophan. Pseudomonads and xanthomonads could be distinguished by the presence of N-acetyltryptophan, which was found only in xanthomonad culture filtrates. These authors suggested that IAA may favour bacterial multiplication *in planta* by promotion of solute uptake due to modification of the electrochemical proton gradient across the host plasmalemma. According to Zapata *et al.* (1988) IAA production alone by strains of *X. c.* pv. *phaseoli* is insufficient to induce callus. In addition, the production of an extracellular component with **cytokinin**-like activity appeared necessary.

The production of large quantities of **ethylene** can be induced in plant tissue after infection with xanthomonads (Lund, 1973), although the bacteria can also produce small amounts of ethylene, as was shown for *X. c.* pv. *vesicatoria* (Ben-David *et al.*, 1986). In citrus leaves infected with *X. c.* pv. *citri* the pathogen-induced ethylene biosynthesis appeared to be closely associated with disease progression. Dutta and Biggs (1991) suggested that ethylene production may be a plant defense response leading to abscission of leaves.

4.6.7 Ice nucleation

Many strains of *X. c.* pv. *translucens* are active in ice nucleation, i.e. have the ability to catalyse ice formation at temperatures above −10°C. Epiphytic populations of these bacteria on wheat plants may, therefore, enhance frost damage which can, in turn, increase disease severity (Azad and Schaad, 1988). The ice-nucleation gene *inaX* from *X. c.* pv. *translucens* was characterized by Zhao and Orse (1990). Its nucleotide sequence revealed a homology of 65.8–68.8% with ice-nucleation genes from other species. Ice-nucleation positive strains of *X. campestris* were also isolated from gemmisphere of tea and phyllosphere of vegetables (Goto *et al.*, 1988).

4.6.8 Pigments

The yellow xanthomonadins (brominated aryl polyenes) are produced by nearly all *Xanthomonas* strains and seemingly do not occur in any non-xanthomonad (Feistner, 1990). Several chemically closely related xanthomonadins usually occur as mixtures. They are found associated with cell membranes. Exceptions are the pigments of the pale-yellow *X. maltophilia* which are not xanthomonadins, and the white colony producers *X. populi*, *X. c.* pvs *ricinicola*, *manihotis* and *mangiferaeindicae*. These naturally occurring white strains, as well as pigment-minus mutants of *X. c.* pv. *campestris* which were virulent in cabbage seedlings (Kawalek *et al.*, 1989), indicate that the xanthomonadins do not play a decisive role during pathogenesis. A biological role may be protection against UV light, because in general *Xanthomonas*-caused diseases are more prevalent in warmer countries than diseases caused by *Pseudomonas* species.

X. c. pv. *phaseoli* var. *fuscans*, the incitant of fuscous blight of beans, produces an additional dark-brown pigment. It was found from several years' field tests and field isolations that fuscous blight is more virulent than common blight under the climatic conditions found in Germany. Similar results were obtained by Ekpo and Saettler (1976). However, it is as yet unknown whether these two variants differ in additional features which effect virulence.

4.6.9 Motility

It has been shown in several cases that flagella comprise one factor of virulence and that chemotaxis may assist phytopathogenic bacteria in locating infection courts. For instance, cells of *X. o.* pv. *oryzae* were more attracted to exudates from susceptible than from resistant rice plants (Feng and Kuo, 1975). Motility seems to play a role in penetrating the plant tissue but not in multiplication and spread *in planta*. Thus, Kamoun and Kado (1990) reported that nine virulent strains of *X. campestris* representing pvs *campestris, armoraciae, translucens, vesicatoria* and *pelargonii* were originally non-motile. During culture on swarmplates a switch occurred from the wild type (exopolysaccharide-producing and non-chemotactic) to the swarmer type (exopolysaccharide-deficient and chemotactic), which was irreversible *ex planta* in bacteriological medium. However, in radish leaves swarmer-type strains of *X. c.* pv. *campestris* were able to revert to the wild type.

Similarly, for *X. c.* pv. *malvacearum* motility is apparently not important in spreading infection within cotton leaves during the week following inoculation, since a motile strain did not produce significantly more watersoaked areas in susceptible leaves (Essenberg *et al.*, 1979). Also, Du Plessis (1984) reported that polar flagella were usually present on *X. c.* pv. *pruni* cells grown on aritificial media, whereas flagella were never observed on the specimens in plum tissue observed by scanning electron microscopy. The available free water in the intercellular spaces of plant tissue (Slatyer, 1967) is obviously not sufficient to permit bacterial motility.

4.6.10 General fitness

It seems obvious that only those bacterial strains which are capable of expressing all their diverse metabolic functions can cause disease. Therefore, apathogenic or low virulent mutants which are selected in genetic experiments are not necessarily deficient in specific virulence or pathogenicity genes. Also, **reduced growth rate** of mutants due to metabolic insufficiencies can result in non-appearance of disease symptoms (Klement, 1990b).

De Crecy-Lagard *et al.* (1990) characterized a DNA fragment from *X. c.* pv. *campestris* with an open reading frame that coded for a CAP (catabolite activation factor of *E. coli*)-like protein, named **CLP**. Genetic experiments indicated that CLP plays a role in the regulation of phytopathogenicity and that CLP effects synthesis of xanthan gum, pigment, and extracellular enzymes. On the other hand, Dzhalilov and Chernyaeva (1989) reported that the respiratory activity of *X. c.* pv.

malvacearum was 1.5–6 times higher in low virulent isolates compared with highly virulent isolates.

4.7 PLANT-RESISTANT REACTIONS TO XANTHOMONADS

The necessity for resistance breeding as one important measure to control diseases caused by xanthomonads is expanded in section 7.4, as well as strategies to obtain durable resistance. The phenomenon of resistance has been described in different terms by pathologists, geneticists, physiologists, epidemiologists, etc., relating to their own point of view (Federation of British Plant Pathologists, 1973). Therefore, these terms do not correspond precisely, e.g. vertical/horizontal, qualitative/quantitative, race-specific/race non-specific, monogenic/polygenic, major gene/minor gene resistance.

Hopefully, a better understanding of these different types of resistance will result when we know more about the decisive plant reactions during the resistant response. At present, seven reaction types effecting resistance against xanthomonads can be differentiated.

4.7.1 General activation of metabolism

During the incompatible interaction with many plant pathogens, including the xanthomonads, a general activation of metabolism occurs. Especially, an increased respiration has been reported. Thus, Fischer (1981) compared respiratory rates of resistant and susceptible pepper fruit tissue inoculated with *X. c.* pv. *vesicatoria* (10^7 cells/ml). An immediate increase in oxygen consumption was detected in resistant tissue, whereas susceptible tissue did not reflect an increase in respiration until about 30 h after inoculation. This observation seemed explicable from the ATP levels detected in tissue during the 80-h course of the experiments. In resistant tissue, ATP levels fell immediately; in susceptible tissue, the decrease in ATP coincided with the delayed increase in respiration. However, it was concluded that the potentiated respiration was similar in both instances, except that it occurred more rapidly in the incompatible combination.

Similar observations have been made with other plant reactions after inoculation with xanthomonads, such as wound healing and suberization, membrane leakage, phytoalexin production, accumulation and oxidation of phenolics, and synthesis of agglutinins. All these responses may also occur during the compatible reaction, but with lower speed or intensity than in the incompatible reaction. Even the increase in ethylene synthesis which appears closely associated with disease progression in citrus leaves infected with *X. c.* pv. *citri* may be a plant defence response as it leads to the abscission of leaves (Dutta and Biggs, 1991). Obviously, a subtle balance of several reactions decides on the final outcome of the host–parasite interaction.

4.7.2 Wound healing and cutinization

The formation of physico-chemical barriers is a well-known resistance reaction against fungi, but has rarely been reported for *Xanthomonas*-caused diseases. McPherson (1981) recorded meristematic activity in tissues of pelargonium inoculated with an avirulent isolate of *X. c.* pv. *pelargonii*; this did not occur in the usual field outbreaks of the disease associated with virulent strains of the pathogen.

Koizumi (1983) studied the relationship between wound-healing of citrus leaves and infection by *X. c.* pv. *citri*. At low relative humidity infection was completely inhibited after formation of meristem tissues composed of thin dividing cells. Wound tissues incubated in a moist environment became less susceptible when the cell walls in the outer tissues were lignified. For several *Xanthomonas*-caused diseases, especially for *X. o.* pv. *oryzae*, it was often reported that plants were more susceptible in the seedling stage than at the adult stage, and that old senescent leaves were more resistant than younger leaves (Kaul and Sharma, 1987; see also section 4.7.9). This effect may be due to continuing cutinization. According to Häusermann (1944), a cutin-containing layer develops on the surfaces of mesophyll cells where they are exposed to intercellular air spaces. Suberization diminished degradation of potato tuber slices by pectate lyases (Zucker and Hankin, 1970). Also, the induction of water-soaking (see section 4.6.5) appears more difficult in cutinized or suberized leaf mesophyll tissue.

4.7.3 Hypersensitive reaction

The hypersensitive reaction (HR) of plant tissue to inoculation with phytopathogenic bacteria is defined as a rapid and irreversible increase of membrane permeability (membrane leakage) leading to loss of structural integrity and confluent tissue necrosis within 24 h. Thus, electrolyte leakage increased in a resistant tomato genotype 12 h after infiltration with 10^8 cfu of *X. c.* pv. *vesicatoria* per ml, whereas none occurred in a susceptible cultivar after 24 h (Jones and Scott, 1986). Compatible bacteria cause appreciable leakage 2–3 days after inoculation. During the development of the hypersensitive and the normosensitive (compatible interaction) necrosis, almost identical physiological and biochemical changes occur in the plant except that they develop more slowly in the normosensitive necrosis. There are also transitional stages between the resistant and susceptible reactions, e.g. *X. c.* pv. *malvacearum* in cotton (Klement, 1982).

The HR is induced by xanthomonads on non-host plants and on certain resistant cultivars. It is generally assumed but not proven that the biochemical mechanisms are the same in both systems. **Histological** observations of Cason *et al.* (1977) showed that, with large inoculum doses of *X. c.* pv. *malvacearum*, damage to cotton leaves occurred within 4–6 h after inoculation. Chloroplast damage, cell compression, and tissue shrinkage followed by necrosis and desiccation were the most obvious features of the HR. Jones and Fett (1985) observed that during the HR, cells of *X. c.* pv. *glycines* were immobilized by fibrillar or electron-dense amorphous material on the surface of leaf mesophyll cells. Other host cells' responses included invagination of host plasmalemma, accumulation of vesicles and amorphous material between host plasmalemma and cell wall, and migration and disorientation of thylakoid membrane systems in chloroplasts.

Stall and Cook (1979) studied the **mechanism** of HR induction in pepper leaves with an incompatible race of *X. c.* pv. *vesicatoria*. Occurrence of HR was blocked if the inoculated leaves were kept watersoaked after inoculation. Bacteria immobilized in water agar in the leaves also did not elicit the HR, but multiplied like a compatible race. Therefore, it was assumed that at least transient bacterium–cell contact is necessary for the HR to occur. This assumption was weakened by Keen and Holliday (1982) because (a) excess water in the intercellular space may dilute possible bacterial elicitors of the HR, and (b) gas exchange was impaired by water-soaking so that metabolism was reduced. Cell division is not absolutely necessary for HR elicitation, as shown in experiments where the inoculum of *X. c.* pv. *vesicatoria* was mixed with nalidixic acid (Meadows and Stall, 1981), but active metabolism of the bacteria is essential for HR induction. *X. c.* pv. *vesicatoria* did not cause cell collapse in an atmosphere devoid of oxygen (Stall *et al.*, 1982a).

Electrophysiological studies of Novacky (1980) indicated that the passive (diffusion) component of the membrane potential is affected during the early phase of HR development. Changes in membrane lipids are associated with the HR. It was suggested by Adám *et al.* (1989) that (oxy) free radicals may play a significant role in inducing damage to membrane lipids. However, the role of active oxygen species in host–pathogen interaction is still an open question (Sutherland, 1991). Jiao *et al.* (1992) analysed the lipid composition of citrus leaves and found that in species resistant to *X. c.* pv. *citri* the degree of fatty acid unsaturation was less than in susceptible species. A more saturated lipid composition makes membranes less fluid and therefore possibly more resistant to bacterial attack.

The mechanism of **inhibition of bacterial growth** during the HR has not been elucidated completely. Three phenomena of the HR were seen in response to concentrated inocula by Essenberg *et al.* (1979a): tissue desiccation, loss of pigments and a drop in number of viable bacteria. However, the latter two did not occur in experiments with *X. c.* pv. *malvacearum* in immune cotton leaves when the first, desiccation, was prevented and are therefore probably not the result of an active host response. It was suggested that desiccation may cause cell rupture leading to loss of pigments, and that bacteria may be killed by lack of water or by toxic substances which concentrate during desiccation. Bacteriostatic compounds which diffuse from the damaged or dead cells were regarded as the main principle which inhibits bacterial growth during the HR (Essenberg *et al.*, 1979a,b). Thus, when immune cotton plants were maintained in continuous dark, bacteria multiplied to levels typically found in susceptible plants, although the ultrastructural reactions resembled the HR (Morgham *et al.*, 1988). However, sesquiterpenoid phytoalexins did not accumulate in immune plants maintained in the dark, because their synthesis is light-dependent.

Genetic analysis of very few *Xanthomonas*-caused diseases indicated that a 'gene-for-gene' interaction occurs in certain race/cultivar combinations (e.g. *X. c.* pvs *vesicatoria* or *malvacearum*). In these cases it is assumed that a resistance gene in a particular host cultivar confers resistance against physiological races that express the matching avirulence gene, so that the HR results. The underlying mechanisms of these interactions have not been analysed at a molecular level, although elicitors, elicitor receptors, signal interactions, and signal transduction mechanisms are being discussed (Dixon and Lamb, 1990) as is the ion channel defence model (Gabriel *et al.*, 1988). The molecular characterization of race-specific defence responses in the host plant is just beginning (Reimers and Leach, 1989), and relatively simple models with different *X. c.* pv. *campestris* strains on *Arabidopsis* have been developed (Daniels *et al.*, 1991).

However, care should be taken not to overstate the HR as a typical resistance reaction against xanthomonads. It was pointed out by Kiraly and Barna (1985) that the HR induced by phytopathogenic fungi in association with resistance is not necessarily the cause of the resistant response. Also, in the case of phytopathogenic bacteria several aspects of the HR indicate that this reaction type may not be of general importance as a resistance mechanism (Rudolph, 1976). Thus, it appears unreal that under natural conditions leaves are infiltrated by highly concentrated bacterial suspensions (10^7–10^8 cfu/ml) which are necessary to induce the typical HR. Although it was reported that individual dead cells also occur in leaves containing only a few incompatible bacteria (Turner and Novacky, 1974; Essenberg *et al.*, 1979a), it has not been elucidated why bacterial multiplication was inhibited in these systems. When Zachowski (1989) inoculated an incompatible race of *X. c.* pv. *malvacearum* in low concentration, no visible symptoms developed. However, infiltration of water 7 days later led to a typical HR, indicating that lack of water had hindered further multiplication of the bacteria. On the other hand, Essenberg *et al.* (1979b) concluded from similar experiments that a localized bacteriostasis was responsible for inhibition of bacterial growth. The advantage of the HR in resistance breeding is that this reaction type is easy to recognize and to incorporate into breeding lines. An example is the HR to *X. c.* pv. *vesicatoria* which is simply inherited (Cook and Stall, 1963; Jones and Scott, 1986). However, a polygenic resistance based on major and minor genes may be more durable.

4.7.4 Phytoalexin production

Phytoalexins have been defined as low-molecular-weight, antimicrobial compounds that are both synthesized by and accumulated in plants after exposure to microorganisms (Paxton, 1981). A vast literature appeared on their role in resistance towards fungal plant pathogens (Mahadevan, 1991), in contrast to very few reports regarding xanthomonads. The most convincing data have been obtained by Essenberg and co-workers on phytoalexins from cotton (*Gossypium*) inhibitory to *X. c.* pv. *malvacearum*. Two sesquiterpenoid naphthols, 2,7-dihydroxycadalene and 2-hydroxy-7-methoxycadalene and their optically active oxidation products, lacinilene C and lacinilene C7-methylether were identified. Accumulating concentrations in resistant cultivars, localization around infection sites, water-solubilities, inhibitory activities toward *X. c.* pv. *malvacearum*, and correlation of resistance to light-dependent synthesis strongly suggest that these compounds play a decisive role in resistance of certain cotton cultivars (Pierce and Essenberg, 1987; Morgham *et al.*, 1988; Essenberg *et al.*, 1990). Under light, cotton phytoalexins can act as photosensitizers, leading to the production of biocidal active oxygen species (Sun *et al.*, 1989).

The soybean isoflavonoids glyceollin and coumestrol showed growth inhibition of *X. c.* pv. *glycines* (Fett and Osman, 1982); Holliday and Keen (1982) concluded that the accumulation of glyceollin may lead to some reduction of growth but that it is not the sole resistance mechanism. However, further studies of Fett (1984) indicated that the soybean isoflavonoids and isoflavone glucosides did not determine resistance to *X. c.* pv. *glycines*. The bush bean isoflavonoids phaseollinisoflavan and kievitone inhibited the bean pathogen *X. c.* pv. *phaseoli* as well as *X. c.* pvs *campestris* and *vesicatoria* (Wyman and Van Etten, 1978), but whether these compounds are inhibitory to the bacteria *in planta* is unknown. Interestingly, these compounds show a selective toxicity to Gram-positive bacteria (Gnanamanickam and Smith, 1980). Ota (1983b) described phytoalexin-like substances which accumulated in citrus leaves after resistance induction by *Pseudomonas* sp. against *X. c.* pv. *citri*. It may be that these substances belong to the less well-specified group of compounds treated in section 4.7.5. The occurrence of phytoalexins was also reported in bacterial blight resistant rice leaves (Nakanishi and Watanabe, 1977).

4.7.5 Accumulation of phenolics and related antibacterial substances

Many pathophysiological studies revealed that the level of phenolics and other secondary substances in plants increases after invasion by pathogens, especially in the resistant reaction. But the enhanced aromatic biosynthesis and the stimulated enzyme activities are characteristic not only of infections but also of injury and accelerated ageing (Goodman *et al.*, 1986). An extended 90-page table of Mehadevan (1991), comprising the inhibition of phytopathogens by diverse phenolics of plant origin, demonstrates that aromatic compounds are structurally non-specifically active substances. Nevertheless, exact data on the role of these substances in resistance against xanthomonads are limited.

In rice leaves antimicrobial compounds with a diterpene skeleton (named oryzalides A and B and oryzalic acid A) were recently identified (Kono *et al.*, 1991). Occurrence in resistant leaves and biological activity suggest a role in resistance. Tannin-like materials have been implicated in resistance to *X. c.* pv. *pelargonii* (Wainwright and Nelson, 1972). Other reports described increased contents of phenolics in resistant plants, e.g. in cotton and cabbage inoculated with *X. c.* pvs *malvacearum* and *campestris*, respectively (Namasivayam *et al.*, 1971; Jalali *et al.*, 1976), without proving a participation in resistance.

In several cases, increased **peroxidase** has been correlated to resistance against xanthomonads. Thus, Urs and Dunleavy (1974) concluded that peroxidase plays an essential role in soybean resistance to bacterial pustule (*X. c.* pv. *glycines*), because of a greater antibacterial activity of peroxidase from the resistant cultivar. Correlations for increased peroxidase activity and resistance were reported for *X. c.* pvs *glycines* (Rama *et al.*, 1974), *phaseoli* (Cinar, 1975), and *malvacearum* (Venere, 1980). Reimers and Leach (1991) showed that lignin-like phenolic compounds were synthesized by peroxidases in resistant rice leaves after inoculation with *X. o.* pv. *oryzae*. It is not known whether lignin actually restricts bacterial pathogenesis in these systems. However, precursors of lignin have been shown to have potent antibacterial activity.

4.7.6 Agglutination of bacteria

The first indications for agglutination of bacteria during the resistant reaction originated from **histological studies** of diseased plant tissue. Several studies showed that incompatible strains of diverse bacterial species and saprophytic bacteria were surrounded by fibrillar or electron-dense material at the surface of plant cell walls. These observations were also made for xanthomonads. Thus, Koizumi (1979) reported that in tissues of resistant *Citrus junos*, fibrillar materials coagulated in intercellular spaces and ensheathed bacteria 96 h after inoculation. In the vessels of resistant rice cultivars the irregular mass of bacterial cells (*X. c.* pv. *citri*) was enveloped by a kind of fibrillar material which seemed to be derived from the inner vessel wall layer (Horino, 1981). The unenveloped bacteria in the compatible interaction were significantly larger. Al-Mousawi *et al.* (1982b) observed that in immune cotton leaves compact masses of bacterial cells of *X. c.* pv. *malvacearum* were associated with fibrillar material, which did not completely envelop all the bacteria. In rice leaf parenchymatous tissue incompatible to *X. o.* pv. *oryzae*, most bacterial cells were morphologically abnormal and surrounded by fibrillar material which appeared to originate from the host cell wall (Tsuno and Wakimoto, 1983). In later stages (14 days post inoculation) the surrounding material changed to have a granular appearance. Cells of an incompatible strain of *X. c.* pv. *glycines* were enveloped to varying degrees by fibrillar or electron-dense amorphous material on the surface of leaf mesophyll cells (Jones and Fett, 1985). The extent of bacterial envelopment was inversely related to bacterial growth. However, in all of these studies a causal relationship could not be finally established. It has also been pointed out that entrapment of bacteria in intercellular spaces may be a passive phenomenon brought about by air–water interphases due to evaporation of intercellular water after infiltration of the bacterial suspension (Hildebrand *et al.*, 1980).

Additional information was obtained from trials to extract and characterize plant substances which agglutinate bacteria (**agglutinins**). Fett and Sequeira (1980a,b) isolated a glycoprotein from soybean seeds (cv. Clark) which in the undissociated state agglutinated cells of *X. c.* pv. *glycines*. The agglutination was inhibited by both purified lipopolysaccharides (LPS) and extracellular polysaccharides (EPS) from *X. c.* pv. *glycines*. LPS and EPS from other bacterial species were ineffective. Takahashi and Doke (1983) extracted proteinaceous agglutinin from citrus leaves which agglutinated unwashed cells of *X. c.* pv. *citri* and its EPS, but not washed cells and unwashed EPS non-producing mutants. A phytohaemagglutinin from bean seeds which agglutinated *X. c.* pv. *phaseoli* and several other bacterial species was purified by Castresana *et al.* (1987). Lima *et al.* (1990) isolated an agglutinin from seeds and stems of cassava cv. Cacau. The protein agglutinated avirulent strains of *X. c.* pv. *manihotis* with greater intensity than the virulent strains, and induced gel precipitation of bacterial LPS and EPS. The interaction between agglutinin and LPS or non-capsulated bacterial cells was inhibited by EPS of virulent strains.

It may be **concluded** from these reports that during the resistance response interactions between bacterial and plant polymers result in densely packed bacterial cells which are therefore unable to colonize the intercellular spaces (Figure 4.8). In susceptible combinations, on the other hand, a synergistic interaction between EPS/LPS and plant polymers leads to a gel-like slime embedding the bacteria (Figure 4.2) and facilitating bacterial growth *in planta* (see section 4.6.5). The diverse complexity of plant cell wall polymers (Wydra and Rudolph, 1992) may explain why only in very rare cases (the compatible interaction) a synergistic interaction between bacterial and plant polymers occurs. It is not known whether earlier reports on common antigens of cotton and *X. c.* pv. *malvacearum* (DeVay *et al.*, 1967) or of rice leaves and *X. o.* pv. *oryzae* (Bordoloi and Addy, 1985) are related to these interactions.

4.7.7 Degradation or inactivation of bacterial virulence or pathogenicity factors

Several bacterial determinants for virulence and/or pathogenicity are described in section 4.6. Any inhibition or degradation of these bacterial products by the host plant should result in reduced susceptibility. This resistance mechanism may be decisive for the phenolics (see section 4.7.5) which have long been known as enzyme inhibitors. With many xanthomonads, dissolution of host cell walls is an essential part of pathogenesis (section 4.3.1), so that inhibition of the responsible enzymes should slow down disease development. For instance, histological studies of Wainwright and Nelson (1972) indicated the occurrence of qualitatively different tannin-like materials in *Pelargonium* spp. resistant to *X. c.* pv. *pelargonii*. It was assumed that tannin-caused inhibition of cellulolytic and pectolytic enzymes accounted for the lack of lateral spread and for bacterial pocket formation in the resistant *Pelargonium* spp.

Non-synergistic interaction between bacterial extracellular polysaccharides and plant polymers (agglutinins) (see section 4.7.6) is another example of this type of resistance mechanism. Since resistance toward xanthomonads was mostly studied by tests for inhibition of bacterial growth, these indirect types of disease resistance may have been overlooked in the past.

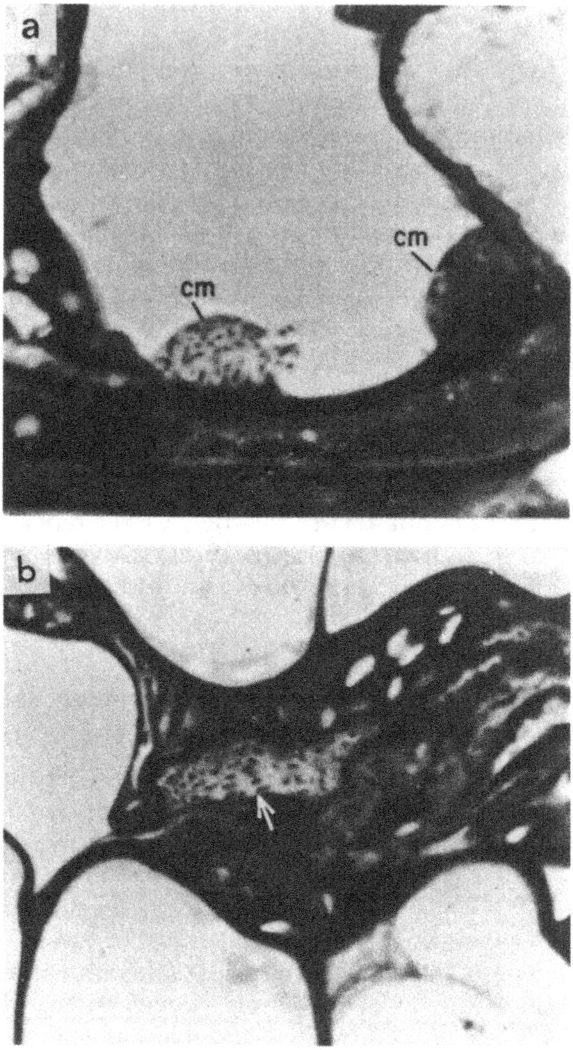

Figure 4.8 Cell colonies of *X. c.* pv. *malvacearum* in resistant cotton leaves 7 days after inoculation: (a) intercellular space between darkly stained host cells showing two compact masses (cm) of bacteria, the left of which is burst; (b) bacteria trapped between collapsed resistant host cells (from Essenberg *et al.*, 1979).

4.7.8 Induced resistance

Pretreatment with inducing agents such as living or killed bacteria, subcellular components of bacteria, or abiotic agents and stress may protect plants against disease incited by pathogens. Klement (1990a), studying pseudomonads that cause leaf spot, differentiated early, late and systemic induced resistance. Xanthomonads appear to be inhibited by the same types of induced resistance. Thus, Hsu and Dickey (1972) reported that in bean and tomato leaves, pre- or simultaneous inoculations with incompatible *X. phaseoli*, *X. vesicatoria*, *X. campestris* and *P. fluorescens* decreased multiplication of compatible pathogens. Zachowski (1989) observed a late induced resistance in cotton cotyledons toward *X. c.* pv. *malvacearum* after a pretreatment with bacterial extracellular polysaccharides 2 days before inoculation. Recently, a search for specific and unspecific elicitors of induced resistance, also against xanthomonads, has attracted many studies in molecular plant pathology (Lamb *et al.*, 1989). The physiological and biochemical mechanisms of induced resistance are obviously the same as those described in sections 4.7.1–4.7.7.

4.7.9 Adult resistance

Adult plant resistance is defined as resistance that is expressed as the plant matures. Many workers reported that rice plants are more susceptible to *X. o.* pv. *oryzae* at seedling stage than at the adult stage (Kaul and Sharma, 1987). According to Goel and Gupta (1990) the growth stage of the plant when adult resistance became operative ranged from maximum tillering (50 days after sowing, DAS) to the booting stage (70 DAS) in different cultivar/isolate combinations. The disease severity in some cultivar/isolate combinations decreased gradually from seedling to boot stage, whereas in others an abrupt decline in disease severity was noticed at/or after maximum tillering. The expression of adult resistance was differential. Similar observations have been made for other xanthomonads, e.g. *X. c.* pvs *cyamopsidis* (Gandi and Chand, 1988) and *citri* (Stall *et al.*, 1982b).

In addition to adult plant resistance to *X. o.* pv. *oryzae* caused by the Xa-3 gene, a change in quantitative resistance (assessed by lesion length) with increasing plant age was determined (Koch and Mew, 1991). Also the restriction of lesion size in the later stages of infection, though the interaction is compatible genetically, may be induced by similar physiological conditions (Lamb *et al.*, 1989). The biochemical or physiological mechanisms for development of resistance during plant ageing are unknown; depletion of free amino acids is one possible explanation (Sridhar *et al.*, 1979).

4.7.10 Tolerance

A host in which there is apparently unrestricted and extensive colonization by a pathogen without symptom development is termed tolerant (Federation of British Plant Pathologists, 1973). In virology the term tolerance has also been taken to describe plants which survive infection and sustain a low yield reduction irrespective of virus acccumulation or symptom expression (Goodman *et al.*, 1986). Because the resistance to some *Xanthomonas*-caused diseases is always incomplete, the terms tolerance or field resistance are also applied here. For instance, Saettler and Mabagala (1987) described two bush bean cultivars with a tolerance to *X. c.* pv. *phaseoli* under field conditions. Because of the predominantly additive gene control of the reaction against this disease, increased levels of tolerance in progeny of some crosses were observed (Valladares-Sanchez *et al.*, 1980). In tolerant bush bean cultivars high levels of bacteria may be reached without causing marked disease symptoms (Bozzano, 1993). It is very difficult to characterize the mechanisms of disease resistance in these cases.

4.7.11 Anatomic features

Only a few studies are available in which resistance to xanthomonads has been shown to be based on the morphology of the host. Evidence for a structurally conditioned resistance of *Lycopersicon* species to *X. c.* pv. *vesicatoria* was obtained by Ramos *et al.* (1992). Stomata frequency as well as stomatal width correlated with the number of bacterial lesions per cm^2. In addition, other morphological features such as the raised stomatal complex in *L. hirsutum* and persistent, hydrophobic coating of the epidermis in *L. peruvianum* appeared to be relevant in disease response.

The resistance of *Leersia japonica* to hydathodal invasion by *X. c.* pv. *oryzae* was attributed to the morphology of the upper surface of the guard cells, where well-developed ledges were believed to prevent invasion by the bacteria (Horino, 1984). The lesion development by *X. c.* pv. *campestris* in white cabbage cultivars tended to be positively correlated with guttation through hydathodes (Ruissen and Gielink, 1993). Similarly, the degree of resistance of hyacinths to *X. c.* pv. *hyacinthi* was correlated with small stomata. But also hyacinths with short and/or narrow leaves were generally more resistant than those with long and/or broad leaves (Van Tuyl, 1982). It was also observed that under natural infection incidence of bacterial blight of rice (*X. o.* pv. *oryzae*) was lower on cultivars with erect leaves than on those with intermediate or horizontal leaves (Philip and Devadath, 1980).

4.7.12 Conclusion

Recent studies on the analysis of resistance against bacterial plant pathogens, including xanthomonads, are dominated by testing for the hypersensitive response (HR), probably because this reaction type is so easy and fast to record. However, it should not be neglected that several other mechanisms of resistance toward xanthomonads exist, some of which may be more durable and more important in practice.

4.8 EFFECTS OF THE ENVIRONMENT

4.8.1 Water and humidity

In contrast to fungi, the xanthomonads, as most other bacteria, cannot colonize dry surfaces. This is probably the main reason why, in general, bacteria are less important as plant pathogens than fungi, and why the situation is reversed in human and veterinary medicine. The mammalian blood and digestion system, epithelial cells, intestinal mucosa, lung tissues, etc., represent ideal media for bacterial growth. In contrast, plant surfaces are generally dry. Even in the intercellular spaces, where most of the xanthomonads multiply, the plant cell walls are more or less cutinized or suberized. Therefore, xanthomonads have developed specific mechanisms to attract water and to create a milieu which allows multiplication in the intercellular space (see section 4.6.5). Xanthomonads need free water for colonization of leaf surfaces (see section 4.1), for penetration (see section 4.2), and also for multiplication within the plant tissue (see section 4.4).

The importance of free water as a vehicle for pathogen **dispersal** is generally recognized as being pre-eminent for bacteria (Goodman, 1982). Thus, wind-driven rainsplash or watersplash of sprinkler irrigation is very important in dispersal of all leaf spot-causing bacteria, including xanthomonads, as was reported for several *X. c.* pathovars, (Venette, 1982). Sprinkler irrigation is more effective in spreading bacterial pathogens than surface irrigation. Even in dry weather, evening dews will furnish enough water for spread of bacterial leaf pathogens. This was also concluded from experiments on the spread of *X. c.* pv. *phaseoli* var. *fuscans* in bean fields (Cinar and Rudolph, 1970). Manicom (1986) reported that the major weather factor affecting mango fruit infection by *X. c.* pv. *mangiferaeindicae* was rainfall.

The **duration** of free moisture on the plant surface is decisive for a successful **infection**, e.g. in cluster bean, infected by *X. c.* pv. *cyamopsidis* (Gandi and Chand, 1988). A 24–48 h wetting period was required for significant infection of peach by *X. c.* pv. *pruni* (Shepard and Zehr, 1987), and Koizumi (1983) found that wounded citrus leaf tissues incubated in a moist environment were twice as susceptible to *X. c.* pv. *citri* than leaves incubated at low relative humidity. On the basis of similar observations, Rakesh *et al.* (1992) developed a model for disease intensity by *X. c.* pv. *vignicola* including minimum and maximum relative humidity.

Continuous high humidity and a film of free water on leaf surfaces can also result in **water-congested areas** in the intercellular spaces, which are often observed after periods of dew. These water-congested spaces may allow the bacteria to establish microcolonies which can initiate a susceptible reaction (Rudolph, 1980). Thus, water-soaking of tomato leaves enhanced infection by *X. c.* pv. *vesicatoria* (Vakili, 1967). Cook and Stall (1977) analysed the effect of water on the hypersensitive reaction (HR) of pepper to *X. c.* pv. *vesicatoria* and found that detached leaves, when watersoaked after infiltration with the bacterium, did not develop a HR. Bacterial multiplication in these leaves was similar to that in susceptible leaves, left *in situ* after inoculation. However, typical development of the HR began 2–3 h after water-soaking was permitted to dissipate.

4.8.2 Temperature

In general, *Xanthomonas*-caused diseases are more destructive in tropical and subtropical regions than in temperate climates, whereas the fluorescent pseudomonads are typically endemic in the temperate climatic zones. The temperature optima for both groups of bacteria are similar (25–30°C). However, the xanthomonads are not cold tolerant, in contrast to most of the pseudomonads (Young *et al.*, 1977). Most of the xanthomonads do not grow at 5°C and some are unable to grow below 9°C (Dye and Lelliott, 1974), whereas many pathovars of *Pseudomonas syringae* are characterized by a minimum growth temperature around 0°C (Stapp, 1956). It was pointed out by Goodman (1982) that a certain growth rate must probably be achieved by each pathogen before infection can be established, because at low temperature, resulting in reduced bacterial replication, plant resistance mechanisms may keep pace with the feeble growth of the pathogen. Numerous examples have been reported, e.g. the bean cultivar Jules and the bean P.I. 207272 are moderately resistant to *X. c.* pv. *phaseoli* in temperate zones but are susceptible in field evaluation in the tropics (Webster *et al.*, 1983). Also in Germany, temperature appeared to be the most important factor for spread of *X. c.* pv. *phaseoli* var. *fuscans* in bean fields (Çinar and Rudolph, 1970). A remarkable spreading was only observed when the average day and night temperature was above 19°C for several days. The vascular wilt pathogen, *X. c.* pv. *pelargonii*, is disseminated primarily during vegetative propagation. Often, symptom expression is suppressed under the cool conditions prevailing at propagation, so that symptomless but latently infected plants are produced and sold. In warm weather, symptom expression occurs and apparently healthy plants suddenly wilt and die. Incidence and severity of disease by *X. c.* pv. *pruni* on peach were greater at 30 than at 24°C (Shepard and Zehr, 1987). Also, the severity of *Syngonium* blight caused by *X. c.* pv. *syngonii* was greatest for plants maintained at 30°C (Chase, 1988). Similarly, 30°C was the optimum temperature of infection of cluster bean by *X. c.* pv. *cyamopsidis* (Gandhi and Chand, 1988).

4.8.3 Wind

Wind-blown soil (especially sandy soil) can cause considerable abrasion of leaves. This injury to the host can facilitate the initial phase of bacterial infection. Thus, strong winds, even in the absence of rainfall, were included as important factors influencing the incidence of citrus bacterial canker (Koizumi, 1983), and the patterns of wind-blown rain were consistent with the predominant spread of citrus bacterial spot in Florida (Gottwald *et al.*, 1989). The waterborne inoculum of *X. c.* pvs *citri* and *citrumelo* must be driven with enough force to cause water-soaking of citrus leaf tissues before disease can develop. Wind speeds of 8–12 m/s were reported to be necessary for infection of citrus foliage by *X. c.* pvs *citri* and *citrumelo* (Gottwald *et al.*, 1992). Manicom (1986) reported significant correlations of a wind/rain index and a storm index with infection levels of *X. c.* pv. *mangiferaeindicae* on mangoes. Sand blasting of tomato leaves prior to spraying with a turbid suspension of *X. c.* pv. *vesicatoria* resulted in much greater infection rates than without wounding (Vakili, 1967).

4.8.4 Light

In general, light intensity and duration is positively correlated with the intensity of the susceptible reaction. However, light can also induce a higher resistance toward xanthomonads. This inconsistent effect is probably due to two different light-dependent reactions. Increased photosynthesis results in higher concentrations of assimilates, so that the bacteria are provided with a rich medium which may stimulate bacterial growth. On the other hand, some plant species, such as cotton (see section 4.7.4), synthesize light-dependent phytoalexins which inhibit bacterial growth. The first possibility was confirmed by experiments of Smith and Kennedy (1970) and Klement *et al.* (1987). When plants were incubated in complete darkness for several days, a normally susceptible reaction was changed into resistance (*X. c.* pvs *phaseoli* and *malvacearum*). Cytological studies showed that in darkened leaves the bacteria were more densely packed and not embedded in large masses of bacterial slime (Zachowski *et al.*, 1990).

On the other hand, an enhanced susceptibility with darkness or low light susceptibility was reported not only in cotton/*X. c.* pv. *malvacearum* during the incompatible combination (Zachowski, 1989), but also in sesame/*X. c.* pv. *sesami* (Thomas, 1965), pepper/*X. c.* pv. *phaseoli*

(Sasser *et al.*, 1974), and rice/*X. o.* pv. *oryzae* (Reddy *et al.*, 1978; Devadath *et al.*, 1987).

4.8.5 Nutrition of the host plant

Many workers have studied the effect of fertilizers on the plant's reaction toward infection with xanthomonads. In most cases increased levels of nitrogen, phosphorus, magnesium, and micronutrients enhanced susceptibility, whereas calcium and potassium decreased susceptibility (Woltz and Jones, 1979; Mohanty and Reddy, 1982; Mohanty *et al.*, 1983; Philip and Devadath, 1984; Rao and Hiremath, 1985; Devadath *et al.*, 1987). Thus, better nutrition of peaches raised susceptibility to *X. c.* pv. *pruni* (Matthee and Daines, 1969), whereas McGuire *et al.* (1991) determined lower epiphytic populations of *X. c.* pv. *vesicatoria* on tomato plants receiving combinations of higher rates of nitrogen and potassium. Notwithstanding that the nutritional requirements of the xanthomonads are relatively simple (Rudolph, 1990c), only one or a few host plant species will allow maximum multiplication of each of the known *Xanthomonas* species or pathovars of *X. campestris*. Therefore, the fascinating phenomenon of very narrow host specificity for most of the xanthomonads (see also Chapter 1) cannot be explained by nutritional requirements.

4.8.6 Air pollutants

Air pollutants can markedly affect the severity of bacterial diseases of plants. The reaction can be shifted to higher susceptibility or to a higher degree of resistance. Thus, the apparent infection rate of bush bean with *X. c.* pv. *phaseoli* increased with higher hydrogen fluoride concentrations, as measured by lesion size and expansion (Reynolds and Laurence, 1990). On the other hand, increasing cumulative sulphur dioxide (SO_2) concentration resulted in a significant decrease in the rate of lesion appearance (Reynolds *et al.*, 1987). The consequences of SO_2 exposure are cuticle abrasion, epidermal cell damage and stomatal closure (Rist and Davis, 1979; Kimmerer and Kozlowski, 1981). Also, ozone significantly reduced injury by *X. c.* pv. *alfalfae* (Howell and Graham, 1977) and inhibited infection by *X. fragariae* (Laurence and Wood, 1978).

4.8.7 Phylloplane microorganisms

Especially in the epiphytic phase (see section 4.1), but also during pathogenesis, xanthomonads can be affected by phylloplane micro-organisms. Most studies have dealt with epiphytic bacteria, and little is known of the effect of other organisms, such as fungi or phages, on phytopathogenic xanthomonads. The majority of bacteria found on aerial plant surfaces are Gram-negative genera (*Erwinia, Pseudomonas, Xanthomonas, Flavobacterium*). Very often, *Erwinia herbicola* and *Pseudomonas fluorescens* have been identified. Four types of interactions by which phytopathogenic bacteria can be affected were differentiated by Blakeman (1982): parasitism, nutrient competition and change of pH, antibiotic production, and stimulation of the host's defence. Most studies with phylloplane bacteria were carried out in order to screen for possibilities of **biological control**.

Studies with *E. herbicola* against *X. o.* pv. *oryzae* (Hsieh and Buddenhagen, 1974; Santhi *et al.*, 1987) and with *Pseudomonas* sp. against *X. c.* pv. *citri* (Ota, 1983a) revealed an inhibition *in vitro* and *in planta*, and indicated competition for nutrients and/or lowering of pH as the mechanisms involved. On the other hand, disease reduction by epiphytic bacterial strains in the case of *X. c.* pvs *graminis* (Schmidt, 1988a,b) and *pruni* (Randhawa and Civerolo, 1986) was not related to inhibition *in vitro*, so that an induced resistance in the host plant was assumed. Most of the epiphytic bacteria in grasses were found beneath the leaf sheath. Applied pruniphages reduced cell numbers of *X. c.* pv. *pruni* on the leaves, but pruniphage-resistant strains developed (Randhawa and Civerolo, 1986). Rao *et al.* (1978) described strains of *Pseudomonas, Erwinia* and *Aspergillus* which reduced disease by *X. o.* pv. *oryzicola* when applied together or 24 h after inoculation with the pathogen. Several phylloplane bacteria also induced protection against *X. c.* pv. *malvacearum* (Sinha and Verma, 1983). Ridé *et al.* (1979) determined an abundant population of pseudomonads and chromogenic bacteria on aerial structures of poplar, especially on resistant clones. Some of the epiphytic strains but not the yeasts were antagonistic toward *X. populi*.

REFERENCES

Adam, A., Farkas, T., Somlyai, G., Hevesi, M. and Kiraly, Z. (1989) Consequence of O_2 generation during a bacterially induced hypersensitive reaction in tobacco: deterioration of membrane lipids. *Physiological and Molecular Plant Pathology*, **34**, 13–26.

Aggour, A.R., Coyne, D.P., Vidaver, A.K. and Eskridge, K.M. (1989) Transmission of the common blight pathogen in bean seed. *Journal of the American Society of Horticultural Science*, **114**, 1002–8.

Al-Mousawi, A.H., Richardson, P.E., Essenberg, M. and Johnson, W.M. (1982a) Ultrastructural studies of a compatible interaction between *Xanthomonas campestris* pv. *malvacearum* and cotton. *Phytopathology*, **72**, 1222–30.

Al-Mousawi, A.H., Richardson, P.E., Essenberg, M. and Johnson, W.M. (1982b) Cotyledon and leaf ultrastructure of a bacterial blight-immune cotton line inoculated with a low level of *Xanthomonas campestris* pv. *malvacearum*. *Phytopathology*, **72**, 1230–4.

Angeles-Ramos, R., Vidaver, A.K. and Flynn, P. (1991) Characterization of epiphytic *Xanthomonas campestris* pv. *phaseoli* and pectolytic xanthomonads recovered from symptomless weeds in the Dominican Republic. *Phytopathology*, **81**, 677–81.

Arlat, M., Gough, C.L., Barber, C.E., Boucher, C. and Daniels, M.J. (1991) *Xanthomonas campestris* contains a cluster of *hrp* genes related to the larger *hrp* cluster of *Pseudomonas solanacearum*. *Molecular Plant–Microbe Interactions*, **4**, 593–601.

Azad, A. and Schaad, N.W. (1988) The relationship of *Xanthomonas campestris* pv. *translucens* to frost and the effect of frost on black chaff development in wheat. *Phytopathology*, **78**, 95–100.

Baker, J.C. (1992) Active oxygen metabolism during plant/bacterial recognition, in *Proceedings of the Fourth International Symposium on Biotechnology and Plant Protection*, College Park, MD, 1991 (eds D. Bills and S.P. Kung), Singapore Publishers, Singapore, in press.

Bashan, Y. and Okon, Y. (1986) Internal and external infections of fruits and seeds of peppers by *Xanthomonas campestris* pv. *vesicatoria*. *Canadian Journal of Botany*, **64**, 2865–71.

Beaulieu, C., Minsavage, G.V., Canteros, B.I. and Stall, R.E. (1991) Biochemical and genetic analysis of a pectate lyase gene from *Xanthomonas campestris* pv. *vesicatoria*. *Molecular Plant–Microbe Interactions*, **4**, 446–51.

Bech, K. and Paludan, N. (1990) Is *Xanthomonas campestris* pathovar *begoniae* transmitted in meristem culture? Inoculation trials with *Begonia elatior* and testing for bacteria in meristem plants. *Tidsskr. Planteavl.*, **94**, 337–44.

Ben-David, A., Bashan, Y. and Okon, Y. (1986) Ethylene production in pepper (*Capsicum annuum*) leaves infected with *Xanthomonas campestris* pv. *vesicatoria*. *Physiological and Molecular Plant Pathology*, **29**, 305–16.

Benedict, A.A., Alvarez, A.M. and Pollard, L.W. (1990) Pathovar-specific antigens of *Xanthomonas campestris* pv. *begoniae* and *X. campestris* pv. *pelargonii* detected with monoclonal antibodies. *Applied and Environmental Microbiology*, **56**(2), 572–4.

Benett, E.L. and Carter, C.E. (1990) Influences of water congestion and *Xanthomonas campestris* pv. *pruni* on proteins of peach leaves. *Phytopathology*, **80**, 1023.

Bent, A., Carland, F., Dahlbeck, D., *et al.* (1991) Gene-for-gene relationships specifying disease resistance in plant–bacterial interactions, in *Advances in Molecular Genetics of Plant–Microbe Interactions*, Vol. 1 (eds H. Hennecke and D.S. Verma), Kluwer Academic Publishers, Dordrecht, pp. 32–6.

Birch, R.G. and Patil, S.S. (1987a) Correlation between albicidin production and chlorosis induction by *Xanthomonas albilineans*, the sugarcane leaf scald pathogen. *Physiological and Molecular Plant Pathology*, **30**, 199–206.

Birch, R.G. and Patil, S.S. (1987b) Evidence that an albicidin-like phytotoxin induces chlorosis in sugarcane leaf scald disease by blocking plastid DNA replication. *Physiological and Molecular Plant Pathology*, **30**, 207–14.

Blakeman, J.P. (1982) Phylloplane interactions, in *Phytopathogenic Prokaryotes*, Vol. 1 (eds M.S. Mount and G.H. Lacey), Academic Press, New York, pp. 307–33.

Bonas, U., Schulte, R., Fenselau, S., Minsavage, G.V., Staskawicz, B.J. and Stall, R.E. (1991) Isolation of gene cluster from *Xanthomonas campestris* pv. *vesicatoria* that determines pathogenicity and the hypersensitive response on pepper and tomato. *Molecular Plant–Microbe Interactions*, **4**, 81–8.

Bonas, U., Stall, R.E. and Staskawicz, B. (1989) Genetic and structural characterization of the avirulence gene avrBs3 from *Xanthomonas campestris* pv. *vesicatoria*. *Molecular and General Genetics*, **218**, 127–36.

Bordoloi, D.K. and Addy, S.K. (1985) Common antigens between *Xanthomonas campestris* pv. *oryzae* and rice leaves to resistance and susceptibility. *Indian Phytopathology*, **38**, 112–15.

Borkar, S.G. and Verma, J.P. (1989) Exopolysaccharide, a water soaking inducing factor produced by bacterial blight of cotton, *Xanthomonas campestris* pv. *malvacearum*. *Cot. Fib. Trop.* **XLIV**, 149–52.

Bozzano, G. (1993) Unterschiedliche Sortenresistenz von Buschbohnen (*Phaseolus vulgaris* L.) gegenüber den beiden Formen des bakteriellen Bohnenbrandes (*Xanthomonas campestris* pv. *phaseoli* und *X. c.* pv. *phaseoli* var. *fuscans*). Magister-Arbeit, Fachbereich Agrarwissenschaften, Universität Göttingen.

Bradbury, J.F. (1986) *Guide to Plant Pathogenic Bacteria*, CAB International, Farnham Royal, UK, 332 pp.

Cafati, C.R. and Saettler, A.W. (1980) Effect of host on multiplication and distribution of bean common blight bacteria. *Phytopathology*, **70**, 675–9.

Cason, Jr, E.T., Richardson, P.E., Brinkerhoff, L.A. and Gholson, R.K. (1977) Histopathology of immune and susceptible cotton cultivars inoculated with *Xanthomonas malvacearum*. *Phytopathology*, **67**, 195–8.

Castresana, M.C., Serra, M.T. and Tejerina, G. (1987) Purification of phytohaemagglutinins from *Phaseolus vulgaris* and their interaction with plant pathogenic bacteria. *Journal of Phytopathology*, **119**, 345–57.

Chase, A.R. (1988) Effects of temperature and preinoculation light level on severity of *Syngonium* blight caused by *Xanthomonas campestris*. *Journal of Environmental Horticulture*, **6**, 61–3.

Çinar, Ö. (1975) Der Einfluß der Infektion mit *Xanthomonas phaseoli* var. *fuscans* (Burkh.) Starr et Burkh. auf die Peroxydase-Aktivität bei anfälligen und resistenten Bohnensorten. *Journal of Turkish Phytopathology*, **4**, 47–52.

Çinar, Ö. and Rudolph, K. (1970) Zur Bedeutung von *Xanthomonas phaseoli* var. *fuscans* (Burkh.) Starr et Burkh., dem Erreger des Bohnenbrandes, für Deutschland. *Z. Pfl.Krankh. Pfl.Schutz*, **77**, 515–24.

Çinar, Ö. and Rudolph, K. (1972) Susceptibility of different bean species and cultivars to *Xanthomonas phaseoli* var. *fuscans*, cause of bacterial blight of beans. *Z. Pfl.Krankh. Pfl.schutz*, **79**, 193–200.

Cook, A.A. (1988) Association of citrus canker pustules with leaf miner tunnels in North Yemen. *Plant Disease*, **72**, 546.

Cook, A.A. and Stall, R.E. (1963) Inheritance of resistance in pepper to bacterial spot. *Phytopathology*, **53**, 1060–2.

Cook, A.A. and Stall, R.E. (1977) Effects of watersoaking on response to *Xanthomonas vesicatoria* in pepper leaves. *Phytopathology*, **67**, 1101–3.

Cooke, D.L., Epton, H.A.S., Waites, W.M. and Leifert, C. (1993) Elimination of plant pathogenic bacteria by plant tissue culture; the problem of latent persistence *in vitro*. *Proceedings of the Eighth International Conference on Plant Pathogenic Bacteria*, 9–12 June, Versailles, France.

Corey, R.R. and Starr M.P. (1957) Colony types of *Xanthomonas phaseoli*. *Journal of Bacteriology*, **74**, 137–40.

Costerton, J.W., Geesey, G.G. and Cheng, K.-J. (1978) How bacteria stick. *Scientific American*, **238**, 86–95.

Daniels, M.J., Fan, M.J., Barber, C.E., *et al.* (1991) Interactions between *Arabidopsis thaliana* and *Xanthomonas campestris*, in *Advances in Molecular Genetics of Plant–Microbe Interactions*, Vol. 1 (eds H. Hennecke and D.S. Verma), Kluwer Academic Publishers, Dordrecht, pp. 84–9.

De Cleene, M. (1989) Scanning electron microscopy of the establishment of compatible and incompatible *Xanthomonas campestris* pathovars on the leaf surface of Italian ryegrass and maize. *Bulletin OEPP/EPPO Bulletin*, **19**, 81–8.
De Crecy-Lagard, V., Glaser, P., Lejeune, P., *et al.* (1990) A *Xanthomonas campestris* pv. *campestris* protein similar to catabolite activation factor is involved in regulation of phytopathogenicity. *Journal of Bacteriology*, **172**, 5877–83.
De Feyter, R. and Gabriel, D.W. (1991) At least six avirulence genes are clustered on a 90-kilobase plasmid in *Xanthomonas campestris* pv. *malvacearum*. *Molecular Plant–Microbe Interactions*, **4**, 423–32.
Dea, I.C.M. and Morris, E.R. (1977) Synergistic xanthan gels, in *Extracellular Microbial Polysaccharides* (eds P.A. Sandford and A. Laskin), ACS Symp. Ser. 45, Washington DC, pp. 174–82.
Dekker, R.F.H. and Candy, G.P. (1979) The ß-mannanases elaborated by the phytopathogen *Xanthomonas campestris*. *Archives of Microbiology*, **122**, 297–9.
Devadath, S., Dath, A.P. and Jain, R.K. (1987) Effect of shade, nitrogen fertilization, waterlogging and inoculum concentration on the incidence of bacterial blight of rice. *Indian Phytopathology*, **40**, 529–30.
DeVay, J.E., Schnathorst, W.C. and Foda, M.S. (1967) Common antigens and host–parasite interactions, in *The Dynamic Role of Molecular Constituents in Plant Parasite Interaction* (eds C.J. Mirocha and I. Uritani), APS Press, St Paul, MN, pp. 313–28.
Diab, S., Bashan, Y. and Okon, Y. (1982a) Studies of infection with *Xanthomonas campestris* pv. *vesicatoria*, causal agent of bacterial scab of pepper in Israel. *Phytoparasitica*, **10**, 183–91.
Diab, S., Bashan, Y., Okon, Y. and Henis, Y. (1982b) Effects of relative humidity on bacterial scab caused by *Xanthomonas campestris* pv. *vesicatoria* on pepper. *Phytopathology*, **72**, 1257–60.
Diachun, S. and Valleau, W.D. (1946) Growth and overwintering of *Xanthomonas vesicatoria* in association with wheat roots. *Phytopathology*, **36**, 277–80.
Dianese, J.C. and Schaad, N.W. (1982) Isolation and characterization of inner and outer membranes of *Xanthomonas campestris* pv. *campestris*. *Phytopathology*, **72**, 1284–9.
Digat, B. (1978a) Mise en évidence de la latence épiphylle du *Xanthomonas pelargonii* (Brown) Starr et Burkholder chez le *Pelargonium*. *Annales of Phytopathology*, **10**, 61–6.
Digat, B. (1978b) Selection sanitaire des boutures de *Pelargonium* et de *Begonia* X Elator 'Rieger' *vis-à-vis* des bacterioses par utilisation de l'immuno-fluorescence. *Annales of Phytopathology*, **10**, 67–78.
Dixon, R.A. and Lamb, C.J. (1990) Molecular communication in interactions between plants and microbial pathogens. *Annual Review of Plant Physiology and Plant Molecular Biology*, **41**, 339–67.
Dow, J.M., Clarke, B.R., Milligan, D.E., Tang, J.-L. and Daniels, M.J. (1990) Extracellular proteases from *Xanthomonas campestris* pv. *campestris*, the black rot pathogen. *Applied and Environmental Microbiology*, **56(10)**, 2994–8.
Dow, J.M., Daniels, M.J., Dums, F., Turner, P.C. and Gough, C. (1989a) Genetic and biochemical analysis of protein export from *Xanthomonas campestris*. *Journal of Cell Science*, Suppl. **11**, 59–72.
Dow, J.M., Milligan, D.E., Jamieson, L., Barber, C.E. and Daniels, M.J. (1989b) Molecular cloning of a polygalacturonate lyase gene from *Xanthomonas campestris* pv. *campestris* and the role of the gene product in pathogenicity. *Physiological and Molecular Plant Pathology*, **35**, 113–20.
Dow, J.M., Scofield, G., Trafford, K., Turner, P.C. and Daniels, M.J. (1987) A

gene cluster in *Xanthomonas campestris* pv. *campestris* required for pathogenicity controls the excretion of polygalacturonate lyase and other enzymes. *Physiological and Molecular Plant Pathology*, **31**, 261–71.

Du Plessis, H.J. (1984) Scanning electron microscopy of *Xanthomonas campestris* pv. *pruni* in plum petioles and buds. *Phytopathologische Zeitschrift*, **109**, 277–84.

Du Plessis, H.J. (1987) Canker development on plum shoots following systemic movement of *Xanthomonas campestris* pv. *pruni* from inoculated leaves. *Plant Disease*, **71**, 1078–80.

Du Plessis, H.J. (1990) Systemic invasion of plum seed and fruit by *Xanthomonas campestris* pv. *pruni* through stalks. *Journal of Phytopathology*, **130**, 37–45.

Ducourneau, C. and Auriol, P. (1986a) Une phytotoxine de *Xanthomonas campestris* pv. *vasculorum*. I. Purification et caractérisation partielle. *Journal of Phytopathology*, **117**, 336–42.

Ducourneau, C. and Auriol, P. (1986b) Une phytotoxine de *Xanthomonas campestris* pv. *vasculorum*. II. Production *in vivo*. *Journal of Phytopathology*, **117**, 343–8.

Durgapal, J.C., Baleshwar, S. and Pandey, K.R. (1980) Mode of infection of rice seeds by *Xanthomonas oryzae*. *Indian Journal of Agricultural Science*, **50**, 624–6.

Dutta, A.K. and Rafey, A. (1982) Effect of pruning on rice bacterial blight. *International Rice Research Newsletter*, **7**, 10.

Dutta, S. and Biggs, R.H. (1991) Regulation of ethylene biosynthesis in citrus leaves infected with *Xanthomonas campestris* pv. *citri*. *Physiol. Plantarum*, **82**, 225–30.

Duveiller, E. (1992) Toward management of bacterial leaf streak of wheat and triticale. Thèse de Doctorat en Sciences Agronomiques, Université Catholique de Louvain.

Dye, D.W. and Lelliott, R.A. (1974) Genus II. *Xanthomonas*, in *Bergey's Manual of Determinative Bacteriology* (eds R.E. Buchanan and N.E. Gibbons), Williams & Wilkins, Baltimore, pp. 243–9.

Dzhalilov, A.U. and Chernyaeva, I.I. (1989) (Respiration of *Xanthomonas malvacearum* bacteria with various degrees of virulence). *Doklady Vsesoyuznoi Ordena Lenina i Ordena Trudovogo Krasnogo Znameni Akademii Sel'skokhozyaistvennykh Nauk Imeni V.I. Lenina*, **9**, 26–9 (in Russian) (*Review of Plant Pathology*, **71**, 158, 1992).

Egawa, H., Yoshii, K. and Ueyama, A. (1968) Phenylacetic acid, a metabolite of *Xanthomonas oryzae* (Uyeda et Ishiyama) Dowson in culture affecting upon the depressive growth of young roots of rice seedlings. *Annals of the Phytopathological Society of Japan*, **34**, 46–50.

Egel, D.S., Graham, J.H. and Riley, T.D. (1991) Population dynamics of strains of *Xanthomonas campestris* differing in aggressiveness on Swingle citrumelo and grapefruit. *Phytopathology*, **81**, 666–71.

Ekpo, E.J.A. and Saettler, A.W. (1976) Pathogenic variation in *Xanthomonas phaseoli* and *X. phaseoli* var. *fuscans*. *Plant Disease Reporter*, **60**, 80–3.

El-Banoby, F.E. and Rudolph, K.W.E. (1989) Multiplication of *Xanthomonas campestris* pvs. *secalis* and *translucens* in host and non-host plants (rye and barley) and development of water soaking. *Bulletin OEPP/EPPO Bulletin*, **19**, 105–11.

Ercolani, G.L. (1984) Infectivity titrations with bacterial plant pathogens. *Annual Review of Phytopathology*, **22**, 35–52.

Essenberg, M., Cason Jr, E.T., Hamilton, B., Brinkerhoff, L.A., Gholson, R.K. and Richardson, P.E. (1979a) Single cell colonies of *Xanthomonas malvacearum* in susceptible and immune cotton leaves and the local resistant response to colonies in immune leaves. *Physiological Plant Pathology*, **15**, 53–68.

Essenberg, M., Hamilton, B., Cason Jr, E.T., Brinkerhoff, L.A., Gholson, R.K. and Richardson, P.E. (1979b) Localized bacteriostasis indicated by water dispersal of colonies of *Xanthomonas malvacearum* within immune cotton leaves. *Physiological Plant Pathology*, **15**, 69–77.

Essenberg, M., Grover Jr, P.B. and Cover, E.C. (1990) Accumulation of antibacterial sesquiterpenoids in bacterially inoculated *Gossypium* leaves and cotyledons. *Phytochemistry*, **29**, 3107–13.

Esterio, M.A. and Latorre, B.A. (1982) Potential sources of inoculum of *Xanthomonas juglandis* in walnut blight outbreaks. *Journal of Horticultural Science*, **57**, 69–72.

Ewbank, E. (1992) Etude du catabolisme d'acides aminés conduisant à la formation de phytotoxines par *Xanthomonas campestris* pv. *manihotis* (agent pathogène de la bactériose du manioc). Thèse de Doctorat en Sciences Agronomiques, Université Catholique de Louvain.

Ewbank, E. and Maraite, H. (1990a) Amino acid catabolism in *Xanthomonas campestris* pathogenesis. *Proceedings of the Seventh International Conference on Plant Pathogenic Bacteria*, Budapest, Hungary, pp. 111–16.

Ewbank, E. and Maraite, H. (1990b) Conversion of methionine to phytotoxic 3-methylthiopropionic acid by *Xanthomonas campestris* pv. *manihotis*. *Journal of General Microbiology*, **136**, 1185–9.

Federation of British Plant Pathologists (1973) A guide to the use of terms in plant pathology. Phytopathology Paper 17, Commonwealth Mycological Institute, Kew, Surrey, England, 55 pp.

Feistner, G. (1990) Pigments, in *Methods in Phytobacteriology* (eds Z. Klement, K. Rudolph and D.C. Sands), Akadémiai Kiadó, Budapest, pp. 233–44.

Feliciano, A. and Daines, R.H. (1970) Factors influencing ingress of *Xanthomonas pruni* through peach leaf scars and subsequent development of spring cankers. *Phytopathology*, **60**, 1720–6.

Feng, T.Y. and Kuo, T.T. (1975) Bacterial leaf blight of rice plants. VI. Chemotactic responses of *Xanthomonas oryzae* to water droplets exudated from water pores on the leaf of rice plants. *Acad. Sin. Inst. Bot. Bull.* (Taipei), **16**, 126–36.

Fett, W.F. (1984) Accumulation of isoflavonoids and isoflavone glucosides after inoculation of soybean leaves with *Xanthomonas campestris* pv. *glycines* and pv. *campestris* and a study of their role in resistance. *Physiological Plant Pathology*, **24**, 303–20.

Fett, W.F. and Osman, S.F. (1982) Inhibition of bacteria by the soybean isoflavonoids glyceollin and coumestrol. *Phytopathology*, **72**, 755–60.

Fett, W.F., Osman, S.F. and Dunn, M.F. (1987) Auxin production by plant-pathogenic pseudomonads and xanthomonads. *Applied and Environmental Microbiology*, **53**, 1839–45.

Fett, W.F. and Sequeira, L. (1980a) A new bacterial agglutinin from soybean. I. Isolation, partial purification, and characterization. *Plant Physiology*, **66**, 847–52.

Fett, W.F. and Sequeira, L. (1980b) A new bacterial agglutinin from soybean. II. Evidence against a role in determining pathogen specificity. *Plant Physiology*, **66**, 853–8.

Fischer, E. (1981) Relation between the transmembrane potential, the energy status, and the ion content in pepper fruit tissue affected by bacterial infection. PhD dissertation, Department of Plant Pathology, University of Missouri-Columbia.

Fujimoto, D.K., Leach, J.E. and Vidaver, A.K. (1991) Identification of proteinase genes from *Xanthomonas campestris* pv. *phaseoli. Phytopathology*, **81**, 1188.

Gabriel, D.W. (1989) Genetics of plant parasite populations and host–parasite specificity, in *Plant–Microbe Interaction III* (eds T. Kosuge and E.W. Nester), MacMillan, New York, pp. 343–79.

Gabriel, D.W., Loschke, D.C. and Rolfe, B.G. (1988) Gene-for-gene recognition: the ion channel defense model, in *Molecular Plant–Microbe Interactions* (eds D.P. Verma and R. Palacios), APS Press, St Paul, MN, pp. 3–14.

Gandhi, S.K. and Chand, J.N. (1988) Plant age, temperature, and free moisture in relation to susceptibility of clusterbean to *Xanthomonas campestris* pv. *cyamopsidis. Proceedings of the Indian Academy of Sciences, Plant Science*, **98**, 49–54.

Gasperini, C., Bazzi, C. and Mazzucchi, U. (1984) Autumn inoculation of *Xanthomonas campestris* pv. *pruni* through leaf scars in plum trees in the Po valley. *Phytopathologia Mediterranea*, **23**, 60–2.

Gholson, R.K., Rodgers, C. and Pierce, M. (1989) Extracellular proteases of *Xanthomonas campestris* pv. *malvacearum. Phytopathology*, **79**, 1199.

Gnanamanickam, S.S. and Smith, D.A. (1980) Selective toxicity of isoflavonoid phytoalexins to Gram-positive bacteria. *Phytopathology*, **70**, 894–6.

Goel, R.K. and Gupta, A.K. (1990) Host age in relation to resistance in rice to bacterial blight caused by *Xanthomonas campestris* pathovar *oryzae. Tropical Agriculture*, **67**, 368–70.

Gonzalez, C.F., Xu, G.W., Li, H.L. and Cosper, J.W. (1991) *Leersia hexandra*, an alternative host for *Xanthomonas campestris* pv. *oryzae* in Texas. *Plant Disease*, **75**, 159–62.

Goodman, C.A. and Hattingh, M.J. (1988) Mechanical transmission of *Xanthomonas campestris* pv. *pruni* in plum nursery trees. *Plant Disease*, **72**, 643.

Goodman, R.N. (1982) The infection process, in *Phytopathogenic Prokaryotes*, vol. 1 (eds M.S. Mount and G.H. Lacy), Academic Press, New York, pp. 31–62.

Goodman, R.N., Kiraly, Z. and Wood, K.R. (1986) *The Biochemistry and Physiology of Plant Disease*, University of Missouri Press.

Goto, M., Huang, B.L., Makino, T., Goto, T. and Inaba, T. (1988) A taxonomic study on ice nucleation-active bacteria isolated from gemmisphere of tea (*Thea sinensis* L.), phyllosphere of vegetables and flowers of *Magnolia denudata* Desr. *Annals of the Phytopathological Society of Japan*, **54**, 189–97.

Goto, M. and Okabe, N. (1959) Studies on the cellulolytic enzymes of phytopathogenic bacteria. Part 4. On the nature of cellulose in the lesions of black rot of cauliflower (*Xanthomonas campestris*) and Citrus canker (*Xanthomonas citri*). *Report of the Faculty of Agriculture, Shizuoka University*, **9**, 21–3.

Goto, M. and Okabe, N. (1960) Studies on the cellulolytic enzymes of phytopathogenic bacteria. Part 3. Some experiments on the mechanism of the decomposition of cellulose on its derivatives due to cellulases. *Report of the Agricultural Research Branch, Shizuoka University Bulletin*, **10**, 27–31.

Goto, M., Tadauchi, Y. and Okabe, N. (1979) Interaction between *Xanthomonas citri* and *Erwinia herbicola in vitro* and *in vivo. Annals of the Phytopathological Society of Japan*, **45**, 618–24.

Gottwald, T.R., Graham, J.H. and Richie, S.M. (1992) Relationship of leaf surface populations of strains of *Xanthomonas campestris* pv. *citrumelo* to development of citrus bacterial spot and persistence of disease symptoms. *Phytopathology*, **82**, 625–32.

Gottwald, T.R., Miller, C., Brlansky, R.H., Gabriel, D.W. and Civerolo, E.L. (1989) Analysis of the spatial distribution of citrus bacterial spot in a Florida citrus nursery. *Plant Disease*, **73**, 297–303.

Gough, C., Genin, S., Zischek, C., *et al.* (1992) Possible functions for *Pseudomonas solanacearum hrp* genes and conservation among Gram-negative phytopathogenic bacteria. *Proceedings of the Eighth International Conference on Plant Pathogenic Bacteria*, 9–12 June, Versailles, France.

Groth, D.E. and Braun, E.J. (1986) Growth kinetics and histopathology of *Xanthomonas campestris* pv. *glycines* in leaves of resistant and susceptible soybeans. *Phytopathology*, **76**, 959–65.

Hammerschlag, F.A. (1984) Optical evidence for an effect of culture filtrates of *Xanthomonas campestris* pv. *pruni* on peach mesophyll cell membranes. *Plant Science Letters*, **34**, 295–304.

Hammerschlag, F.A. and Bottino, P. (1978) *In vitro* activity of a low molecular weight toxic fraction from culture filtrates of *Xanthomonas pelargonii* and *X. pruni*. *Phytopathology News*, **12**, 155–6.

Häusermann, E. (1944) Über die Benetzungsgröße der Mesophyllinterzellularen. *Ber. Schweiz Bot. Ges.*, **54**, 541–78.

Hayward, A.C. (1977) Occurrence of glycoside hydrolases in plant pathogenic and related bacteria. *Journal of Applied Bacteriology*, **43**, 407–11.

Helander, I., Salkinoja-Salonen, M. and Rylander, R. (1980) Chemical structure and inhalation toxicity of lipopolysaccharides from bacteria on cotton. *Infection and Immunity*, **29**, 859–62.

Hildebrand, D.C., Alosi, M.C. and Schroth, M.N. (1980) Physical entrapment of pseudomonads in bean leaves by films formed at air–water interfaces. *Phytopathology*, **70**, 98–101.

Hirano, S.S. and Upper, C.D. (1983) Ecology and epidemiology of foliar bacterial plant pathogens. *Annual Review of Phytopathology*, **21**, 243–69.

Hokawat, S. (1988) Pathophysiological characterization of the susceptible and resistant reaction of soybeans against the pathogen of pustule disease (*Xanthomonas campestris* pv. *glycines* Nakano, Dye). Dissertation, Universität Göttingen.

Hokawat, S. and Rudolph, K. (1988a) Effect of extracellular polysaccharides of *Xanthomonas campestris* pv. *glycines* on multiplication and survival of the pathogen, in *Proceedings of the 26th Science Conference, Plant Division*, 3–5 February 1988, Kasetsart, Thailand, pp. 23–30.

Hokawat, S. and Rudolph, K. (1988b) Virulence of several strains of *Xanthomonas campestris* pv. *glycines* and reaction of different soybean cultivars against the pathogen, in *Proceedings of the 26th Science Conference, Plant Division*, 3–5 February 1988, Kasetsart, publ. by Kasetsart University and Ministr. Agricult. & Cooperative and Minstr. Technol. & Energy, Kasetsart, Thailand, pp. 49–56.

Holliday, M.J. and Keen, N.T. (1982) The role of phytoalexins in the resistance of soybean leaves to bacteria: effect of glyphosate on glyceollin accumulation. *Phytopathology*, **72**, 1470–4.

Horino, O. (1981) Ultrastructural histopathology of rice leaves infected with *Xanthomonas campestris* var. *oryzae* on Kogyoku group rice varieties with different levels of resistance at seedling stage. *Annals of the Phytopathological Society of Japan*, **47**, 501–9.

Horino, O. (1984) Ultrastructure of water pores in *Leersia japonica* Makino and *Oryza sativa* L.: its correlation with the resistance to hydathodal invasion of *Xanthomonas campestris* pv. *oryzae*. *Annals of the Phytopathological Society of Japan*, **50**, 72–6.

Howell, R.K. and Graham, J.H. (1977) Interaction of ozone and bacterial leafspot of alfalfa. *Plant Disease Reporter*, **61**, 565–7.

Hseu, S.H., Hsu, S.T., Tzeng, K.C. and Tsai, T.W. (1990) Light and electron microscopy of tomato leaves infected with *Xanthomonas campestris* pathovar *vesicatoria*. *Plant Protection Bulletin*, **32**(3), 218–28.

Hsieh, S.P.Y. and Buddenhagen, I.W. (1974) Suppressing effects of *Erwinia herbicola* on infection by *Xanthomonas oryzae* and on symptom development in rice. *Phytopathology*, **64**, 1182–5.

Hsu, S.-T. and Dickey, R.S. (1972) Interaction between *Xanthomonas phaseoli, Xanthomonas vesicatoria, Xanthomonas campestris* and *Pseudomonas fluorescens* in bean and tomato leaves. *Phytopathology*, **62**, 1120–6.

Huang, J.-S. (1986) Ultrastructure of bacterial penetration in plants. *Annual Review of Phytopathology*, **24**, 141–57.

Jalali, B.L., Singh, G. and Grover, R.K. (1976) Role of phenolics in bacterial blight resistance in cotton. *Acta Phytopath. Acad. Scient. Hung.*, **11**, 81–3.

Jiao, H.J., Wang, S.Y. and Civerolo, E.L. (1992) Lipid composition of citrus leaves from plants resistant and susceptible to citrus bacterial canker. *Journal of Phytopathology*, **135**, 48–56.

Jones, J.B., Pohronezny, K.L., Stall, R.E. and Jones, J.P. (1986) Survival of *Xanthomonas campestris* pv. *vesicatoria* in Florida on tomato crop residue, weeds, seeds, and volunteer tomato plants. *Phytopathology*, **76**, 430–4.

Jones, J.B. and Scott, J.W. (1986) Hypersensitive response in tomato to *Xanthomonas campestris* pv. *vesicatoria*. *Plant Disease*, **70**, 337–9.

Jones, S.B. and Fett, W.F. (1985) Fate of *Xanthomonas campestris* infiltrated into soybean leaves: an ultrastructural study. *Phytopathology*, **75**, 733–41.

Jones, S.B. and Fett, W.F. (1987) Bacterial pustule disease of soybean: microscopy of pustule development in a susceptible cultivar. *Phytopathology*, **77**, 266–74.

Kaiser, W.J. and Ramos, A.H. (1979) Two bacterial diseases of cowpea in East Africa. *Plant Disease Reporter*, **63**, 304–8.

Kaiser, W.J. and Vakili, N.G. (1978) Insect transmission of pathogenic xanthomonads to bean and cowpea in Puerto Rico. *Phytopathology*, **68**, 1057–63.

Kamiunten, H. and Fujita, T. (1990) Electrophoretic analysis of the lipopolysaccharide from *Xanthomonas campestris* pv. *oryzae* and *Pseudomonas avenae*. *Kyushu Byogaichu Kenkyukaiho*, **36**, 14–15.

Kamoun, S. and Kado, C.I. (1990) Phenotypic switching affecting chemotaxis, xanthan production, and virulence in *Xanthomonas campestris*. *Applied and Environmental Microbiology*, **56**, 3855–60.

Kamoun, S., Kamdar, H. and Kado, C.I. (1991) Genetic and molecular analysis of *Xanthomonas campestris* pathogenesis. *Journal of Cell Biochemistry*, Suppl 15, Part A, p. 44.

Kaul, M.L.H. and Sharma, K.K. (1987) Bacterial blight in rice – a review. *Biol. Zentr.bl.*, **106**, 141–67.

Kawalek, M.D., Poplawsky, A.R. and Schaad, N.W. (1989) Molecular cloning of xanthomonadin pigment synthesis genes from *Xanthomonas campestris* for identification of xanthomonads. *Phytopathology*, **79**, 1182.

Kearney, B. and Staskawicz, B. (1990) Widespread distribution and fitness contribution of *Xanthomonas campestris* avirulence gene avrBs2. *Nature*, **346**, 385–6.

Keen, N.T. and Holliday, M.J. (1982) Recognition of bacterial pathogens by plants, in *Phytopathogenic Prokaryotes II* (eds M.S. Mount and G.H. Lacy), Academic Press, New York, pp. 179–217.

Kelman, A. and Hruschka, J. (1973) The role of motility and aerotaxis in the selective increase of avirulent bacteria in still broth cultures of *Pseudomonas solanacearum*. *Journal of General Microbiology*, **76**, 177–88.

Kim, H.K., Orser, C., Lindow, S.E. and Sands, D.C. (1987) *Xanthomonas campestris* pv. *translucens* strains active in ice nucleation. *Plant Disease*, **71**, 994–7.

Kim, S.-G. and Yoo, J.-Y. (1992) Changes in protein patterns resulting from infection of rice leaves with *Xanthomonas oryzae* pv. *oryzae*. *Molecular Plant–Microbe Interactions*, **5**, 356–60.

Kimmerer, T.W. and Kozlowski, T.T. (1981) Stomatal conductance and sulfur uptake of five clones of *Populus tremuloides* exposed to sulfur dioxide. *Plant Physiology*, **67**, 990–5.

Kingsley, M.T. and Gabriel, D.W. (1991) A mutation in *Xanthomonas campestris* pv. *citrumelo* affects both host-specific virulence and exopolysaccharide production. *Phytopathology*, **81**, 1195.

Király, Z. and Barna B. (1985) Hypersensitivity revisited. *Acta Phytopath. Academiae Scientiarum Hungaricae*, **20**, 3–6.

Klement, Z. (1982) Hypersensitivity, in *Phytopathogenic Prokaryotes*, Vol. II (eds M.S. Mount and G.H. Lacy), Academic Press, New York, pp. 149–77.

Klement, Z. (1990a) Induced resistance, in *Methods in Phytobacteriology* (eds Z. Klement, K. Rudolph and D.C. Sands), Adadémiai Kiadó, Budapest, pp. 477–9.

Klement, Z. (1990b) Sources of mistakes in pathological assays with genetic mutants, in *Methods in Phytobacteriology* (eds Z. Klement, K. Rudolph and D.C. Sands), Akadémiai Kiadó, Budapest, pp. 526–8.

Klement, Z., Gross, M. and Rudolph, K. (1987) Leaf necrosis instead of water-soaking due to light deficiency after inoculation with *Pseudomonas* and *Xanthomonas*, in *Plant Pathogenic Bacteria* (eds E.L. Civerolo, A. Collmer, R.E. Davis and A.G. Gillaspie), *Proceedings of the Sixth International Conference on Plant Pathogenic Bacteria*, Maryland, 2–7 June 1985, Martinus Nijhoff Publishers, Dordrecht, pp. 430–6.

Kloepper, J.W. and Schroth, M.N. (1981) Plant growth-promoting rhizobacteria and plant growth under gnotobiotic conditions. *Phytopathology*, **71**, 642–4.

Koch, M.F. and Mew, T.W. (1991) Effect of plant age and leaf maturity on the quantitative resistance of rice cultivars to *Xanthomonas campestris* pv. *oryzae*. *Plant Disease*, **75**, 901–4.

Koizumi, M. (1977) Behaviour of *Xanthomonas citri* (Hasse) Dowson and histological changes of diseased tissue in the process of lesion extension. *Annals of the Phytopathological Society of Japan*, **43**, 129–36.

Koizumi, M. (1979) Ultrastructural changes in susceptible and resistant plants of citrus following artificial inoculation with *Xanthomonas citri* (Hasse) Dowson. *Annals of the Phytopathological Society of Japan*, **45**, 635–44.

Koizumi, M. (1983) Relationship between wound-healing process of citrus leaf tissues and successful infection through wounds by *Xanthomonas campestris* pv. *citri* (Hasse) Dye. *Annals of the Phytopathological Society of Japan*, **49**, 352–60.

Kono, Y., Uzawa, J., Kobayashi, K., *et al.* (1991) Structures of oryzalides A and B, and oryzalic acid A, a group of novel antimicrobial diterpenes, isolated from healthy leaves of a bacterial leaf blight-resistant cultivar of rice plant. *Agricultural and Biological Chemistry*, **55**, 803–11.

Kumar, J. and Prasad, M. (1993) Enzymatic activities in relation to carbohydrate breakdown during pathogenesis of tomato fruits infected with *Xanthomonas campestris* pv. *vesicatoria*. *Proceedings of the Eighth International Conference on Plant Pathogenic Bacteria*, 9–12 June, Versailles, France.

Kuo, T.-T., Lin, B.-C. and Li, C.-C. (1970) Bacterial leaf blight of rice plant. III. Phytotoxic polysaccharides produced by *Xanthomonas oryzae*. *Botanical Bulletin Academia Sinica*, **11**, 46–54.

Lamb, C.J., Lawton, M.A., Dron, M. and Dixon, R.A. (1989) Signals and transduction mechanisms for activation of plant defenses against microbial attack. *Cell*, **56**, 215–24.

Laurence, J.A. and Wood, F.A. (1978) Effect of ozone on infection of wild strawberry by *Xanthomonas fragariae*. *Phytopathology*, **68**, 689–92.

Lawson, C.J. and Symes, K.C. (1977) Xanthan gum-acetolysis as a tool for elucidation of structure, in *Extracellular Microbial Polysaccharides* (eds P.A. Sandford and A. Laskin), American Chemical Society, Washington DC, pp. 183–91.

Lawson, R.H., Dienelt, M.M. and Civerolo, E.L. (1989) Histopathology of *Xanthomonas campestris* pv. *citri* from Florida and Mexico in wound-inoculated detached leaves of *Citrus aurantifolia*: light and scanning electron microscopy. *Phytopathology*, **79**, 329–35.

Leach, J.G., Lilly, V.G., Wilson, H.A. and Purvis Jr, M.R. (1957) Bacterial polysaccharides: the nature and function of the exudate produced by *Xanthomonas phaseoli*. *Phytopathology*, **47**, 113–20.

Leben, C. and Daft, G.C. (1967) Population variations of epiphytic bacteria. *Canadian Journal of Microbiology*, **13**, 151–6.

Leyns, F., De Cleene, M., Van Bogaert, G., Van De Wiele, A. and De Ley, J. (1988) Preliminary investigations about the mode of transmission and spread of *Xanthomonas campestris* pv. *graminis* on forage grasses. *Journal of Phytopathology*, **122**, 76–88.

Liao, C.H. (1989) Analysis of pectate lyases produced by soft rot bacteria associated with spoilage of vegetables. *Applied and Environmental Microbiology*, **55**, 1677–83.

Liao, C.H. and Wells, J.M. (1987) Association of pectolytic strains of *Xanthomonas campestris* with soft rots of fruits and vegetables at retail markets. *Phytopathology*, **77**, 418–22.

Lima, M.I.P.M., Romeiro, R.S. and Muchovej, J.J. (1990) Cassava agglutination factor against *Xanthomonas campestris* pathovar *manihotis*. *Fitopatologia Brasileira*, **15**, 48–52.

Lüderitz, O., Freudenberg, M.A., Galanos, C., Lehmann, V., Rietschel, E.T. and Shaw, D.H. (1982) Lipopolysaccharides of Gram-negative bacteria, in *Current Topics in Membranes and Transport*, Vol. 17, Academic Press, New York, pp. 79–151.

Lund, B.M. (1973) The effect of certain bacteria on ethylene production by plant tissue, in *Fungal Pathogenicity and the Plant's Response* (eds R.J.W. Byrde and C.V. Cuting), Academic Press, New York, pp. 69–86.

Mahadevan, A. (1991) *Post-infectional Defence Mechanisms*, Today and Tomorrows Printers and Publishers, New Delhi, 871 pp.

Manicom, B.Q. (1986) Factors affecting bacterial black spot of mangoes caused by *Xanthomonas campestris* pv. *mangiferaeindicae*. *Annals of Applied Biology*, **109**, 129–35.

Maraite, H. and Meyer, J.A. (1975) *Xanthomonas manihotis* (Arthaud-Berthet) Starr, causal agent of bacterial wilt, blight and leaf spots of cassava in Zaire. *PANS*, **21**, 27–37.

Martin Jr, W.R., Sterling, W.L., Kenerley, C.M. and Morgan, P.W. (1988) Transmission of bacterial blight of cotton, *Xanthomonas campestris* pv. *malvacearum*, by feeding of the cotton fleahopper: implications for stress ethylene-induced square loss in cotton. *Journal of Entomological Science*, **23**, 161–8.

Mason, M.M. (1935) A comparison of the maximal growth rates of various bacteria under optimal conditions. *Journal of Bacteriology*, **29**, 103–10.

Matthee, F.N. and Daines, R.H. (1969) The influence of nutrition on susceptibility of peach foliage to water congestion and infection by *Xanthomonas pruni*. *Phytopathology*, **59**, 285–7.

Mazzucchi, U., Bazzi, C. and Pupillo, P. (1979) The inhibition of susceptible and hypersensitive reactions by protein–lipopolysaccharide complexes from phytopathogenic pseudomonads: relationship to polysaccharide antigenic determinants. *Physiological Plant Pathology*, **14**, 19–30.

McGuire, R.G., Jones, J.B., Stanley, C.D. and Csizinszky, A.A. (1991) Epiphytic populations of *Xanthomonas campestris* pv. *vesicatoria* and bacterial spot of tomato as influenced by nitrogen and potassium fertilization. *Phytopathology*, **81**, 656–60.

McPherson, G.M. (1981) *Xanthomonas pelargonii* infections of *Pelargonium*. PhD thesis, The University of Leeds, UK.

Meadows, M.E. and Stall, R.E. (1981) Different induction periods for hypersensitivity in pepper to *Xanthomonas vesicatoria* determined with antimicrobial agents. *Phytopathology*, **71**, 1024–7.

Mew, T.W., Mew, I.P.C. and Huang, J.S. (1984) Scanning electron microscopy of virulent and avirulent strains of *Xanthomonas campestris* pv. *oryzae* on rice leaves. *Phytopathology*, **74**, 635–41.

Miles, M.J., Lee, I. and Atkins, E.D.T. (1991) Molecular resolution of polysaccharides by scanning tunneling. *J. Vac. Sci. Technol., B*, **9**, 1206–9.

Miles, W.G., Daines, R.H. and Rue, J.W. (1977) Presymptomatic egress of *Xanthomonas prunii* from infected peach leaves. *Phytopathology*, **67**, 895–7.

Misawa, T. and Miyazaki, E. (1972) Studies on the leaf blight of rice plants (I) Alteration of content of carbohydrates, nitrogenous and phosphorus compounds in the diseased leaves. *Annals of the Phytopathological Society of Japan*, **38**, 375–80.

Mitchell, R.E. (1991) Coronatine analogues produced by *Xanthomonas campestris* pv. *phormiicola*. *Phytochemistry*, **30**, 3917–20.

Mohanty, S.K. and Reddy, P.R. (1982) Effect of calcium and magnesium on the susceptibility of rice plants to bacterial leaf blight. *Current Science*, **51**, 298–9.

Mohanty, S.K., Reddy, P.R. and Sridhar, R. (1983) Effect of major nutrients on the susceptibility of rice plants to bacterial leaf blight. *Z. Pfl.Krankh. Pfl.Schutz*, **90**, 50–4.

Morgham, A.T., Richardson, P.E., Essenberg, M. and Cover, E.C. (1988) Effects of continuous dark upon ultrastructure, bacterial populations and accumulation of phytoalexins during interactions between *Xanthomonas campestris* pv. *malvacearum* and bacterial blight susceptible and resistant cotton. *Physiological and Molecular Plant Pathology*, **32**, 141–62.

Morris, E.R. (1979) Polysaccharide structure and conformation in solutions and gels, in *Polysaccharides in Food* (eds J.M.V. Blanshard and J.R. Mitchell), Butterworth, London, pp. 15–31.

Morris, E.R., Rees, D.A., Young, G., Walkinshaw, M.D. and Darke, A. (1977) Order–disorder transition for a bacterial polysaccharide in solution. A role for polysaccharide conformation in recognition between *Xanthomonas* pathogen and its plant host. *Journal of Molecular Biology*, **110**, 1–16.

Mulrean, E.N. and Schroth, M.N. (1982) Ecology of *Xanthomonas campestris* pv. *juglandis* on Persian (English) walnuts. *Phytopathology*, **72**, 434–8.

Nakanishi, K. and Watanabe, M. (1977) Studies on the mechanisms of resistance of rice plants against *Xanthomonas oryzae*. IV. Extraction and partial purification of antibacterial substances from infected leaves. *Annals of the Phytopathological Society of Japan*, **43**, 449–54.

Namasivayam, L., Hedge, R.K. and Balasubramanian, A. (1971) Studies on the biochemical changes in cabbage plants infected with *Xanthomonas campestris* (Pam.) Dow. *Phytopathologia Mediterranea*, **10**, 63–7.

Noda, T., Sato, Z., Kobayashi, H., Iwasaki, S. and Okuda, S. (1980) Isolation and structural elucidation of phytotoxic substances produced by *Xanthomonas campestris* pv. *oryzae* (Ishiyama) Dye. *Annals of the Phytopathological Society of Japan*, **46**, 663–6.

Novacky, A. (1980) Disease-related alteration in membrane function, in *Plant Membrane Transport: current conceptual issues* (eds R.M. Spanswick, W.J. Lucas and J. Dainty), Elsevier/North-Holland Biomedical Press, Amsterdam, pp. 369–80.

Novacky, A. and Ullrich-Eberius, C.I. (1982) Relationship between membrane potential and ATP level in *Xanthomonas campestris* pv. *malvacearum* infected cotton cotyledons. *Physiological Plant Pathology*, **21**, 237–49.

Ota, T. (1983a) Interactions *in vitro* and *in vivo* between *Xanthomonas campestris* pv. *citri* and antagonistic *Pseudomonas* sp. *Annals of the Phytopathological Society of Japan*, **49**, 308–15.

Ota, T. (1983b) Production of phytoalexin-like substances in the citrus leaves inoculated with a bacterium (*Pseudomonas* sp.) antagonistic against *Xanthomonas campestris* pv. *citri*. *Annals of the Phytopathological Society of Japan*, **49**, 676–82.

Pandian, S.K., Somasundaram, T., Jayaraman, J. and Lakshmanan, M. (1982) A new toxin isolated from *Xanthomonas malvacearum* which inhibits mitochondrial ATP-ADP translocase. *Archives of Biochemistry and Biophysics*, **216**, 768–70.

Paxton, J.D. (1981) Phytoalexins – a working redefinition. *Phytopathologische Zeitschrift*, **101**, 106–9.

Perreaux, D., Maraite, H. and Meyer, J.A. (1982) Identification of 3(methylthio) propionic acid as a blight inducing toxin produced by *Xanthomonas campestris* pv. *manihotis in vitro*. *Physiological Plant Pathology*, **20**, 313–19.

Persley, G.J. (1979) Epiphytic survival of *Xanthomonas manihotis* in relation to the disease cycle of cassava bacterial blight, in *Proceedings of the Fourth International Conference on Plant Pathogenic Bacteria*, Angers, France (ed. Station de Pathologie Végétale et Phytobactériologie), **2**, 773–7.

Philip, R. and Devadath, S. (1980) Relation between some leaf characters and the occurrence of bacterial blight of rice. *Riso* **29**, 317–22.

Philip, R. and Devadath, S. (1981) Studies on the physiology of bacterial blight infected tolerant and susceptible rice cultivars. *Phytopathologische Zeitschrift*, **101**, 65–71.

Philip, R. and Devadath, S. (1984) Effect of micronutrients on the development of bacterial blight of rice. *Phytopathologische Zeitschrift*, **111**, 91–5.

Pierce, L., Schroth, M.N. and McCain, A.H. (1990) Viscosity test for preliminary identification of strains of *Xanthomonas campestris*. *Plant Disease*, **74**, 646–7.

Pierce, M. and Essenberg, M. (1987) Localization of phytoalexins in fluorescent mesophyll cells isolated from bacterial blight-infected cotton cotyledons and separated from other cells by fluorescence-activated cell sorting. *Physiological and Molecular Plant Pathology*, **31**, 273–90.

Prasad, M. (1987) Capability of phytopathogenic bacteria to secrete wall degrading enzymes – a criterion for their identification, in *Plant Pathogenic Bacteria* (eds E.L. Civerolo, A. Collmer, R.E. Davis and A.G. Gillaspie), Martinus Nijhoff Publishers, Dordrecht, pp. 862–75.

Prasad, M. and Bhushan, A. (1987) The influence of carbon nutrition on activation and activities of pectolytic enzyme in cultures of *Xanthomonas campestris* pv. *vignicola*. *Zentralblatt Mikrobiologie*, **142**, 631–8.

Preece, T.F. (1982) The progression of bacterial diseases in plants, in *Bacteria and Plants* (eds M.E. Rhodes-Roberts and F.A. Skinner), Academic Press, New York, pp. 71–83.

Pruvost, O. and Gardan, L. (1988) Dynamics of epiphytic survival of *Xanthomonas campestris* pathovar *corylina, Xanthomonas campestris* pathovar *juglandis, Erwinia herbicola* and *Pseudomonas paucimobilis* on hazelnut leaves. *Agronomie,* **8**, 925–32.

Rakesh, S., Bhatnagar, M.G. and Dashora, P.K. (1992) Metereological factors influencing bacterial blight of cowpea. *Proceedings of the Eighth International Conference on Plant Pathogenic Bacteria*, Versailles, France, 9–12 June.

Rama Raje Urs, N.V. and Dunleavy, J.M. (1974) Function of peroxidase in resistance of soybean to bacterial pustule. *Crop Science*, **14**, 740–4.

Ramirez, M.E., Fucikovsky, L., Garcia-Jiminez, F., Quintero, R. and Galindo, E. (1988) Xanthan gum production by altered pathogenicity variants of *Xanthomonas campestris. Applied Microbiology and Biotechnology*, **29**, 5–10.

Ramos, L.J., Narayanan, D.R. and McMillan Jr, R.T. (1992) Association of stomatal frequency and morphology in *Lycopersicon* species with resistance to *Xanthomonas campestris* pv. *vesicatoria. Plant Pathology*, **41**, 157–64.

Randhawa, P.S. and Civerolo, E.L. (1986) Interaction of *Xanthomonas campestris* pv. *pruni* with pruniphage and epiphytic bacteria on detached peach leaves. *Phytopathology*, **76**, 549–53.

Rao, C.S., Devadath, S. and Dath, A.P. (1978) Effect of three antagonists on the development of bacterial leaf streak of rice. *Canadian Journal of Microbiology*, **24**, 1010–12.

Rao, K.S.P. and Hiremath, R.V. (1985) Influence of host nutrition on the severity of bacterial blight of cowpea. *Indian Phytopathology*, **38**, 81–4.

Rao, N.S.R.K. and Nayudu, M.V. (1979) Enzymological and permeability changes involved in bacterial leaf blight of rice. *Phytopathologische Zeitschrift*, **96**, 77–82.

Reddy, M.N., Stuteville, D.L. and Sorensen, E.L. (1971) Protease production during pathogenesis of bacterial leaf spot of alfalfa and by *Xanthomonas alfalfae in vitro. Phytopathology*, **61**, 361–5.

Reddy, M.N., Stuteville, D.L. and Sorensen, E.L. (1974) Xylanase activity of *Xanthomonas alfalfae* in culture and during pathogenesis of bacterial leaf spot of alfalfa. *Phytopathologische Zeitschrift*, **80**, 215–23.

Reddy, P.R., Nayak, P. and Sridhar, R. (1978) Physiology of bacterial leaf blight of rice: influence of light intensity on some biochemical changes associated with the disease development. *Indian Phytopathology*, **30**, 51–4.

Reimers, P.J. and Leach, J.E. (1989) Molecular characterization of race-specific defense responses in the rice–*Xanthomonas campestris* pv. *oryzae* interaction. *Journal of Cell Biochemistry* Suppl. p. 325.

Reimers, P.J. and Leach, J.E. (1991) Race-specific resistance to *Xanthomonas oryzae* pv. *oryzae* conferred by bacterial blight resistance gene Xa-10 in rice (*Oryza sativa*) involves accumulation of a lignin-like substance in host tissues. *Physiological and Molecular Plant Pathology*, **38**, 39–55.

Reynolds, K.L. and Laurence, J.A. (1990) Development of common blight and accumulation of fluoride in Red Kidney bean plants exposed continuously or intermittently to hydrogen fluoride. *Phytopathology*, **80**, 211–16.

Reynolds, K.L., Zanelli, M. and Laurence, J.A. (1987) Effects of sulfur dioxide exposure on the development of common blight in field-grown Red Kidney beans. *Phytopathology*, **77**, 331–4.

Ridé, J.P. (1983) Cell walls and other structural barriers in defence, in *Biochemical Plant Pathology* (ed. J.A. Callow), Wiley, London, pp. 215–36.

Ridé M., Ridé, S. and Poutier, J.-C. (1979) Factors affecting the survival of *Xanthomonas populi* on aerial structures of poplar, in *Proceedings of the Fourth International Conference on Plant Pathogenic Bacteria*, Angers, France, pp. 803–14.

Rietschel, E.Th., Lüderitz, O. and Volk, W.A. (1975) Nature, type of linkage, and absolute configuration of (hydroxy) fatty acids in lipopolysaccharides from *Xanthomonas sinensis* and related strains. *Journal of Bacteriology*, **122**, 1180–8.

Rist, D.L. and Davis, D.D. (1979) The influence of exposure temperature and relative humidity on the response of pinto bean foliage to sulfur dioxide. *Phytopathology*, **69**, 231–5.

Robeson, F.J. and Cook, D.R. (1985) Production of low molecular weight carboxylic acids by *Xanthomonas campestris* pv. *campestris* in relation to the amino acid composition of the medium and their possible involvement in pathogenesis. *Physiological Plant Pathology*, **26**, 219–30.

Rolfs, F.M. (1915) A bacterial disease of stone fruits. *Cornell University Agriculture Experimental Station Memoir*, **8**, 377–436.

Romantschuk, M. (1993) Fimbriae (pilus) mediated attachment of *Pseudomonas syringae*, *Xanthomonas campestris* and *Erwinia rhapontici* to plant surfaces, in *Molecular Mechanisms of Bacterial Virulence* (eds C.I. Kado and J. Crosa), Kluwer Academic Publishers, Dordrecht, in press.

Rudolph, K. (1976) Models of interaction between higher plants and bacteria, in *Specificity in Plant Diseases* (eds R.K.S. Wood and A. Graniti), Plenum Press, New York, pp. 109–26.

Rudolph, K. (1980) Multiplication of bacteria in leaf tissue. *Angew. Botanik*, **54**, 1–9.

Rudolph, K. (1990a) Toxins as taxonomic features, in *Methods in Phytobacteriology* (eds Z. Klement, K. Rudolph and D.C. Sands), Akadémiai Kiadó, Budapest, pp. 251–67.

Rudolph, K. (1990b) Bacterial growth assessment in the plant to determine degree of resistance, in *Methods in Phytobacteriology* (eds Z. Klement, K. Rudolph and D.C. Sands), Akadémiai Kiadó, Budapest, pp. 356–62.

Rudolph, K. (1990c) Special requirements for certain bacterial species, in *Methods in Phytobacteriology* (eds Z. Klement, K. Rudolph and D.C. Sands), Akadémiai Kiadó, Budapest, pp. 57–9.

Rudolph, K., Ebrahim-Nesbat, F., Mendgen, K. and Thiele, C. (1987) Cytological observations of leaf spot-causing bacteria in susceptible and resistant hosts, in *Plant Pathogenic Bacteria* (eds E.L. Civerolo, A. Collmer, R.E. Davis and A.G. Gillaspie), *Proceedings of the Sixth International Conference on Plant Pathogenic Bacteria*, Maryland, 2–7 June 1985, Martinus Nijhoff Publishers, Dordrecht, pp. 604–12.

Rudolph, K.W.E., Gross, M., Neugebauer, M., *et al.* (1989) Extracellular polysaccharides as determinants of leaf spot diseases caused by pseudomonads and xanthomonads, in *Phytotoxins and Plant Pathogenesis* (eds A. Graniti, R.D. Durbin and A. Ballio), NATO ASI Series, Vol. H27, Springer Verlag, Berlin, pp. 177–218.

Ruissen, M.A. and Gielink, A.J. (1993) The development of black rot in cabbage as a result of difference in guttation between cultivars. *Proceedings of the Eighth International Conference on Plant Pathogenic Bacteria*, 9–12 June, Versailles, France.

Saettler, A.W. and Mabagala R.B. (1987) Studies on bacterial blight resistance. *Annual Report of the Bean Improvement Cooperation*, **30**, 46–7.

Saettler, A.W. and Mount, M.S. (1982) Manipulation of plant–prokaryote interaction, in *Phytopathogenic Prokaryotes*, Vol. 2 (eds M.S. Mount and G.H. Lacy), pp. 329–42.

Santhi, D.P., Unnamalai, N. and Gnanamanickam, S.S. (1987) Epiphytic association of *Erwinia herbicola* with rice leaves infected by *Xanthomonas campestris* pv. *oryzae* and its interaction with pathogen. *Indian Phytopathology*, **40**, 327–32.

Sasser, M., Andrews, A.K. and Doganay, Z.V. (1974) Inhibition of photosynthesis diminishes antibacterial action of pepper plants. *Phytopathology*, **64**, 770–2.

Schaad, N.W. and Dianese, J.C. (1981) Cruciferous weeds as sources of inoculum of *Xanthomonas campestris* in black rot of crucifers. *Phytopathology*, **71**, 1215–20.

Schaad, N.W. and White, W.C. (1974) Survival of *Xanthomonas campestris* in soil. *Phytopathology*, **64**, 1518–20.

Schmidt, D. (1988a) *Pseudomonas fluorescens* and *Erwinia herbicola* reduce wilt of grasses caused by *Xanthomonas campestris* pv. *graminis*. *Journal of Phytopathology*, **122**, 245–52.

Schmidt, D. (1988b) Prevention of bacterial wilt of grasses by phylloplane bacteria. *Journal of Phytopathology*, **122**, 253–60.

Schmidt, D. (1989) Epidemiological aspects of bacterial wilt of fodder grasses. *Bulletin OEPP*, **19**, 89–95.

Schulte, R., Herbers, K., Fenselau, S., Balbo, I., Stall, R.E. and Bonas, U. (1991) Characterization of genes from *Xanthomonas campestris* pv. *vesicatoria* that determine avirulence and pathogenicity on pepper and tomato, in *Advances in Molecular Genetics of Plant–Microbe Interactions*, Vol. 1 (eds H. Hennecke and D.S. Verma), Kluwer Academic Publishers, Dordrecht, pp. 61–4.

Schuster, M.L. and Coyne, D.P. (1974) Survival mechanisms of phytopathogenic bacteria. *Annual Review of Phytopathology*, **12**, 199–221.

Schuster, M.L. and Sayre, R.M. (1967) A coryneform bacterium induces purple-colored seed and leaf hypertrophy of *Phaseolus vulgaris* and other Leguminosae. *Phytopathology*, **57**, 1064–6.

Sequeira, L. (1980) Defenses triggered by the invader: recognition and compatibility phenomena, in *Plant Disease, an Advanced Treatise*, Vol. 5 (eds J.G. Horsfall and E.B. Cowling), pp. 179–200.

Sharon, E., Bashan, Y., Okon, Y. and Hewis, Y. (1982) Presymptomatic multiplication of *Xanthomonas campestris* pv. *vesicatoria* on the surface of pepper leaves. *Canadian Journal of Botany*, **60**, 1041–5.

Shekhawat, G.S. and Patel, P.N. (1979) Histology of pigeon pea plant infected with *Xanthomonas cajani*. *Zeitschrift für Pflanzenkrankheiten und Pflanzenschütz*, **86**, 18–24.

Shekhawat, G.S., Patel, P.N. and Raj, S. (1977) Histology of cowpea plant infected with *Xanthomonas vignicola*. *Zeitschrift für Pflanzenkrankheiten und Pflanzenschütz*, **84**, 547–58.

Shekhawat, G.S. and Srivastava, D.N. (1972) Mode of infection in bacterial leaf streak of rice and histology of the diseased leaf. *Phytopathologische Zeitschrift*, **74**, 84–90.

Shelton, A.M. and Hunter, J.E. (1985) Evaluation of the potential of the flea beetle *Phyllotreta cruciferae* to transmit *Xanthomonas campestris* pv. *campestris*, causal agent of black rot of crucifers. *Canadian Journal of Plant Pathology*, **7**, 308–10.

Shepard, D.P. and Zehr, E.I. (1987) Temperature and moisture effects on bacterial spot of peach caused by *Xanthomonas campestris* pv. *pruni*. *Phyto-*

pathology, **77**, 1709.

Sinha, P.P. and Verma, J.P. (1983) Role of phylloplane bacteria in bacterial blight disease of cotton. *International Journal of Tropical Plant Diseases*, **1**, 125–33.

Slatyer, R.O. (1967) Plant–water relationships. *Experimental Botany*, **2**, 148–9.

Smith, M.A. and Kennedy, B.W. (1970) Effect of light on reactions of soybean to *Pseudomonas glycinea*. *Phytopathology*, **60**, 723–5.

Sridhar, R., Mohanty, S.K. and Reddy, P.R. (1979) Physiology of bacterial leaf streak of rice: relationship between leaf senescence and post-infectional changes in detached rice leaves. *Acta Phytopathol. Acad. Scient. Hungar.*, **14**, 95–101.

Srivastava, A.K. and Prasad, M. (1989) The production of protease in cultures of *Xanthomonas campestris* pv. *phaseoli* and its activities in bean leaf tissues of varied responses. *Zentralblatt Mikrobiologie*, **144**, 23–7.

Stall, R.E. and Cook, A.A. (1979) Evidence that bacterial contact with the plant cell is necessary for the hypersensitive reaction but not the susceptible reaction. *Physiological Plant Pathology*, **14**, 77–84.

Stall, R.E., Hall, C.B. and Hicks, J.R. (1982a) Development of the cell collapse phase of the hypersensitive reaction in pepper in atmospheres varying in humidity, oxygen, and carbon dioxide. *Phytopathologische Zeitschrift*, **103**, 211–17.

Stall, R.E., Marco, G.M. and Echenique, B.I.C. (1982b) Importance of mesophyll in mature-leaf resistance to cancrosis of citrus. *Phytopathology*, **72**, 1097–100.

Stapp, C. (1956) Bakterielle Krankheiten, in *Handbuch der Pflanzenkrankheiten*, Vol. 2 (eds O. Appel and H. Richter), Paul Parey Verlag, Berlin.

Starr, M.P. and Nasuno, S. (1967) Pectolytic activity of phytopathogenic xanthomonads. *Journal of General Microbiology*, **46**, 425–33.

Stead, D.E. (1990) Differentiation of commonly isolated phytopathogenic species, in *Methods in Phytobacteriology* (eds Z. Klement, K. Rudolph and D.C. Sands), Akadémiai Kiadó, Budapest, pp. 65–75.

Sun, T.J., Essenberg, M. and Melcher, U. (1989) Photoactivated DNA nicking, enzyme inactivation, and bacterial inhibition by sesquiterpenoid phytoalexins from cotton. *Molecular Plant–Microbe Interactions*, **2**, 139–47.

Sun, X.P. and Ritchie, D.F. (1990) Detection of protease produced by *Xanthomonas campestris* pv. *zinniae* in zinnia leaf tissue using the substrate β-casein. *Phytopathology*, **80**, 1023.

Suslow, T.V. and Schroth, M.N. (1982) Role of deleterious rhizobacteria as minor pathogens in reducing crop growth. *Phytopathology*, **72**, 111–15.

Sutherland, I.W. (ed.) (1977) *Surface Carbohydrates of the Prokaryotic Cell*. Academic Press, London.

Sutherland, I.W. (1985) Biosynthesis and composition of Gram-negative bacterial extracellular and wall polysaccharides. *Annual Review of Microbiology*, **39**, 243–70.

Sutherland, I.W. and Kennedy, A.F.D. (1986) Comparison of bacterial lipopolysaccharides by high-performance liquid chromatography. *Applied and Environmental Microbiology*, **52**, 948–50.

Sutherland, M.W. (1991) The generation of oxygen radicals during host plant responses to infection. *Physiological and Molecular Plant Pathology*, **39**, 79–93.

Swarup, S., Yang, Y., Kingsley, M.T. and Gabriel, D.W. (1992) A *Xanthomonas citri* pathogenicity gene, *pthA*, pleiotropically encodes gratuitous avirulence on nonhosts. *Molecular Plant–Microbe Interactions*, **5**, 204–13.

Tabei, H. (1967) Anatomical studies of rice plant affected with bacterial leaf blight, with special reference to stomatal infection at the coleoptile and the foliage leaf sheath of rice seedling. *Annals of the Phytopathological Society of Japan*, **33**, 12–16.

Takahashi, T. and Doke, N. (1983) Agglutination of *Xanthomonas campestris* pv. *citri*, a causal pathogen of citrus canker, by proteinaceous components from citrus leaves. *Annals of the Phytopathological Society of Japan*, **49**, 600–9.

Takahashi, T. and Doke, N. (1984) A role of extracellular polysaccharides of *Xanthomonas campestris* pv. *citri* in bacterial adhesion to citrus leaf tissues in preinfectious stage. *Annals of the Phytopathological Society of Japan*, **50**, 565–73.

Tako, M. (1991a) Synergistic interaction between deacylated xanthan and galactomannan. *Journal of Carbohydrate Chemistry*, **10**, 619–33.

Tako, M. (1991b) Synergistic interaction between xanthan and tara bean gum. *Carbohydrate Polymers*, **16**, 239–52.

Tamura, K., Takikawa, Y., Tsuyumu, S., Goto, M. and Kijima, T. (1988) Coronatine production by *Xanthomonas campestris* pv. *phormiicola*. *Proceedings of the International Congress on Plant Pathology*, Kyoto, 1988.

Thomas, C.A. (1965) Effect of photoperiod and nitrogen on reaction of sesame to *Pseudomonas sesami* and *Xanthomonas sesami*. *Plant Disease Reporter*, **49**, 119–20.

Timmer, L.W. (1989) Inoculum production from Asiatic citrus canker lesions and epiphytic survival of *Xanthomonas campestris* pv. *citri* in Argentina. *Phytopathology*, **79**, 1182.

Timmer, L.W., Gottwald, T.R. and Zitko, S.E. (1991) Bacterial exudation from lesions of Asiatic citrus canker and citrus bacterial spot. *Plant Disease*, **75**, 192–5.

Timmer, L.W., Marois, J.J. and Achor, D. (1987) Growth and survival of xanthomonads under conditions nonconducive to disease development. *Phytopathology*, **77**, 1341–5.

Tripathi, H.H.S. and Vyas, K.M. (1990) Purification of glycopeptide toxin produced by *Xanthomonas campestris* pv. *betlicola*. *National Academy of Science Letters*, **13**, 263–6 (*Review of Plant Pathology*, **71**, 231).

Trujillo, P.G.E. (1989) The role of non-host plants and weeds in the epidemiological cycle of *Xanthomonas campestris* pv. *phaseoli* which causes bacterial blight in beans. *Revista de la Facultad de Agronomia, Universidad Central de Venezuela*, **15**, 69–84.

Tseng, Y.-H., Weng, S.-F., Lin, H.-M. and Lai, Y.-F. (1983) Pleiotropic effects in non-mucoid mutants of *Xanthomonas campestris* pv. *oryzae*. *Proceedings of the National Science Council*, Part B: Basic Science, National Science Council, Taipei, Taiwan, Republic of China, Vol. 7, pp. 44–50.

Tsuno, K. and Wakimoto, S. (1983) Ultrastructural investigations on the host–parasite interactions in the rice leaf parenchymatous tissue infiltrated with bacterial leaf blight and leaf streak pathogens. *Annals of the Phytopathological Society of Japan*, **49**, 659–69.

Turner, J.G. and Novacky, A. (1974) The quantitative relation between plant and bacterial cells involved in the hypersensitive reaction. *Phytopathology*, **64**, 885–90.

Ullrich, W.R. and Novacky, A. (1982) Ammonia accumulation and its effect on the membrane potential in bacteria-inoculated cotton and tobacco. *Phytopathology*, **72**, 981.

Urs, N.V.R. and Dunleavy, J.M. (1974) Function of peroxidase in resistance of soybean to bacterial pustule. *Crop Science*, **14**, 740–4.

Vakili, N.G. (1967) Importance of wounds in bacterial spot (*Xanthomonas vesicatoria*) of tomatoes in the field. *Phytopathology*, **57**, 1099–103.

Valladares-Sanchez, N.E., Coyne, D.P. and Schuster, M.L. (1980) Diallele analysis of genetics of reaction to *Xanthomonas phaseoli* in dry beans (*Phaseolus vulgaris*). *Annual Report of the Bean Improvement Cooperation*, **23**, 21–2.

Van Tuyl, J.M. (1982) Breeding for resistance to yellow disease of hyacinths. II. Influence of flowering time, leaf characters, stomata and chromosome number on the degree of resistance. *Euphytica*, **31**, 621–8.

Venere, R.J. (1980) Role of peroxidase in cotton resistant to bacterial blight. *Plant Science Letters*, **20**, 47–56.

Venette, J.R. (1982) How bacteria find their hosts, in *Phytopathogenic Prokaryotes*, Vol. 2 (eds M.S. Mount and G.H. Lacy), Academic Press, New York, pp. 4–30.

Verdier, V., Schmit, J. and Lemattre, M. (1990) Etude en microscopie électronique a balayage de l'installation de deux souches de *Xanthomonas campestris* pv. *manihotis* sur feuilles de vitroplants de manioc. *Agronomie*, **10**, 93–102.

Verma, J.P., Formanek, H., Singh, R.P., Trivedi, B.M. and Jindal, J.K. (1983) Chemistry and ultrastructure of surface layers of phytopathogenic bacteria. *Indian Journal of Plant Pathology*, **1**, 68–74.

Vidhyasekaran, P., Alvenda, M.E. and Mew, T.W. (1989) Physiological changes in rice seedlings induced by extracellular polysaccharide produced by *Xanthomonas campestris* pv. *oryzae*. *Physiological and Molecular Plant Pathology*, **35**, 391–402.

Volk, W.A. (1966) Cell wall lipopolysaccharides from *Xanthomonas* species. *Journal of Bacteriology*, **91**, 39–42.

Volk, W.A. (1968) Quantitative assay of polysaccharide components obtained from cell wall lipopolysaccharides of *Xanthomonas* species. *Journal of Bacteriology*, **95**, 980–2.

Wainwright, S.H. and Nelson, P.E. (1972) Histopathology of *Pelargonium* species infected with *Xanthomonas pelargonii*. *Phytopathology*, **62**, 1337–47.

Wallis, F.M., Rijkenberg, F.H.J., Joubert, J.J. and Martin, M.M. (1973) Ultrastructural histopathology of cabbage leaves infected with *Xanthomonas campestris*. *Physiological Plant Pathology*, **3**, 371–8.

Watanabe, M. and Asaumi, T. (1975) Studies on the physiological changes in the rice plants infected with *Xanthomonas oryzae*. I. Relationship between the increase in respiration rate and multiplication of the bacteria in the infected leaves. *Annals of the Phytopathological Society of Japan*, **41**, 364–72.

Webster, D.M., Temple, S.R. and Galvez, G. (1983) Expression of resistance to *Xanthomonas campestris* pv. *phaseoli* in *Phaseolus vulgaris* under tropical conditions. *Plant Disease*, **67**, 394–6.

Weller, D.M. and Saettler, A.W. (1980) Colonization and distribution of *Xanthomonas phaseoli* and *Xanthomonas phaseoli* var. *fuscans* in field-grown navy beans. *Phytopathology*, **70**, 500–6.

Wilson, H.A., Lilly, V.G. and Leach, J.G. (1965) Bacterial polysaccharides. IV. Longevity of *Xanthomonas phaseoli* and *Serratia marcescens* in bacterial exudates. *Phytopathology*, **55**, 1135–8.

Woltz, S.S. and Jones, J.P. (1979) Effects of magnesium on bacterial spot of pepper and tomato and on the *in vitro* inhibition of *Xanthomonas vesicatoria* by streptomycin. *Plant Disease Reporter*, **63**, 182–4.

Wrather, J.A., Sappenfield, W.P. and Baldwin, C.H. (1986) Colonization of cotton buds by *Xanthomonas campestris* pv. *malvacearum*. *Plant Disease*, **70**, 551–2.

Wydra, K. and Rudolph, K. (1992) Analysis of toxic extracellular polysaccharides, in *Modern Methods of Plant Analysis*, New Series, Vol. 13, *Plant Toxin Analysis* (eds H.F. Linskens and J.F. Jackson), Springer-Verlag, Berlin, pp. 113–83.

Wyman, J.G. and VanEtten, H.D. (1978) Antibacterial activity of selected isoflavonoids. *Phytopathology*, **68**, 583–9.

Xu, G.-W. and Gonzalez, C.F. (1989) Evaluation of TN4431-induced protease mutants of *Xanthomonas campestris* pv. *oryzae* for growth *in planta* and pathogenicity. *Phytopathology*, **79**, 1210.

Yang, P., Vauterin, L., Vancanneyt, M., Swings, J. and Kersters, K. (1992) Establishment of a fatty acid data base for automated and rapid identification of strains from the genus *Xanthomonas*. *Proceedings of the Eighth International Conference on Plant Pathogenic Bacteria*, 9–12 June, Versailles, France.

Young, J.M. (1974) Effect of water on bacterial multiplication in plant tissue. *New Zealand Journal of Agricultural Research*, **17**, 115–19.

Young, J.M., Luketina, R.C. and Marshall, A.M. (1977) The effects of temperature on growth *in vitro* of *Pseudomonas syringae* and *Xanthomonas pruni*. *Journal of Applied Bacteriology*, **42**, 345–54.

Yuen, J.E. (1989) Modeling the growth and fusion of lesions of *Xanthomonas campestris* pv. *dieffenbachiae* on *Anthurium*. *Phytopathology*, **79**, 1218.

Zachowski, M.A. (1989) Zur Pathophysiologie von *Xanthomonas campestris* pv. *malvacearum* (Smith) Dye, dem Erreger der eckigen Blattfleckenkrankheit, und zur Resistenz verschiedener Baumwollsorten (*Gossypium hirsutum* L.). Dissertation, Universität Göttingen.

Zachowski, M.A., Çinar, Ö. and Rudolph, K. (1989) Characterization of *Xanthomonas campestris* pv. *malvacearum* (angular leaf spot of cotton) isolates from Turkey and resistance tests of Turkish cotton cultivars. *Phytopathologia Mediterranea*, **28**, 10–15.

Zachowski, M.A., Ebrahim-Nesbat, F. and Rudolph, K. (1990) Effect of continuous darkness on production of exopolysaccharides by *Xanthomonas campestris* pv. *malvacearum* in susceptible cotton cotyledons, in *Plant Pathogenic Bacteria, Proceedings of the Seventh International Conference on Plant Pathogenic Bacteria*, Budapest, 11–16 June 1989 (ed. Z. Klement), Akadémiai Kiadó, Budapest, pp. 173–8.

Zachowski, M.A. and Rudolph, K. (1991) Reduction of bacterial blight infestation of cotton seeds by treatment with sodium hypochlorite. *Journal of Phytopathology*, **313**, 55–8.

Zapata, M. and Vidaver, A.K. (1991) Polypeptide changes detected in compatible and incompatible interactions of the tepary bean upon inoculation with different bacteria. *Phytopathology*, **81**, 700.

Zapata, M., Vidaver, A.K. and Dam, R. (1988) Extracellular cytokinin-like activity and production of IAA detected in filtrates of *Xanthomonas campestris* pv. *phaseoli*. *Phytopathology*, **78**, 1610.

Zaragoza, B.A. and Mew, T.W. (1979) Relationship of root injury to the 'Kresek' phase of bacterial blight of rice. *Plant Disease Reporter*, **63**, 1007–11.

Zhao, J.L. and Orse, C.S. (1990) Conserved repetition in the ice nucleation gene *ina*X from *Xanthomonas campestris* pv. *translucens*. *Molecular and General Genetics*, **223**, 163–6.

Zoina, A. and Volpe, E. (1992) Epidemiological aspects of lettuce bacterial spot induced by *Xanthomonas campestris* pv. *vitians*. *Proceedings of the Eighth International Conference on Plant Pathogenic Bacteria*, 9–12 June, Versailles, France.

Zucker, M. and Hankin, L. (1970) Physiological basis for a cycloheximide-induced soft rot of potatoes by *Pseudomonas fluorescens*. *Ann. Bot.*, **34**, 1047–62.

Zvoutete, P., Claflin, I.E. and Ramundo, B.A. (1991) Seedborne role and longevity of *Xanthomonas campestris* pv. *holcicola* in grain sorghum. *Phytopathology*, **81**, 1194.

5

Ecology of plant pathogenic xanthomonads

R.E. Stall[1], *T.R. Gottwald*[2], *M. Koizumi*[3] *and N.C. Schaad*[4]
[1] *Department of Plant Pathology, University of Florida*
[2] *USDA-ARS, Orlando*
[3] *Fruit Tree Research Station, Shimizu*
[4] *USDA-ARS, Foreign Disease-Weed Science Research, Frederick*

5.1 INTRODUCTION

The ecology of *Xanthomonas* involves the mutual relationships of the bacterium with biotic and abiotic environments. Increase, stationary and decline phases of growth of the bacterium result from the relationships. Special environmental conditions are also involved in the epidemiology of the bacterium. The purpose of this chapter is to review the interaction of *Xanthomonas* with various environmental factors.

Only a few of the more than 130 pathogenic variants have been studied ecologically. Those that are harmful to the most economically important plants have received the most attention. Most variations of the ecology of those studied seem to be related to the plant to which it is pathogenic. In general, however, the pathogenic phase of the bacterium contributes to growth, and in the absence of the living host plant, the bacterial population is usually stationary or declining.

Xanthomonas.
Edited by J.G. Swings and E.L. Civerolo.
Published in 1993 by Chapman & Hall, London.
ISBN 0 412 434202

Of particular importance in the studies of ecology of strains of plant pathogenic *Xanthomonas* is the accurate identification of the xanthomonads. Yellow-pigmented bacteria that are similar to *Xanthomonas* are commonly associated with leaves and seeds of many plants (Lahman and Schaad, 1985; Ou, 1985; Gitaitis *et al.*, 1987; Jones *et al.*, 1989; Angeles-Ramos *et al.*, 1991). Such bacteria are easily mistaken for *Xanthomonas.* Although the genus *Xanthomonas* has been characterized physiologically (Bradbury, 1984), few of the characteristics are diagnostic for the genus. The presence of xanthomonadins, which can be identified spectrophotometrically (Starr, 1981) or by thin-layer chromatography (Irey and Stall, 1982), is useful for identification of the pigmented strains. Recently, fatty acid profiles of xanthomonads have been extremely beneficial for identification of bacteria in the genus (Sasser, 1990).

Plant pathogenicity tests can sometimes be an aid in identification. However, caution must be used when pathogenicity tests are undertaken. Proper pathogenicity tests are particularly important when dealing with quarantined organisms such as *X. campestris* pv. *citri* and *X. oryzae* pv. *oryzae*. Appropriate inoculum concentrations and inoculation methods should be used and plants should be kept under normal growing conditions. Even saprophytic bacteria such as *Escherichia coli, Pseudomonas fluorescens* and *Enterobacter aerogenes* can produce symptoms in tobacco leaves when high levels of inoculum are used and plants are held under free moisture (Johnson, 1937). When yellow-pigmented saprophytic bacteria are present on bean seeds in large numbers, disease symptoms will develop in bean seedlings when such seeds germinate and grow under very high humidity (Lahman and Schaad, 1985).

Xanthomonads that may be saprophytic in nature have been isolated from plant tissues. An 'unusual form' of *X. campestris* was isolated from apple explants in tissue culture (Maas *et al.*, 1985). The bacterium did not cause disease in apple plants or any of several other plants that were inoculated. 'Opportunistic' xanthomonads were isolated from tomato plants and weeds in tomato production areas, and were not pathogenic on tomato seedlings (Gitaitis *et al.*, 1987). The xanthomonads were pectolytic and caused soft-rot of tomato fruit under very favourable conditions for disease. These opportunistic xanthomonads were identified as *X. campestris* by typical biochemical tests, fatty acid profiles, and by the presence of xanthomonadin pigments. Pectolytic xanthomonads isolated in and near bean fields produced red-brown and occasionally even watersoaked lesions on bean leaves and pods, but the symptoms could be distinguished from those produced by *X. c.* pv. *phaseoli* (Angeles-Ramos *et al.*, 1991). Of 17 presumptive xanthomonads isolated from weeds in bean fields, 11 had xanthomonadins and were pathogenic, whereas six had xanthomonadins, but were not pathogenic.

The ecology of these organisms with respect to their increase and decline in the environment has not been investigated.

5.2 CLIMATIC CONDITIONS FAVOURABLE FOR *XANTHOMONAS*

Plant disease in nature is markedly influenced by the environment, especially by weather conditions. Since populations of *Xanthomonas* increase during disease development, weather conditions affect the bacterium either directly or indirectly. High temperatures during early spring seasons, lower temperatures with cloudy days in mid-summer, and high temperatures in early autumn are most favourable for bacterial leaf blight of rice caused by X. *oryzae* pv. *oryzae* in Japan (Tagami and Mizukami, 1962; Mizukami and Wakimoto, 1969). Citrus canker, caused by *X. c.* pv. *citri*, only occurs when citrus growing seasons coincide with rainy seasons (Peltier and Frederich, 1926). In South America, the incidence of citrus canker increases when average temperature exceeds 20°C and precipitation in one month is over 100 mm (Alcaraz, 1978), or when mean temperatures are around 30°C and relative humidity is above 75% (Pallazzo *et al.*, 1984).

Citrus canker incidence in sweet orange can fluctuate markedly. During 1976–78, the disease was epidemic when the annual rainfall was over 1500 mm. Citrus growth occurs when the daily maximum temperatures are 20°C or more. The disease increased during the growing season when rainfall occurred frequently, i.e. 200 mm or more in one month (from June to October 1978 and from April to June 1977). In contrast, canker incidence decreased when monthly rainfall was below 100 mm in summer and autumn. Abundant rainfall in spring when new sprouts develop and occurrence of typhoons in autumn were directly related to disease prevalence. In general, fluctuations of the diseases caused by these xanthomonads are closely related to temperature, rainfall, wetness, and strong winds. Each factor influences not only the pathogen, but also the susceptibility of the host plant. In addition, interactions of two or more factors are also important.

5.2.1 Effects of temperature

Few differences occur in the effects of temperature on growth *in vitro* of *Xanthomonas* strains. For minimum temperatures, none grow at 5°C, most grow at 7°C, but some are unable to grow below 9°C. For maximum temperatures, all grow at 30°C, most grow at 35°C, and none grow at 40°C (Dye and Lelliott, 1974).

Temperature effects on growth of *Xanthomonas in vitro* may be different than growth *in vivo*. Multiplication of *X. c.* pv. *citri* inoculated into

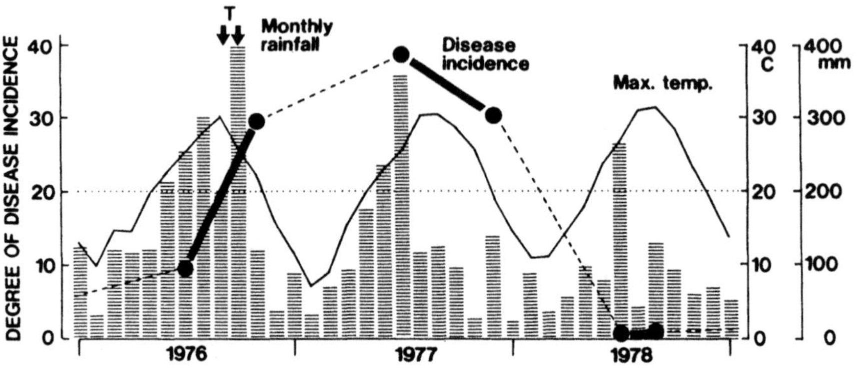

Figure 5.1 Weather conditions and the prevalence of citrus canker caused by *X. c.* pv. *citri* in the field at Kuchinotsu, Nagasaki, Japan. T indicates typhoon.

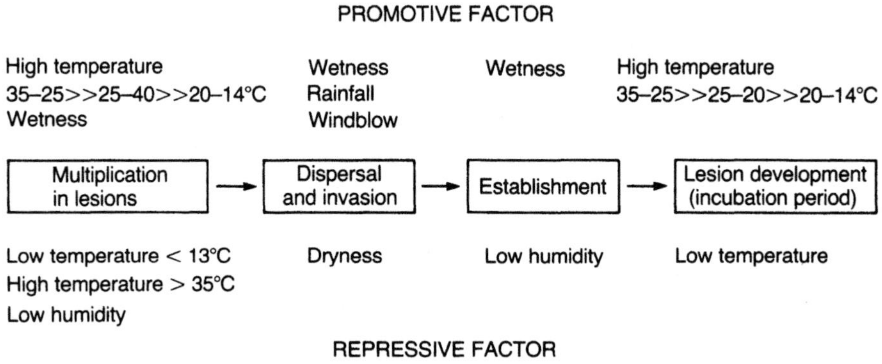

Figure 5.2 Climatic factors influencing the prevalence of citrus canker caused by *X. c.* pv. *citri.*

citrus tissues occurs in two phases, initial growth and secondary growth (Koizumi, 1976a,b). The former occurs just after inoculation into the intercellular spaces of leaves and mild decomposition of host cell walls. Growth occurs even if the temperature is 6°C and may establish bacteria in the host tissues. Secondary growth occurs abundantly after the degeneration of the plasmalemma and tonoplast adjacent to causal bacteria and results in formation of bacterial masses which move to surrounding intercellular spaces (Koizumi, 1976a,b, 1979). When citrus leaves are inoculated via wounds and kept in a moist chamber, secondary growth begins with lesion appearance. Growth *in planta* under these conditions occurs at 15–36°C, but not at 6 and 40°C (Koizumi, 1976a,c). Based on the relationship between symptom appearance and secondary growth, effective temperatures for secondary growth ranged from 13 to 36°C. The time periods between inoculation and occurrence of secondary growth are as follows: 216 h at 16°C, 102 h at 20°C, 60 h at 24°C, 43 h at 28°C, 36 h at 32°C, and 33 h at 36°C (Koizumi, 1976c).

An empirical formula ($f(x)$) of the period (h) dependent on temperature (x) can be derived from the above data. An index of disease progress (IDP) is based on the integral, $1/f(x)$, from zero to a time (h) after inoculation. Thus, host cells degenerate and bacteria multiply when the IDP reaches 1.0 (Koizumi, 1976c). This can be used to estimate the incubation period under temperature-shift conditions. Thus, lesions appear when the IDP reaches 3.0 or more if inoculation was made through stomata, and bacteria can overwinter at high population levels in the tissues where IDP has reached 1.0 in autumn (Koizumi, 1976c, 1977b,c). Lesions appear within 3–5, 6–10 and 10–15 days when temperature shifts occur around 30, 25 and 20°C, respectively. The optimum temperature for lesion development of citrus canker is about 30°C and the maximum temperature is 35°C (Peltier, 1920; Loucks, 1930).

Bacterial leaf spot of peach caused by *X. c.* pv. *pruni* developed 3, 5 and 6–8 days after inoculation when incubated at 28/23°C (day/night), 23/18°C and 18/13°C, respectively (Takanashi, 1978). Therefore, disease developed more rapidly in warm seasons. Under those high temperatures lesions turned brown and necrotic, followed by formation of shot holes. The shot holes resulted in rapid extinction of bacteria from the tree, so that the bacteria were not prevalent if occasional rainfalls occurred. Bacteria could not overwinter in cankers which developed in high temperatures in summer. In contrast, infections which occurred in autumn, from mid-September to November, produced bacteria that overwintered at high populations. The autumn infections were followed by formation of spring cankers which became important as inoculum sources in the next growing season (Thornberry and Anderson, 1933; Adam *et al.*, 1955; Feliciano and Daines, 1970; Takanashi, 1978).

Lesion extension of rice bacterial leaf blight is also influenced by temperature (Tagami and Mizukami, 1962). Alternate temperatures of 35/27°C (day/night) and 33/25°C are most favourable and those of 29/21°C or lower are markedly less favourable (Horino *et al.*, 1982). Marked extension of lesions provides abundant multiplication of bacteria and further prevalence of the disease. Symptoms of black rot, caused by *X. c.* pv. *campestris*, are not normally expressed in cabbage plants growing at temperatures below 18–20°C (Schaad, 1982). Atypical black spots commonly occur in infected cauliflower growing in the cooler coastal California areas. *Phleum pratense* is highly susceptible to *X. c.* pv. *graminis* during periods of hot dry weather, but no symptoms occur during cool, wet weather (Sletten, 1989).

A treatment of 53°C for 10 min is lethal for cells of *X. oryzae* pv. *oryzae* in the growth stage, and a treatment of 56°C for 10 min is lethal for cells in a bacterial mass that has dried out (Tagami and Mizukami, 1962). The number of *X. c.* pv. *citri* cells in a leaf lesion was markedly decreased when dipped into hot water at 48°C; the lethal point is at 48°C for 60 min, or 51°C for 15 min. Bacteria became non-detectable within 48 h when incubated in inoculated leaves at 40°C (Koizumi, 1976a). In contrast, lower temperatures such as freezing were not lethal for the bacteria.

5.2.2 Effects of moisture

Moisture as provided by rainfall, irrigation, dew, fog, and plant water congestion is necessary for bacterial survival, dispersion, invasion into host plants, and growth in the plants. When raindrops hit a lesion, bacteria are easily suspended into water. Water falling against the suspension raises abundant drops containing bacteria and provides easy transportation through the air and deposition onto a new host. When bacterial plant pathogens suspended in water are inoculated onto leaves, their establishment is markedly enhanced when water congestion occurs in the leaves which are then kept under high humidity (Wiles and Walker, 1952; Williams and Keen, 1967). In addition, movement of bacterial masses (zooagleae) in leaf tissues to surrounding intercellular spaces is promoted by wetness. When atmospheric humidity is low, bacterial masses in leaf tissues lose their mobility and are rapidly enclosed within a protective cell layer of host tissues. This induces entrapment of bacteria within limiting tissues and such lesions lose their significance as an inoculum source (Koizumi, 1977a, 1979).

5.3 LIFE CYCLE

5.3.1 Pathogenic phase

Strains of xanthomonads, unlike several of their pseudomonad relatives, such as *Pseudomonas solanacearum* and *P. syringae* pv. *syringae*, do not normally cause disease in botanically unrelated plants (Wernham, 1948; Klement, 1968; Starr, 1981). In a test of pathogenicity of 37 strains belonging to 17 *Xanthomonas* nomenspecies (currently named pathovars) to 16 'taxonomically distinct' host plants, each xanthomonad caused disease only in the plant species from which it was originally isolated (Wernham, 1948). In contrast, Dye (1958) reported infection of bean seedlings with 20 *Xanthomonas* nomenspecies after several serial passages, and concluded that *Xanthomonas* spp. were not host-specific. Under more natural conditions (Schnathorst, 1966a) host specificity of several xanthomonads after serial passage through bean plants remained unaltered. The non-specific results of Dye may have been due to too much inoculum, abnormally high humidities, and misinterpretation of true symptoms. In a few cases, a xanthomonad may infect many plant species in the same plant family, including many weeds. A typical example of a xanthomonad with a large host range within the same plant family is *X. c.* pv. *campestris*.

Disease symptoms caused by xanthomonads consist of leaf spots, streaks, wilts, cankers, chloroses and soft rots (Walker, 1969; Moffet and Croft, 1983). Disease may occur on foliage (hypocotyls, stems and leaves) and reproductive growth (flowers, fruits and seeds). Bacteria typically enter plants through injuries, stomates, hydathodes and possibly trichomes. Those xanthomonads that cause wilt, including *X. oryzae* pv. *oryzae* and *X. c.* pvs *pelargonii, manihotis* and *campestris*, enter hydathodes or stomates of leaves and then move into the vascular system (Walker, 1969; Watanabe, 1975). Such vascular infection normally leads to seed infection.

Most xanthomonads are very competent in initiation of infection. Theoretically, one cell in a favourable location in a susceptible plant can multiply to a population that will cause necrosis. However, not all *X. c.* pv. *citri* cells that enter citrus leaves cause lesions. One factor that influences the proportion of cells that initiate infection in citrus leaves is development of resistance in plant tissues (Stall *et al.*, 1982). Additional cells may be needed to insure that bacteria enter plant tissues. When applied to gutation droplets, fewer than 10 viable *X. c.* pv. *campestris* cells are needed for infection of cabbage. Rice seedlings are easily infected by dipping injured roots into a suspension containing as few as 100 cfu/ml of *X. oryzae* pv. *oryzae* (Watanabe, 1975).

Bacterial populations *in vivo* generally increase exponentially under conditions favourable for multiplication, then level out as lesions near their maximum size or host tissues become limiting. Xanthomonads grow relatively slowly in host tissues. The doubling time for *X. c.* pv. *phaseoli* in leaf tissue is 18.8 h, and 5×10^6 cfu/20 cm^2 is needed before symptoms are expressed (Weller and Saettler, 1980b). Similarly, cabbage does not normally express symptoms until the concentration of *X. c.* pv. *campestris* reaches 10^5-10^6 cfu/40 cm^2 (Schultz and Grabrielson, 1986). Populations *in vivo* are subject to wide variations due to host plant resistance, tissue age and plant vigour. With some diseases populations remain at high levels in diseased tissue for long periods and with other diseases the populations fall dramatically after necrosis occurs (Egel *et al.*, 1991). As bacterial lesions age or host tissues senesce, *in vivo* populations may decline.

5.3.2 Epiphytic phase

Like many plant pathogenic bacteria, xanthomonads may be associated with symptomless plants as epiphytes. Populations of cells may increase on the plants in which case they are called residents according to Leben's (1974) definition. This is in contrast to the short-term existence of bacteria on symptomless plants as the result of some dispersal event. The latter bacteria are termed casuals (Leben, 1965). The resident and the casual phases are difficult to distinguish in nature and most studies of the epiphytic potential of xanthomonads do not do so (Hirano and Upper, 1983). In addition, most studies on the resident nature of xanthomonads do not distinguish between growth on surfaces of plants or within the plant. Growth in either location would qualify the bacteria as residents (Leben, 1974). Xanthomonads as epiphytes are affected by foliar age, host physiology, other microflora, crop architecture, as well as meterological conditions which can range from favourable to harsh. Recoverable plant surface populations can fluctuate 10- to 1000-fold over very short time periods (Timmer *et al.*, 1987). Populations of epiphytic xanthomonads on leaves have been described by log-normal and, more recently, Weibull distributions (Weller and Saettler, 1980b; Hirano *et al.*, 1982; Ishimaru *et al.*, 1991).

X. c. pv. *vesicatoria* cells moved from the surface of germinating tomato seeds to stems of developing seedlings when held under high humidity (Leben, 1963). No symptoms of disease occurred. *X. c.* pv. *manibotis*, the causal agent of bacterial blight of cassava, normally causes leaf spots, wilting, severe leaf defoliation, and tip die-back during the rainy season, but few or no symptoms during the dry season (Daniel and Boher, 1985). The pathogen survived as an epiphyte on leaves of infected plants (without symptoms) for 8 months during the dry season. When the rainy season started, infection and symptoms occurred. The number of recoverable colony forming units (cfu) per leaf ranged from 1×10^3 to 1×10^4 or less during the dry season. *X. c.* pv. *juglandis* cells survive as epiphytes on buds and leaves of walnut trees without symptoms (Mulrean and Schroth, 1981). Populations ranged up to 5.9×10^5 and 1×10^9 cfu per bud and leaf, respectively. Such epiphytic populations serve as a source of secondary inoculum for late season infections of nuts.

X. c. pv. *phaseoli* survives as an epiphyte on bean seeds and leaves (Cafati and Saettler, 1980b; Weller and Saettler, 1980b). Up to 10^5 cfu of *X. c.* pv. *phaseoli* were recovered from healthy bean leaves of diseased plants (Cafati and Saettler, 1980b). Populations of *X. c.* pv. *phaseoli* occurred as residents in or near bean fields on eight weed species from six plant families. *X. c.* pv. *campestris* survived as an epiphyte on symptomless leaves of naturally infected *Brassica campestris* (common mustard) along roadsides of coastal California (Kuan *et al.*, 1986). Natural populations decreased after heavy rains and low night-time temperatures of 0.6–6°C. Populations ranged up to 10^6 cfu/cm^2 on symptomless leaves and non-lesion areas of leaves with symptoms and up to 10^8 cfu/cm^2 in lesion areas. In field studies, *X. c.* pv. *campestris* was recovered on the surface of leaves of cabbage for up to 73 days after being sprayed with inoculum (Schultz and Gabrielson, 1986). *X. c.* pv. *campestris* apparently resides as an epiphyte on crucifer leaves in high numbers and for relatively long periods. *X. c.* pv. *translucens* is epiphytic on many grasses, including *Bromus inermis, Agropyon repens, Poa pratensis, Festuca arundinaceae, F. ruba, Hordeum leporinum* and *Medicago sativa,* which are common in wheat and barley production areas in south central and south eastern Idaho where black chaff is endemic (Thompson *et al.*, 1989). Populations of *X. c.* pv. *translucens* ranged from 13 to 7.5 $\times$ 10^6 cfu/g fresh weight. No symptoms occurred on those weeds which are common along borders of breeder plots in Aberdeen, Idaho. Sletten (1989) used immunofluorescence (IF) and enzyme-linked immunosorbent assays (ELISA) to detect *X. c.* pv. *graminis* on symptomless *Phleum pratense.* Several other xanthomonads, including *X. c.* pvs *pruni, populi, citri, pelargonii* and *malvacearum* have resident phases on leaves of host plants.

Perhaps the most thorough study of epiphytic survival of xanthomonads on host tissue was that of Timmer *et al.* (1987). *X. c.* pv. *vesicatoria* survived on tomato leaves held under high humidity (90%) but not under low humidity (25%). Moreover, other pathovars and a yellow-pigmented xanthomonad-like saprophyte from citrus survived on tomato as well as *X. c.* pv. *vesicatoria* when the plants were held under high humidities. Using scanning electron microscopy, xanthomonad cells occurred in small aggregates on leaves of both host and non-host species. When dry conditions prevailed, these microcolonies became embedded in an exopolysaccharide matrix and tightly adhered to the plant surface. When the plant surface became wet, these dried bacterial colonies rehydrated to some extent and surface bacterial populations were renewed in part. Such results may help explain why bacterial spot of tomato is a common disease problem in humid subtropical Florida, but not in the arid West.

The role of resident or epiphytic populations in disease epidemiology is poorly understood for most diseases caused by xanthomonads. It is likely that epiphytic populations can provide inoculum for disease. For instance, common blight of bean, caused by *X. c.* pv. *phaseoli* var. *fuscans,* was always detected after epiphytic populations exceeded 5 $\times$ 10^6 cfu/leaflet (Hirano and Upper, 1983). Alternatively, the majority of surface bacteria simply may be survivors following epidemiologically significant dissemination events.

5.3.3 Saprophyte phase

Although most xanthomonads are not considered to be nutritionally fastidious in artificial culture (Bradbury, 1984), there is no evidence for their increase in populations in nature except in association with plants. Even those strains that are not pathogenic may increase as residents on living plants in nature (Timer *et al.*, 1987). However, populations of xanthomonads become static or decline in plant tissues that are killed during disease development (Stall and Cook, 1966; Barton-Willis *et al.*, 1989; Egel *et al.*, 1991). In addition, xanthomonad populations progressively decline in crop residues and in soil environments (see below).

5.3.4 Survival phase

Most bacterial plant pathogens do not form special survival structures and are generally intolerant of desiccation and sunlight (Smith, 1905). However, xanthomonads, like many bacteria, produce an exopolysaccharide (EPS) slime which behaves as a hydrophyllic colloid. The water-holding capacity of the EPS may help bacteria survive during unfavourable conditions. Aggregates of cells in plant tissues are embedded in the EPS which constitutes most of the exudates from lesions. Leben (1974) further suggested that bacteria survive more readily in a state of reduced metabolism which he termed hypobiotic. In that state nutrients are not needed and the cells can better withstand adverse stresses. The hypobiotic condition in bacteria is induced naturally as lesions age, but can be induced artificially by loss of water, unfavourable temperatures, and a variety of other means. The survival methods of xanthomonads in specific environments are given below.

5.4 EPIDEMIC DEVELOPMENT

Xanthomonad diseases often appear suddenly. The first questions that are posed in such circumstances usually pertain to the source of the bacterium. This has led to numerous studies on the means of survival and dissemination of plant pathogenic bacteria in many environmental conditions. This is important because there is a direct relationship between the concentration of inoculum and the mean disease grade (severity) (Hunter *et al.*, 1975). Therefore, the means and ability for survival and the modes of dissemination of inoculum are preludes to ensuing epidemics.

5.4.1 Initial inoculum

(a) Soil, crop refuse and roots

The dynamics of xanthomonad populations in soil have been investigated with strains of many pathovars. These experiments give only an approximation of populations in soil because the complexity of microorganisms and other edaphic factors in soil often interfere with recovery of the xanthomonads. Probably the most extensive and complete investigation of the survival of xanthomonads in the soil environment has been with *X. c.* pv. *citri* because of its relevance to procedures for eradication of citrus canker. When cells of *X. c.* pv. *citri* were added directly to soil, survival was limited to no more than one week (Fulton, 1920; Lee, 1920; Loucks, 1930). Soil pH and sterilization had minor effects on survival of *X. c.* pv. *citri* (Vasudeva, 1961, 1963). Graham *et al.* (1989) could not detect *X. c.* pv. *citri* in soil under diseased citrus seedlings 21 days after removal of the seedlings. There was a relationship between populations on symptomless leaves and in the soil under the trees on several sampling dates when the seedling trees were present. Thus, lesions on the seedlings were a constant source of bacteria for soil contamination while diseased trees were present, but once the diseased trees were removed the bacterium quickly disappeared from the soil. On the other hand, *X. c.* pv. *citri* could be detected in diseased leaves placed in soil for 85 days. The bacterium was still present in diseased leaves placed on the surface of the soil at 120 days. The more rapid decline of the bacterium in leaves in soil rather than on the soil surface may be related to the presence of moisture in the soil, because populations of bacteria in leaves placed in moist soil declined more rapidly than those placed in dry soil (Graham *et al.*, 1987). The pathogen survived considerably longer in diseased leaves in soils kept at 5–15°C than at 30°C (Vasudeva, 1963).

X. c. pv. *citri* cells survived in close association with citrus roots and roots of 17 various weed species collected in citrus orchards before new shoots developed in the spring (Goto, 1970; Goto *et al.*, 1975a,b). In association with living roots, the bacterial population did not decline during the winter months. In these studies no evidence was presented that the bacteria could cause infection of host plants. Also, the presence of xanthomonads in the rhizosphere of a host plant or a non-host plant near diseased plants may not be significant, because bacteria may be washed into the soil each time rains or dews occur on the diseased plants. Of epidemiological significance, however, may be the survival of xanthomonads in the rhizospheres of plants unrelated to the host plant and in the absence of host plants. For example, *X. c.* pv. *vesicatoria*, a pathogen of tomato and pepper, was found in association with roots of wheat plants (Diachun and Valleau, 1946). Other reports of this nature are difficult to find, however. A better understanding of the survival of xanthomonads in rhizospheres of plants unrelated to host plants is needed.

Investigations of survival in the soil environment with *X. oryzae* pv. *oryzae* (Tagami *et al.*, 1964; Horino, 1986) and *X. c.* pvs *malvacearum* (Brinkerhoff and Fink, 1964), *glycines* (Fett, 1979; Groth and Braun,

1989), *phaseoli* (Zaumyer and Thomas, 1957), *alfalfae* (Claflin and Stuteville, 1973), *campestris* (Schaad and White, 1974; Schultz and Gabrielson, 1986) and *vesicatoria* (Peterson, 1963; Jones *et al.*, 1986) have been similar to those with *X. c.* pv. *citri*. Xanthomonads survive in soil only for days when added as cells, for months when added in plant residues, and for several seasons in association with living roots.

Debris from infected plants is an excellent reservoir for survival. Overwintered rice stubble in unploughed paddies can be a source of inoculum for *X. oryzae* pv. *oryzae* (Tagami *et al.*, 1964). Cotton plant debris from fields infested with *X. c.* pv. *malvacearum* can be a reservoir for primary inoculum the following season (Brinkerhoff and Fink, 1964). In Tanganyika, diseased cotton debris left on the surface maintained pathogen populations longer than debris that had been buried (Arnold and Arnold, 1961). *X. c.* pv. *malvacearum* survived on dried cotton plants in the laboratory for up to 6 years (Ark, 1958). Although seed free of *X. c.* pv. *phaseoli* is commonly planted, common blight of bean occurs where crop rotation is not practised. The bacterium overwinters in bean straw and survives best when the debris is dry (Schuster and Coyne, 1974).

(b) Perennial plants

Xanthomonads may survive for long periods of time in perennial plants in both active and dormant tissues, especially in temperate climates (Schuster and Coyne, 1974). Walnut blight, caused by *X. c.* pv. *juglandis*, overwinters primarily in holdover cankers or blighted twigs which produce bacterial exudates in the spring. The bacterium also overwinters in leaf and catkin buds. Dusts of pollen from diseased catkins transmitted the bacterium to pistillate flowers resulting in reduced nut set and disease in leaves (Ark, 1944). This perennial source of inoculum explains why walnut blight epidemics can be so severe when only a few cankers are evident. Similarly, *X. c.* pv. *pruni* survives in dormant peach buds (Dhanvantari, 1973). *X. c.* pv. *citri* survives during cool weather in foliar and twig lesions which resume expansion when temperatures begin to rise and the host resumes its growth in the spring. Bacteria exude from these lesions onto the leaf surface when wetted by heavy dew or rain.

(c) Volunteer plants

The residue of many annual crops is either left in the field or incorporated into the soil after final harvest. Seeds of the host plant within the crop residue will often germinate before the crop residue has decomposed. The resulting volunteer seedlings are difficult to find among weeds that also grow, but they can be infected with cells of *Xanthomonas* that might be in the crop residue. These volunteer plants are particularly important for survival of xanthomonads in tropical or subtropical regions and may be important in temperate regions where seeds can survive the winter. Volunteer tomato plants along with the crop residues are the major source of survival of *X. c.* pv. *vesicatoria* in Florida (Jones *et al.*, 1986). Volunteer pepper plants also were sources of survival of the same bacterium. Volunteer rice seedlings aid in the survival of *X. oryzae* pv. *oryzae* (Thri-Murty and Devadath, 1981a).

(d) Weeds

Many xanthomonads infect weeds closely related to host plants, but symptoms in such weeds are often absent or very atypical of those on cultivated species (Schaad and Dianese, 1981; Schultz and Gabrielson, 1986; Thompson *et al.*, 1989; Angeles-Ramos *et al.*, 1991). Even though symptoms often are not clearly evident on weeds, inoculum can build up to high levels in weeds near or in production fields. Such plants as hedge mustard (*Sismbrium officinale* (L.) Scop.), *Alyssium*, Shepardspurse (*Capsella bursapastoris* (L.) Medik.), *Lepidium virginicum*, white top, wild turnip (*Brassica campestris*) and marsh yellowcress (*Rorippa islandica*) do not generally have symptoms, but can have high populations of the black rot pathogen (*X. c.* pv. *campestris*). *X. c.* pv. *phaseoli* can also survive for considerable periods on weed species, which can be primary inoculum sources for bean blight epidemics (Cafati and Saettler, 1980b).

Wild grasses, including *Leersia oryzoides* and its variety *japonica*, and *Zizamia latifolia*, can be infected with *X. oryzae* pv. *oryzae* under natural conditions and serve as overwintering hosts for the pathogen in rice paddies and, thus, sources of initial inoculum for epidemics the following season (Goto *et al.*, 1953; Yoshimura *et al.*, 1956; Inoue *et al.*, 1957; Tagami *et al.*, 1963, 1964; Yoshimura, 1963). *X. oryzae* pv. *oryzae* survives on 11 grasses and 16 other weeds commonly found near rice paddies (Isaka, 1969). An additional five more weed species sustain populations of *X. oryzae* pv. *oryzae* (Wakimoto, 1956; Mizukami, 1961). *X. c.* pv. *translucens* overwintered in winter wheat, winter rye, quack grass and brome grass (Boosalis, 1952). Many of the weeds are perennial plants and thus can harbour xanthomonads for many years.

(e) Seed contamination

Bacteria of many pathovars of *Xanthomonas* are associated with seeds of their host plants (Neergaard, 1977). Seed can have both external contamination and internal infection by xanthomonads. *X. c.* pv. *campestris* cells normally enter siliques of crucifer plants through the vascular system and infect seeds internally (Cook *et al.*, 1952). Infection occurs in the parenchyma tissue of the seed coat and endosperm. Based on histological studies of naturally contaminated cabbage seeds, *X. c.* pv. *campestris* cells are limited to xylem vessels in the endosperm. The pathogen often infects seeds even though disease symptoms in the flowering plant are not evident. Seed infection is much higher in plants receiving overhead irrigation compared to plants receiving surface water.

Bacteria, such as those of *X. c.* pvs *carotae*, *incanae* and *campestris* commonly adhere to the seed coat and then invade cotyledons of germinating seeds through stomata (Baker, 1972). *X. c.* pv. *malvacearum* cells overwinter principally in cotton-seed fuzz that remains on the seed coat after delinting (Brinkerhoff and Hunter, 1963; Hunter and Brinkerhoff, 1964). *X. oryzae* pv. *oryzae* may overwinter on rice grains stored in farmhouses. Infection of rice seeds as high as 100% has been encountered in India. Both the husk and endosperm of rice seeds were found to be infected and the pathogen was transmitted to seedlings raised under high humidity (Srivastava and Rao, 1964). Seedling infection in barley and wheat caused by *X. c.* pv. *cerealis* results from seed transmission (Wallin, 1946). *X. c.* pvs *phaseoli* and *phaseoli* var. *fuscans* can survive and initiate disease from internally infected as well as externally contaminated seed (Cafati and Saettler, 1980a; Weller and Saettler, 1980a).

Some bacteria may die before seeds lose their viability, but others survive beyond the time of seed germinability. Xanthomonads, such as *X. c.* pvs *phaseoli* and *phaseoli* var. *fuscans* have been isolated from 15-year-old bean seed (Schuster and Coyne, 1974). *X. c.* pv. *translucens* survived for more than 5 years in naturally infected wheat seeds stored outside in Idaho, but less than 5 years in seeds stored in the laboratory under constant temperature and humidity (Forster and Schaad, 1989).

5.4.2 Dissemination

Xanthomonads share the abilities and mechanisms for dissemination of most other phytopathogenic bacteria. Several mechanisms have been implicated in passive dispersal of phytopathogenic bacteria over both short and long distances. These include people, mammals, birds, reptiles, as well as insects, mites, agricultural tools and equipment, aircraft, soil particles, and water sources. Phytopathogenic xanthomonads have been isolated from nearly all of these sources (Bashan, 1986); however, the epidemiological significance of many of the mechanisms has not been demonstrated.

(a) Seeds

Epidemics can start from contaminated seed which can act as an effective long-distance dispersal mechanism for xanthomonads. Such long-distance dispersal can be not only responsible for the development of disease in previously disease-free areas, but strains of bacteria not previously important in an area may be established by seed transmission (Pohronezny *et al.*, 1992). The frequency of seed contamination and spatial distribution of infected seed will ultimately dictate disease progress and crop loss. In Canada, *X. c.* pv. *phaseoli* var. *fuscans* reached epidemic levels from contaminated seed soon after the introduction of a new cultivar (Sutton and Williams, 1970). Contamination of less than 0.5% of bean seed with this pathogen can be disastrous when the crop is grown under favourable conditions for disease development (Wallen and Sutton, 1965). The incidence of infested seeds was directly correlated to field incidence of black rot of cabbage and broccoli caused by *X. c.* pv. *campestris* (Schaad *et al.*, 1980). In India, the percentage of cowpea seeds infested with *X. c.* pv. *vignicola* was directly related to subsequent disease progress although midsummer rains and winter monsoons mitigated the differences of initial seedborne inoculum (Shekhawat and Patel, 1977).

(b) Vegetative propagating material

Xanthomonads can also be disseminated long distances by contaminated or diseased transplants, nursery stock, or host propagating materials. Citrus canker in southeast Asia and South America is common in nurseries. Subsequently, transfer of the plants to newly established orchards or to existing orchards as replants introduce new sources of inoculum or augment existing inoculum. Movement of contaminated citrus budwood disperses citrus canker bacteria for long distances and establishes new infections in previously canker-free nurseries. Citrus canker has been unintentionally transported from Asian sources to citrus-producing areas on different continents such as South Africa, South America and the USA (Stall and Seymour, 1983). Such intercontinental dissemination of exotic pathogens is increasing in direct relationship to increasing international commerce and travel (Gottwald and Garnsey, 1990). Likewise, bacterial spot of tomato becomes established in seedbeds and the diseased plants are a source of initial inoculum when transplanted to new tomato fields (Goode and Sasser, 1980). Often transplants are grown in one state or region and transported to another. Xanthomonads associated with citrus bacterial spot were disseminated when contaminated or diseased rootstock seedlings were transported to new areas and transplanted to nurseries (Gottwald and Graham, 1990).

(c) Wind and rain

Wind is a very important agent for short-distance dissemination of xanthomonads. Wind-blown dust from cotton fields infested with *X. c.* pv. *malvacearum* disseminated the angular leaf spot of cotton pathogen resulting in epidemics in adjacent cotton fields (Brown, 1942). Infested crop debris blown into new fields is an important source of *X. c.* pv. *malvacearum* inoculum, especially during the dry season when whirlwinds are prevalent (Mohan, 1983). For instance, a single whirlwind scattered infected, dry-leaf material and inoculum-laden dust over an area of 100 acres in 20 min (Rolfs, 1935).

Raindrops also accomplish short-distance bacterial dispersion. Concentrations of *X. c.* pv. *citri* in rainwater collected from infected foliage were as high as 10^5–10^6 cfu/ml and were 10^4 cfu/ml in rainwater collected under diseased plants. Detectable levels of bacteria occur in rainwater up to 32 m from diseased foliage (Stall *et al.*, 1980).

Strong winds that accompany rainfall enhance the distance of bacterial dispersion by rain. Dispersion of *X. c.* pv. *citri* from lesions was demonstrated under artificial conditions of wind and rain (Serizawa, 1981). The ratio of water-drops having bacterial cells increased with the increase in wind velocity. Winds over approximately 6 m/s, especially those over 8.5 m/s, also caused abundant injuries in leaves and enhanced infection (Serizawa and Inoue, 1975). *X. oryzae* pv. *oryzae* was also dispersed at least 64 m from infected rice plants. However, the effective dispersion that resulted in new infections was within 1 m distance under winds of 28 m/s maximum and accompanied with rainfall. Similar dispersion of *X. c.* pv. *malvacearum* occurs in cotton fields in the southern USA (Faulwetter, 1917). Epidemics of bacterial leaf stripe of arecanut palm caused by *X. c.* pv. *arecae* in India were directly related to the monsoon season; the disease was of little significance at other times of the year. As rainfall diminished at the end of the monsoon season, disease incidence decreased (Kumar, 1983). Wind-blown water droplets were thought to carry bacteria from transplants diseased with *X. c.* pv. *campestris* downwind to disease-free transplants of Brussels sprouts and cabbage (Hunter *et al.*, 1975).

(d) Irrigation

In Israel, the effect of overhead sprinkler irrigation combined with wind direction was associated with disease severity of bacterial spot of tomato. Infection by *X. c.* pv. *vesicatoria* was significantly more severe in the direction of the prevailing winds which likely dispersed inoculum-laden droplets (Volcani, 1969). The irrigation method directly affected the severity of bacterial spot of tomato in Florida as well. The rate of bacterial spot disease progress was much slower under drip than overhead sprinkler irrigation. Reducing surface moisture on fruit and foliage significantly reduced crop losses (Moss *et al.*, 1987). The use of furrow irrigation, as opposed to sprinkler irrigation, combined with improved sanitation practices has greatly reduced the significance of angular leaf spot of cotton caused by *X. c.* pv. *malvacearum* in California (Schnathorst, 1966b). Flooding and flood irrigation can also affect pathogen dispersal. Flooding in Brussels sprout and cabbage seedbeds disseminated *X. c.* pv. *campestris* by providing optimum conditions (Hunter *et al.*, 1975). Rice seedlings grown in deeply irrigated or easily flooded nurseries become more contaminated with *X. oryzae* pv. *oryzae* than seedlings grown in uplands or semi-arid conditions. The former condition leads to higher levels of primary inoculum in newly established paddies and more severe epidemics (Mizukami and Wakimoto, 1969). Once paddies are established, the depth of irrigation water and flooding can influence disease development.

(e) Aerosols

Aerosol particles are often naturally generated when water droplets hit wet, bacteria-covered leaf surfaces. Aerosols remain suspended in air for varying periods of time depending upon particle size, rain and wind. Aerosols laden with bacteria are epidemiologically significant because they can be disseminated further than splashed raindrops. Aerial bacterial populations of *X. c.* pv. *campestris* recovered with an Anderson air sampler were highest during rainfall, but were still detectable at low levels during dry conditions. Aerosol dispersal from cruciferous weeds was an important source of primary inoculum for cabbage black rot epidemics and a short-distance dispersal mechanism (Kuan *et al.*, 1982, 1986). Airborne concentrations of the cabbage black rot bacteria were as high as 0.002 cells/l of air sampled near cruciferous weeds in California (Kuan *et al.*, 1982). Thus, airborne bacteria from weeds near fields may act as an inoculum source for black rot epidemics. In contrast, airborne concentrations of *X. c.* pv. *vesicatoria* were reduced by rainfall and irrigation and were highest during dry conditions in the afternoon in tomato fields established from transplants. Aerosols with *X. c.* pv. *vesicatoria* were increased by clipping and harvesting operations (McInnes *et al.*, 1986, 1988).

(f) Insects

Insect transmission of phytopathogenic xanthomonads has been confirmed in a few cases. Three coleoptera species, *Cerotoma ruficornis* Oliv., *Diaprepes abbreviatus* L. and *Chalcodemus eceninus* Boheman, and two hemiptera species, *Nezara viridula* L. and *Empoasca* sp. transmit *X. c.* pvs *phaseoli*, *phaseoli* var. *fuscans* and *vignicola* in bean and cowpea (Kaiser and Vakili, 1978). Transmission was directly related to wounding due to insect feeding, and thus chewing insects were more efficient transmitters of the bacteria than sucking insects. Flea bettles (*Phyllotreta cruciferae*) have a limited potential for transmitting *X. c.* pv. *campestris* to crucifers in the greenhouse, but transmission could not be confirmed under field conditions. Leaf hoppers and grasshoppers were capable of transmitting *X. oryzae* pv. *oryzae*, but the bacterium could not be isolated from within the insect bodies. The bacteria were transmitted on the chewing mouthparts (Thri-Murty and Devadath, 1981b). The angular leaf blight of cotton pathogen is transmitted by several insect species. Spotted bollworm caterpillars feeding on infected cotton leaves acquired the bacterium within 15–20 min of feeding activity. *X. c.* pv. *malvacearum* cells were detected in faecal material and remained viable for up to 48 h (Borkar *et al.*, 1980). In growth chamber experiments, cotton fleahoppers (*Pseudatomoscelis seriatus* Reuter) artificially infested with *X. c.* pv. *malvacearum* caused 51% of plants they attacked to be infected. In addition, the bacterium was re-isolated from 60% of the salivary glands, 95% of the alimentary tracts, and 76% of the faecal spots examined (Martin *et al.*, 1988). In greenhouse tests, the cotton bug (*Lygus vosseleri* Popp.) transmitted *X. c.* pv. *malvacearum*. In Uganda, tests with bactericides against *X. c.* pv. *malvacearum* and insecticides against *L. vosseleri*, separately and in combination, led researchers to believe that the cotton bug and bacterial infection are related (Logan and Coaker, 1960).

5.4.3 Plant ingress

(a) Wounds

Injuries open wounds for bacterial invasion unless protective layers are formed in the surrounding tissues. *X. c.* pv. *citri* cells invade through wounds for some periods after injury and subsequently develop lesions (Koizumi, 1983). The susceptible period is dependent on temperature: 2–3 days at 25–35°C, 6–8 days at 20°C, 11–15 days at 15°C and 40 days or more at 10°C. In the field the susceptible periods are 20 days or more in winter and about 10 days in early spring. Mechanical transmission of *X. c.* pv. *pruni* during pruning and budding resulted in rapid canker development in new shoots of pruned plum nursery trees (Goodman and Hattingh, 1988) and reduced bud take and infection in plum and apricot trees (Goodman and Hattingh, 1986). *X. c.* pv. *graminis* in fodder grasses was spread extensively when grasses were mowed, especially when wet (Schmidt, 1989). Less aggressive strains of the xanthomonads causing bacterial spot of citrus were spread only by mechanical transmission such as planting, budding, stripping leaves by hand during budding, and hedging, but aggressive strains were capable of spreading naturally, e.g. via rain splash (Gottwald and Graham, 1990). Bacterial spot of tomato is also favoured by wounding of plants. Winds can cause scraping of tomato leaves leading to small wounds or breakage of epidermal hairs, or can cause blasting of leaves by small soil particles. The size of the resulting lesions is directly related to the extent of wound damage (Vakili, 1967). Leaf scars are important infection sites for bacterial leaf spot pathogen of stone fruits, but leaf scars rapidly form protective cell layers. Ingress only occurs within 24 h after leaves drop. Therefore, forced removal of leaves during windy rains, such as occurs in hurricanes, is favourable for ingress through leaf scars (Adam *et al.*, 1955; Feliciano and Daines, 1970).

(b) Natural openings

Xanthomonads also enter natural openings in plants, such as stomates, hydathodes and glands. Entry into these openings, however, is dependent upon the presence of water in the pores. Water may flood the pores through force of rainfall, overhead irrigation, or by gutation of the plant. If water is present in the pores, xanthomonads may enter plants through chemotaxic movement of the bacteria (Feng and Kuo, 1975), or be drawn into the plant by uptake of the water by the plant tissues (Diachun *et al.*, 1944). Inoculations of tomato with *X. c.* pv. *vesicatoria* was more successful when plants were kept at high humidity prior to inoculation rather than after inoculation (Davis and Halmos, 1958). This was probably because of water congestion of leaves prior to inoculation and rapid entry of the pathogen into the leaves through congested stomatal pores.

The size and number of stomatal pores per square centimetre of leaf tissue may affect the severity of diseases caused by xanthomonads. The relative susceptibility of mandarin and grapefruit leaves to citrus canker was related to the size of the stomatal pores (McLean, 1921). Mandarin leaves, being more resistant to *X. c.* pv. *citri* have smaller stomatal pores than grapefruit leaves which are more susceptible. Greater pressure was needed to force water through the stomatal pores of mandarins than grapefruit (McLean and Lee, 1922). Ramos and Volin (1987) related the relative disease response of tomato lines to bacterial spot caused by *X. c.* pv. *vesicatoria* to stomatal frequency. Physiologically induced stomatal closure before inoculations reduced disease incidence and severity.

5.4.4 Secondary inoculum

The dynamics of bacterial populations *in vivo* play a significant role in secondary inoculum production and availability. Rapidly expanding lesions or newly expanded infections are more epidemiologically important as continual inoculum sources. Although overwintering lesions or infections usually have fewer viable bacteria, these low populations often increase rapidly in the spring when favourable conditions for multiplication occur. Lesions caused by *X. c.* pv. *citri* were active for a longer period of time for susceptible cultivars of citrus compared with more resistant cultivars (Koizumi, 1977a). Thus, the inoculum produced per lesion on susceptible cultivars is greater and available for a longer period of time than that produced on resistant cultivars. Similarly, bacterial multiplication and spread within plant lesions caused by *X. oryzae* pv. *oryzae* were more extensive for compatible than for incompatible host-race combinations (Barton-Willis *et al.*, 1989). Such resistance can have a profound effect on disease development.

The presence of antagonistic microbes can also affect epidemic development. Co-infection of *X. c.* pv. *citri* and strains of a *Pseudomonas* species greatly reduced the multiplication of *X. c.* pv. *citri* (Ota, 1983). Such co-infection may thus have some mitigating influence on pathogen spread by reducing available inoculum throughout the epidemic. Lesions of citrus canker are usually contaminated with saprophytic organisms, but the role of the saprophytes in the rapid decline of *X. c.* pv. *citri* in the lesions was not determined (Goto *et al.*, 1979).

Xanthomonad phages are abundant in nature and occur together with host bacteria in soil and lesions of diseased plants (Hayward, 1964; Stolp and Starr, 1964). However, data regarding the effect of bacteriophages on natural populations of xanthomonads do not exist. Phages can reflect bacterial populations in some cases, however. To this end, the quantity of phages in rice paddy water is related to eventual severity of *X. oryzae* pv. *oryzae* epidemics. Epidemics of bacterial leaf blight of rice can be forecast based on the amount of phage in rice paddy water combined with rainfall during critical growing stages (Yoshimura and Tagami, 1968).

Mineral nutrition of plants may affect development of diseases caused by xanthomonads, but the effects seem to be dependent upon the disease. Excess nitrogen and fertilizers with unsuitable combinations of nitrogen, phosphate and potassium increased rice leaf blight severity as did too-high applications of silicate, magnesium and shale (Mizukami and Wakimoto, 1969). High epiphytic populations of *X. c.* pv. *vesicatoria* developed on leaves of tomato plants receiving low rates of nitrogen and potassium. Defoliation from disease was greater with low rates of nitrogen. Potassium did not affect defoliation from disease (McGuire *et al.*, 1991). The percentage of blighted leaf area of *Syngonium podophyllum* caused by *X. c.* pv. *dieffenbachiae* decreased linearly as amounts of a complete fertilizer (20:20:20) increased (Chase, 1989).

5.5 QUANTITATIVE EPIDEMIOLOGY

The spatial and/or temporal dynamics of xanthomonad diseases were examined in relationship to the environment in only a few cases in early work. For instance, spatial patterns of angular leaf spot of cotton in Uganda were related to soil wash channels (Hansford *et al.*, 1933). Re-examination of original distribution maps of angular leaf spot of cotton in a field in South Carolina (Faulwetter, 1917) revealed clusters of infected plants and significant within-row aggregation, as one would expect in cultivated row crops.

5.5.1 Rates of disease progress

Recently, citrus canker epidemiology was extensively studied in Argentina during an eradication effort. Rates (k) of disease increase were calculated from linearized Gompertz transformed data from orchards to be 0.04–0.06, 0.1, 0.18 and 0.13–0.24 per day for mandarin, satsuma, navel and sweet orange trees, respectively. Slopes of Gompit (y) versus $\log_{10}$ (m) linearlized disease gradients ranged from −0.21 to −4.13, but discrete foci of inoculum were not determined in the blocks of trees studied (Danos *et al.*, 1984). Spread of *X. c.* pv. *citri* from discrete inoculum sources and early phases of canker epidemics in concentric circles of trees planted around an inoculated focal tree at 5 m increments were best described by the Gompertz model. Gompertz rates (k) of disease increase were 0.005 and 0.009 per day for orange and grapefruit plots, respectively (Gottwald *et al.*, 1988b). Citrus canker epidemics in nurseries and newly established orchards, where initial inoculum consisted of a single inoculated tree surrounded by nursery or orchard trees, were most appropriately described by the Gompertz model when exponential, monomolecular, logistic, Gompertz and Weibull models were fitted to disease progress data by non-linear regression analysis (Gottwald *et al.*, 1989b). Gompertz rates (k) of citrus canker increase were 0.029, 0.029 and 0.032 per day for Swingle, orange and grapefruit cultivars, respectively. Slopes, compared by t-test, demonstrated a significantly higher rate of increase for grapefruit trees compared to orange and Swingle trees. The higher rates of increase in nurseries compared to the orchards reflected more conducive environmental conditions present during the tests and the faster rate of increase than can be achieved in plants in very dense nursery-planting patterns versus the greater spacings of orchards (Gottwald *et al.*, 1989b).

The rates of increase of Gompertz-transformed incidence of bacterial canker of four major cowpea cultivars were compared to examine their relative susceptibilities in Georgia (Gitaitis and Bell, 1986). Gompertz rate (*k*) helped to distinguish the least susceptible, and, therefore, the superior choice of black-eyed pea cultivars to be grown. For these tests, lesions were categorized into different sizes but this had no apparent effect on final ratings.

Other temporal models for increase of disease caused by phyto-pathogenic xanthomonads have been useful. Logistic rates (*r*) of increase of black rot of cabbage in Florida were estimated to be 0.57 per day, nearly 10 times that recorded in a later study in Hawaii where logistic rates were 0.04–0.06 per day (Strandberg, 1973; Yuen *et al.*, 1987). Comparison of monomolecular, logistic, and Gompertz disease progress models for disease increase of black rot in fields in Hawaii indicated that the monomolecular model was the most appropriate followed by the Gompertz and then the logistic. The rates of increase of black rot of cabbage for different strains of *X. c.* pv. *campestris* differed markedly under identical field conditions when monitored by serotyping of recovered strains with monoclonal antibodies. In this case, differences in strain virulence were not demonstrable under laboratory conditions, but inherent differences were demonstrable under field conditions (Yuen *et al.*, 1987). Co-inoculation of strains in the same field resulted in reduced disease progress for each strain compared to the rates of disease progress of individually inoculated strains. Such reduced disease progress of co-inoculated strains demonstrates the effect of competition between strains of *X. c.* pv. *campestris* (Yuen *et al.*, 1987).

5.5.2 Disease gradients

The direction of spread of the pathogen of citrus canker over time has been studied and related to environmental conditions. Simple water-splash dispersal of inoculum is quite effective in the close spacings of plants in citrus nurseries, whereas the greater distances between trees in citrus orchards require wind to carry the inoculum-laden rain from tree to tree (Gottwald *et al.*, 1988a,b). Normally, disease gradients tend to flatten over time as pathogens spread further from the inoculum source due to increases in disease incidence at distances farther from the inoculum source. However, with citrus canker, infected trees often partially defoliate when disease becomes severe (Gottwald *et al.*, 1988b, 1989b). Therefore, during citrus canker epidemics, disease gradients often fluctuate over time due to disease-induced defoliation, which lowers disease incidence, followed by new flushes, which become diseased, causing rises in disease incidence again (Gottwald *et al.*, 1988b).

The ratio of upwind to downwind disease gradients can be compared to examine preferential direction of spread of a disease. Based on upwind and downwind disease gradients in citrus orchard plots in Argentina a northeast preference of pathogen spread was related to the blowing rainstorms from the southwest (Gottwald *et al.*, 1988b, 1989b). For citrus canker in nurseries, no such directionality was associated with disease increase, indicating the strong influence of overhead irrigation compared to wind-blown rain (Gottwald *et al.*, 1989b). During a 1990 outbreak of citrus canker in south Florida, three major foci of infection were detected, separated by 100 m or more. No significant difference occurred among the slopes of directional disease gradients, indicating a general spread from a point source by rain splash and mechanical dissemination by machinery combined.

In two outbreaks of citrus bacterial spot wind direction during rainstorms was associated with spatial dissemination of the citrus bacterial spot pathogen from nursery edges. These rainstorm-related dissemination patterns were substantiated by analysis of directional disease gradients in which gradients were calculated for fan-shaped segments of the nurseries where the primary focus of disease was considered to have been the most severely infected trees on the edges of nurseries. Citrus bacterial spot disease gradients calculated in this way were best fit by a probit (x) versus probit (y) model, where x and y are distance in metres and disease incidence, respectively (Gottwald *et al.*, 1989a; Gottwald and Graham, 1990).

5.5.3 Disease distribution

The presence and severity of primary and secondary foci of infection were clearly visualized when isopathic maps of citrus canker were examined. The maps are particularly useful in visualizing the regions of discontinuity of disease gradients from the primary focal points of infection. The presence of these discontinuities raises questions concerning the validity or usefulness of disease gradient models for splash-dispersed bacterial pathogens such as *X. c.* pv. *citri*. Perhaps more important than disease gradients is a more thorough understanding of the spatial relationships of these aggregates of disease that are formed by splash dispersal.

Most diseases caused by splash-dispersed phytopathogenic bacteria are aggregated under field conditions, owing to the formation of secondary foci or clusters of disease plants. Bacterial diseases are usually most aggregated early in the epidemic. This aggregation becomes less discernible as the individual aggregates of disease plants continue to grow and start to coalesce later in the epidemic. This has been documented for citrus canker, citrus bacterial spot, and a few other diseases caused by phytopathogenic xanthomonads (Gottwald and Graham, 1990; Gottwald *et al.*, 1988b, 1989b). Crop spacing plays a crucial role in disease aggregation. Closer within-row spacing, when crop canopies touch or intertwine, most often results in preferential spread down the crop row. This within-row spatial aggregation is often augmented by mechanical spread of the disease down rows which is influenced by traffic patterns of humans and machinery within the crop. The spatial distribution of bacterial spot of tomato, caused by *X. c.* pv. *vesicatoria*, is a good example (Pohronezny *et al.*, 1990). Normally, workers proceed down field rows while thinning plants to proper field spacing. Bacterial spot incidence, examined by ordinary runs analysis, was highly aggregated within rows which had been thinned. Transmission of *X. c.* pv. *vesicatoria* by workers was diminished from 87% to 44% when thinning operations were restricted to the afternoon when plants were dry compared to thinning conducted when the plants were wet with dew. Pathogen transmission was further reduced to 3.7% and 0%, respectively, when workers periodically decontaminated their hands with ethanol or a povidone-iodine solution (Pohronezny *et al.*, 1990).

Data for cabbage black rot from 47 field plots in Florida were subjected to analyses of goodness-of-fit to seven different standard distributions (Strandberg, 1973). As expected, the disease was highly aggregated. This aggregation was demonstrated by the predominance of fit of the 31 out of 47 data sets to the negative binomial distribution. This aggregation was further demonstrated by doublet analysis in which significantly higher-than-expected within-row and across-row aggregation was encountered with only slightly higher within- than across-row aggregation. This slightly higher within- than across-row aggregation was directly related to mechanical spread by farm equipment, which always entered the field from one end and travelled down the crop rows. Spatial analyses of black rot infections were also highly aggregated for many plots in Hawaii as indicated by ordinary runs analysis (Yuen *et al.*, 1987). For those cases where random spatial patterns of black rot existed, high winds, rain and flooding may have mitigated the expected aggregated distribution of the pathogen.

Bacterial leaf blight of rice, caused by *X. oryzae* pv. *oryzae*, appears in small patches in rice paddies. Even though this is a vascular disease, the pathogen can spread from plant to plant by contact of foliage and through irrigation water. These patches originate from a few infected plants which serve as focal centres of disease. Leaf blight frequency data of infected plants were best described by the negative binomial distribution. The *k*-parameter of the negative binomial distribution indicated extreme clustering of infected plants in the field. Patchiness of bacterial leaf blight was further confirmed by ordinary runs and doublet analysis (Nayak and Reddy, 1987).

Citrus canker is highly aggregated in citrus nurseries. Lloyd's Index of Patchiness was very high during the initial stages of epidemics in three nurseries of different cultivars in Argentina, then declined as the epidemic progressed and clusters of diseased plants (secondary foci) began to coalesce (Gottwald *et al.*, 1989b). Based on ordinary runs analysis of citrus trees in these non-disturbed nurseries, aggregation was equivalent within versus across rows due to the close spacing of plants. In commercial nurseries with normal traffic, a greater within-row aggregation is expected due to the influence of mechanical spread (Gottwald *et al.*, 1989b). The spatial distribution of bacterial spot in citrus nurseries in Florida was also aggregated as indicated by variance-to-mean ratios, Lloyd's Index of Patchiness, and distribution fitting, as well as ordinary runs analysis. Greater within- than across-row aggregation occurred in all cases studied (Gottwald *et al.*, 1989a; Gottwald and Graham, 1990). This greater within- than across-row aggregation was directly related to mechanical transmission in four out of six of the nurseries examined (Gottwald *et al.*, 1989; Gottwald and Graham, 1990).

The general cluster size, shape and orientation of bacterial spot within citrus nurseries as determined by spatial autocorrelation analysis (SAA) related directly to the mechanisms of dissemination, e.g. mechanical transmission or blowing rainstorms. Cluster shape was longest in the direction of mechanical spread with minimal width across nursery rows (Gottwald and Graham, 1990). In many instances, non-contiguous spatial proximity patterns existed for both citrus canker in nurseries and orchards and citrus bacterial spot in nurseries. The spatial lag distance between the main proximity pattern and the non-contiguous portions of the pattern relates to the general distances and orientation of secondary foci to the primary focus and to each other and reflects the dissemination of the xanthomonad by splash dispersal (Gottwald and Graham, 1990). Spatial proximity patterns were also examined by SAA for a 1990 outbreak of citrus canker in south Florida. These patterns indicated a strong within- and across-row relationship between diseased individuals as well as association between individuals at oblique angles to each other. These same spatial patterns indicated numerous non-contiguous diseased tree associations. These results reflected spread of *X. c.* pv. *citri* which was influenced by blowing rain, the effect of air-blast sprayers, and mechanical spread by a herbicide applicator.

Spatio-temporal autoregressive moving average models (STARMA) were calculated for the data of citrus canker in the orchards and nurseries discussed above. This analysis examines the relationship between the incidence and spatial location of diseased trees at one point in time to those trees that develop disease over time. The spatial temporal disease increase for citrus nurseries infected with citrus canker was well represented by a STARMA (0, 4, 1, 1) model, whereas the appropriate model forms for orange and grapefruit orchards were STARMA (0, 1, 4, 1) and STARMA (0, 2, 1, 2), respectively. For the citrus nursery STARMA model, this means that spatio-temporal partial autocorrelations decayed rapidly over spatial lags (lags in this case represent the distance between 10-tree groups of trees within a row) but remained significant out to four temporal lags (0, 4, 1, 1). The spatio-temporal autocorrelations were absent after one temporal lag and decayed rapidly within the first temporal lag (0, 4, 1, 1). These results are indicative of a common underlying process for citrus canker epidemics that can be manifest in various temporal trends of spatial autocorrelation. This process is reflected in the close lattice spacing of citrus trees in nurseries which influences spread primarily by splash dissemination of bacteria to immediately adjacent neighbouring plants.

New pathogen methodologies are coming into use to monitor pathogen dissemination and progress of diseases. Field spread of citrus bacterial spot has been monitored by the use of a genomic DNA probe, by immunofluorescent microscopy, and by the use of monoclonal antibodies (Gottwald *et al.*, 1988a). A panel of monoclonal antibodies was utilized to differentiate 323 strains of *X. c.* pv. *dieffenbachiae* into 12 serotypes. The monoclonal antibodies were used to follow anthurium blight in fields and farms in Hawaii (Alvarez *et al.*, 1990). Cabbage black rot epidemics resulting from different strains of *X. c.* pv. *campestris* were monitored by serotyping recovered strains via monoclonal antibodies (Yuen *et al.*, 1987). Such field detection probes will allow the detection and monitoring of phytopathogenic xanthomonads in advance of symptom manifestation and thus help identify new, and possibly more accurate, associations of xanthomonad populations and the environment.

REFERENCES

Adam, A.V., Powell, D. and Anderson, H.W. (1955) Time of peach twig infection by *Xanthomonas pruni* in relation to spring canker incidence. *Phytopathology*, **45**, 285–7.

Alcaraz, L.M.F. (1978) Factores climaticos relacionados con la cancrosis de los citrus (*Xanthomonas citri* (Hasse) Dow.) en Corrientes. *J. Fitosan. Argentina*, **3**, 647–59.

Alvarez, A.M., Lipp, R.L. and Benedict, A.A. (1990) Use of monoclonal antibodies to study the epidemiology of bacterial blight of anthurium. *Phytopathology*, **80**, 886.

Angeles-Ramos, R., Vidaver, A.K. and Flyn, P. (1991) Characterization of epiphytic *Xanthomonas campestris* pv. *phaseoli* and pectolytic xanthomonads recovered from symptomless weeds in the Dominican Republic. *Phytopathology*, **81**, 677–81.

Ark, P.A. (1944) Pollen as a source of walnut bacterial blight infection. *Phytopathology*, **34**, 330–4.

Ark, P.A. (1958) Longevity of *Xanthomonas malvacearum* in dried cotton plants. *Plant Disease Reporter*, **42**, 1293.

Arnold, M.H. and Arnold, K.M. (1961) Bacterial blight of cotton: trash-borne infection. *The Empire Cotton Growing Review*, **38**, 258–70.

Baker, K.F. (1972) Seed pathology, in *Seed Biology* (ed. T.T. Kozlowski), Academic Press, New York, pp. 318–416.

Barton-Willis, P.A., Roberts, P.D., Guo, A. and Leach, J.E. (1989) Growth dynamics of *Xanthomonas campestris* pv. *oryzae* in leaves of rice differential cultivars. *Phytopathology*, **79**, 573–8.

Bashan, Y. (1986) Field dispersal of *Pseudomonas syringae* pv. *tomato*, *Xanthomonas campestris* pv. *vesicatoria*, and *Alternaria macrospora* by animals, people, birds, insects, mites, agricultural tools, aircraft, soil particles, and water sources. *Canadian Journal of Botany*, **64**, 276–81.

Blank, L.M. and Hunter, R.E. (1954) Soil carry-over of *Xanthomonas malvacearum* in New Mexico. *Phytopathology*, **44**, 331.

Boosalis, M.G. (1952) The epidemiology of *Xanthomonas translucens* (J.J. et R.) Dowson on cereals and grasses. *Phytopathology*, **42**, 387–95.

Borkar, S.G., Verma, J.P. and Singh, R.P. (1980) Transmission of *Xanthomonas malvacearum* (Smith) Dowson, the incitant of bacterial blight of cotton through spotted bollworms. *Indian Journal of Entomology*, **42**, 390–7.

Bradbury, J.F. (1984) Genus II. *Xanthomonas* Dowson 1939, in *Bergey's Manual of Systemic Bacteriology*, Vol. 1 (eds N.R. Kreig and G. Holt), Williams & Wilkins, Baltimore, pp. 199–210.

Brinkerhoff, L.A. and Fink, G.B. (1964) Survival and infectivity of *Xanthomonas malvacearum* in cotton plant debris and soil. *Phytopathology*, **54**, 1198–201.

Brinkerhoff, L.A. and Hunter, R.E. (1963) Internally infected seed as a source of inoculum for the primary cycle of bacterial blight of cotton. *Phytopathology*, **53**, 1397–401.

Brown, J.G. (1942) Wind dissemination of angular leaf spot of cotton. *Phytopathology*, **32**, 81–90.

Cafati, C.R. and Saettler, A.W. (1980a) Transmission of *Xanthomonas phaseoli* in seed of resistant and susceptible *Phaseolus* genotypes. *Phytopathology*, **70**, 638–40.

Cafati, C.R. and Saettler, A.W. (1980b) Role of nonhost species as alternate inoculum sources of *Xanthomonas phaseoli*. *Plant Disease*, **64**, 194–6.

Chase, A.R. (1989) Effect of nitrogen and potassium fertilizer rates on severity of *Xanthomonas* blight of *Syngonium podophyllum*. *Plant Disease*, **73**, 972–5.

Claflin, L.E. and Stuteville, D.L. (1973) Survival of *Xanthomonas alfalfae* in alfalfa debris and soil. *Plant Disease Reporter*, **57**, 52–3.

Cook, A.A., Larson, R.H. and Walker, J.C. (1952) Relation of the black rot pathogen to cabbage seed. *Phytopathology*, **42**, 316–20.

Daniel, J.F. and Boher, B. (1978) Ecology of cassava bacterial blight: epiphytic survival of *Xanthomonas manihotis* on aerial parts of the cassava plant, in *Proceedings of the Fourth International Conference on Plant Pathogenic Bacteria* (ed. Station de Pathologie Vegetale et Phytobacteriology), INRA, Angers, France, pp. 763–71.

Daniel, J.-F. and Boher, B. (1985) Some aspects of survival of *Xanthomonas campestris* pv. *manihotis*, the causal agent of cassava bacterial blight. *Agronomie*, **5**, 339–46.

Danos, E., Berger, R.D. and Stall, R.E. (1984) Temporal and spatial spread of citrus canker within groves. *Phytopathology*, **74**, 904–8.

Davis, D. and Halmos, S. (1958) The effect of air moisture on the predisposition of tomato to bacterial spot. *Plant Disease Reporter*, **42**, 110–11.

Dhanvantari, B.N. (1973) Population dynamics of *Xanthomonas pruni* on peach bud and leaf surface and its relation to other microflora. Abstract 0812, in *Abstracts of Papers, Second International Congress on Plant Pathology*, The American Phytopathological Society, St Paul, MN.

Diachun, S. and Valleau, W.D. (1946) Growth and overwintering of *Xanthomonas vesicatoria* in association with wheat roots. *Phytopathology*, **36**, 277–80.

Diachun, S., Valleau, W.D. and Johnson, E.M. (1944) Invasion of watersoaked tobacco leaves by bacteria, solutions, and tobacco mosaic virus. *Phytopathology*, **24**, 250–3.

Dye, D.W. (1958) Host specificity in *Xanthomonas*. *Nature*, **182**, 1813–14.

Dye, D.W. and Lelliott, R.A. (1974) Genus II. *Xanthomonas* Dowson 1939, in *Bergey's Manual of Determinative Bacteriology* (eds R.E. Buchanan and N.E. Gibbons), Williams & Wilkins, Baltimore, MD, pp. 243–9.

Egel, D.S., Graham, J.W. and Riley, T.D. (1991) Population dynamics of strains of *Xanthomonas campestris* differing in aggressiveness on Swingle citrumelo and grapefruit. *Phytopathology*, **81**, 666–71.

Faulwetter, R.C. (1917) Dissemination of the angular leafspot of cotton. *Journal of Agricultural Research*, **8**, 457–75.

Feliciano, A. and Daines, R.H. (1970) Factors influencing ingress of *Xanthomonas pruni* through peach leaf scars and subsequent development of spring cankers. *Phytopathology*, **60**, 1720–6.

Feng, T.Y. and Kuo, T.T. (1975) Bacterial leaf blight of rice plant. VI. Chemotactic responses of *Xanthomonas oryzae* to water pores on the leaf of rice plants. *Botanical Bulletin Academia Sinica*, **16**, 126–36.

Fett, W.F. (1979) Survival of *Pseudomonas glycinea* and *Xanthomonas phaseoli* var. *sojensis* in leaf debris and soybean seed in Brazil. *Plant Disease Reporter*, **63**, 79–83.

Forster, R.L. and Schaad, N.W. (1989) Longevity of *Xanthomonas campestris* pv. *translucens* in wheat seed under two storage conditions, in *Proceedings of the Seventh International Conference on Plant Pathogenic Bacteria* (ed. Z. Klement), Akademiai Kiado, Budapest, pp. 329–31.

Fulton, H.R. (1920) Decline of *Pseudomonas citri* in the soil. *Journal of Agricultural Research*, **19**, 207–23.

Gitaitis, R.D. and Bell, D.K. (1986) Epidemiology and control of bacterial blight and canker of cowpea. *Plant Disease*, **70**, 187–98.

Gitaitis, R.D., Sasser, M.J., Beaver, R.W., McInnes, T.B. and Stall, R.E. (1987) Pectolytic xanthomonads in mixed infection with *Pseudomonas syringae* pv. *syringae*, *P. syringae* pv. *tomato*, and *Xanthomonas campestris* pv. *vesicatoria* in tomato and pepper transplants. *Phytopathology*, **76**, 430–4.

Goode, M.J. and Sasser, M. (1980) Prevention – the key to controlling bacterial spot and bacterial speck of tomato. *Plant Disease*, **64**, 831–4.

Goodman, C.A. and Hattingh, M.J. (1986) Transmission of *Xanthomonas campestris* pv. *pruni* in plum and apricot nursery trees by budding. *HortScience*, **21**, 995–6.

Goodman, C.A. and Hattingh, M.J. (1988) Mechanical transmission of *Xanthomonas campestris* pv. *pruni* in plum nursery trees. *Plant Disease*, **72**, 643.

Goto, K., Fukatsu, R. and Ohata, K. (1953) Relationship between the occurrence of bacterial leaf blight disease on rice and that on wild grasses in endemic areas. *Plant Protection, Tokyo*, **7**, 365–8 (Japanese).

Goto, M. (1970) Studies on citrus canker. III. Survival of *Xanthomonas citri* (Hasse) Dowson in soils and on the surface of weeds. *Bulletin of the Faculty of Agriculture Shizuoka University*, **20**, 21–9 (Japanese, English summary).

Goto, M., Ohta, K. and Okabe, N. (1975a) Studies on saprophytic survival of *Xanthomonas citri* (Hasse) Dowson. 2. Longevity and survival density of the bacterium on artificially infested weeds, plant residues, and soils. *Annals of the Phytopathological Society of Japan*, **41**, 141–7.

Goto, M., Ohta, K. and Okabe, N. (1975b) Studies on saprophytic survival of *Xanthomonas citri* (Hasse) Dowson. 1. Detection of the bacterium from a grass (*Zoysia japonica*). *Annals of the Phytopathological Society of Japan*, **41**, 9–14.

Goto, M., Tadauchi, Y. and Okabe, N. (1979) Interaction between *Xanthomonas citri* and *Erwinia herbicola in vitro* and *in vivo*. *Annals of the Phytopathological Society of Japan*, **45**, 618–24.

Gottwald, T.R., Civerolo, E.L., Garnsey, S.M., Brlansky, R.H., Graham, J.H. and Gabriel, D.W. (1988a) Dynamics and spatial distribution of *Xanthomonas campestris* pv. *citri* group E strains in simulation nursery and new grove situations. *Plant Disease*, **72**, 781–7.

Gottwald, T.R. and Garnsey, S.M. (1990) Major citrus diseases in Southeast Asia. *Citrus Industry*, **7**, 58–63, 78.

Gottwald, T.R. and Graham, J.H. (1990) Spatial pattern analysis of epidemics of citrus bacterial spot in Florida citrus nurseries. *Phytopathology*, **80**, 181–90.

Gottwald, T.R., McGuire, R.G. and Garran, S. (1988b) Asiatic citrus canker: spatial and temporal spread in simulated new grove situations in Argentina. *Phytopathology*, **78**, 739–45.

Gottwald, T.R., Miller, C., Brlansky, R.H., Gabriel, D.W. and Civerolo, E.L. (1989a) Analysis of the spatial distribution of citrus bacterial spot in a Florida citrus nursery. *Plant Disease*, **73**, 297–303.

Gottwald, T.R., Timmer, L.W. and McGuire, R.G. (1989b) Analysis of disease progress of citrus canker in nurseries in Argentina. *Phytopathology*, **79**, 1276–83.

Graham, J.H., Gottwald T.R., Civerolo, E.L. and McGuire, R.G. (1989) Population dynamics and survival of *Xanthomonas campestris* in soil in citrus nurseries in Maryland and Argentina. *Plant Disease*, **73**, 423–7.

Graham, J.H., McGuire, R.G. and Miller, J.W. (1987) Survival of *Xanthomonas campestris* pv. *citri* in citrus plant debris and soil in Florida and Argentina. *Plant Disease*, **71**, 1094–8.

Groth, D.E. and Braun, E.J. (1989) Survival, seed transmission, and epiphytic development of *Xanthomonas campestris* pv. *glycines* in the North-Central United States. *Plant Disease*, **73**, 326–30.

Hansford, C.G., Hosking, H.R., Stoughton, R.H. and Yates, F. (1933) An experiment on the incidence and spread of angular leaf-spot disease of cotton in Uganda. *Annals of Applied Biology*, **20**, 404–20.

Hayward, A.C. (1964) Bacteriophage sensitivity and biochemical group in *Xanthomonas malvacearum*. *Journal of General Microbiology*, **35**, 287–98.

Hirano, S.S., Nordheim, E.V., Arny, D.C. and Upper, C.D. (1982) Lognormal distribution of epiphytic bacterial populations on leaf surfaces. *Applied and Environmental Microbiology*, **44**, 695–700.

Hirano, S.S. and Upper, C.D. (1983) Ecology and epidemiology of foliar bacterial plant pathogens. *Annual Review of Phytopathology*, **21**, 243–69.

Horino, O. (1986) Epidemiology and control of bacterial leaf blight and other bacterial diseases of rice. *Japan Pesticide Information*, **49**, 3–6.

Horino, O., Mew, T.W. and Yamada, T. (1982) The effect of temperature on the development of bacterial leaf blight on rice. *Annals of the Phytopathological Society of Japan*, **48**, 72–5.

Hunter, J.E., Abawi, G.S. and Becker, R.F. (1975) Observations on the source and spread of *Xanthomonas campestris* in an epidemic of black rot in New York. *Plant Disease Reporter*, **59**, 384–7.

Hunter, R.E. and Brinkerhoff, L.A. (1964) Longevity of *Xanthomonas malvacearum* on and in cotton seed. *Phytopathology*, **54**, 617.

Inoue, Y., Goto, K. and Ohata, K. (1957) Overwintering and mode of infection of leaf blight bacteria of rice plant. *Bulletin of the Horticultural Division of Tokai-Kinki Agricultural Experimental Station*, **4**, 74–82 (Japanese, English summary).

Irey, M.S. and Stall, R.E. (1982) Value of xanthomonadins for identification of pigmented *Xanthomonas campestris* pathovars, in *Proceedings of the Fifth International Conference on Plant Pathogenic Bacteria* (ed. J.C. Lozano), CIAT, Cali, Columbia, pp. 85–95.

Isaka, M. (1969) Studies on bacterial leaf blight of rice plant. On some grasses and weeds as carriers of the pathogen. *Proceedings of the Association of Plant Protection, Kyushu*, **17**, 14–19 (in Japanese).

Ishimaru, C., Eskridge, K.M. and Vidaver, A.K. (1991) Distribution analyses of naturally occurring epiphytic populations of *Xanthomonas campestris* pv. *phaseoli* on dry beans. *Phytopathology*, **81**, 262–8.

Johnson, J. (1937) Relation of water-soaked tissues to infection by *Bacterium angulatum* and *B. tabacum* and other organisms. *Journal of Agriculture Research*, **55**, 599–618.

Jones, J.B., Pohronezny, K.L., Stall, R.E. and Jones, J.P. (1986) Survival of *Xanthomonas campestris* pv. *vesicatoria* in Florida on tomato crop residue, weeds, seeds, and volunteer tomato plants. *Phytopathology*, **76**, 430–4.

Jones, R.K., Barnes, L.W., Gonzalez, C.F., Leach, J.E., Alvarez, A.M. and Benedict, A.A. (1989) Identification of low-virulence strains of *Xanthomonas campestris* pv. *oryzae* from rice in the United States. *Phytopathology*, **79**, 984–90.

Kaiser, W.J. and Vakili, N.G. (1978) Insect transmission of pathogenic xanthomonads to bean and cowpea in Puerto Rico. *Phytopathology*, **68**, 1057–63.

Klement, Z. (1968) Pathogenicity factors in regard to relationships of phyto-pathogenic bacteria. *Phytopathology*, **58**, 1218–21.

Koizumi, M. (1976a) Behavior of *Xanthomonas citri* (Hasse) Dowson in the infection process. I. Multiplication of the bacteria and histological changes following needle-prick inoculation. *Annals of the Phytopathological Society of Japan*, **42**, 407–16.

Koizumi, M. (1976b) Behavior of *Xanthomonas citri* (Hasse) Dowson in the infection process. II. Multiplication of the bacteria and histological changes of the host plant following rubber-block press or infiltration inoculation. *Annals of the Phytopathological Society of Japan*, **42**, 517–25.

Koizumi, M. (1976c) Incubation period of citrus canker in relation to temperature. *Bulletin of the Fruit Tree Research Station, Series B, Okitsu*, **3**, 33–46.

Koizumi, M. (1977a) Behavior of *Xanthomonas citri* (Hasse) Dowson and histological changes of diseased tissues in the process of lesion extension. *Annals of the Phytopathological Society of Japan*, **43**, 129–36.

Koizumi, M. (1977b) Factors related to the occurrence of spring canker caused by *Xanthomonas citri* (Hasse) Dows. *Bulletin of the Fruit Tree Research Station Series B, Okitsu*, **4**, 115–29.

Koizumi, M. (1977c) Relation of temperature to the development of citrus canker lesions in the spring. *Proceedings of the International Society of Citriculture*, **3**, 924–8.

Koizumi, M. (1979) Ultrastructural changes in susceptible and resistant plants of citrus following artificial inoculation with *Xanthomonas citri*. *Annals of the Phytopathological Society of Japan*, **45**, 635–44.

Koizumi, M. (1983) Relationship between wound-healing process of citrus leaf tissues and successful infection through wounds by *Xanthomonas campestris* pv. *citri* (Hasse) Dye. *Annals of the Phytopathological Society of Japan*, **49**, 352–60.

Kuan, T.-L., Minsavage, G.V. and Schaad, N.W. (1982) Airborne dispersal of *Xanthomonas campestris* pv. *campestris*. *Phytopathology*, **72**, 945.

Kuan, T.-L., Minsavage, G.V. and Schaad, N.W. (1986) Aerial dispersal of *Xanthomonas campestris* pv. *campestris* from naturally infected *Brassica campestris*. *Plant Disease*, **70**, 409–13.

Kumar, S.N.S. (1983) Epidemiology of bacterial leaf stripe disease of arecanut palm. *Tropical Pest Management*, **29**, 249–52.

Lahman, L.K. and Schaad, N.W. (1985) Evaluation of the 'Dome Test' as a reliable assay for seedborne bacterial blight pathogens of beans. *Plant Disease*, **69**, 680–3.

Leben, C. (1963) Multiplication of *Xanthomonas vesicatoria* on tomato seedlings. *Phytopathology*, **53**, 778–81.

Leben, C. (1965) Epiphytic microorganisms in relation to plant disease. *Annual Review of Phytopathology*, **3**, 209–30.

Leben, C. (1974) Survival of plant pathogenic bacteria. Special Circular 100, Ohio Agricultural Research and Development Center, Wooster, Ohio.

Lee, H.A. (1920) Behavior of the citrus canker organism in the soil. *Journal of Agricultural Research*, **19**, 189–206.

Logan, C. and Coaker, T.H. (1960) The transmission of bacterial blight of cotton (*Xanthomonas malvacearum* (E.F. Smith) Dowson) by the cotton bug, *Lygus vosseleri* Popp. *The Empire Cotton Growing Review*, **37**, 26–9.

Loucks, K.W. (1930) Some physiological studies of *Phytomonas citri*. *Journal of Agricultural Research*, **41**, 247–58.

Maas, J.L., Finney, M.M., Civerolo, E.L. and Sasser, M. (1985) Association of an unusual strain of *Xanthomonas campestris* with apple. *Phytopathology*, **75**, 438–45.

Martin, Jr, W.R., Sterling, W.L., Kenerley, C.M. and Morgan, P.W. (1988) Transmission of bacterial blight of cotton, *Xanthomonas campestris* pv. *malvacearum*, by feeding of the cotton fleahopper: implications for stress ethylene-induced square loss in cotton. *Journal of Entomological Science*, **23**, 161–8.

McGuire, R.G., Jones, J.B., Stanley, C.D. and Csizinsky, A.A. (1991) Epiphytic populations of *Xanthomonas campestris* pv. *vesicatoria* and bacterial spot of tomato as influenced by nitrogen and potassium fertilization. *Phytopathology*, **81**, 656–60.

McInnes, T.B., Gitaitis, R.D. and McCarter, S.M. (1986) The effect of cultural practices on the generation of bacterial aerosols in tomato transplants. *Phytopathology*, **76**, 1070.

McInnes, T.B., Gitaitis, R.D., McCarter, S.M., Jaworski, C.A. and Phatak, S.C. (1988) Airborne dispersal of bacteria in tomato and pepper transplant fields. *Plant Disease*, **72**, 575–9.

McLean, F.T. (1921) A study of the structure of the stomata of two species of citrus in relation to citrus canker. *Bulletin of the Torrey Botanical Club*, **48**, 101–6.

McLean, F.T. and Lee, A. (1922) Pressures required to cause stomatal infections with the citrus canker organism. *Phillipine Journal of Science*, **20**, 309–20.

Mizukami, T. (1961) Studies on the ecological properties of *Xanthomonas oryzae* (Uyeda et Ishiyama) Dowson, the causal organism of bacterial leaf blight of rice plant. *Science Bulletin of the Faculty of Agriculture Saga University*, **13**, 1–85.

Mizukami, T. and Wakimoto, S. (1969) Epidemiology and control of bacterial leaf blight of rice. *Annual Review of Phytopathology*, **7**, 51–72.

Moffett, M.L. and Croft, B.J. (1983) *Xanthomonas* sp., in *Plant Bacterial Diseases: a diagnostic guide* (eds P.C. Fahy and G.J. Persley), Academic Press, Sydney, pp. 189–228.

Mohan, S.K. (1983) Seed transmission and epidemiology of *Xanthomonas campestris* pv. *malvacearum*. *Seed Science and Technology*, **11**, 859–65.

Moss, M.A., Pohronezny, K., Schenk, J. and Dankers, W. (1987) Effect of irrigation method on the development of bacterial spot epidemics in tomato. *Phytopathology*, **77**, 1712.

Mulrean, E.N. and Schroth, M.N. (1981) Ecology of *Xanthomonas campestris* pv. *juglandis* on Persian (English) walnuts. *Phytopathology*, **71**, 434–8.

Nayak, P. and Reddy, P.R. (1987) The pattern of bacterial leaf blight disease development and spread in rice. *Journal of Phytopathology*, **119**, 255–61.

Neergaard, P. (1977) *Seed Pathology*, Vol. 1, John Wiley, New York.

Ota, T. (1983) Interactions *in vitro* and *in vivo* between *Xanthomonas campestris* pv. *citri* and antagonistic *Pseudomonas* sp. *Annals of the Phytopathological Society of Japan*, **49**, 308–15.

Ou, S.H. (1985) *Rice Diseases*, Commonwealth Mycology Institute, Kew, Surrey, England.

Palazzo, D.A., Malavolta, Jr, V.A., Nogueira, E.M., Carvalho, M.L.V. and Rossetti, V. (1984) Epidemiological observations on *Xanthomonas campestris* pv. *citri* (Hasse) Dye in relation to different citrus species and varieties. *Proceedings of the International Society of Citriculture 1984*, **1**, 382–5.

Peltier, G.L. (1920) Influence of temperature and humidity on the growth of *Pseudomonas citri* and its host plants and on infection and development of the disease. *Journal of Agricultural Research*, **20**, 447–506.

Peltier, G.L. and Frederich, W.J. (1926) Effects of weather on the world distribution and prevalence of citrus canker and citrus scab. *Journal of Agricultural Research*, **32**, 147–64.

Peterson, G.H. (1963) Survival of *Xanthomonas vesicatoria* in soil and diseased tomato plants. *Phytopathology*, **53**, 765–7.

Pohronezny, K., Moss, M.A., Dankers, W. and Schenk, J. (1990) Dispersal and management of *Xanthomonas campestris* pv. *vesicatoria* during thinning of direct-seeded tomato. *Plant Disease*, **74**, 800–5.

Pohronezny, K., Stall, R.E., Canteros, B.I., Kegley, M., Datnoff, L. and Subramanya, R. (1992) Sudden shift in the prevalent race of *Xanthomonas campestris* pv. *vesicatoria* in pepper fields in Southern Florida. *Plant Disease*, **76**, 118–20.

Ramos, L.J. and Volin, R.B. (1987) Role of stomatal opening and frequency on infection of *Lycopersicon* spp. by *Xanthomonas campestris* pv. *vesicatoria*. *Phytopathology*, **77**, 1311–17.

Rolfs, F.M. (1935) Dissemination of the bacterial leaf spot organism. *Phytopathology*, **25**, 971.

Sasser, M. (1990) Identification of bacteria through fatty acid analysis, in *Methods in Phytobacteriology* (eds Z. Klement *et al.*), Akademiai Kiado, Budapest, pp. 199–204.

Schaad, N.W. (1982) Detection of seedborne plant pathogens. *Plant Disease*, **66**, 885–90.
Schaad, N.W. and Dianese, J.C. (1981) Cruciferous weeds as sources of inoculum of *Xanthomonas campestris* in black rot of crucifers. *Phytopathology*, **71**, 1215–20.
Schaad, N.W., Sitterly, W.R. and Humaydan, H. (1980) Relationship of incidence of seedborne *Xanthomonas campestris* to black rot of crucifers. *Plant Disease*, **64**, 91–2.
Schaad, N.W. and White, W.C. (1974) Survival of *Xanthomonas campestris* in soil. *Phytopathology*, **64**, 1518–20.
Schmidt, D. (1989) Epidemiological aspects of bacterial wilt of fodder grasses. *Bulletin OEPP/EPPO Bulletin*, **19**, 89–95.
Schnathorst, W.C. (1966a) Unaltered specificity in several xanthomonads after repeated passage through *Phaseolus vulgaris*. *Phytopathology*, **56**, 58–60.
Schnathorst, W.C. (1966b) Eradication of *Xanthomonas malvacearum* from California through sanitation. *Plant Disease Reporter*, **50**, 168–71.
Schultz, T. and Gabrielson, R.L. (1986) *Xanthomonas campestris* pv. *campestris* in western Washington crucifer seed fields: occurrence and survival. *Phytopathology*, **76**, 1306–9.
Schuster, M.L. and Coyne, D.P. (1974) Survival mechanisms of phytopathogenic bacteria. *Annual Review of Phytopathology*, **12**, 199–221.
Serizawa, S. (1981) Recent studies on the behavior of the causal bacterium of the citrus canker. *Proceedings of the International Society of Citriculture, 1981*, **1**, 395–7.
Serizawa, S. and Inoue, K. (1975) Studies on citrus canker III. The influence of wind blowing on infection. *Bulletin of the Schizuoka Prefecture Citrus Experimental Station*, **11**, 54–67.
Shekhawat, G.S. and Patel, P.N. (1977) Seed transmission and spread of bacterial blight of cowpea and leaf spot of green gram in summer and monsoon seasons. *Plant Disease Reporter*, **61**, 390–2.
Sletten, A. (1989) Detection methods for *Xanthomonas campestris* pv. *graminis* on forage grasses. *Bulletin OEPP/EPPO Bulletin*, **19**, 69–72.
Smith, E.F. (1905) *Bacteria in Relation to Plant Diseases*, Vol. 1, Carnegie Institution of Washington, Washington DC.
Srivastava, D.N. and Rao, Y.P. (1964) Seed transmission and epidemiology of the bacterial blight disease of rice in North India. *Indian Phytopathology*, **17**, 77–8.
Stall, R.E. and Cook, A.A. (1966) Multiplication of *Xanthomonas vesicatoria* and lesion development in resistant and susceptible pepper. *Phytopathology*, **56**, 1152–4.
Stall, R.E., Marco, G.M. and Canteros de Echenique, B.I. (1982) Importance of mesophyll in mature-leaf resistance to cancrosis of citrus. *Phytopathology*, **72**, 1097–100.
Stall, R.E., Miller, J.W., Marco, G.M. and Cauteros, B.I. (1980) Population dynamics of *Xanthomonas citri* causing canerosis of citrus in Argentina. *Proceedings of the Florida State Horticultural Society*, **93**, 10–14.
Stall, R.E. and Seymour, C.P. (1983) Canker, a threat to citrus in the Gulf-Coast States. *Plant Disease*, **67**, 581–5.
Starr, M.P. (1981) The genus *Xanthomonas*, in *The Prokaryotes: a handbook on habits, isolation and identification of bacteria* (eds M.P. Starr, H. Stolp, H.G. Truper, A. Balows and H.G. Schlegel), Springer-Verlag, Berlin, pp. 742–63.
Stolp, H. and Starr, M. (1964) Bacteriophage reactions and speciation of phytopathogenic xanthomonads. *Phytopathologische Zeitschrift*, **51**, 442–78.

Strandberg, J. (1973) Spatial distribution of cabbage black rot and the estimation of diseased plant populations. *Phytopathology*, **63**, 998–1003.

Sutton, J.C. and Williams, P.H. (1970) Comparison of extracellular polysaccharide of *Xanthomonas campestris* from culture and from infected cabbage leaves. *Canadian Journal of Botany*, **48**, 645–51.

Tagami, Y., Kuhara, S., Kurita, T., *et al.* (1963) Epidemiological studies on the bacterial leaf blight of rice, *Xanthomonas oryzae* (Uyeda et Ishiyama) Dowson. 1. The overwintering of the pathogen. *Bulletin of the Kyushu Agricultural Experimental Station*, **9**, 89–122.

Tagami, Y., Kuhara, S., Kurita, T., Fujii, H., Sekiya, N. and Sato, T. (1964) Epidemiological studies on the bacterial leaf blight of rice, *Xanthomonas oryzae* (Uyeda et Ishiyama) Dowson. II. Successive change in the population of the pathogen in paddy fields during rice growing period. *Bulletin of the Kyushu Agricultural Experimental Station*, **10**, 23–50.

Tagami, Y. and Mizukami, T. (1962) Review of rice bacterial leaf blight. Spec. Bull. Pest Forecast 10. Ministry of Agriculture and Forestry, Japan, pp. 10–13.

Takanashi, K. (1978) Ecological studies on bacterial shot hole caused by *Xanthomonas pruni* (E.F. Smith) Dowson. *Bulletin of the Fruit Tree Research Station, Series A*, **5**, 1–71.

Thompson, D.C., Schaad, N.W. and Forster, R.L. (1989) New perennial hosts of epiphytic populations of *Xanthomonas campestris* pv. *translucens*. *Phytopathology*, **79**, 1168.

Thornberry, H.H. and Anderson, H.W. (1933) Overwintering of *Phytomonas pruni* on peach. *Phytopathology*, **23**, 787–801.

Thri-Murty, V.S. and Devadath, S. (1981a) Role of volunteer rice seedlings in the perpetuation of *Xanthomonas campestris* pv. *oryzae*. *Oryza*, **18**, 59–60.

Thri-Murty, V.S. and Devadath, S. (1981b) Studies on the transmission and survival of *Xanthomonas campestris* pv. *oryzae* through insects. *Indian Phytopathology*, **34**, 162–3.

Timmer, L.W., Marois, J.J. and Achor, D. (1987) Growth and survival of xanthomonads under conditions nonconducive to disease development. *Phytopathology*, **77**, 1341–5.

Vakili, N.G. (1967) Importance of wounds in bacterial spot (*Xanthomonas vesicatoria*) of tomatoes in the field. *Phytopathology*, **57**, 1099–103.

Vasudeva, R.S. (1963) Report of the Division of Mycology and Plant Pathology. *Review of Applied Mycology*, **40**, 10.

Vasudeva, R.S. (1963) Report of the Division of Mycology and Plant Pathology. *Review of Applied Mycology*, **42**, 593.

Volcani, Z. (1969) The effect of mode of irrigation and wind direction on disease severity caused by *Xanthomonas vesicatoria* on tomato in Israel. *Plant Disease Reporter*, **53**, 459–61.

Wakimoto, S. (1956) Considerations on the overwintering and the infection mechanism of *Xanthomonas oryzae*. *Plant Protection, Japan*, **10**, 421–4.

Walker, J.C. (1969) *Plant Pathology*, 3rd edn, McGraw-Hill, New York.

Wallen, V.R. and Sutton, M.D. (1965) *Xanthomonas phaseoli* var. *fuscans* (Burkh.) Starr & Burkh. in field beans in Ontario. *Canadian Journal of Botany*, **43**, 437–46.

Wallin, J.R. (1946) Seed and seedling infection of barley, bromegrass, and wheat by *Xanthomonas translucens* var. *cerealis*. *Phytopathology*, **36**, 446–57.

Watanabe, Y. (1975) Ecological studies on kresek phase of bacterial leaf blight of rice. *Bulletin of Tokaikini National Agricultural Experimental Station*, **28**, 50–123.

Weller, D.M. and Saettler, A.W. (1980a) Evaluation of seed-borne *Xanthomonas phaseoli* and *X. phaseoli* var. *fuscans* as primary inocula in bean blights. *Phytopathology*, **70**, 148–52.

Weller, D.M. and Saettler, A.W. (1980b) Colonization and distribution of *Xanthomonas phaseoli* and *Xanthomonas phaseoli* var. *fuscans* in field-grown navy beans. *Phytopathology*, **70**, 500–6.

Wernham, C.C. (1948) The species value of pathogenicity in the genus *Xanthomonas*. *Phytopathology*, **38**, 283–91.

Wiles, A.B. and Walker, J.C. (1952) Epidemiology and control of angular leaf spot of cucumber. *Phytopathology*, **42**, 105–8.

Williams, P.H. and Keen, N.T. (1967) Relation of cell permeability alterations to water congestion in cucumber angular leaf spot. *Phytopathology*, **57**, 1378–85.

Yoshimura, S. (1963) Diagnostic and ecological studies of rice bacterial leaf blight, caused by *Xanthomonas oryzae* (Uyeda et Ishiyama) Dowson. *Bulletin of the Hokuriki Agricultural Experimental Station*, **5**, 27–182.

Yoshimura, S., Morihashi, T. and Suzuki, Y. (1956) Distribution of the important grasses for overwintering of *Xanthomonas oryzae* and their diseased situations in the field. *Annals of the Phytopathological Society of Japan*, **24**, 6.

Yoshimura, S. and Tagami, Y. (1968) Forecasting and control of bacterial leaf blight of rice in Japan. *Proc. Symp. Rice*, pp. 25–38.

Yuen, G.Y., Alvarez, A.M., Benedict, A.A. and Trotter, K.J. (1987) Use of monoclonal antibodies to monitor the dissemination of *Xanthomonas campestris* pv. *campestris*. *Phytopathology*, **77**, 366–70.

Zaumeyer, W.J. and Thomas, H.R. (1957) A monographic study of bean diseases and methods for their control. *Technical Bulletin 868*, US Department of Agriculture, Washington DC.

6

Genetics of *Xanthomonas*

M.J. Daniels
The Sainsbury Laboratory, John Innes Centre, Norwich Research Park

and

J.E. Leach
Department of Plant Pathology, Kansas State University

6.1 INTRODUCTION

The title of this chapter is not wholly accurate. Our primary purpose is not to describe *Xanthomonas* genetics as a topic *per se* but rather to illustrate the use of genetic approaches to analyse aspects of *Xanthomonas* biology. Most of the studies reported so far have been concerned either with phytopathogenicity or with the production of polysaccharide. One of the most striking features of modern science is the impact which genetics (including molecular genetics) has had upon the development of almost all aspects of biology. In the 1950s and 1960s classical bacterial genetics consisted mainly of isolation and mapping of mutations (i.e. genetic manipulation *in vivo*) and was applied with great success to investigating the properties of so-called 'metabolic black boxes', pathways or processes about which little was known. Frequently the genetics suggested models which could be tested biochemically. Most of the genetics of this era was carried out with *Escherichia coli* and its relatives and a small number of other

Xanthomonas.
Edited by J.G. Swings and E.L. Civerolo.
Published in 1993 by Chapman & Hall, London.
ISBN 0 412 434202

bacteria (e.g. *Bacillus subtilis, Streptomyces coelicolor*) (Hayes, 1968). The main limitation of working with other bacteria was the need for methods of transferring DNA between strains, a problem which persists to this day. From the early 1970s, classical genetics has been supplemented by molecular genetics, allowing manipulation of genetic material *in vitro* and its propagation in surrogate hosts (e.g. *E. coli*). The analytical power of the range of techniques now available for isolating and characterizing genes is well known. Moreover, the techniques and approaches used are very similar for *E. coli*, yeast, animals, plants or *Xanthomonas*. These technical matters are discussed in numerous books and we do not describe standard methodology here. One consequence of the generality of the techniques has been an expansion in the range of organisms studied. A recent book illustrates some of the diverse members and aspects of the prokaryotic world which have received attention from geneticists (Hopwood and Chater, 1989).

Pathogenicity or, more generally, host–pathogen interaction can be thought of as a series of sequential events involving the pathogen and the host, and it is tempting to believe that genetic approaches will help us to unravel the poorly-understood interconnected pathways. The history of genetic studies of bacterial plant pathogens has been comprehensively reviewed by Chatterjee and Vidaver (1986). Although *Agrobacterium, Erwinia* and *Pseudomonas* were increasingly studied during the 1970s, *Xanthomonas* was largely ignored. Early reports of transformation of *X. campestris* pv. *phaseoli* and *X. oryzae* pv. *oryzae* have not been followed up and are difficult to evaluate (Corey and Starr, 1957a,b; Yamasaki *et al.*, 1966). Lai *et al.* (1977) showed that broad host-range plasmids (R factors) of incompatibility group P could be transferred to xanthomonads. This demonstrated the feasibility of using gene cloning vectors derived from IncP plasmids to transfer DNA into the bacteria. Certain derivatives of IncP plasmids, such as R68.45 (Haas and Holloway, 1976), can also be used as chromosome-mobilizing agents and several examples of genetic linkage were demonstrated in *X. c.* pv. *campestris*. However, the convenience of using recombinant DNA techniques has meant that conventional genetic techniques have not been exploited further for the study of pathogenicity functions.

Research during the last decade shows that xanthomonads are in general amenable to molecular genetic analysis. However, as discussed below, certain pathovars or strains may have frustrating properties such as an inability to receive cloning vectors by conjugation, lack of stable maintenance of plasmids, or inability to be mutagenized with transposons. Anybody starting a research programme with a previously unstudied xanthomonad strain should be prepared to invest some effort into screening many independent wild-type strains in order to find the one which offers the best combination of attributes.

Our discussion of methodology below is limited to some key techniques which have been specially modified for use in *Xanthomonas* research, and most of this chapter is devoted to consideration of the classes of gene needed for pathogenicity.

6.2 METHODOLOGY

6.2.1 Mutation

Isolation and study of mutations are two of the cornerstones of genetics. Mutagenesis has been used as part of the strategy for isolating *Xanthomonas* pathogenicity and other genes. The loose term of convenience, 'pathogenicity gene', will be used to denote any gene which is required for normal pathogenicity, but is not known to be essential for growth of the bacteria *ex planta*. Mutagenesis has also been widely used to characterize cloned DNA, particularly to define the size and location of genes and to allow assessment of their roles in pathogenicity.

Mutations may be induced by irradiation, mutagenic chemicals or by insertion of DNA fragments (transposons and insertion sequences) into genes. The properties of commonly used mutagenic agents and the types of mutations induced by them have been discussed (Carlton and Brown, 1981) and summarized (Chatterjee and Vidaver, 1986). Experimental protocols are given by Miller (1972). Physicochemical agents have been used for mutagenesis of xanthomonads (Whitfield *et al.*, 1981; Tsuchiya *et al.*, 1982; Daniels *et al.*, 1984a; Atkins *et al.*, 1987; Harding *et al.*, 1987; Thorne *et al.*, 1987; Mellano and Cooksey, 1988; Bonas *et al.*, 1991).

Transposons are mobile genetic elements which carry as part of their structure a selectable antibiotic resistance gene. They have many advantages as mutagenic agents and they can be exploited for many types of genetic manipulation *in vivo* and *in vitro* (Berg and Howe, 1989). Mills (1985) reviewed the use of transposons with plant pathogenic bacteria. Three main properties of transposon mutagenesis have been exploited by pathologists. (a) Mutations are absolutely linked to a selectable antibiotic resistance marker, enabling mutants to be easily identified and manipulated *in vivo*, irrespective of the phenotypes conferred by the wild type and mutant genes. (b) They are also linked to a known DNA sequence, so that DNA fragments carrying the mutant gene can be located by hybridization to the transposon sequence. This is exploited in some cloning strategies. (c) Transposon-induced mutations are highly polar, giving complete loss of function of the mutant gene and of downstream genes in the same operon. This means that phenotypes are usually clear and non-leaky. However, there may be instances when other types of mutation are more informative and, despite the undoubted value of transposons, one should not neglect the useful properties of more traditional mutagenic agents.

Many DNA transfer systems are available for delivering transposons to target bacteria, generally based on 'suicide' plasmid vectors or other replicons which cannot maintain themselves in the bacterium (Berg and Howe, 1989). The choice of the most appropriate transposon and delivery system is usually a matter of trial and error. There is considerable variation in transposon–vector–bacterium interaction both between and within species or pathovars and much preliminary screening may be necessary before a useful combination is found. For unknown reasons many xanthomonads seem to be refractory to mutagenesis with commonly used transposons (Turner *et al.*, 1984). However, successful mutagenesis has been achieved with several systems (Mellano and Cooksey, 1988; Shaw *et al.*, 1988; Joseph-Liauzun *et al.*, 1989; Hötte *et al.*, 1990; Seal *et al.*, 1990). Transposon mutagenesis of cloned DNA is described in a later section.

6.2.2 Gene exchange

(a) Barriers to gene transfer

A major constraint on genetic analysis of *Xanthomonas* spp. has been the difficulty of introducing or maintaining cloning vectors in the bacteria. One reason for this is the presence of restriction–modification systems in *Xanthomonas* which recognize and cleave specific DNA sequences depending on the modification or lack of modification of certain bases. Recently, DeFeyter and Gabriel (1991a) described an approach to overcome the interference of restriction–modification systems with gene transfer to *X. c.* pv. *malvacearum*. DNA modification genes encoding *Xma*I and *Xma*III methylases were cloned from *X. c.* pv. *malvacearum* and introduced into the *E. coli* cloning host to enhance the transfer frequency of constructs containing other *X. c.* pv. *malvacearum* DNA. Presumably, the *X. c.* pv. *malvacearum* methylase genes modified the DNA in *E. coli* such that it was not destroyed by the corresponding endonucleases upon subsequent introduction into *X. c.* pv. *malvacearum* strains. The improvement in transfer efficiency of plasmids was significant (up to 10^5-fold higher), suggesting that previous problems with poor transfer frequency into *X. c.* pv. *malvacearum* were caused primarily by the organism's endogenous restriction systems.

Another method to circumvent restriction–modification systems is to obtain mutants defective in these systems for use as recipients. A mutant of *X. c.* pv. *campestris* was selected which could be transformed with plasmid DNA isolated either from *E. coli* or *X. c.* pv. *campestris* at equal efficiencies (Atkins *et al.*, 1987). Strains of *X. o.* pv. *oryzae* that were selected for growth on medium containing the cytosine methylase inhibitor 5-azacytidine were more efficient recipients of plasmid DNA from *E. coli* either by conjugation or by electroporation.

(b) Cloning vectors

Mobilizable broad host-range plasmids are used to introduce foreign DNA into bacteria. Cloning vectors suitable for use with phytopathogenic bacteria were made using plasmids from the IncP (e.g. pLAFR1, Friedman *et al.*, 1982; pLAFR3, Staskawicz *et al.*, 1987; pLAFR5, Keen *et al.*, 1988; pIJ3200, Liu *et al.*, 1990), IncQ (pKT230, Bagdasarian *et al.*, 1981; pIJ3100, Osbourn *et al.*, 1987) and IncW (pSa747, Tait *et al.*, 1983; pUFR027 and pUFR034, DeFeyter *et al.*, 1990; pRI40, Innes *et al.*, 1988) incompatibility groups because members of these groups were known to replicate in a wide variety of Gram-negative species. Several of these broad host-range vectors have useful features for cloning, such as a range of unique restriction enzyme sites in a localized 'polylinker' region, one or more selectable markers, and a complementation system (e.g. *lacZ*α) for distinguishing the vector from derivatives containing cloned DNA inserts. Some of the cloning vectors that have been used in *Xanthomonas* spp. and their features are listed in Table 6.1.

Generalizations about the usefulness of particular vectors for analysis of *Xanthomonas* spp. are difficult to make. DeFeyter *et al.* (1990) evaluated plasmids from the three incompatibility groups (IncP, IncQ and IncW) for use as cloning vectors in *X. c.* pvs *malvacearum, campestris, vesicatoria, translucens, phaseoli* and *citri*. Their findings underline the importance of testing various vectors for suitability with particular *Xanthomonas* strains. For example, the IncP plasmids, such as pLAFR1 and pLAFR3, have been used successfully for gene transfer in several *Xanthomonas* strains (Daniels *et al.*, 1984b; Gough *et al.*, 1988; Swanson *et al.*, 1988; Whalen *et al.*, 1988; Fujimoto *et al.*, 1991). However, for some species, such as *X. c.* pv. *malvacearum* (DeFeyter *et al.*, 1990) and *X. o.* pv. *oryzae* (Leach *et al.*, 1989), the IncP plasmids were not sufficiently transferred or stably maintained to be useful. Plasmids from the IncW group are the most useful for transfer into *X. c.* pv. *malvacearum* and *X. o.* pv. *oryzae* (Leach *et al.*, 1989; DeFeyter *et al.*, 1990). Some strains of *X. c.* pv. *campestris* are excellent recipients of IncW plasmids (10^{-3} transconjugants per recipient bacterium; DeFeyter *et al.*, 1990), whereas transfer of IncW vectors into other strains has not been achieved (Liu *et al.*, 1990).

Table 6.1 Broad host-range gene cloning vectors and their use in *Xanthomonas*

Vector	*Relevant characteristics*	*Unique cloning sites*	*Reference*	*Used with Xanthomonas*
pLAFR1	IncP, Mob^+, Tc^R, *cos*, RK2 replicon, 21.6 kb	*Eco*RI	Friedman *et al.* (1982)	*X. c.* pv. *campestris* (Daniels *et al.*, 1984b)
pLAFR3	Derivative of pLAFR1, 23 kb, polylinker from pUC8, *lacZ*α^+	*Bam*HI, *Hin*dIII, *Pst*I, *Eco*RI	Staskawicz *et al.* (1987)	*X. c.* pv. *vesicatoria* (Swanson *et al.*, 1988; Whalen *et al.*, 1988); *X. c.* pv. *campestris* (Gough *et al.*, 1988); *X. c.* pv. *glycines* (Hwang *et al.*, submitted)
pIJ3200	Derivative of pLAFR1, 22 kb, polylinker from Bluescript, T3 and T7 promoters	*Sst*I, *Bst*XI, *Xba*I, *Bam*HI, *Pst*I, *Hin*dIII, *Xho*I, *Kpn*I, *Cla*I, *Spe*I	Liu *et al.* (1990)	*X. c.* pv. *campestris* (Liu *et al.*, 1990; *X. c.* pv. *raphani* (Parker *et al.*, in preparation)
pSa747	IncW, Mob^+, Km^R, Sp^R, *cos*, pSa replicon, 15 kb, cloning into unique *Eco*RI site inactivates Sp^R or *Kpn*I site inactivates Km^R	*Eco*RI and *Kpn*I	Tait *et al.* (1983)	*X. o.* pv. *oryzae* (Kelemu and Leach, 1990); *X. c.* pv. *malvacearum* (Gabriel *et al.*, 1986); *X. c.* pv. *vesicatoria* (DeFeyter *et al.*, 1990)
pHMI	pRI40 (Innes *et al.*, 1988) derivative, IncW, Mob^+, Sm^R, Sp^R, *cos*, pSa replicon, 13.3 kb, *parA*, polylinker from pUC19, *lacZ*α^+	*Hin*dIII, *Pst*I, *Sal*I, *Kpn*I, *Sac*I, *Eco*RI	Innes (personal communication)	*X. o.* pv. *oryzae* (Choi and Leach, unpublished)
pUFR034	IncW, Mob^+, Nm^R, *cos*, pSa replicon, 8.7 kb, *parA*, polylinker from pUC18, *lacZ*α^+	*Bam*HI, *Kpn*I, *Sac*I, *Eco*RI	DeFeyter *et al.* (1990)	*X. c.* pvs *campestris*, *malvacearum*, *vesicatoria*, *phaseoli*, *citri* (DeFeyter *et al.*, 1990); *X. o.* pv. *oryzae* (Choi and Leach, unpublished)

pUFR043	IncW, Mob^+, Nm^R, Gm^R, *cos*, pSa replicon, 10.3 kb, *parA*, polylinker from pUC18, *lacZα*$^+$	*Bam*HI, *Kpn*I, *Eco*RI, *Sst*I, *Sal*I	DeFeyter and Gabriel (1991b)	*X. c.* pv. *malvacearum* (DeFeyter and Gabriel, 1991b), *X. o.* pvs *oryzae* (Choi and Leach, unpublished) and *oryzicola* (C. Raymundo, unpublished)
pUFR047	IncW, Mob^+, Gm^R, Ap^R, pSa replicon, 8.6 kb, *parA*, polylinker from pUC18, *lacZα*$^+$	*Bam*HI, *Kpn*I, *Eco*RI, *Sst*I, *Sma*I, *Sal*I, *Pst*I, *Hin*dIII	DeFeyter and Gabriel (1991b)	*X. c.* pvs *malvacearum* (DeFeyter and Gabriel, 1991b), *translucens* (Waney *et al.*, 1991) and *oryzae* (Choi and Leach, unpublished)
pUFR027	IncW, Mob^+, Nm^R, pSa replicon, 8.7 kb, *parA*, polylinker from pUC18, *lacZα*$^+$	*Bam*HI, *Kpn*I, *Sac*I, *Eco*RI	DeFeyter *et al.* (1990)	*X. c.* pvs *malvacearum, campestris, phaseoli, citri, translucens, vesicatoria* (DeFeyter *et al.*, 1990), *X. o.* pv. *oryzae* (Choi and Leach, unpublished)
pKT230	IncQ, Mob^+, Km^R, Sm^R, RSF1010 replicon, 11.9 kb	*Bam*HI, *Bst*EII, *Eco*RI,*Sst*I, *Xma*I, *Xho*I, *Hin*dIII	Bagdasarian *et al.* (1981)	*X. c.* pvs *campestris, malvacearum, vesicatoria, translucens, phaseoli, citri* (DeFeyter *et al.*, 1990)
pIJ3100	IncQ, Mob^+, Sm^R, Cm^S, promoter probe plasmid, RSF1010 replicon, 13.2 kb	*Sma*I, *Bam*HI, *Sal*I, *Hin*dIII	Osbourn *et al.* (1987)	*X. c.* pv. *campestris* (Osbourn *et al.*, 1987)

Note: Inc refers to the incompatibility group (P, Q or W) of the parent plasmid; Mob^+ is the ability to be mobilized in conjugation between strains by appropriate *trans*-acting genes; *cos* is a DNA fragment originating from phage λ permitting the DNA to be packaged into phage capsids; *lacZα*$^+$ indicates a sequence permitting production of β-galactosidase by intragenic complementation in suitable *E. coli* strains; *parA* controls partition of plasmids between cells at division; superscript^R or ^{S} refers to conferring of resistance or susceptibility, respectively, to the antibiotics ampicillin (Ap), chloramphenicol (Cm), gentamycin (Gm), kanamycin (Km), neomycin (Nm), spectinomycin (Sp), streptomycin (Sm) and tetracycline (Tc).

In addition to a high transfer frequency, a suitable cloning vector should be inherited stably in the recipient *Xanthomonas*. This is especially important when transconjugants of plant pathogenic bacteria are screened for interaction with host plants in the absence of antibiotic selection. Plasmids replicating in a low or moderate number of copies per cell are often not inherited stably if they lack an efficient partition system for distribution of plasmid copies at cell division (Nordstrom, 1985). These problems have been overcome by the incorporation of the *parA* locus, isolated from the *Agrobacterium* plasmid pTAR (Gallie and Kado, 1987), into vectors such as pUFR027, pUFR034 (DeFeyter *et al.*, 1990), pRI40 (Innes *et al.*, 1988) and pHMI. The presence of the *parA* locus stabilizes these plasmids in *Xanthomonas* species.

(c) Gene exchange and introduction by conjugation

Broad host-range vectors are not self-transmissible; efficient transfer of the vectors between bacteria is possible by triparental matings (Ditta *et al.*, 1980). In triparental matings, *trans*-acting transfer functions which mobilize the cloning vector are provided by helper plasmids such as pRK2013 (Figurski and Helinski, 1979) or pSa322 (Tait *et al.*, 1983).

Biparental matings, in which the genes encoding transfer functions are incorporated into the genome of *E. coli* S17-1 (Simon *et al.*, 1983), have been shown to be a more efficient means of mobilizing vectors into *X. c.* pv. *phaseoli* and *X. o.* pv. *oryzae* than triparental matings.

(d) Transformation and electrotransformation

The introduction of plasmid DNA into bacterial cells by transformation is an important element of recombinant DNA technology. Many xanthomonads do not respond to chemical treatments that induce competence necessary for transformation in other bacteria. As a result, transformation systems involving the chemical induction of competence have been reported for only a few *Xanthomonas* spp. (Corey and Starr, 1957a,b; Yamasaki *et al.*, 1966; Gabriel, 1984; Atkins *et al.*, 1987; Murooka *et al.*, 1987). The reported transformation systems, in general, are not highly efficient or reproducible.

Electrotransformation, a technique by which bacteria are transformed by electroporation, has been used to successfully introduce plasmid DNA into a number of bacterial genera, including *Xanthomonas* (Kamoun and Kado, 1990a; White and Gonzalez, 1991). Electroporation is the application of high-intensity electric fields of short duration to reversibly permeabilize biomembranes. Specific parameters for electroporation have been determined for a few *Xanthomonas* strains and transformation efficiencies between 10^6 and 10^9 transformants/μg DNA have been reported, indicating that electrotransformation is indeed useful for gene transfer in this genus (Kamoun and Kado, 1990a; White and Gonzalez, 1991). Variation in responses between strains of a pathovar or between pathovars, however, indicates that for optimal DNA transfer electrical parameters may need to be determined for each bacterial strain (White and Gonzalez, 1991).

(e) Transduction

Bacteriophages, which infect and replicate in bacteria, are commonly used to transfer DNA efficiently into Gram-negative bacteria. However, the use of such systems requires that the bacteriophage infection and replication cycle be well characterized and, usually, that its DNA be modified to contain selectable markers and cloning sites. Although phages have been described and characterized for many *Xanthomonas* strains (see Chapter 2), none of them have been modified for use in genetic manipulation. It may be possible to use the powerful *E. coli* lambda technology for the genetic analysis of *Xanthomonas*. Strains of *Rhizobium meliloti* and *A. tumefaciens* that were sensitive to infection by phage lambda were created by the introduction of a broad host-range plasmid, pTROY, carrying *lamB*^{c} (a constitutive derivative of *lamB*, the gene encoding the receptor for lambda (De Vries *et al.*, 1984)). Although a derivative of pTROY was successfully introduced into *X. o.* pv. *oryzae*, infection with a lambda phage carrying Tn5 (Km^r) was not detected. Because lambda is potentially so useful, the application of this technology should be attempted with other *Xanthomonas* strains.

6.2.3 Analysis of cloned genes

Gene cloning exercises generally lead to the isolation of a fragment of *Xanthomonas* DNA which can either direct some biological function, such as complementation of a mutation, or contains sequences hybridizing to a particular probe. The task facing the experimenter is to narrow down and further characterize the DNA region of interest. Two general preliminary approaches are subcloning and transposon mutagenesis. The aim of subcloning is to find the smallest subfragment of the original cloned DNA which retains the property of interest. Efficient subcloning demands a detailed map of sites of cleavage by restriction endonucleases, and it can be a time-consuming operation to construct a map. Nevertheless, subcloning is usually needed for sequencing of the genes of interest. As an alternative to subcloning, it may be possible to locate genes by preparing 'nested' deletions *in vitro*, extending into the cloned DNA fragment from each end (Liu *et al.*, 1990).

Transposon mutagenesis of cloned DNA has proved to be a most efficient procedure for defining regions of interest. The recombinant plasmid containing the cloned DNA is first introduced into a suitable strain of *E. coli* containing a transposon. Random transposition occurs in the cells, and in some cases the transposon will insert into the cloned DNA. By exploiting the antibiotic resistance markers of the transposon and the cloning vector it is possible to isolate genetically a sample of transposon-carrying recombinant plasmids. The position of the transposon in each clone is determined by restriction mapping, and a selection of mutant clones with transposon insertions scattered at intervals of not more than 1 kb is taken for further study. First, the effect of each insertion on the biological activity of the cloned DNA is observed (e.g. insertions which abolish the ability to complement an original mutation). This should locate the position of the gene of interest. Second, the transposon-containing recombinant plasmids are used as vehicles for localized or 'site-directed' mutagenesis of the wild-type *Xanthomonas* genome. This technique was introduced for *Rhizobium* (Ruvkun and Ausubel, 1981), and adapted for *X. c.* pv. *campestris* (Turner *et al.*, 1985). Essentially, the technique involves transferring each mutant recombinant plasmid into the wild-type *Xanthomonas* strain. In a small fraction of the population recombination between the bacterial genome and the homologous cloned DNA results in 'marker-exchange', i.e. transfer of the transposon to the corresponding site in the genome. Genetic procedures, often based on plasmid incompatibility, are used to isolate the marker-exchange mutants. Thus a series of new mutants is produced, each containing a single transposon insertion pre-mapped with respect to the original cloned DNA. There are two main uses of marker exchange: (a) linked or clustered genes can be detected and (b) each marker-exchange mutant is isogenic with the wild-type bacterium, apart from the single transposon insertion. Thus one can be confident that phenotypic alterations are caused by the mutation rather than differences in the genetic backgrounds.

Transposons have also been used to monitor gene expression. Several derivatives have been constructed containing promoterless 'reporter' genes giving easily assayable phenotypes adjacent to one end. If the transposon inserts in the correct orientation into a gene which is being actively expressed the reporter gene is also expressed via transcriptional or translational fusion. Examples of reporter genes are *lacZ* (β-galactosidase; Kroos and Kaiser, 1984), *uid* (=*gus*, β-glucuronidase; Bonas *et al.*, 1991), *lux* (luminescence; Shaw *et al.*, 1988) and *phoA* (alkaline phosphatase; Manoil and Beckwith, 1985).

Transposon mutagenesis has played a key role in the characterization of all the classes of *Xanthomonas* genes discussed below, and the papers cited give details of technical adaptations to particular organisms and circumstances.

We have discussed the preliminary characterization of cloned *Xanthomonas* genes. More detailed studies, such as transcription analysis, elucidation of the nature of gene products, nucleotide sequencing, etc., follow standard protocols described elsewhere (e.g. Sambrook *et al.*, 1989).

6.2.4 Pathogenicity testing: the geneticist's perception

(a) General considerations

To identify cloned genes or specific mutations which affect a host–pathogen interaction, it is frequently necessary to screen large numbers of individual bacterial strains. For such a screen to be manageable, limitations in time and space as well as screen reliability and authenticity must be considered. Factors such as plant age and cultivar, pathogen inoculum level, and environmental conditions after inoculation also can influence the outcome of the screen. For example, seedling assays offer the advantages of minimal space, supply, and time requirements; however, they may not accurately represent the disease syndrome. Different levels of screening may be required to assure that initially selected clones or mutants are actually involved in disease symptom development. Thus, the type of plant assay chosen is critical because it will determine the accuracy and correctness of the conclusions reached about the role of a gene in fitness, pathogenicity, or avirulence.

(b) Screening for genes involved in resistance induction

Plant assays for incompatibility usually rely on the appearance of the hypersensitive response (HR), i.e. the rapid necrosis of plant tissues induced by an avirulent pathogen that results in the inhibition of pathogen multiplication and spread. A technique successfully used to observe the HR in various *Xanthomonas*–plant interactions requires infiltration of leaves with the blunt end of a syringe against the leaf surface and gentle pressure infiltration of a bacterial suspension (*ca* 10^7–10^9 cfu/ml) through the stomata and into the leaf (Gabriel *et al.*, 1986; Swanson *et al.*, 1988; Whalen *et al.*, 1988; Kamoun and Kado, 1990a; Arlat *et al.*, 1991; Bonas *et al.*, 1991; Reimers and Leach, 1991). The HR is usually visible within 16–48 h, although the timing and appearance may vary in different *Xanthomonas*–plant interactions. Compatible interactions are usually typified by spreading watersoaked lesions. After the primary rapid screen for HR induction, the multiplication of selected strains in host plants is compared with that of wild type to confirm the correlation between the HR and resistance. In some interactions, for example that between *X. c.* pv. *campestris* and *Arabidopsis*, a typical HR (browning) does not occur, and the absence of water-soaking or other symptoms characteristic of a compatible reaction after infiltration of leaves is used as a measure of incompatibility (Daniels *et al.*, 1991; Tsui *et al.*, 1991).

Incompatible interactions involving certain bacterial blight resistance genes in rice do not result in a typical HR or browning response (Bai, 1990). In addition, several resistance genes are not expressed in the seedling stage of growth (Mew, 1987). In these cases, mature plants are inoculated by pricking leaves with a bacteria-laden needle or clipping with scissors dipped in bacterial suspensions. Quantitative differences in lesion length are used to distinguish between virulent and avirulent strains of *X. o.* pv. *oryzae* (Bai, 1990; Kelemu and Leach, 1990). In interactions such as these, it is very important to compare bacterial populations with wild-type strains to be certain that the genes affected are indeed involved in the host–pathogen interaction.

(c) Screening for genes involved in pathogenicity

Testing for alterations in pathogenicity are, in general, more troublesome than tests for the HR because differences between strains may be very subtle if the pathogen's aggressiveness (a quantitative difference) is affected rather than a total loss of pathogenicity (a qualitative difference). Inoculation of seedlings by wounding or by infiltration of bacterial suspensions with syringes has been used to identify genes involved in pathogenesis from *X. c.* pv. *translucens* (Mellano and Cooksey, 1988; Sawczyc *et al.*, 1989; Waney *et al.*, 1991), *X. c.* pv. *citri* (Swarup *et al.*, 1991a) and *X. c.* pv. *campestris* (Daniels *et al.*, 1984a). Direct application of inoculum by wounding is more simple and rapid than infiltration, and usually results in reproducible expression of symptoms. For example, Daniels *et al.* (1984a) inoculated 4-day-old turnip seedlings grown on agar in plastic dishes by stabbing the stem with a needle touched to a colony of *X. c.* pv. *campestris*. Of the seedlings inoculated with virulent bacteria, 90–95% developed characteristic spreading translucent lesions, which gradually turned brown, within 2–3 days.

Although seedling assays which involve wounding have been used successfully to identify several pathogenicity genes from *Xanthomonas* spp., genes involved in some aspects of disease development may be missed (Daniels, 1987). Natural infection of crucifers by *X. c.* pv. *campestris* usually results from epiphytic colonization followed by entry of the bacteria through the hydathodes, although entry can also occur through wounds (Cook *et al.*, 1952). Inoculation of turnip seedlings by stabbing with a bacteria-infested needle would clearly bypass several plant barriers and thus, bacterial genes required for the early stages of infection may not be detected (Daniels, 1987).

Several assays have been developed which attempt to mimic natural plant infections more precisely. However, they often require mature plants and therefore need more space, time and resources. An assay where seeds are soaked in bacterial suspensions prior to planting alleviates the problem of breaching natural plant barriers and mimics infection via seed transmission (Gough *et al.*, 1988). Mist inoculation of mature plants (Mew *et al.*, 1984; Shaw and Kado, 1988) mimics infection through hydathodes. Wounding leaves of mature plants with an infested needle or pipette tip (Birch and Patil, 1987; Shaw and Kado, 1988; Kamoun and Kado, 1990a) imitates infection through naturally occurring wounds. If the leaves are submerged in bacterial suspensions prior to wounding (Gough *et al.*, 1988; Dow *et al.*, 1990), the suspension is drawn into the vessel ends and resembles entry through hydathodes. Dow *et al.*, (1990) clearly demonstrated that protease-deficient strains of *X. c.* pv. *campestris* were less aggressive than wild-type strains if bacteria were introduced into mature leaves of turnip variety 'Just Right' by small wounds at the vein endings. Differential responses between the strains also were observed when different bacterial inoculum levels, ranging from 10^4 to 10^8 cfu/ml, were used (Dow *et al.*, 1990). The same mutants showed little or no reduction in virulence from wild-type strains in seedling bioassays with the variety 'Just Right', although there was a slight delay in the onset of rotting in seedlings inoculated with the protease-deficient strains (Tang *et al.*, 1987). Symptom development was the same between the protease-deficient mutants and wild-type strains after inoculation of seedlings of another turnip variety, Golden Ball, suggesting there may be different sensitivities to the proteases between host varieties (Tang *et al.*, 1987).

Another important consideration in selecting an assay is that the symptoms monitored are characteristic of the disease. Shaw and Kado (1988) compared several techniques for inoculation of *X. c.* pv. *campestris* into crucifers. These included a seedling assay (Daniels *et al.*, 1984a) and mature plant assays involving infiltration of attached and detached leaves (Osbourn *et al.*, 1987), mist inoculation (hydathode infections), and wound inoculations. They concluded that inoculation of mature plant tissues resulted in symptoms more typical of black rot. Furthermore, they observed that bacteria normally not pathogenic to crucifers could cause necrosis or maceration when infiltrated into detached leaves or inoculated to seedling stems. Thus, selection of an adequate plant assay for genes involved in plant–pathogen interactions requires an understanding of the disease syndrome.

6.3 PROPERTIES OF THE *XANTHOMONAS* GENOME

The general properties of the *Xanthomonas* genome are mentioned in Chapter 2. Repetitive DNA elements have been described in *X. c.* pv. *vesicatoria* (Bonas *et al.*, 1989; Kearney and Staskawicz, 1990a), *X. o.* pv. *oryzae* (Leach *et al.*, 1990) and *X. c.* pv. *citri* (Tu *et al.*, 1989). Complex hybridization patterns observed when cosmid clones were used as probes to total genomic DNA suggest the existence of repeated DNA sequences in other xanthomonads (Lazo *et al.*, 1987; Gabriel *et al.*, 1988; Leach *et al.*, 1990). The presence of these repetitive sequences has been used to analyse the genetic interrelationships between *Xanthomonas* species and pathovars (Leach *et al.*, 1990; Quobela *et al.*, 1990, 1991) and to investigate pathogen population structures (Leach and White, 1992).

Several repetitive sequences are insertion sequences (Kearney *et al.*, 1988; Tu *et al.*, 1989; Kearney and Staskawicz, 1990a; Yun, 1991). Tu *et al.* (1989) characterized two mobile elements which had inserted into the broad host-range plasmid RP4 after introduction into *X. c.* pv. *citri*. The elements also were capable of transposition in *E. coli*.

An insertion sequence (IS*476*) in *X. c.* pv. *vesicatoria* can insert into and inactivate an avirulence gene, *avrBs1*, suggesting such elements might play a role in race evolution (Kearney *et al.*, 1988). IS*476* is 1225 bp in length, has 26 bp inverted repeats and causes a 4-bp target site duplication upon insertion into *avrBs1* (Kearney and Staskawicz, 1990a). The element is related to IS*3* from *E. coli* and IS*51* from *Pseudomonas syringae* pv. *savastanoi*. The mobility of the element was demonstrated using a transposition assay based on the insertional inactivation of the *sacRB* locus of *Bacillus subtilis* (Kearney and Staskawicz, 1990a).

A repetitive element originally cloned from the genome of *X. o.* pv. *oryzae* (Leach *et al.*, 1990) later was found to be a mobile element based on insertional inactivation of *sacRB* (Yun, 1991). The element, called IS*112*, is 1055 bp in length with 25 bp terminal inverted repeats and causes a 3-bp target site duplication at the site of insertion into *sacRB*. Is*112* is related to IS*4351* from *Bacteroides fragilis*.

6.4 GENES WHICH ARE IMPORTANT FOR PATHOGENICITY

6.4.1 Extracellular polysaccharide biosynthesis

Most plant pathogenic bacteria produce extracellular polysaccharide (EPS) in culture, giving colonies on agar a slimy, glistening appearance. Variants which no longer produce EPS are frequently less virulent than the EPS$^+$ wild type. This has led to suggestions that EPS is a pathogenicity factor. Although several hypotheses have been advanced for EPS action, little is known about the mechanisms involved (Coplin and Cook, 1990).

The major polysaccharide produced by *Xanthomonas* is xanthan gum. Because of the unusual rheological properties of its solutions, xanthan has many industrial applications (see Chapter 9). This fact has led to xanthan becoming one of the most extensively studied microbial polysaccharides (Sutherland, 1985). The biosynthesis of xanthan in *X. c.* pv. *campestris* was studied using cells permeabilized with toluene so that sugar nucleotide intermediates could be taken up (Ielpi *et al.*, 1981). A pathway involving sequential additions to intermediates bound to a lipid carrier molecule was deduced. Whitfield *et al.* (1981) isolated EPS mutants of *X. c.* pv. *campestris*. Some of these and other mutants were used by Barrère *et al.* (1986) to isolate cosmids which restored EPS production from a genomic library of *X. c.* pv. *campestris*. Using transposon mutagenesis some clustering of genes was found. These genes could restore function to certain EPS$^-$ *Rhizobium* mutants (Borthakur *et al.*, 1986). Using a similar strategy, Harding *et al.* (1987) found a cluster of genes spanning 13.5 kb containing at least five complementation groups. Thorne *et al.* (1987) similarly found three unlinked clusters of genes affecting xanthan synthesis. Cosmids which could restore EPS production to three *X. c.* pv. *campestris* mutants were subjected to Tn*5lac* mutagenesis (Hötte *et al.*, 1990), revealing 12 complementation groups in a 35 kb region of DNA. The most detailed studies of xanthan biosynthesis have been conducted by a group of researchers at Synergen Inc. (Vanderslice *et al.*, 1989). A strategy similar to those just described was used to isolate cosmid clones carrying EPS genes. Not only was an operon (*gum*) defined by mutagenesis, but the whole region was sequenced. Using the procedure of Ielpi *et al.* (1981) to study mutants blocked in each gene, it was possible to ascribe a biochemical function to each gene. Regrettably, these invaluable data are not readily available in the public domain.

The regulation of xanthan biosynthesis is poorly understood, but probably very complex (e.g. Kamoun and Kado, 1990b). Examples of regulatory genes which affect xanthan biosynthesis are given in a subsequent section.

As with other classes of EPS produced by plant pathogens, the role of xanthan in pathogenesis is not clear. Experiments reported by Barrère *et al.* (1986) indicated that EPS$^-$ mutants of *X. c.* pv. *campestris* produced symptoms in turnip seedlings indistinguishable from those produced by the wild type. However, this pathogenicity test is not a good model of the natural disease, and probably simulates only the later, necrotic stages of the disease (see section 2.4.3). When a more stringent pathogenicity test was used in which bacteria were introduced into veins at the leaf margins (Dow *et al.*, 1990), EPS^- mutants were severely reduced in virulence. Thus, it is possible that xanthan is required at an early stage of infection, but once the disease is well advanced it becomes dispensable.

6.4.2 Extracellular enzymes

Certain *X. campestris* pathovars produce extracellular enzyme activities such as protease, pectinase, glucanase, amylase and lipase (Chapter 2). Since these enzymes have the capacity to degrade components of plant tissues, they are obvious candidates as pathogenicity factors. Genetic studies have contributed to an understanding of the role of certain enzymes.

Simple plate tests are available for screening bacterial colonies which produce one or more of the enzymes listed above. It is possible to clone structural genes for the enzymes by strategies which depend on the detection of the enzyme produced from the cloned gene.

(a) Protease

E. coli does not produce extracellular protease activity. When *E. coli* colonies harbouring clones of a genomic library of *X. c.* pv. *campestris* were grown on agar containing skimmed milk, a clone which produced a zone of clearing characteristic of protease-producing bacteria was found (Tang *et al.*, 1987). It is unusual for *X. c.* pv. *campestris* genes to be efficiently expressed in *E. coli,* and attempts to clone pectinase, glucanase and amylase genes by analogous methods were unsuccessful. The protease gene was located by subcloning, Tn*5* mutagenesis, deletion analysis and sequencing (Tang *et al.*, 1987; Liu *et al.*, 1990). The sequence showed strong relatedness to several known serine proteases, and biochemical studies by Dow *et al.* (1990) confirmed that the enzyme belonged to the family of serine metalloproteases. *X. c.* pv. *campestris* also produces a second protease (a metalloprotease). The enzymes could be 'fingerprinted' by the pattern of peptides produced by digestion of purified β-casein. Although all strains of *X. c.* pv. *campestris* tested gave the same activity fingerprints, other pathovars yielded different patterns, suggesting that each produced protease(s) with characteristic peptide bond cleavage specificities. The cloned *X. c.* pv. *campestris* protease gene was used to construct Tn*5* marker exchange mutants which had lost both protease activities. When these mutants were inoculated into turnip seedlings both symptoms and bacterial growth *in planta* were not significantly different from the wild type (Tang *et al.*, 1987). However, when inoculated into the margins of mature turnip leaves, the protease-deficient mutants were much less virulent (Dow *et al.*, 1990). Thus, the requirement for protease in pathogenicity is reminiscent of that for EPS described above, suggesting a role for the enzyme in the early stages of infection of the plant.

Protease-deficient mutants of *X. o.* pv. *oryzae* have been isolated which showed reduced aggressiveness to rice, and impaired ability to grow *in planta* (G. Wu and C.F. Gonzalez, personal communication). The protease gene of *X. c.* pv. *campestris* is known to have a homologue in the *X. o.* pv. *oryzae* genome (Todd *et al.*, 1990), but it is not yet known whether this is the gene which was mutated in the experiments of Wu and Gonzalez.

(b) Endoglucanase

X. c. pv. *campestris* endoglucanase genes are not expressed in *E. coli*. However, when a cosmid library of *X. c.* pv. *campestris* DNA was introduced into the non-glucanolytic *X. c.* pv. *translucens*, clones which directed the production of glucanase activity (detected on carboxymethylcellulose-containing agar) were obtained. Two genes were found, one encoding an intracellular activity, and the other the extracellular enzyme which is the most abundant secreted protein in *X. c.* pv. *campestris* cultures (Gough *et al.*, 1988). The protein was purified and sequenced at the N-terminus and the gene was sequenced. Comparison of the two sequences indicated that maturation and export of the enzyme through the inner membrane of the bacteria involves removal of a signal sequence of 23 amino acids from the N-terminus of the protein. The sequence of the gene indicates that the enzyme is a member of the most common class of prokaryotic glucanases ('cellulases'; Gough *et al.*, 1990). No role for the extracellular glucanase in pathogenesis has been demonstrated. Mutants devoid of activity showed no significant loss in virulence (Gough *et al.*, 1988).

(c) Amylase

Cosmids carrying the *X. c.* pv. *campestris* amylase gene(s) were isolated by their ability to produce amylase activity in a strain of *X. c.* pv. *translucens* which did not otherwise degrade starch. The genes have not been characterized.

(d) Pectinase

Pectin degradation by *X. campestris* pathovars depends on pectinmethylesterase and polygalacturonate lyase (PGL; Starr and Nasuno, 1967). A PGL structural gene of *X. c.* pv. *campestris* was cloned by virtue of its ability to be expressed at a detectable level in non-pectolytic strains of *X. c.* pv. *translucens* and *X. c.* pv. *vesicatoria* (Dow *et al.*, 1989). *X. c.* pv. *campestris* produces three isoforms of PGL (Dow *et al.*, 1987; Beaulieu *et al.*, 1991). Strains carrying Tn*5* insertions in the chromosomal copy of the PGL gene retained virulence (Dow *et al.*, 1989). However, the mutants still produced the other isoforms; to determine the role of PGL in pathogenesis it would be necessary to mutate all three structural genes. Some *X. c.* pv. *vesicatoria* strains are also pectolytic, but in this pathovar only one PGL species is produced. Beaulieu *et al.* (1991) cloned this gene and produced PGL^- mutants. No reduction in virulence could be detected in the mutants. The cloned gene was found to hybridize to DNA of all xanthomonads tested, whether or not they were pectolytic. At the moment the significance of pectolysis in pathogenicity is not understood.

(e) Enzyme secretion

Mutations in the structural genes coding for extracellular enzymes affect only the one enzyme. However, two classes of mutation have pleiotropic effects, giving reductions in all the enzymes. One class lies in regulatory genes (discussed in the next section), and the other interferes with secretion of enzymes from the bacterial cell into the medium. A recombinant cosmid from a genomic library of *X. c.* pv. *campestris* which could restore pathogenicity to a mutant was isolated (Daniels *et al.*, 1984b). Turner *et al.* (1985) showed that the cloned DNA contained a cluster of genes spanning more than the 10 kb needed for pathogenicity, and Dow *et al.* (1987) found that mutations in this region resulted in failure to secrete PGL, protease and endoglucanase. The enzymes were synthesized at near wild-type levels, but accumulated in the bacterial periplasm. Beaulieu *et al.* (1991) found mutants of *X. c.* pv. *vesicatoria* which had simultaneously lost PGL and protease activity. These may be secretion mutants. The secretion mutants of *X. c.* pv. *campestris* are avirulent in all types of plant test. Since loss of several of the individual enzyme activities by mutation of structural genes has relatively little effect on virulence, it is possible that the enzymes act synergistically. Alternatively, the secretion system may also transport an as yet unidentified protein which is essential for virulence.

Part of the secretion gene cluster of *X. c.* pv. *campestris* has been sequenced (Dums *et al.*, 1991). Six putative genes, *xpsE–J*, organized in an operon were found, and the deduced protein products showed striking relatedness to the pullulanase secretion gene products, PulE–J, of *Klebsiella*. Further secretion genes were characterized by Hu *et al.*, (1992) and one of these was related to *pulD* of *Klebsiella*. It is becoming apparent that secretion systems of many Gram-negative bacteria are highly related (Pugsley *et al.*, 1990). De Groot *et al.* (1991) found that secretion genes (*xcp*) of *Pseudomonas aeruginosa* could restore secretion of protease and amylase to a *xps* mutant of *X. c.* pv. *campestris*, and, on the other hand, *xps* genes gave some restoration of elastase secretion to a *P. aeruginosa xcp* mutant. In contrast to the process of protein export through the inner membrane of bacteria, little is known about the mechanisms of secretion through the outer membrane.

6.4.3 Regulatory genes

Several independent systems pleiotropically regulate the synthesis of extracellular enzymes and polysaccharide in *X. c.* pv. *campestris*. A cosmid clone which restored pathogenicity and protease and PGL production to a mutant (Daniels *et al.*, 1984b) was found to contain a cluster of genes which behave as positive regulators of synthesis of not only protease and PGL, but also endoglucanase, amylase and EPS (Tang *et al.*, 1991). Sequencing of part of the cluster revealed that at least one gene, designated *rpfC*, is a member of the family of prokaryotic 'two-component' regulatory genes (Stock *et al.*, 1989). Usually the products of these gene pairs act to regulate expression of some target gene(s) in response to environmental changes or stimuli. One of the components, the sensor, is usually anchored by N-terminal domains to the inner membrane, with portions of the molecule exposed to the periplasm probably acting as a receiver of the external stimulus. Transduction of the signal is accomplished by phosphorylation, firstly of a histidinyl residue in the conserved C-terminus of the sensor protein, and secondly by transfer of the phosphate moiety to an aspartyl residue in a conserved N-terminal region of the second component, the regulator. Activation of the regulator by the phosphate transfer reaction allows it to modulate expression of the target genes. Usually the sensor and regulator are encoded by separate genes (often linked), but in the case of the *X. c.* pv. *campestris rpfC* gene the two components are encoded by one gene. It is not known whether the system functions as a single protein, or whether post-translational cleavage generates separate functional entities. Since the only common feature of the genes regulated by *rpfC* (and other genes of the cluster) is that they are required for virulence, it is possible that the system acts to allow the bacteria to adjust their biosynthetic activities to the changes in their environment (i.e. the plant) as disease progresses. There is evidence that *rpfC* homologues are found in *X. o.* pv. *oryzae* and *X. c.* pv. *translucens* (Sawczyc *et al.*, 1989; Todd *et al.*, 1990) and, based on its sequence, the *X. o.* pv. *oryzae* gene is highly related to *rpfC*. Little is known about other genes in the cluster and how they interact with *rpfC*.

The activation or positive regulation of synthesis of extracellular enzymes and EPS in *X. c.* pv. *campestris* is 'balanced' by a negative regulatory system (Tang *et al.*, 1990). Increasing the copy number of the regulatory gene gives coordinate repression of synthesis of enzymes and EPS, and conversely mutation of the chromosomal copy of the gene results in coordinate overproduction. It is likely that negative regulation acts via a protein which binds to a conserved motif upstream of the promoter of cellulase and protease genes.

In many bacteria cyclic AMP and its receptor protein CAP are known to modulate transcription of many genes. A DNA fragment of *X. c.* pv. *campestris* partially restored function to an *E. coli* mutant defective in the cAMP–CAP system, and a gene designated *clp* was thereby identified coding for a protein similar to CAP. A *clp* mutant of *X. c.* pv. *campestris* was constructed and showed reduced virulence. The mutant overproduced protease, but produced less cellulase, pectinase, amylase and EPS. The EPS produced had different properties to the xanthan of the wild-type strain (De Crécy-Lagard *et al.*, 1990).

Thus, the synthesis of extracellular enzymes in *X. c.* pv. *campestris* is regulated in a complex manner by at least three independent systems. The significance of this complexity is not understood.

Since *rpfC*, a positive regulatory gene, is a 'two-component' regulatory gene, it was of interest to determine whether *X. c.* pv. *campestris* possessed other environment-sensing systems which could be important for host–pathogen interaction. Using oligonucleotides which would code for conserved domains of two-component proteins, Osbourn *et al.* (1990a) isolated a number of DNA fragments containing genes for putative two-component systems. In most cases, mutagenesis of the corresponding genomic region produced no obvious phenotypic change relevant to pathogenicity, but mutations in one gene caused a 50% reduction in EPS production. Sequencing confirmed that the fragment contained linked genes encoding a sensor-regulator pair. However, the reduction in EPS levels in the mutant was not enough to give detectably lower virulence.

6.4.4 *hrp* genes

The *h*ypersensitive *r*eaction and *p*athogenicity (*hrp*) genes are required for several bacterial–plant interaction phenotypes. Inactivation of the *hrp* genes has pleiotropic effects; mutants are attenuated in their ability to colonize and initiate disease symptoms in host plants and cannot elicit the hypersensitive reaction (HR) in non-host plants (Lindgren *et al.*, 1988). In addition, Hrp^- mutants cannot elicit the HR in incompatible interactions with resistant host plants (Huynh *et al.*, 1989; Bonas *et al.*, 1991).

(a) *Identification and characterization of* hrp *genes in* Xanthomonas

Identification of *hrp* gene clusters in plant pathogenic bacteria has been accomplished by complementation of Hrp^- mutants or through the use of heterologous probes (see Willis *et al.*, 1991, for review). The presence of *hrp* clusters in xanthomonads was implicated by hybridization with clones containing *Pseudomonas solanacearum hrp* sequences (Boucher *et al.*, 1987). Arlat *et al.* (1991) used *P. solanacearum hrp* sequences to identify cosmid clones carrying *hrp* gene clusters from libraries of *X. c.* pv. *campestris* and *X. c.* pv. *vitians*. The *hrp* gene cluster from *X. c.* pv. *campestris* was about 25 kb in size, and was homologous and colinear over a continuous 19 kb DNA region with the *P. solanacearum hrp* cluster.

Hrp^- mutants of *X. c.* pv. *vesicatoria* that were non-pathogenic to pepper and did not elicit an HR on resistant pepper cultivars or on tobacco (a non-host plant), were complemented by clones from a wild-type genomic library (Bonas *et al.*, 1991). The *hrp* gene region identified covers about 25 kb and is organized into at least six different complementation groups, suggesting a number of operons. Mutants in all six complementation groups failed to grow to wild-type levels *in planta*. DNA sequences homologous to a 23 kb region from the *hrp* cluster hybridized with genomes of other pathovars of *X. campestris*. Cross-hybridization between the *hrp* sequences from *X. c.* pv. *vesicatoria* and genomic DNA from different species of *Pseudomonas* was not observed. However, weak hybridization was observed with *P. solanacearum*.

Most *hrp* genes that have been described thus far are part of large clusters (see Willis *et al.*, 1991, and Mindrinos *et al.*, 1990, for reviews). It is not clear if there are several *hrp* clusters in a bacterial genome, or if all genes which function to confer the Hrp phenotype are part of large clusters. Kamoun and Kado (1990a) described a locus (*hrpXc*) of about 2.1 kb which complemented a transposon-induced *hrp* mutant of *X. c.* pv. *campestris*. A locus which functionally complements the *X. c.* pv. *campestris* mutant also was cloned from *X. o.* pv. *oryzae* using *hrpXc* as a probe. The *hrpXc* sequences do not hybridize with the *hrp* cluster cloned from *X. c.* pv. *vesicatoria* or from *P. solanacearum*, suggesting that the locus is not part of the described clusters (Kamoun and Kado, 1990a). It is not certain whether or not the *hrpXc* locus is part of a cluster.

(b) *Conservation of* hrp *genes and their function*

The *hrp* genes are structurally conserved within a pathovar, and between several pathovars of *X. campestris* (Kamoun and Kado, 1990a; Bonas *et al.*, 1991; Arlat *et al.*, 1991). In some pathovars, homology between *hrp* genes was sufficient to allow for recombination, giving marker exchange (Bonas *et al.*, 1991). Non-pathogenic mutants of *X. c.* pv. *citri*, *X. c.* pv. *dieffenbachiae* and *X. c.* pv. *vignicola* were obtained by marker exchange with transposon-tagged alleles of the *hrp* region from *X. c.* pv. *vesicatoria*.

Stall and Minsavage (1990) demonstrated hybridization of the cloned *hrp* sequences from *X. c.* pv. *vesicatoria* to 33 plant pathogenic strains from 17 *X. campestris* pathovars. An interesting observation was that opportunistic xanthomonads either did not hybridize with the probe, or hybridized weakly. However, conclusions about the pathogenicity of the opportunistic xanthomonads in relation to the *hrp* genes cannot be drawn until more is known about the extent of variation that can occur in a *hrp* gene cluster with no change in function of gene products.

Functional complementation of *hrp* sequences has been described between pathovars, species and genera. Cross-complementation experiments have shown the *hrp* sequences from *X. c.* pv. *vesicatoria* and *X. c.* pv. *phaseoli* (Bonas *et al.*, 1991), *X. c.* pv. *vitians* and *X. c.* pv. *campestris* (Arlat *et al.*, 1991), and *X. c.* pv. *campestris* and *X. o.* pv. *oryzae* are functionally interchangeable. A Hrp^- mutant of *X. c.* pv. *campestris* was partially complemented by *hrp* sequences from *P. solanacearum*, indicating that some functional relatedness exists between the *hrp* genes of different genera (Arlat *et al.*, 1991).

(c) Regulation and function of hrp *genes*

Expression of *hrp* genes has been shown to be regulated by nutrient conditions (Huynh *et al.*, 1989; Arlat *et al.*, 1991), osmolyte concentrations (Mindrinos *et al.*, 1990) and plant-inducing factors (Lindgren *et al.*, 1989; Rahme *et al.*, 1991; Schulte *et al.*, 1991; Schulte and Bonas, 1992a,b). Using fusions to a luciferase (*lux*) or chloramphenicol acetyltransferase (*cat*) reporter gene, the expression of the *hrp* locus *hrpXc* from *X. c.* pv. *campestris* was found to be strongly induced when the bacteria are in radish leaves (Kamoun and Kado, 1990a). The level of expression of the *hrpXc* promoter in rich or minimal medium was about two to three orders of magnitude lower than that of a constitutive tetracycline resistance gene promoter, whereas in radish leaves the two promoters displayed similar activities. In contrast, the homologue of *hrpXc* cloned from *X. o.* pv. *oryzae* is constitutively expressed, and does not appear to be plant inducible. Similarly, Schulte *et al.* (1991) and Schulte and Bonas (1992b) observed no induction of genes (measured using *gus* fusions) in any of the six complementation groups of the *hrp* cluster cloned from *X. c.* pv. *vesicatoria* after growth in rich or minimal media. However, after growth *in planta*, induction of the genes was observed.

Variations of carbon sources or other nutrients in minimal media was not reported in the above examples. Schulte and Bonas (1992a) reported that sucrose and sulphur induced the *hrp* locus of *X. c.* pv. *vesicatoria*. Expression of the genes (studied by *lacZ* fusions) in the *hrp* cluster from *X. c.* pv. *campestris* was repressed in rich medium; expression in minimal medium was dependent on the carbon source supplied to the cells (Arlat *et al.*, 1991). Transcription of the genes was not regulated by genes which control the synthesis of extracellular enzymes (which are required for pathogenicity), suggesting the *hrp* genes are regulated independently of other pathogenicity genes.

The biological role of *hrp* gene products in the HR and pathogenicity are not yet known. A regulatory function has been suggested for one of the *hrp* gene products from *P. syringae* pv. *phaseolicola*; it contains a conserved domain characteristic of certain regulatory proteins of the NtrC family (Grimm and Panopoulos, 1989). Kamoun and Kado (1990a) demonstrated that a Hrp^- mutant could be complemented *in planta* by co-inoculation with the wild-type strain, suggesting either a diffusible or extracellular product could cross-feed and correct the mutant phenotype.

6.4.5 Avirulence genes

Cultivar-/race-specific interactions characterized by gene-for-gene complementarity have been described for several *Xanthomonas* spp. and their host plants. In such interactions, the resistance response is proposed to be controlled by an avirulence (*avr*) gene from the pathogen and a corresponding resistance gene in the host (Flor, 1955; Ellingboe, 1976). The first microbial avirulence gene was cloned from *Pseudomonas syringae* pv. *glycinea* by Staskawicz *et al.* (1984). Since then, avirulence genes have been cloned from several *Pseudomonas* and *Xanthomonas* spp. (see Keen, 1990, and Keen and Staskawicz, 1988, for reviews). Although several *avr* genes have been indentified and characterized, it is not yet known how the genes function to elicit host resistance.

(a) Indentification and characterization of avirulence genes

The gene-for-gene hypothesis predicts that *avr* genes confer a positive function, i.e. the ability to elicit resistance. This allows for the identification of *avr* genes by introduction of libraries from avr^+ strains (those which elicit resistance in a particular cultivar) into virulent strains; the transconjugants are screened on the cultivar of interest for changes in the interaction phenotype from compatible (susceptible) to incompatible (resistant). To date, avirulence genes have been cloned from *X. c.* pv. *vesicatoria* (Swanson *et al.*, 1988; Whalen *et al.*, 1988; Bonas *et al.*,

Table 6.2 Properties of avirulence genes cloned from *Xanthomonas* strains

Pathogen	*Avirulence gene*	*Plant resistance gene recognized*	*Size and location*	*Relatedness to other genes*	*Other characteristics*	*Interaction phenotype*	*Reference*
Xcv	*avrBs1*	*Bs1*, pepper	5.3 kb, plasmid	None	49.8 kDa protein, homology to carboxy terminus of AvrA protein from *P. s.* pv. *glycinea*	Dark, papery brown, confluent necrosis by 8–12 h; electrolyte leakage	Swanson *et al.* (1988), Ronald and Staskawicz (1988)
Xcv	*avrBs2*	*Bs2*, pepper	2.4 kb, chromosome	None		Medium brown necrosis by 12–36 h; electrolyte leakage	Minsavage *et al.* (1990), Kearney and Staskawicz (1990b)
Xcv	*avrBs3*	*Bs3*, pepper	3.6–3.7 kb, plasmid	*Xoo* and *Xcm avr* genes, *avrBsP*	Constitutive; 102 bp direct repeat 17.5 times; 112 kDa protein; intracellular location	Medium brown, confluent necrosis by 24–36 h; electrolyte leakage	Bonas *et al.* (1989), Schulte *et al.* (1991), Knoop *et al.* (1991)
Xcv	*avrBsT*	*BsT*, pepper	4.3 kb, plasmid	None		Light brown, non-confluent necrosis by 8–12 h, electrolyte leakage	Minsavage *et al.* (1990)
Xcv	*avrBsP*, also *avrBs3-i*	Gene in tomato	1.7 kb, plasmid	*avrBs3*	Carboxy terminus not cloned, but still functional; six 102 bp repeats	Electrolyte leakage, reduced bacterial growth, small flecks or no lesions	Canteros *et al.* (1991)
Xcv	*avrRxv*	*Rxv* and equivalent genes in many plants	2.1 kb, chromosome	None to other pvs; only to *Xcv*		Dark necrotic HR to *Phaseolus vulgaris* cv. *sprite*	Whalen *et al.* (1988)

Xcm	*avrB4*	*B4*, cotton	Plasmid	*avrBs3*	Also functions in *B2*	HR	DeFeyter and Gabriel (1991b)
Xcm	*avrb6*	*b6*, cotton	Plasmid	*avrBs3*	Also functions in *B2*	HR	DeFeyter and Gabriel (1991b)
Xcm	*avrb7*	*b7*, cotton	Plasmid	*avrBs3*	Also functions in *B2*	HR	DeFeyter and Gabriel (1991b)
Xcm	*avrBIn*	*BIn*, cotton	Plasmid	*avrBs3*	Also functions in *B2*	HR	DeFeyter and Gabriel (1991b)
Xcm	*avrB101*	*B?*, cotton	Plasmid	*avrBs3*	Also functions in *B2*	HR	DeFeyter and Gabriel (1991b)
Xcm	*avrB102*	Unknown gene, cotton	Plasmid	*avrBs3*	Also functions in *B2*	HR	DeFeyter and Gabriel (1991b)
Xoo	*avrxa5*	*xa-5*, rice	Unknown	*avrBs3*		Tan-brown at 48–72 h, reduced lesion length	Leach *et al.* (in preparation)
Xoo	*avrXa7*	*Xa-7*, rice	4.3 kb	*avrBs3*	Approx. 25 copies of 102 bp repeat	Dark brown ring by 48–72 h, peroxidase by 48 h, reduced lesion length and bacterial numbers	Leach *et al.* (in preparation)
Xoo	*avrXa10*	*Xa-10*, rice	3.2 kb	*avrBs3*	Approx. 15 copies of 102 bp repeat	Dark brown at 24–48 h, peroxidase by 24 h, reduced lesion length and bacterial numbers	Leach *et al.* (in preparation)
Xcr	*avrXca*	Gene(s) in *Arabidopsis* (most ecotypes), not *Brassica*	1.8 kb	None		No visible symptoms, bacterial growth inhibited *in planta*	Parker *et al.* (in preparation)

The pathogen abbreviations *Xcm*, *Xcr*, *Xcv* and *Xoo* denote, respectively, *Xanthomonas campestris* pvs *malvacearum*, *raphani* and *vesicatoria*, and *X. oryzae* pv. *oryzae*.
HR = hypersensitive reaction.

1989; Minsavage *et al.*, 1990; Canteros *et al.*, 1991), *X. c.* pv. *malvacearum* (DeFeyter and Gabriel, 1991b), *X. c.* pv. *campestris* and *X. o.* pv. *oryzae* (Table 6.2).

A few avirulence genes from *Xanthomonas* spp. have been completely or partially sequenced (Ronald and Staskawicz, 1988; Bonas *et al.*, 1989; Canteros *et al.*, 1991). The transcription start site of the *avrBs3* gene has been mapped (Knoop *et al.*, 1991). Unlike the *hrp* genes from *X. c.* pv. *vesicatoria*, which are plant-inducible (Schulte *et al.*, 1991), the *avrBs3* gene is expressed constitutively in cells grown in minimal or complex medium and *in planta* (Knoop *et al.*, 1991). In some cases, the protein products or fusion proteins have been produced (Ronald and Staskawicz, 1988; Knoop *et al.*, 1991). Knoop *et al.* (1991) used polyclonal antibodies raised against a fusion protein (β-galactosidase fused with the *avrBs3*-encoded protein) to identify the 122 kDa protein product in extracts from *X. c.* pv. *vesicatoria* cells. Their findings indicate that AvrBs3 is localized intracellularly in *X. c.* pv. *campestris,* and is mainly in the soluble fraction. Thus far, nothing about the characterized *Xanthomonas* genes has provided clues as to how the genes function in resistance expression.

(b) Relationships among avr *genes*

A generalization about conservation of *avr* gene sequence and function between pathovars, species and genera is not possible. Several *avr* genes share no homology with other *avr* genes (*avrBs2, avrBsT, avrRxv, avrXca*). On the other hand, increasing numbers of avirulence genes have been identified which share some degree of sequence relatedness. Ronald and Staskawicz (1988) found that the *avrBs1* gene from *X. c.* pv. *vesicatoria* encodes a 49.8 kDa protein with a significant amount of homology to the C-terminal region of the protein encoded by *avrA*, a *P. syringae* pv. *glycinea* avirulence gene (Staskawicz *et al.*, 1987). The significance of the relatedness is not known.

Bonas *et al.* (1989) probed Southern blots of DNA from different *X. c.* pv. *vesicatoria* strains with an internal fragment of *avrBs3*, an avirulence gene cloned from *X. c.* pv. *vesicatoria,* and reported that additional homologous sequences were present in a number of strains. When a 1.7 kb subclone containing the homologous sequence (called *avrBsP*) was introduced into *X. c.* pv. *vesicatoria* strains normally pathogenic to tomato, the transconjugants induced an HR in tomato (Canteros *et al.*, 1991; Schulte *et al.*, 1991). Like *avrBs3*, *avrBsP* is plasmid-borne. The nucleotide sequences of *avrBsP* and *avrBs3* are nearly identical (Canteros *et al.*, 1991).

The *avrBs3* gene also hybridized with six additional pathovars of *X. campestris*, including *X. c.* pv. *malvacearum* (Bonas *et al.*, 1989). Swarup *et al.* (1991b) reported that sequences of six avirulence genes from *X. c.*

pv. *malvacearum* and a pathogenicity gene cloned from *X. c.* pv. *citri* (Swarup *et al.*, 1991a) hybridized with *avrBs3*. Like *avrBs3* and *avrBsP*, the avirulence genes are located on a plasmid (DeFeyter and Gabriel, 1991b). Leach *et al.* (in preparation) found at least three *avrBs3*-related sequences in *X. o.* pv. *oryzae* which functioned as avirulence genes (*avrxa5*, *avrXa7* and *avrXa10*) in the rice–*X. o.* pv. *oryzae* interaction.

The structure of *avrBs3* and its homologues is interesting. The *avrBs3* gene (3.6 kb) contains 17.5 direct repeats of 102 bp which have 91–100% identity with each other (Bonas *et al.*, 1989). The *avrBsP* gene contains six of the 102 bp repeats. In addition, the sequences of *avrBs3* and *avrBsP* differ only in a few bases of the repeat sequences (Canteros *et al.*, 1991). Two of the homologues from *X. o.* pv. *oryzae* (*avrXa7* and *avrXa10*) contain similar repeat structures, but with a different copy number (15 and 25, respectively). The significance of the repeat structures in avirulence function is not known. No differences between the two *avr* genes from *X. c.* pv. *vesicatoria* (*avrBs3* and *avrBsP*) were found in the sequence of the N-terminal region; however, differences were observed in the organization and sequences of the internal repetitive region (Canteros *et al.*, 1991). Thus, it is tempting to speculate that either the copy number or sequence of these repeat sequences is important to the Avr protein function. In support of this, Schulte *et al.* (1991) and Herbers *et al.* (1992) constructed derivatives of *avrBs3* containing varying numbers of repeat units. In some cases the avirulence-determining ability was lost, but in other cases new plant specificities were generated.

The apparently large family of *avrBs3*-related genes is unique to *Xanthomonas* spp., since although many *avr* genes have been cloned from *Pseudomonas* spp., none have shown such relatedness. The *avrBs3* family functions to confer avirulence in pathogens of both dicots (*X. c.* pvs *vesicatoria*, *citri* and *malvacearum*) and monocots (*X. o.* pv. *oryzae*). Whether or not the *avrBs3* homologues from a dicot pathogen can confer specific resistance in a monocot is not yet known. Another avirulence gene (*avrRxv*) cloned from *X. c.* pv. *vesicatoria* and unrelated to *avrBs3* does confer the ability to inhibit disease production in both dicots and monocots (Whalen *et al.*, 1988).

(c) Role of avirulence genes in non-host resistance

Whalen *et al.* (1988) addressed the question of whether avirulence genes could account for non-host resistance (pathovar specificity). A gene (*avrRxv*) from one pathovar (*X. c.* pv. *vesicatoria*), which elicits an HR on bean, functioned in another pathovar (*X. c.* pv. *phaseoli*) to elicit resistance in bean in a race-/cultivar-specific manner. In addition, the presence of the avirulence gene (*avrRxv*) resulted in the inhibition of disease production by several *X. campestris* pathovars on their normally susceptible hosts (viz. pv. *glycines* on soybean, pv. *vignicola* on cowpea, pv. *holcicola* on corn, pv. *alfalfae* on alfalfa, pv. *malvacearum* on cotton). Thus, the genetic mechanisms governing non-host resistance may be similar to those governing host resistance, and in some cases, avirulence genes may account for pathovar specificity.

(d) Inactivation of avirulence genes in nature

To understand how avirulence genes influence cultivar/race specificity, it is important not only to determine how they function to elicit resistance, but also how that function is modulated in the bacterium to allow for adaptation to changes in the host population. Populations of avirulent *X. c.* pv. *vesicatoria* strains on pepper plants with the *Bs1* resistance gene changed to virulent populations at high frequencies (Dahlbeck and Stall, 1979). Kearney *et al.* (1988) demonstrated that *avrBs1*, the avirulence gene corresponding to *Bs1*, was inactivated in virulent strains by insertion of IS*476*, an insertion sequence present in *X. c.* pv. *vesicatoria* (Kearney and Staskawicz, 1990a). Thus, insertion elements such as IS*476* might contribute to the adaptability of bacterial pathogens to changing host genotypes.

(e) Contribution of avirulence genes to pathogen fitness

In several cases, inactivation of resident avirulence genes has not obviously reduced the overall pathogenic fitness of the pathogen (for review, see Gabriel, 1989). This may account for the instability of the resistance genes in the field, that is, avirulence gene functions may be readily lost if they are not essential to the pathogen. Kearney and Staskawicz (1990b) noticed that the *Bs2* resistance gene, which confers resistance in pepper to *X. c.* pv. *vesicatoria*, had not been overcome in the field. Sequences related to *avrBs2*, the avirulence gene corresponding to the pepper resistance gene *Bs2*, occur in all *X. c.* pv. *vesicatoria* races and in many other pathovars of *X. campestris*. Marker exchange mutants containing an inactivated *avrBs2* gene were less aggressive to a susceptible cultivar of pepper than the wild-type *X. c.* pv. *vesicatoria* strain. Complementation of the mutant with a plasmid-borne copy of *avrBs2* allowed the growth of the mutant to approach bacterial numbers comparable to the wild-type strain. These results suggest that *avrBs2* may be essential for the pathogen. Similarly, spontaneous mutants in *avrb6*, an avirulence gene cloned from *X. c.* pv. *malvacearum*, were reduced in ability to watersoak susceptible cotton plants (DeFeyter and Gabriel, 1991b). Together these studies suggest that plant resistance genes may be more durable in crops if mutations in the corresponding avirulence genes reduce the pathogen's virulence.

6.4.6 Other genes

The classes of gene so far described are concerned primarily with symptom production and determination of host specificity. However, the disease cycle involves many more aspects of bacterial physiology, e.g. pre-infection colonization of plants, survival in soil and on plant debris, preference for certain portals of entry into plants, etc. It would not be surprising if many genes were important in determining these aspects of ecological adaptation.

Additional 'pathogenicity' genes are known. For example, Osbourn *et al.* (1990b) described and sequenced a gene from *X. c.* pv. *campestris* which when mutated gave a non-pathogenic phenotype. However, the production of enzymes and EPS and the ability of the bacteria to grow *in planta* were not impaired, and the reason for virulence loss is unknown. The gene does not hybridize with other known genes required for pathogenicity.

Early stages of research on pathogenicity are often based on the implicit assumption that pathogenicity can be dissociated from normal processes of bacterial growth. The fact that expression of *hrp* genes is dependent upon the nutritional status of the bacteria clearly indicates that pathogenicity and 'housekeeping' functions are in reality intimately connected. It will be interesting to determine whether mutants altered in metabolic functions, such as the mutants defective in fructose metabolism described by De Crécy-Lagard *et al.* (1991), are affected in plant interaction.

The genes involved in biosynthesis of the yellow pigment xanthomonadin are also probably important in the total process of plant–bacterial interaction. Although non-pigmented mutants of *X. c.* pv. *campestris* retain pathogenicity in laboratory tests, mutants of *X. c.* pv. *juglandis* were found to be light-sensitive (Jenkins and Starr, 1982). Thus, pigment production is probably essential for survival of xanthomonads as epiphytes, which is an important factor in the life cycle of many pathovars. Genes involved in xanthomonadin biosynthesis may be easily cloned by complementation of white mutants (Daniels *et al.*, 1984b), and their study would be of interest not only because of the ecological importance of the pigment but also because xanthomonadin is one of the few natural products containing bromine.

6.5 THE CONTRIBUTION OF GENETICS TO THE UNDERSTANDING OF BACTERIA–PLANT INTERACTIONS

The literature is full of reports that various enzymes or other molecules are involved in pathogenesis. These implications are largely based on correlative evidence, i.e. the presence and amount of the factor correlates with aggressiveness or virulence. The ability to clone and inactivate genes in *Xanthomonas* spp. has enabled the careful analysis of the necessity or importance of individual factors in interactions between pathogens and plants. As we have emphasized above, it is possible to produce near-isogenic bacterial strains differing from one another only in a single mutation in a precisely known position in the genome. Comparison of such strains provides the simplest (or, as some enthusiasts would say, the only) method for evaluating the role of genes and their products in pathogenicity.

The *hrp* genes are required for both pathogenicity and induction of the HR, indicating key factors are similar and required in both interactions. The analysis of pathogenicity genes facilitates our understanding of the various phases of disease development. For example, protease-deficient mutants of *X. c.* pv. *campestris* are less aggressive to turnip than wild-type strains only if the method of inoculation resembles a hydathode infection, suggesting a role for the proteases in the early stages of infection (Dow *et al.*, 1990).

Understanding the regulation of bacterial genes involved in plant pathogenesis goes hand-in-hand with determining the role of the genes in pathogenesis. For example, the *rpfC* locus in *X. c.* pv. *campestris* regulates genes required for virulence (Tang *et al.*, 1991), suggesting that common regulation of virulence genes allows the pathogen to adjust to changes in environment as disease progresses. Alternatively, some genes, such as the avirulence gene *avrBs3*, are produced constitutively (Knoop *et al.*, 1991). Perhaps the products of these genes are required not only for early recognition by the host plant, but possibly for other biochemical processes.

Access to cloned pathogen avirulence genes and host cultivars near-isogenic for single resistance genes has enabled researchers to test the gene-for-gene hypothesis at the molecular level. The evidence is convincing that avirulence genes specify incompatibility only on host cultivars carrying the complementary resistance gene (for review, see Keen, 1990). An exception to this pattern was reported for the interactions between *X. c.* pv. *malvacearum* and cotton (DeFeyter and Gabriel, 1991b). All six *avr* genes cloned from *X. c.* pv. *malvacearum* induced a strong HR on cotton cultivar 101-102B (which contains angular leaf spot resistance genes B_2, B_3 and B_{sm}) and all but *avrBIn* induced a weak HR on cultivar Acala-B2 (which contains B_2). The authors suggest that in this interaction a gene-for-genes recognition occurs, i.e. the resistance gene B_2 interacts with more than one pathogen avirulence gene to confer resistance. Further, they suggest that the stability of B_2-derived resistance in cotton is the result of the resistance gene-determined recognition of more than one pathogen *avr* gene product. The avirulence genes cloned from the cotton pathogen are homologues of *avrBs3*; it will be interesting to see if other members of that gene family (*avr* genes from *X. c.* pv. *vesicatoria* and *X. o.* pv. *oryzae*) are recognized by the cotton gene B_2, and if there are comparable resistance genes in pepper, tomato or rice.

Many plant pathologists have modified their research programmes in recent years to take advantage of developments in molecular genetic technology. They are surely wise to have done so. Without the application of genetics, *hrp* genes, avirulence genes, regulatory genes and protein secretion genes would not have been discovered, and we would have no clear idea of the importance of extracellular enzymes and EPS. The genetic results have opened up new areas of investigation for biochemical and physiological plant pathologists by suggesting possible mechanisms of action, and collaboration between pathologists from different backgrounds promises to yield new insights into disease processes.

REFERENCES

Arlat, M., Gough, C.L., Barber, C.E., Boucher, C. and Daniels, M.J. (1991) *Xanthomonas campestris* contains a cluster of *hrp* genes related to the larger *hrp* cluster of *Pseudomonas solanacearum*. *Molecular Plant–Microbe Interactions*, **4**, 593–601.

Atkins, D.T., Barber, C.E. and Daniels, M.J. (1987) Transformation of *Xanthomonas campestris* pathovar *campestris* with plasmid DNA. *Journal of General Microbiology*, **133**, 2727–31.

Bagdasarian, M., Lurz, R., Ruckert, B., *et al.* (1981) Specific-purpose plasmid cloning vectors, II. Broad-host-range, high copy number, RSF1010 derived vectors and a host-vector system for gene cloning in *Pseudomonas*. *Gene*, **16**, 237–47.

Bai, J. (1990) Infection types of rice bacterial blight caused by *Xanthomonas campestris* pv. *oryzae*. M.S. thesis, University of the Philippines, Los Banos, The Philippines.

Bak, A.L., Christiansen, C. and Stenderup, A. (1970) Bacterial genome sizes determined by DNA renaturation studies. *Journal of General Microbiology*, **64**, 377–80.

Barrère, G.C., Barber, C.E. and Daniels, M.J. (1986) Molecular cloning of genes involved in the biosynthesis of the extracellular polysaccharide xanthan by *Xanthomonas campestris* pv. *campestris*. *Int. J. Biol. Macromol.*, **8**, 372–4.

Beaulieu, C., Minsavage, G.V., Canteros, B.I. and Stall, R.E. (1991) Biochemical and genetic analysis of a pectate lyase gene from *Xanthomonas campestris* pv. *vesicatoria*. *Molecular Plant–Microbe Interactions*, **4**, 446–51.

Berg, D.E. and Howe, M.M. (eds) (1989) *Mobile DNA*, American Society for Microbiology, Washington DC.

Birch, R.G. and Patil, S.S. (1987) Correlation between albicidin production and chlorosis induction by *Xanthomonas albilineans*, the sugarcane leaf scald pathogen. *Physiological and Molecular Plant Pathology*, **30**, 199–206.

Bonas, U., Schulte, R., Fenselau, S., Minsavage, G.V., Staskawicz, B.J. and Stall, R.E. (1991) Isolation of a gene cluster from *Xanthomonas campestris* pv.

vesicatoria that determines pathogenicity and the hypersensitive response on pepper and tomato. *Molecular Plant–Microbe Interactions*, **4**, 81–8.

Bonas, U., Stall, R.E. and Staskawicz, B. (1989) Genetic and structural characterization of the avirulence gene *avrBs3* from *Xanthomonas campestris* pv. *vesicatoria*. *Molecular and General Genetics*, **218**, 127–36.

Borthakur, D., Barber, C.E., Lamb, J.W., Daniels, M.J., Downie, J.A. and Johnston, A.W.B. (1986) A mutation that blocks exopolysaccharide synthesis prevents nodulation of peas by *Rhizobium leguminosarum* but not of beans by *R. phaseoli* and is corrected by cloned DNA from *Rhizobium* or the phytopathogen *Xanthomonas*. *Molecular and General Genetics*, **203**, 320–3.

Boucher, C., Van Gijsegem, F., Barberis, P.A., Arlat, M. and Zischek, C. (1987) *Pseudomonas solanacearum* genes controlling both pathogenicity on tomato and hypersensitivity on tobacco are clustered. *Journal of Bacteriology*, **169**, 5626–32.

Canteros, B., Minsavage, G., Bonas, U., Pring, D. and Stall, R. (1991) The cloning and characterization of a gene from *Xanthomonas campestris* pv. *vesicatoria* that determines avirulence in tomato. *Molecular Plant–Microbe Interactions*, **4**, 628–32.

Carlton, B.C. and Brown, B.J. (1981) Gene mutation, in *Manual of Methods for General Microbiology*, American Society for Microbiology, Washington DC, pp. 222–42.

Chatterjee, A.K. and Vidaver, A.K. (1986) Genetics of pathogenicity factors: application to phytopathogenic bacteria. *Advances in Plant Pathology*, **4**, 1–218.

Cook, A.A., Walker, J.C. and Larson, R.H. (1952) Studies on the disease cycle of black rot of crucifers. *Phytopathology*, **42**, 162–7.

Coplin, D.L. and Cook, D. (1990) Molecular genetics of extracellular polysaccharide biosynthesis in vascular phytopathogenic bacteria. *Molecular Plant–Microbe Interactions*, **3**, 271–9.

Corey, R.R. and Starr, M.P. (1957a) Genetic transformation of colony type in *Xanthomonas phaseoli*. *Journal of Bacteriology*, **74**, 141–5.

Corey, R.R. and Starr, M.P. (1957b) Genetic transformation of streptomycin resistance in *Xanthomonas phaseoli*. *Journal of Bacteriology*, **74**, 146–50.

Dahlbeck, D. and Stall, R.E. (1979) Mutations for change of race in cultures in *Xanthomonas vesicatoria*. *Phytopathology*, **69**, 634–6.

Daniels, M.J. (1987) Molecular genetics of pathogenicity of *Xanthomonas campestris*: technical aspects, in *Genetics and Plant Pathogenesis* (eds P.R. Day and G.J. Jellis), Blackwell, Oxford, pp. 45–9.

Daniels, M.J., Barber, C.E., Turner, P.C., Cleary, W.G. and Sawczyc, M.K. (1984a) Isolation of mutants of *Xanthomonas campestris* pv. *campestris* showing altered pathogenicity. *Journal of General Microbiology*, **130**, 2447–55.

Daniels, M.J., Barber, C.E., Turner, P.C., Sawczyc, M.K., Byrde, R.J.W. and Fielding, A.H. (1984b) Cloning of genes involved in pathogenicity of *Xanthomonas campestris* pv. *campestris* using the broad host range cosmid pLAFR1. *EMBO Journal*, **3**, 3323–8.

Daniels, M.J., Fan, M.J., Barber, C.E., Clarke, B.R. and Parker, J.E. (1991) Interaction between *Arabidopsis thaliana* and *Xanthomonas campestris*, in *Advances in Molecular Genetics of Plant–Microbe Interactions*, Vol. 1 (eds H. Hennecke and D.P.S. Verma), Kluwer, Dordrecht, pp. 84–9.

De Crécy-Lagard, V., Glaser, P., Lejeune, P., *et al.* (1990) A *Xanthomonas campestris* pv. *campestris* protein similar to catabolite activation factor is involved in regulation of phytopathogenicity. *Journal of Bacteriology*, **172**, 5877–83.

De Crécy-Lagard, V., Lejeune, P., Bouvet, O.M.M. and Danchin, A. (1991) Identification of two fructose transport and phosphorylation pathways in

Xanthomonas campestris pv. *campestris*. *Molecular and General Genetics*, **227**, 465–72.

DeFeyter, R. and Gabriel, D.W. (1991a) Use of cloned DNA methylase genes to increase the transfer frequency of foreign genes into *Xanthomonas campestris* pv. *malvacearum*. *Journal of Bacteriology*, **173**, 6421–7.

DeFeyter, R. and Gabriel, D.W. (1991b) At least six avirulence genes are clustered on a 90-kilobase plasmid in *Xanthomonas campestris* pv. *malvacearum*. *Molecular Plant–Microbe Interactions*, **4**, 423–32.

DeFeyter, R., Kado, C.I. and Gabriel, D.W. (1990) Small, stable shuttle vectors for use in *Xanthomonas*. *Gene*, **88**, 65–72.

De Groot, A., Filloux, A. and Tommassen, J. (1991) Conservation of *xcp* genes, involved in the two-step protein secretion process, in different *Pseudomonas* species and other Gram-negative bacteria. *Molecular and General Genetics*, **229**, 278–84.

De Vries, G.E., Raymond, C.K. and Ludwig, R.A. (1984) Extension of bacteriophage (lambda) host range: selection, cloning, and characterization of a constitutive (lambda) receptor gene. *Proceedings of the National Academy of Sciences, U.S.A.*, **81**, 6080–4.

Ditta, G., Stanfield, S., Corbin, D. and Helinski, D. (1980) Broad host range DNA cloning system for Gram-negative bacteria: construction of a gene bank of *Rhizobium meliloti*. *Proceedings of the National Academy of Sciences, U.S.A.*, **77**, 7347–51.

Dow, J.M., Clarke, B.R., Milligan, D.E., Tang, T.L. and Daniels, M.J. (1990) Extracellular proteases from *Xanthomonas campestris* pv. *campestris*, the black rot pathogen. *Applied and Environmental Microbiology*, **56**, 2994–8.

Dow, J.M., Milligan, D.E., Jamieson, L., Barber, C.E. and Daniels, M.J. (1989) Molecular cloning of a polygalacturonate lyase gene from *Xanthomonas campestris* pv. *campestris* and role of the gene product in pathogenicity. *Physiological and Molecular Plant Pathology*, **35**, 113–20.

Dow, J.M., Scofield, G., Trafford, K., Turner, P.C. and Daniels, M.J. (1987) A gene cluster in *Xanthomonas campestris* pv. *campestris* required for pathogenicity controls the excretion of polygalacturonate lyase and other enzymes. *Physiological and Molecular Plant Pathology*, **31**, 261–71.

Dums, F., Dow, J.M. and Daniels, M.J. (1991) Structural characterisation of protein secretion genes of the bacterial phytopathogen *Xanthomonas campestris* pathovar *campestris*: relatedness to secretion systems of other Gram-negative bacteria. *Molecular and General Genetics*, **229**, 357–64.

Ellingboe, A.H. (1976) Genetics of host–parasite interactions, in *Encyclopedia of Plant Pathology, New Series*, Vol. 4, *Physiological Plant Pathology* (eds R. Heitefuss and P.H. Williams), Springer-Verlag, New York, pp. 761–78.

Figurski, D.H. and Helinski, D.R. (1979) Replication of an origin-containing derivative of plasmid RK2 dependent on a plasmid function provided in *trans*. *Proceedings of the National Academy of Sciences, U.S.A.*, **76**, 1648–52.

Flor, H.H. (1955) Host–parasite interaction in flax rust – its genetics and other implications. *Phytopathology*, **45**, 680–5.

Friedman, A.M., Long, S.R., Brown, S.E., Buikema, W.J. and Ausubel, F.M. (1982) Construction of a broad host range cosmid cloning vector and its use in the genetic analysis of *Rhizobium* mutants. *Gene*, **18**, 289–96.

Fujimoto, D.K., Leach, J.E. and Vidaver, A.K. (1991) Identification of proteinase genes from *Xanthomonas campestris* pv. *phaseoli*. *Phytopathology*, **81**, 1188 (abstr.).

Gabriel, D.W. (1984) Plasmid transformation of *Xanthomonas campestris* pv. *malvacearum*. *Phytopathology*, **74**, 837.

Gabriel, D.W. (1989) The genetics of plant pathogen population structure and host–parasite specificity, in *Plant–Microbe Interactions: molecular and genetic perspectives*, vol. 3 (eds T. Kosuge and E.W. Nester), Macmillan, New York, pp. 343–79.

Gabriel, D.W., Burges, A. and Lazo, G.R. (1986) Gene-for-gene interactions of five cloned avirulence genes from *Xanthomonas campestris* pv. *malvacearum* with specific resistance genes in cotton. *Proceedings of the National Academy of Sciences, U.S.A.*, **83**, 6415–19.

Gabriel, D.W., Hunter, J.E., Kingsley, M.T., Miller, J.W. and Lazo, G.R. (1988) Clonal population structure of *Xanthomonas campestris* and genetic diversity among citrus canker strains. *Molecular Plant–Microbe Interactions*, **1**, 59–65.

Gallie, D. and Kado, C.I. (1987) *Agrobacterium tumefaciens* pTAR *parA* promoter region involved in autoregulation, incompatibility and plasmid partitioning. *Journal of Molecular Biology*, **193**, 465–78.

Gough, C.L., Dow, J.M., Barber, C.E. and Daniels, M.J. (1988) Cloning of two endoglucanase genes of *Xanthomonas campestris* pv. *campestris*: analysis of the role of the major endoglucanase in pathogenesis. *Molecular Plant–Microbe Interactions*, **1**, 275–81.

Gough, C.L., Dow, J.M., Keen, J., Henrissat, B. and Daniels, M.J. (1990) Nucleotide sequence of the *engXCA* gene encoding the major endoglucanase of *Xanthomonas campestris* pv. *campestris*. *Gene*, **89**, 53–9.

Grimm, C. and Panopoulos, N.J. (1989) The predicted protein product of a pathogenicity locus from *Pseudomonas syringae* pv. *phaseolicola* is homologous to a highly conserved domain of several procaryotic regulatory proteins. *Journal of Bacteriology*, **171**, 5031–8.

Haas, D. and Holloway, B.W. (1976) R-factor variants with enhanced sex factor activity in *Pseudomonas aeruginosa*. *Molecular and General Genetics*, **144**, 243–51.

Harding, N.E., Cleary, J.M., Cabanas, D.K., Rosen, I.G. and Kang, K.S. (1987) Genetic and physical analyses of a cluster of genes essential for xanthan gum biosynthesis in *Xanthomonas campestris*. *Journal of Bacteriology*, **169**, 2854–61.

Hayes, W. (1968) *The Genetics of Bacteria and their Viruses*, 2nd edn, Blackwell, Oxford.

Herbers, K., Conrads-Strauch, J. and Bonas, U. (1992) Race specificity of plant resistance to bacterial spot disease determined by repetitive motifs in a bacterial avirulence protein. *Nature*, **356**, 172–4.

Hopwood, D.A. and Chater, K.F. (eds) (1989) *Genetics of Bacterial Diversity*, Academic Press, London.

Hötte, B., Ruth-Arnold, I. and Pühler, A. (1990) Cloning and analysis of a 35.3-kilobase DNA region involved in exopolysaccharide production in *Xanthomonas campestris* pv. *campestris*. *Journal of Bacteriology*, **172**, 2804–7.

Hu, N.T., Hung, M.N., Chiou, S.J., *et al.* (1992) Cloning and characterisation of a gene required for the secretion of extracellular enzymes across the outer membrane by *Xanthomonas campestris* pv. *campestris*. *Journal of Bacteriology*, **174**, 2679–87.

Huynh, T.V., Dahlbeck, D. and Staskawicz, B.J. (1989) Bacterial blight of soybean: regulation of a pathogen gene determining host cultivar specificity. *Science*, **245**, 1374–7.

Ielpi, L., Couso, R. and Dankert, M. (1981) Lipid-linked intermediates in the biosynthesis of xanthan gum. *FEBS Letters*, **130**, 253–6.

Innes, R.W., Hirose, M.A. and Kuempel, P.L. (1988) Induction of nitrogen-fixing nodules on clover requires only 32 kilobase pairs of DNA from the *Rhizobium trifolii* symbiosis plasmid. *Journal of Bacteriology*, **170**, 3793–802.

Jenkins, C.L. and Starr, M.P. (1982) The brominated aryl-polyene (xanthomonadin) pigments of *Xanthomonas juglandis* protect against photobiological damage. *Current Microbiology*, **7**, 323–6.

Joseph-Liauzun, E., Fellay, R. and Chandler, M. (1989) Transposable elements for efficient manipulation of a wide range of Gram-negative bacteria: promoter probes and vectors for foreign genes. *Gene*, **85**, 83–9.

Kamoun, S. and Kado, C.I. (1990a) A plant-inducible gene of *Xanthomonas campestris* pv. *campestris* encodes an exocellular component required for growth in the host and hypersensitivity on nonhosts. *Journal of Bacteriology*, **172**, 5165–72.

Kamoun, S. and Kado, C.I. (1990b) Phenotypic switching affecting chemotaxis, xanthan production, and virulence in *Xanthomonas campestris*. *Applied and Environmental Microbiology*, **56**, 3855–60.

Kearney, B. and Staskawicz, B. (1990a) Characterization of IS*476* and its role in bacterial spot disease of tomato and pepper. *Journal of Bacteriology*, **172**, 143–8.

Kearney, B. and Staskawicz, B. (1990b) Widespread distribution and fitness contribution of *Xanthomonas campestris* avirulence gene *avrBs2*. *Nature*, **346**, 385–6.

Kearney, B., Ronald, P.C., Dahlbeck, D. and Staskawicz, B.J. (1988) Molecular basis for evasion of plant host defence in bacterial spot disease of pepper. *Nature*, **332**, 541–3.

Keen, N.T. (1990) Gene-for-gene complementarity in plant–pathogen interactions. *Annual Review of Genetics*, **24**, 447–63.

Keen, N.T. and Staskawicz, B. (1988) Host range determinants in plant pathogens and symbionts. *Annual Review of Microbiology*, **42**, 421–40.

Keen, N.T., Tamaki, S., Kobayashi, D. and Trollinger, D. (1988) Improved broad-host-range plasmids for DNA cloning in Gram-negative bacteria. *Gene*, **70**, 191–7.

Kelemu, S. and Leach, J.E. (1990) Cloning and characterization of an avirulence gene from *Xanthomonas campestris* pv. *oryzae*. *Molecular Plant–Microbe Interactions*, **3**, 59–65.

Knoop, V., Staskawicz, B. and Bonas, U. (1991) Expression of the avirulence gene *avrBs3* from *Xanthomonas campestris* pv. *vesicatoria* is not under the control of *hrp* genes and is independent of plant factors. *Journal of Bacteriology*, **173**, 7142–50.

Kroos, L. and Kaiser, D. (1984) Construction of Tn*5lac*, a transposon that fuses *lacZ* expression to exogenous promoters, and its introduction in *Myxococcus xanthus*. *Proceedings of the National Academy of Sciences, U.S.A.*, **81**, 5816–20.

Lai, M., Panopoulos, N.J. and Shaffer, S. (1977) Transmission of R plasmids among *Xanthomonas* spp. and other plant pathogenic bacteria. *Phytopathology*, **67**, 1044–50.

Lazo, G.R., Roffey, R. and Gabriel, D.W. (1987) Pathovars of *Xanthomonas campestris* are distinguishable by restriction fragment-length polymorphism. *International Journal of Systemic Bacteriology*, **37**, 214–21.

Leach, J.E., Kelemu, S., Ardales, E. and Leung, H. (1989) The genetics of *Xanthomonas campestris* pv. *oryzae*, in *Bacterial Blight of Rice, Proceedings of the International Workshop on Bacterial Blight of Rice, 1988*, pp. 199–206.

Leach, J.E. and White, F.W. (1992) Molecular probes for disease diagnosis and monitoring, in *Rice Biotechnology* (eds G. Khush and G. Toenniessen), CAB International, Wallingford, UK, in press.

Leach, J.E., White, F.W., Rhoads, M.L. and Leung, H. (1990) A repetitive

DNA sequence differentiates *Xanthomonas campestris* pv. *oryzae* from other pathovars of *X. campestris*. *Molecular Plant–Microbe Interactions*, **3**, 238–46.

Lindgren, P.B., Fredrick, R., Govindarajan, A.G., Panopoulos, N.J., Staskawicz, B.J. and Lindow, S.E. (1989) An ice-nucleation reporter gene system: identification of inducible pathogenicity genes in *Pseudomonas syringae* pv. *phaseolicola*. *EMBO Journal*, **8**, 1291–301.

Lindgren, P.B., Panopoulos, N.J., Staskawicz, B.J. and Dahlbeck, D. (1988) Genes required for pathogenicity and hypersensitivity are conserved and interchangeable among pathovars of *Pseudomonas syringae*. *Molecular and General Genetics*, **211**, 499–506.

Liu, Y.N., Tang, J.L., Clarke, B.R., Dow, J.M. and Daniels, M.J. (1990) A multipurpose broad host range cloning vector and its use to characterise an extracellular protease gene of *Xanthomonas campestris* pathovar *campestris*. *Molecular and General Genetics*, **220**, 433–40.

Manoil, C. and Beckwith, J. (1985) Tn*phoA*, a transposon probe for protein export signals. *Proceedings of the National Academy of Sciences, U.S.A.*, **82**, 8129–32.

Mellano, V.J. and Cooksey, D.A. (1988) Development of host range mutants of *Xanthomonas campestris* pv. *translucens*. *Applied and Environmental Microbiology*, **54**, 884–9.

Mew, T.W. (1987) Current status and future prospects of research on bacterial blight of rice. *Annual Review of Phytopathology*, **25**, 359–82.

Mew, T.W., Mew, I.C. and Huang, J.S. (1984) Scanning electron microscopy of virulent and avirulent strains of *Xanthomonas campestris* pv. *oryzae* on rice leaves. *Phytopathology*, **74**, 635–41.

Miller, J. (1972) *Experiments in Molecular Genetics*, Cold Spring Harbor Laboratory, Cold Spring Harbor, New York.

Mills, D. (1985) Transposon mutagenesis and its potential for studying virulence genes in plant pathogens. *Annual Review of Phytopathology*, **23**, 297–320.

Mindrinos, M.N., Rahme, L.G., Fredrick, R.D., Hatziloukas, E., Grimm, C. and Panopoulos, N.J. (1990) Structure, function, regulation, and evolution of genes involved in pathogenicity, the hypersensitive response, and phaseolotoxin immunity in the bean halo blight pathogen, in *Pseudomonas: biotransformations, pathogenesis, and evolving biotechnology* (eds S. Silver, A.M. Chakrabarty, B. Iglewski and S. Kaplan), American Society for Microbiology, Washington DC, pp. 74–81.

Minsavage, G.V., Dahlbeck, D., Whalen, M.C., *et al.* (1990) Gene-for-gene relationships specifying disease resistance in *Xanthomonas campestris* pv. *vesicatoria* – pepper interactions. *Molecular Plant–Microbe Interactions*, **3**, 41–7.

Murooka, Y., Iwamoto, H., Hamamoto, A. and Yamauchi, T. (1987) Efficient transformation of phytopathogenic strains of *Xanthomonas* species. *Journal of Bacteriology*, **169**, 4406–9.

Nordstrom, K. (1985) Replication, incompatibility and partition, in *Plasmids in Bacteria* (eds D.R. Helinski, S.N. Cohen, D.B. Clewell, D.A. Jackson and A. Hollaender), Plenum, New York, pp. 119–23.

Osbourn, A.E., Barber, C.E. and Daniels, M.J. (1987) Identification of plant-induced genes of the bacterial pathogen *Xanthomonas campestris* pathovar *campestris* using a promoter-probe plasmid. *EMBO Journal*, **6**, 23–8.

Osbourn, A.E., Clarke, B.R., Stevens, B.J.H. and Daniels, M.J. (1990a) Use of oligonucleotide probes to identify members of two-component regulatory systems in *Xanthomonas campestris* pv. *campestris*. *Molecular and General Genetics*, **222**, 145–51.

Osbourn, A.E., Clarke, B.R. and Daniels, M.J. (1990b) Identification and DNA sequence of a pathogenicity gene of *Xanthomonas campestris* pv. *campestris*. *Molecular Plant–Microbe Interactions*, **3**, 280–5.

Pugsley, A.P., d'Enfert, C., Reyss, I. and Kornacker, M.G. (1990) Genetics of extracellular protein secretion by Gram-negative bacteria. *Annual Review of Genetics*, **24**, 67–90.

Quobela, M., Claflin, L.E. and Nowell, D.C. (1990) Evidence that *Xanthomonas campestris* pv. *zeae* can be distinguished from other pathovars capable of infecting maize by restriction fragment length polymorphism of genomic DNA. *Canadian Journal of Plant Pathology*, **12**, 183–6.

Quobela, M., Leach, J.E., Pearson, D.L. and Claflin, L.E. (1991) Characterization of strains of *Xanthomonas campestris* pv. *holcicola* by PAGE of membrane proteins and by REA and RFLP analysis of genomic DNA. *Plant Disease*, **75**, 32–6.

Rahme, L.G., Mindrinos, M.N. and Panopoulos, N.J. (1991) The genetic and transcriptional organization of the *hrp* cluster of *Pseudomonas syringae* pathovar *phaseolicola*. *Journal of Bacteriology*, **173**, 575–86.

Reimers, P.J. and Leach, J.E. (1991) Race-specific resistance to *Xanthomonas oryzae* pv. *oryzae* conferred by bacterial blight resistance gene *Xa-10* in rice (*Oryza sativa*) involves lignification of host tissues. *Physiological and Molecular Plant Pathology*, **38**, 39–55.

Ronald, P.C. and Staskawicz, B.J. (1988) The avirulence gene *avrBs1* from *Xanthomonas campestris* pv. *vesicatoria* encodes a 50-kD protein. *Molecular Plant–Microbe Interactions*, **1**, 191–8.

Ruvkun, G.B. and Ausubel, F.M. (1981) A general method for site-directed mutagenesis in prokaryotes. *Nature*, **289**, 85–8.

Sambrook, J., Fritsch, E.F. and Maniatis, T. (1989) *Molecular Cloning: a Laboratory Manual*, 2nd edn, Cold Spring Harbor Laboratory, Cold Spring Harbor, New York.

Sawczyc, M.K., Barber, C.E. and Daniels, M.J. (1989) The role in pathogenicity of some related genes in *Xanthomonas campestris* pathovars *campestris* and *translucens*: a shuttle strategy for cloning genes required for pathogenicity. *Molecular Plant–Microbe Interactions*, **2**, 249–55.

Schulte, R. and Bonas, U. (1992a) A *Xanthomonas* pathogenicity locus is induced by sucrose and sulfur-containing amino acids. *Plant Cell*, **4**, 79–86.

Schulte, R. and Bonas, U. (1992b) The expression of the *hrp* gene cluster from *Xanthomonas campestris* pv. *vesicatoria* that determines pathogenicity and hypersensitivity on pepper and tomato is plant-inducible. *Journal of Bacteriology*, **174**, 815–23.

Schulte, R., Herbers, K., Fenselau, S., Balbo, I., Stall, R.E. and Bonas, U. (1991) Characterization of genes from *Xanthomonas campestris* pv. *vesicatoria* that determine avirulence and pathogenicity on pepper and tomato, in *Advances in Molecular Genetics of Plant–Microbe Interactions*, Vol. 1 (eds H. Hennecke and D.P.S. Verma), Kluwer, Dordrecht, pp. 61–4.

Seal, S.E., Cooper, R.M. and Clarkson, J.M. (1990) Identification of a pathogenicity locus in *Xanthomonas campestris* pv. *vesicatoria*. *Molecular and General Genetics*, **222**, 452–6.

Shaw, J.J. and Kado, C.I. (1988) Whole plant wound inoculation for consistent reproduction of black rot of crucifers. *Phytopathology*, **78**, 981–6.

Shaw, J.J. Settles, L.G. and Kado, C.I. (1988) Transposon Tn*4431* mutagenesis of *Xanthomonas campestris* pv. *campestris*: characterisation of a non-pathogenic mutant and cloning of a locus for pathogenicity. *Molecular Plant–Microbe Interactions*, **1**, 39–45.

Simon, R., Priefer, U. and Puhler, A. (1983) A broad host range mobilization system for *in vivo* genetic engineering: transposon mutagenesis in Gram-negative bacteria. *Biotechnology*, **1**, 784–91.

Stall, R.E. and Minsavage, G.V. (1990) The use of *hrp* genes to identify opportunistic xanthomonads, in *Plant Pathogenic Bacteria, Proceedings of the Seventh International Conference on Plant Pathogenic Bacteria, Budapest, Hungary, 1989* (ed. Z. Klement), Akadémiai Kiadó, Budapest, pp. 369–74.

Starr, M.P. and Nasuno, S. (1967) Pectolytic activity of phytopathogenic xanthomonads. *Journal of General Microbiology*, **46**, 425–33.

Staskawicz, B., Dahlbeck, D., Keen, N. and Napoli, C. (1987) Molecular characterization of cloned avirulence genes from race 0 and race 1 of *Pseudomonas syringae* pv. *glycinea*. *Journal of Bacteriology*, **169**, 5789–94.

Staskawicz, B.J., Dahlbeck, D. and Keen, N.T. (1984) Cloned avirulence gene of *Pseudomonas syringae* pv. *glycinea* determines race-specific incompatibility on *Glycine max* (L.) Merr. *Proceedings of the National Academy of Sciences, U.S.A.*, **81**, 6024–8.

Stock, J.B., Ninfa, A. and Stock, A.M. (1989) Protein phosphorylation and regulation of adaptive responses in bacteria. *Microbiological Reviews*, **53**, 450–90.

Sutherland, I.W. (1985) Biosynthesis and composition of Gram-negative bacterial extracellular and wall polysaccharides. *Annual Review of Microbiology*, **39**, 243–70.

Swanson, J., Kearney, B., Dahlbeck, D. and Staskawicz, G. (1988) Cloned avirulence gene of *Xanthomonas campestris* pv. *vesicatoria* complements spontaneous race change mutants. *Molecular Plant–Microbe Interactions*, **1**, 5–9.

Swarup, S., DeFeyter, R., Briansky, R.H. and Gabriel, D.W. (1991a) A pathogenicity locus from *Xanthomonas citri* enables strains from several pathovars of *X. campestris* to form canker-like lesions on citrus. *Phytopathology*, **81**, 802–9.

Swarup, S., Yang, Y. and Gabriel, D.W. (1991b) *A Xanthomonas citri* pathogenicity gene, *pthA*, has pleiotropic *avr* functions and is similar in DNA sequence to members of a multigene family, widely distributed in *Xanthomonas* spp. *Phytopathology*, **81**, 1189 (abstr.).

Tait, R., Close, T., Lundquist, R., Hagiya, M., Rodriguez, R. and Kado, C. (1983) Construction and characterization of a versatile broad host range DNA cloning system for Gram-negative bacteria. *Biotechnology*, **1**, 269–75.

Tang, J.-L., Gough, C., Barber, C.E., Dow, J.M. and Daniels, M.J. (1987) Molecular cloning of protease gene(s) from *Xanthomonas campestris* pv. *campestris*: expression in *Escherichia coli* and role in pathogenicity. *Molecular and General Genetics*, **210**, 443–8.

Tang, J.-L., Gough, C.L. and Daniels, M.J. (1990) Cloning of genes involved in negative regulation of production of extracellular enzymes and polysaccharide of *Xanthomonas campestris* pathovar *campestris*. *Molecular and General Genetics*, **222**, 157–60.

Tang, J.-L., Liu, Y.-N., Barber, C.E., Dow, J.M., Wootton, J.C. and Daniels, M.J. (1991) Genetic and molecular analysis of a cluster of *rpf* genes involved in positive regulation of synthesis of extracellular enzymes and polysaccharide in *Xanthomonas campestris* pathovar *campestris*. *Molecular and General Genetics*, **226**, 409–17.

Thorne, L., Tansey, L. and Pollock, T.J. (1987) Clustering of mutations blocking synthesis of xanthan gum by *Xanthomonas campestris*. *Journal of Bacteriology*, **169**, 3593–600.

Todd, G.A., Daniels, M.J. and Callow, J.A. (1990) *Xanthomonas campestris* pv. *oryzae* has DNA sequences containing genes isofunctional with *X. campestris* pv. *campestris* genes required for pathogenicity. *Physiological and Molecular Plant Pathology*, **36**, 73–87.

Tsuchiya, K., Mew, T.W. and Wakimoto, S. (1982) Bacteriological and pathological characterisation of wild types and induced mutants of *Xanthomonas campestris* pv. *oryzae*. *Phytopathology*, **72**, 43–6.

Tsui, J., Somerville, S.C. and Hammerschmidt, R. (1991) Identification of a gene in *Arabidopsis thaliana* that controls resistance to *Xanthomonas campestris* pv. *campestris*. *Physiological and Molecular Plant Pathology*, **38**, 57–65.

Tu, J., Wang, H.-R., Chang, S.-F., Charng, Y.-C., Lurz, R. and Dobrinski, B. (1989) Transposable elements of *Xanthomonas campestris* pv. *citri* originating from indigenous plasmids. *Molecular and General Genetics*, **217**, 505–10.

Turner, P., Barber, C. and Daniels, M. (1984) Behaviour of the transposons Tn*5* and Tn*7* in *Xanthomonas campestris* pv. *campestris*. *Molecular and General Genetics*, **195**, 101–7.

Turner, P., Barber, C. and Daniels, M. (1985) Evidence for clustered pathogenicity genes in *Xanthomonas campestris* pv. *campestris*. *Molecular and General Genetics*, **199**, 338–43.

Vanderslice, R.W., Doherty, D.H., Capage, M.A., *et al.* (1989) Genetic engineering of polysaccharide in *Xanthomonas campestris*, in *Recent Developments in Industrial Polysaccharides: biomedical and biotechnological advances* (eds V. Crescenzi, I.C.M. Dea and S.S. Stivola), Gordon and Breach, New York, pp. 145–56.

Waney, V.R., Kingsley, M.T. and Gabriel, D.W. (1991) *Xanthomonas campestris* pv. *translucens* genes determining host-specific virulence and general virulence on cereals identified by Tn*5-gusA* insertion mutagenesis. *Molecular Plant–Microbe Interactions*, **4**, 623–7.

Whalen, M.C., Stall, R.E. and Staskawicz, B.J. (1988) Characterization of a gene from a tomato pathogen determining hypersensitive resistance in non-host species and genetic analysis of this resistance in bean. *Proceedings of the National Academy of Sciences, U.S.A.*, **85**, 6743–7.

White, T.J. and Gonzalez, C.F. (1991) Application of electroporation for efficient transformation of *Xanthomonas campestris* pv. *oryzae*. *Phytopathology*, **81**, 521–4.

Whitfield, C., Sutherland, I.W. and Cripps, R.E. (1981) Surface polysaccharides in mutants of *Xanthomonas campestris*. *Journal of General Microbiology*, **124**, 385–92.

Willis, D.K., Rich, J.J. and Hrabak, E.M. (1991) *hrp* genes of phytopathogenic bacteria. *Molecular Plant–Microbe Interactions*, **4**, 132–8.

Yamasaki, Y., Murata, N. and Suwa, T. (1966) Restoration of lost virulence of leucine-requiring mutants of *Xanthomonas oryzae* by DNA-mediated transformation. *Proceedings of the Japan Academy*, **42**, 946–9.

Yun, C.H. (1991) Molecular characterization of a repetitive element of *Xanthomonas oryzae* pv. *oryzae*. PhD thesis, Department of Plant Pathology, Kansas State University, Manhattan, Kansas.

7

Management of *Xanthomonas* diseases

T.W. Mew[1] *and M.P. Natural*[2]
[1]*International Rice Research Institute, Los Banos*
[2]*Department of Plant Pathology, University of the Philippines at Los Banos College*

7.1 INTRODUCTION

Effective disease control must be based on knowledge of the life cycle of the pathogen and the interaction between the pathogen and the host plant (Lozano and Wholey, 1974). The weakest link in the disease cycle, when control strategies are most effective, is during the survival phase of the pathogen. The relationship of a pathogen's life cycle to disease control, is based on how and where the bacterial pathogen survives – in soil, water, plant debris, or seed. The mechanism(s) of pathogen dissemination affects the type of epidemic the pathogen may cause.

Transmission to a new host is critical in the life cycle of xanthomonads, especially if there is a period of survival in the absence of the primary host. In general, bacteria do not survive well outside of their primary host because of desiccation and solar radiation. Phytopathogenic bacteria survive in several ways, such as in seed, plant residues, perennial hosts, epiphytically, saprophytically in soil, and in insects. Some important seedborne xanthomonads include *X. c.* pv. *campestris* (Schaad *et al.*, 1980; Harman *et al.*, 1987), *X. c.* pv. *vesicatoria* (Moffet and Croft,

Xanthomonas.
Edited by J.G. Swings and E.L. Civerolo.
Published in 1993 by Chapman & Hall, London.
ISBN 0 412 434202

1983), *X. c.* pv. *translucens* (Forster and Schaad, 1988; Fourest *et al.*, 1990), *X. c.* pv. *glycines* (Singh and Jain, 1988; Groth and Braun, 1989), *X. c.* pv. *phaseoli*, *X. c.* pv. *malvacearum* (Moffet and Croft, 1983), *X. c.* pv. *manihotis* (Persley, 1979; Gonzales, 1987), *X. c.* pv. *carotae* (Kuan and Minsavage, 1985), *X. c.* pv. *pelargonii* (Srivastava and Bais, 1985; Stephens and Tuinier, 1989), and *X. c.* pv. *zinniae* (Moffet and Croft, 1983). Epiphytic survival of xanthomonads is well established, e.g. *X. c.* pvs *manihotis* (Gonzales, 1987), *glycines* (Moffet and Croft, 1983), *pelargonii* (Tuinier and Stephens, 1989) and *pruni* (Young, 1978). Saprophytic survival in soil and in plant debris is usually short-lived.

Crop diseases caused by *Xanthomonas* spp. can be managed by proper cultural practices and crop management, including: adjusting plant density and spacing; rate, time and number of fertilizer applications; cropping systems; field sanitation; seed health/quality; resistant crop cultivars; and water management. Chemicals (broad-spectrum fungicides) and antibiotics are available for the control of some diseases caused by xanthomonads; however, these have been used only in more developed countries. In developing countries, it is not economically feasible for resource-poor farmers or growers to apply chemicals to control bacterial diseases. In addition, chemicals and antibiotics are usually unavailable. Moreover, only a few farmers or growers in the developing countries understand that diseases of crops are caused by different causal agents such as bacteria, fungi, nematodes and viruses. Although general crop management does not target specific diseases, general farming activities such as weeding do contribute to disease management. Disease control is an integral part of the crop production system. Nevertheless, the bacterial plant diseases are notoriously difficult to control whether in the developing or developed countries.

More than one control method can be used for a plant disease. Whether and to what extent a disease should be controlled is largely determined by the effect of the disease on production, and how fast and to what extent it may become widespread. Furthermore, availability of the disease control technology to minimize disease effects with minimal input needs to be considered. Technically, a disease can be controlled by host plant resistance, by cultural practices (e.g. crop rotation, sanitation), by biological and chemical control, through quarantine regulations for intercountry or interregion movement of plant material, and by an integrated management system approach. However, the most effective strategy is dependent on specific host–pathogen interactions.

The problem of controlling xanthomonad diseases is often not a lack of technology, but a lack of vigorous and uniform application technology for large areas. Knowledge of survival mechanisms, methods of dissemination or transmission, means of penetration, and interaction with other microorganisms is vital in recommending control strategies. Also important is a thorough understanding of factors affecting disease development, such as cultural practices and the atmospheric and soil environment, that reduce inoculum potential and development of disease. Effective disease control recommendations are therefore a combination of research-based information, experience-based knowledge and good judgement. The ultimate objective is to avoid disease losses and to maintain profitability.

Management of bacterial diseases has been reviewed recently in the volumes of *Phytopathogenic Prokaryotes* (Mount and Lacy, 1982).

7.2 SEED HEALTH

The seeds of many agronomic crops are either infected by and/or infested with xanthomonads. In the case of *X. c.* pv. *malvacearum* (Persley, 1979, 1987) and perhaps other *X. campestris* pathovars, there is ample opportunity to contaminate the surface of uninfested seed during threshing with dust from dried, infected plant materials. Elimination or avoidance of plant pathogens by exclusion and eradication has been achieved through stringent seed production systems and regulations. All xanthomonad plant pathogens can be seedborne, seed infested or carried by planting materials. Germplasm exchange and movement has accounted for the major part of the success of modern agricultural development. One of the control strategies for seedborne bacterial pathogens is seed health management. Seed should be thoroughly cleaned to remove associated dust and debris, and treated in hot water at 60°C for 20 min to avoid the inadvertent introduction of pathogens into new areas. Persley (1979) recommended that cassava seed for export should be produced in an area free of cassava blight caused by *X. c.* pv. *manihotis*. The production of bacteria-free seed has been actively practised by crucifer seed growers in the USA through a combination of disease management actions, including seed production in areas non-conducive for infection and disease development. In the USA, collaborative efforts exist between different states in which crucifer seeds, transplants or crops are grown. This committee of researchers has developed and established guidelines for minimizing the threat of black rot and blackleg of crucifers. The guidelines cover all aspects of crucifer production from seed health to crop health. The emphasis is on sanitation in seed production and transplant seedbed operations. This effort has reduced the spread of the pathogens and minimized the losses by epidemics.

For the development of plant diseases caused by *Xanthomonas* spp., seeds are the most important sources of primary inoculum. Seedborne inoculum can be minimized through regulations and seed treatments.

The objective is to insure that seeds for trade and germplasm movement are free from bacterial pathogens. The aim of plant quarantine regulations is to prevent the introduction of new bacterial pathogens into areas where they are absent. Seed treatment is designed to eradicate the bacteria from seeds used both for movement of germplasm and trade, and as disease management to reduce the source of inoculum.

Clean planting material is essential. Many laboratories have established techniques for detecting bacteria in seeds and other vegetative planting materials. Kuan and Minsavage (1985) developed a procedure for detecting *X. c.* pv. *carotae* in carrot seed based on a low temperature (5°C) stationary aqueous soak of the seed in darkness for 18 h followed by a vigorous shaking of the seed in water containing Tween 20, concentration of the bacterial cells by centrifugation and plating of a dilution series of these cells onto a modified Kado and Heskett's D5 medium. Monoclonal antibodies are used to detect *X. c.* pv. *dieffenbachiae* after isolating the bacterium from symptomless plants using a semi-selective medium (Nishijima, 1989). A semi-selective medium was developed by McGuire *et al.* (1987) for isolation of *X. c.* pv. *vesicatoria* from soil and plant materials.

7.2.1 Plant quarantine restrictions

The entry of seedborne pathogens into a country is subject to control under the plant quarantine regulations of that country. The entry status of each pathogen is determined on the basis of its (a) presence or absence in the country, (b) epidemic potential, (c) local distribution, and (d) economic importance. Regulating the entry of seedborne pathogens entails the imposition of regulatory measures not only on the pathogen itself but also on the entry of seeds and other plant parts which may act as carriers of the pathogen concerned. Citrus canker caused by *X. c.* pv. *citri* has been eradicated in northern Australia, New Zealand, South Africa and Florida (USA). All citrus-exporting countries without canker maintain a strict prohibition on import of citrus plants and fruits unless from canker-free countries (Moffet and Croft, 1983). Anthurium and cassava blights are disseminated in vegetative cuttings, thus, import of vegetative materials is strictly controlled by quarantine. Cassava cuttings should never enter international commerce (Lozano, 1987). Vegetative materials should be introduced only as meristem cultures and then only after sufficient tests have been performed to determine the sanitary condition of the introduced materials. Vegetative material of sugarcane is grown under quarantine in Australia for 2 years to avoid the introduction of gumming disease caused by *X. c.* pv. *vasculorum*, and other diseases of sugarcane (Moffet and Croft, 1983).

Exclusion of a pathogen through regulation by legislation is not always successful. *X. c.* pv. *citri* can be detected in soils, citrus roots and various weeds collected in citrus groves before the new shoots develop in spring (Goto, 1972). These weeds may not necessarily be hosts for saprophytic survival of the bacterial pathogens. Bacterial pathogens may be present at low population levels either in association with non-host plants or in the soil, making eradication more difficult.

Fry (1982) suggested plant quarantine regulations should target pathogens which are absent from a region or country. It is desirable to establish tolerance levels depending upon the objective of the seed health and crop health programmes. Schaad *et al.* (1980) reported that in the case of cabbage infected with *X. c.* pv. *campestris*, one seed in 10 000 is tolerated for heading cabbage, whereas for seed production no infection is tolerated. The same applies to seed movement within and between countries and regions.

7.2.2 Seed treatments

Several techniques are used to eradicate pathogenic bacteria from seeds. No single seed treatment is effective for all bacterial pathogens. Success depends on several key factors, such as inoculum that infects or infests the seed, the quality of the seed, the kind of seed, and a combination of treatments.

Fang *et al.* (1957, 1982) proposed that the source of primary infection of bacterial blight of rice caused by *X. oryzae* pv. *oryzae* likely originated from seeds. Subsequently, they developed and tested a bacteriophage technique to specifically detect the pathogen (Fang *et al.*, 1982). The presence of bacteriophage is an indirect indication of the presence of the bacterial pathogen. Once this is confirmed, the next step is to quantify the amount of inoculum present in the seed lots. This is done by removing the hulls from rice seed and grinding them into small pieces. A suspension is made by adding water, and allowed to incubate at room temperature for 48 h. Then an adequate amount of antiphage serum is added to neutralize excess bacteriophages not adsorbed to the bacterial cells, followed by plating on nutrient agar for plaque count. The number of plaques is directly related to the number of bacteria. Once the seed lots are known to carry the bacteria, the next step is to treat the seeds with chemicals or by other means. For chemical treatment seed lots are soaked in 0.5% hydrochloric acid for 72 h or in a solution containing 500 ppm of chloromycetin for 48 h. With both treatments, only 95% control is achieved.

Seed treatments can reduce or minimize the pathogen inoculum level but are often not effective for eradication of the pathogen.

(a) Hot water treatment

A standard procedure for eradicating pathogens from either infected or infested seed is hot water treatment (50°C for 20–30 min). However, this treatment does not result in uniform eradication. A more serious problem, however, is that the treatment frequently results in reduced seed germination and seedling vigour.

This technique has been used for many years. The temperatures used range from 50°C to about 58°C, depending on the plant species. For example, to eradicate *X. c.* pv. *campestris* cabbage seeds can be soaked at 50°C for 25 min while only 18 min is required at the same temperature for cauliflower and collards. Cassava stem pieces are treated at 51°C for 25 min to eradicate *X. c.* pv. *manihotis* (Gonzales, 1987) while sugarcane stalk pieces are subjected to similar thermotherapy to eradicate *X. albilineans* (Ryan and Lopez, 1988). Hot water treatment at 56°C for 10 min effectively eradicated *X. c.* pv. *cyamopsidis* in cluster bean (*Cyamopsis tetragonoloba;* Singh and Swarup, 1986). The efficiency of hot water treatment for eradicating *X. o.* pv. *oryzae* from rice seed is only about 30% (Fang *et al.*, 1982).

(b) Dry heat treatment

Dry heat eliminated seedborne *X. c.* pv. *translucens*, the causal agent of bacterial black chaff and leaf streak of cereal grains, from barley seeds. The seeds were treated at 71–72°C for 7 days without an appreciable reduction in germination (Fang *et al.*, 1957). Dry heat was recommended by Lozano (1987) to treat cassava seeds after careful visual and density selection. This technique may be an alternative way to produce viable bacteria-free seeds.

(c) Chemical seed treatment

Calcium hypochlorite at 10–20 g active ingredient per kg seed as slurry and left sealed in containers for 16 h controlled *X. c.* pv. *campestris* (Schultz *et al.*, 1986). Slurry treatments had no effect on germination but reduced it in some lots after 6 months.

Nyolate, a general disinfestant composed of 10:3:90 sodium hypochlorite (2.73%) : lactic acid (15.1%) : water effectively reduced the detectable level of *X. c.* pv. *campestris* without reducing the germination of kale, collards, turnip, cabbage or Chinese cabbage (Harman *et al.*, 1987). However, the stands of cauliflower and broccoli were reduced.

X. c. pv. *campestris* was eradicated from seeds by soaking in 0.5% cupric acetate dissolved in 0.005 N acetic acid (acidified cupric acetate) for 20 min at 35–50°C but not at 25°C (Schaad *et al.*, 1980). This technique was used by Forster and Schaad (1988) to control black chaff of wheat caused by *X. c.* pv. *translucens*. Wheat seeds were soaked for 20 min at 25°C and rinsed twice with water. Phytotoxicity occurred when the temperature was raised to 45°C. Acidified cupric acetate treatment of wheat seeds is recommended as a component of a Foundation Seed Health Programme.

Antibiotics such as streptocycline (Agrimycin 100) have not been used extensively in seed treatment. However, Singh and Jain (1988) reported that treatment with 100 ppm streptocycline gave the best control for *X. c.* pv. *glycines*, while Singh and Swarup (1986) reported that 200 ppm plantomycin (streptomycin sulphate) as a seed soak for 2 h eradicated *X. c.* pv. *cyamopsidis* in cluster beans. Streptomycin control of pepper or tomato leaf spot caused by *X. c.* pv. *vesicatoria* was quite effective until the bacterium became resistant to the antibiotics. Phenazine appears to be quite effective against *X. o.* pv. *oryzae* on rice.

Other chemicals are reported to control seedborne *Xanthomonas* diseases. In dense glasshouse plantings, leaf spot of zinnia caused by *X. c.* pv. *zinniae* was reduced from 100 to 2% when seeds were soaked for 30 min in 360 g active ingredient Captan per litre of water containing Tween 80 (Strider, 1980). The best control of *X. c.* pv. *glycines* was by seed treatment with Thiram combined with foliar sprays of streptocycline or Agrimycin 100 (Srivastava and Bais, 1985).

7.3 CULTURAL PRACTICES

A primary disease management strategy is compatible cultural practices that can be integrated into crop management. Disease can be managed either by removing the source of inoculum, or delaying and reducing the spread and development of the pathogen from the resting stage to an infective stage. Crop rotation or fallowing has been a common practice since ancient times. This disease practice has been successful for pathogens which have a life cycle largely dependent on the host plant for survival. Many xanthomonad bacterial pathogens are of this type.

7.3.1 Crop residue management

Crop residues and debris serve as a major source of inoculum of many diseases caused by *Xanthomonas* spp. In the tropics, diseased straw, stubble or volunteer plants are often the primary sources of *X. o.* pv. *oryzae* inoculum. Even in temperate regions, infected plant tissues that remain in the fields can be a major source of primary inoculum for infection. In 1982, severe kresek occurred in Korea. The source of inoculum was infected straw left in the field the previous year. Some farmers also used the infected straw to cover their seedbed. As a result, severe kresek occurred a few weeks after transplanting. To control bacterial blight of rice it is, therefore, essential to manage the crop residue. Seedlings should not come into contact with infected rice straw. Straw left from previous crops should be removed as soon as possible, especially if the crops were diseased. Piling rice straw near the source of irrigation water, or ditches of secondary or tertiary irrigation canals, should also be avoided.

All infected plant debris should be incorporated into the soil, where survival of xanthomonads is very poor. Alternatively, infected plant debris should be removed or burned. Six months fallow is sufficient to prevent carryover of *X. c.* pv. *manihotis* in the soil (Persley, 1987), while 2 months fallow was recommended for *X. c.* pv. *dieffenbachiae* (Nishijima, 1989). After removal of infected plants, *X. c.* pv. *citri* could not be recovered from the nursery after 21 days (Graham *et al.*, 1989). In Georgia, *X. c.* pv. *campestris* survived 8 months in buried cabbage stems, but only 42 days in infested soil (Schaad and White, 1974). Soil is manipulated more frequently in multiple cropping systems than in monocrop systems, and it is possible that infected cabbage stems might decay more rapidly in multiple cropping systems and allow more frequent cropping of crucifers. In the arid climate of Sudan, *X. c.* pv. *malvacearum* in infested debris is a threat to subsequent cotton plants, but in the wetter climate of Tanzania the pathogen barely survives between cotton crops in surface debris. Since many bacterial pathogens survive in debris but not in soil, reduced tillage might increase the incidence of bacterial diseases. Pathogens that survive in plant debris can cause severe epidemics (Williams, 1980). A special effort is made in many countries not to grow crucifer crops frequently in a rotation because of black rot.

Weeds must be eliminated because many xanthomonads survive a long time in association with many weeds that grow naturally in plantations. However, the epidemiological significance of this is not completely understood in all cases.

7.3.2 Planting time

Planting time may be manipulated to reduce losses. Rain splash is an important means of pathogen spread and accounts for the increase in disease severity during the rainy season. Delaying planting toward the end of the rainy season increases the likelihood that plants will escape infection or that infection will be minimal. By planting cassava toward the end of the rainy season, a relatively healthy plantation can be established and maintained in good condition (Lozano, 1987).

7.3.3 Crop mixture

Mixed planting of different crop cultivars or species may result in reduced disease incidence. This practice appears attractive especially for vegetable crops in small land areas in the tropics or in less developed regions. The spread of *X. c.* pv. *vesicatoria* was reduced when thinning operations of direct seeded tomato were performed in the afternoon when foliage was dry (Pohronezny *et al.*, 1990). Periodic hand disinfestation with ethanol or providone-iodine as prophylactic hand washes further reduced disease incidence.

7.3.4 Host nutrition

The role of host nutrition in xanthomonad disease management is not well understood and generalizations cannot be readily applied to disease control. Harkness and Marlatt (1970), as cited by Chase (1988), reported that as the nitrogen rate increased, severity of red-edge disease of *Philodendron scandens* subsp. *oxycardium* caused by *X. c.* pv. *dieffenbachiae* decreased. Neither potassium nor phosphorus affected severity of red-edge disease. Likewise, severity of leaf spot of Schefflera and dwarf Schefflera caused by *X. c.* pv. *hederae* decreased as fertilizer level increased up to 5 times that recommended for these plants (Chase and Poole, 1987). Increasing fertilizer rates have also resulted in significant and dramatic decreases in severity of leaf spot of hibiscus caused by *X. c.* pv. *malvacearum* and ficus caused by *X. c.* pv. *fici*. Symptoms of syngonium blight caused by *X. c.* pv. *syngonii* decreased linearly as the rate of complete fertilizer (liquid 20:20:20 or slow release 19:6:12) increased. Increased rates of nitrogen or potassium were equally effective in reducing symptom expression. On the other hand, disease severity of bacterial blight of anthuriums, as well as overall plant quality, increased initially as fertilizer rate increased, then decreased (Chase, 1989).

7.4 CHEMICAL CONTROL

There are only a few bactericides among the vast and ever increasing number of pesticides manufactured. The primary reason for this lack of bactericides is, perhaps, due to the emphasis on control of fungal diseases compared to that of bacterial diseases. Another reason is perhaps related to the general concern caused by rapid development of resistant strains of the bacterial pathogens.

7.4.1 Antibiotics

Antibiotics, such as streptomycin and oxytetracycline, are formulated as agricultural bactericides. However, the agricultural use of antibiotics with medical applications is discouraged because of the potential hazard of transfer of drug resistance from bacteria associated with plants to those associated with animals and humans.

Streptomycin-resistant strains of *X. c.* pv. *dieffenbachiae* were detected on several farms in Hawaii 2 years after the antibiotics were registered for use to control anthurium blight (Nishijima, 1988). Since resistance to streptomycin became so common, some growers stopped using antibiotics. No oxytetracycline-resistant strain has so far been detected on anthurium farms. In the Philippines, about 44% of *X. c.* pv. *citri* isolates from a citrus orchard in the northern part of the country were resistant to 200 ppm of streptomycin. The grower of this orchard used heavy doses of antibiotics to control citrus canker.

Best results for controlling bacteria plant diseases with chemicals are often obtained by combining copper bactericides with antibiotics which are applied either as seed treatments or foliage sprays. Tribasic copper sulphate (0.5%) was effective against *X. c.* pv. *citri* and its effectiveness was further improved by the addition of Agrimycin 100 (streptomycin) at 0.12% (Falico de Alcaraz, 1982). Copper oxychloride (CuOCl) sprayed at 2.5 g/l, two to three times, gave better control of citrus canker than applying the total dosages (5 or 7.5 g/l) in a single application to young shoots of Mexican lime (Medina-Urrutia and Stapleton, 1987). The same combination gave satisfactory control against *Xanthomonas* leaf spot of cotton (Sharma and Chauhan, 1985; Singh and Singh, 1988), and leaf spot of chili (Raju *et al.*, 1980). A combination of Agrimycin 100 (0.025%) plus 0.1% Bavistin (carbendazim) used three times as foliar spray controlled bacterial blight of mungbean (*Vigna radiata*) caused by *X. c.* pv. *phaseoli* (Gaur and Ahmed, 1983).

7.4.2 Fungicides

The most widely used bactericides are copper-based, such as copper oxychloride, copper hydroxide, cuprous oxide and Bordeaux mixture. Copper is, however, phytotoxic to some plants and should be used with caution. Copper resistance of plant pathogenic bacteria, including xanthomonads, is well known (Adaskaveg and Hine, 1985; Bender *et al.*, 1990; Ritchie and Dittapongpitch, 1991). Some fungicides have been identified to have a bactericidal activity. Recently, fosetyl aluminium (Aliette 80 WP) has been used for control of a variety of bacterial diseases on ornamentals, tomatoes, citrus and apples. While fosetyl aluminium provided control of some of these diseases, it was not consistently more effective than the standard bactericides (Chase,

1988). On foliage plants, fosetyl alumimium is equally or more effective than cupric hydroxide (Kocide 101) for control of *Xanthomonas* diseases of pilea, ficus, fittonia and syngonium. It only slightly reduced *Xanthomonas* leaf spot of shefflera. The three forms of the bacterial blight of cotton (i.e. angular leaf spot, black arm and boll rot) can be controlled by treatment with dichlorophen, first at flowering, again at first boll stage, and then every 7–14 days according to the degree of infection and weather conditions (Swanepoel, 1990).

Metham-sodium at 17 ml/kg (32.7% active compound) of air-dried soil killed over 99.9% and 99% of *X. c.* pv. *campestris* in soil in the lab and field tests, respectively, in a soil infested with the pathogen. In addition, it significantly reduced the levels of the pathogen in stem, siliques and seeds, with a concomitant reduction in the incidence of plants with black rot symptoms. The treatment was more successful in reducing disease incidence when applied to the soil at a depth of 30 or 40 cm rather than 20 cm. Seed treatment is recommended in combination with soil fumigation with Methan-sodium for the production of *X. c.* pv. *campestris*-free seed.

7.4.3 Other compounds

New bactericides are being tested against a variety of xanthomonads. For example, the joint action of hydrogen fluoride and sulphur dioxide on the development of common blight of red kidney beans was investigated by Laurence and Reynolds (1986). Pre-inoculation and post-inoculation exposure to sulphur dioxide caused smaller lesions and longer latent periods. Post-inoculation exposure to hydrogen fluoride resulted in longer latent periods. In general, their joint action resulted in smaller lesions and longer latent periods. Based on preliminary tests in Fujian, China, a monomolecular film-forming substance gave good control of citrus canker in leaves and in fruits (Qui and Ni, 1988). Greenhouse and field trials with a suspended colloid (15% TF-130 at 300–1000 ppm) reduced black chaff (*X. c.* pv. *undulosa*) by 81–99% when sprayed two to three times at 7–10 day intervals starting at booting. This resulted in a 10–20% yield increase (Sun and He, 1988). A stable bleaching powder at 75 μg/ml artificial inoculation as seed treatment, 7.25 mg/kg as a soil treatment and a combination of these treatments gave maximum germination and seedling stand in cotton (Gupta and Chauhan, 1988) with good control of *X. c.* pv. *malvacearum*.

7.5 HOST RESISTANCE

In view of the increasing concern about the quality of our living environment, host resistance is likely to play a key role in the management of plant pathogens below economic threshold levels. Incorporation of disease resistance into improved genotypes is of prime importance, not only in conferring stability to crop production, but also in providing the farmers and growers with the desired technology. To develop resistance genetic sources of resistance are needed. For incorporation into resistant varieties, information about host–pathogen interactions, types of diseases and epidemics caused by the pathogens, and the variability of the target pathogen in the ecosystem are needed. As more genetic mechanisms conveying resistance are uncovered due to the rapid advancement in biotechnology, genes for resistance from different sources, including wild or related species, may be incorporated into commercial cultivars. It is reasonable to assume that crop host resistance to xanthomonad diseases will play a strategic role in disease management. The issue, however, is durability of resistance which is related to the rapid life cycles of the bacterial pathogens. In the past, many resistant cultivars have succumbed to new virulent strains/races, in some cases within one or two seasons of usage.

7.5.1 Methods of evaluation and inoculations

Several techniques have been used to evaluate crop cultivars for resistance to xanthomonads. These techniques are used to identify resistance among cultivars and to screen breeding materials for resistance to disease. Mew *et al.* (1987) summarized the methods used for rice bacterial blight and provided details for screening or for identification of resistance to *X. o.* pv. *oryzae*. The methods of screening/assessment of resistance depend largely on the types of resistance. Resistance to certain virulent races can be either specific or non-specific. Host resistance can be evaluated under controlled conditions such as the greenhouse, or under field conditions, based on either natural infection or artificial inoculation. The adequacy of the screening depends on the inoculum pressure used and the conditions for disease development. The scoring systems used to measure the amount of infection and, subsequently, the rogueing of segregating populations in the early generations of the breeding materials play a major role in developing the type of resistance desired. All these require the appropriate sources of resistance to be incorporated in the progenies.

For resistance to xanthomonads in dicot plants, such as tomato and pepper to *X. c.* pv. *vesicatoria*, cowpea and bean to *X. c.* pv. *phaseoli* and cotton to *X. c.* pv. *malvacearum*, hypersensitivity is often detected in the germplasm. Gitaitis (1983) described 'the typical hypersensitive reaction in a plant leaf to phytopathogenic bacteria as a zone of confluent necrosis within 24 hr of inoculation with inocula of 10^8 cfu/ml, whereas under the same conditions, a susceptible reaction requires 3–5 days or more.' For identification of hypersensitivity to the cowpea bacterial blight pathogen, *X. c.* pv. *vignicola*, Gitaitis (1983) identified and characterized two types of resistance, one with the typical hypersensitive reaction (HR), and one represented by a brown-red discoloration without complete collapse of the tissues. Both types of resistance appear to be valuable in improving the resistance to *X. c.* pv. *vignicola*.

In disease screening and resistance identification, it is relatively simple to establish that resistance is a HR. For bacterial diseases, it is not clear if the HR in dicot plants is comparable to vertical resistance (as in the case of many fungal diseases) and if the non-HR is comparable to polygenic, horizontal or partial resistance – different terms used to denote the resistance conditioned by minor genes. Crill *et al.* (1972) used disease damage ratings and actual lesion numbers to estimate the relative susceptibility of tomato cultivars to *X. c.* pv. *vesicatoria*. Based on these parameters, differences among cultivars were concluded to be real and not artifactual, because both methods of determining relative susceptibility produced similar results. The number of lesions was not always associated with the extent of defoliation. They assumed that this type of 'tolerance' to the disease was not worth considering in a varietal development programme. In the case of bacterial blight of rice, lesion length can be used to assess qualitative or quantitative resistance (Mew, 1987), the former conditioned by major genes while the latter is conditioned by minor genes.

Capsicum plants inoculated by atomization under controlled conditions with 10^8 cfu/ml had significantly higher disease severity than plants inoculated with lower inoculum concentrations. Old plants produced less lesions than younger plants (Hartman *et al.*, 1990). This technique was adequate to assess the resistance of *Capsicum* cultivars to *X. c.* pv. *vesicatoria*.

Most screening techniques for bacterial diseases require wounding of the host tissues in order to introduce the inoculum to the infection site. Some scientists are concerned that the level of resistance among cultivars may not be clearly differentiated following wound inoculation. Robeson *et al.* (1989) introduced a hydathode inoculation technique for the simulation of natural black rot infection of cabbage by *X. c.* pv. *campestris*. By introducing the inoculum into guttation droplets on the leaf margin, the bacteria are taken into the plant via the hydathodes during periods of dryness. This technique avoids mechanical injury to the plant and the possibility of producing artifacts. Using this technique, the authors were able to introduce low levels of inoculum to induce high levels of infection. The differences among cultivars in response to the black rot pathogen in the absence of host injury can be readily discerned. The technique also allowed adjustment in the quantity of inoculum applied to each infection court, resulting in proper assessment of resistance. It would be interesting to apply this technique on a large scale to field test cabbage cultivars for the various types of resistance to *X. c.* pv. *campestris*.

7.5.2 Pathogen variation: races and differentials

The problems of deploying resistant cultivars for bacterial disease control are perhaps no different from those encountered in other crop diseases. The emergence of new and virulent strains or races of the target pathogen is of prime concern. The use of resistant cultivars is considered to be the most effective method to control bacterial blight of cotton caused by *X. c.* pv. *malvacearum*. But immediately after the introduction and cultivation of resistant cultivars new races of the pathogen become evident (Brinkerhoff, 1970).

Generally, differential cultivars possessing different resistance genes are required to distinguish variation among pathogen isolates. Methods used to differentiate the responses vary according to the pathogen–host systems. In the case of rice bacterial blight, variability of *X. o.* pv. *oryzae* was based on a set of rice cultivars carrying different R-genes. Plants were inoculated by either the clipping or the pricking method. Strong differential interactions were observed and races were identified (Mew, 1987). Based on the R-genes identified, near-isogenic lines were developed through backcrosses to a common recurrent parent. This set of differential lines are now being used to compare races distributed in the different rice-growing countries in Asia. By monitoring the race frequency distribution change over the years and geographical localities, it was concluded that the emergence of new races was closely related to the cultivation of resistant cultivars (Mew *et al.*, 1992). In a recent study using genetic probes, the spatial distribution of *X. o.* pv. *oryzae* was determined. In any given field, many restriction fragment length polymorphism (RFLP) types exist (Nelson *et al.*, 1991). It is, therefore, reasonable to assume that different genotypes are present in a single race identified by the conventional method using host differentials. The bacterial blight pathogen of rice is, therefore, highly variable.

Races of *X. c.* pv. *malvacearum* were designated on the basis of different disease reactions on a set of upland cotton differentials (Hunter *et al.*, 1968). In India, races were designated based on reactions to the four species of cotton cultivated in that country (Nayudu, 1974). Different strains within each race were designated as biotypes on the basis of disease reactions to cultivars. Singh *et al.* (1977) described and differentiated 14 races in cotton-growing areas around Delhi. Race 21, virulent to polygenes B2 and BN, was reported for the first time. More than 50% of the isolates were virulent to polygenes and to at least two or more other major genes responsible for resistance in cotton in India. It was interesting to note from these surveys that the gene combination of B2 and B3 in an Acala genetic background was immune to all the races of *X. c.* pv. *malvacearum* individually or in mixed inocula.

In Pakistan, losses of cotton to bacterial blight are very high in the Punjab area where summer rains favour disease development (Hussain and Brinkerhoff, 1978). Race 18 was the common race identified in the major cotton-growing regions. Cotton cultivars with single genes and polygenes for resistance appeared susceptible to this race, which accounted for several epidemics in Pakistan. The bacterial blight resistant upland cotton lines introduced from the USA and Africa were homozygous immune (no visible symptoms) to race 18. These are very useful sources of resistance in the varietal improvement programme (Brinkerhoff and Hussain, 1978).

The selection pressure for more virulent races is directly related to the widespread commercial use of xanthomonad-resistant crops. In India, where a wide acreage of resistant indigenous cotton species (*G. arboreum* and *G. herbaceaum*) were grown, more virulent races became predominant. A shift towards increased virulence has also been noticed at all locations in the USA due to the increased cultivation of resistant cultivars. In New Mexico, a new variant of *X. c.* pv. *malvacearum* emerged which was virulent to the resistant cultivar Acalas earlier developed for resistance against race 2 and race 7 (Chew and Booth, 1974). It is also evident that more than one virulent race of the pathogen can exist in a given locality of crop growth (Singh *et al.*, 1977; Lozano, 1987). In cotton, several races of *X. c.* pv. *malvacearum* were present at one place and race 32 biotype A, which attacks the maximum number of blight-resistant genes, is widely distributed (Verma and Singh, 1974; Mew, 1987). It was the most widely distributed race in India, comprising 59% of the races in the area. This race is most virulent because it attacks at least four major bacterial blight resistance genes, namely B_7, B_2, B_{1n} and B_N. The less virulent races which attack major genes and polygenes have limited distribution.

In cowpea (*Vigna unguiculata*) two strains of *X. c.* pv. *vignicola* are responsible for two separate disease syndromes. The first is cowpea bacterial blight and canker, which appears as large, irregularly shaped, necrotic areas on the foliage, similar to those of common blight of bean. Stem cankers are often associated with common blight of bean and can severely affect certain cultivars, causing wilt or plant lodging. The second disease syndrome is bacterial pustule. Initial symptoms are small, raised, watersoaked pustules on the underside of the leaves, and the pustules remain circular throughout their development. Bacterial pustule appears to be present only in Africa while common blight appears in cowpea-growing areas (Ekpo and Saettler, 1977).

In another legume disease, pathogenic variation in *X. c.* pv. *phaseoli* and *X. c.* pv. *phaseoli* var. *fuscans* is based on disease reactions in six *Phaseolus vulgaris* cultivars (Ekpo and Saettler, 1977). *X. c.* pv. *fuscans* isolates are more virulent on the cultivars than *X. c.* pv. *phaseoli* isolates. Plants in the reproductive stage of development (35 days after planting) were often more susceptible to a particular *X. c.* pv. *phaseoli* or *X. c.* pv. *phaseoli* var. *fuscans* isolate than were plants in the vegetative stage of development (21 days after planting).

X. c. pv. *vesicatoria* isolated from tomato are designated as 'tomato' strains based on the fact that the tomato cultivars were susceptible whereas pepper cultivars were hypersensitive (Cook and Stall, 1979). A strain to which pepper is susceptible and tomato hypersensitive has not been found. Hence, based on the hypersensitivity to tobacco and susceptibility to tomato, all isolates were pathogenic. Isolates were differentiated by hypersensitive or susceptible symptoms on pepper cultivars and classified into three groups. All isolates that produced a hypersensitive reaction on pepper cultivars were included in one group. The second group of isolates were those which were virulent to all pepper cultivars. However, selections derived from PI 173192 were hypersensitive. The third group of isolates were those virulent to all pepper cultivars and the selections.

7.5.3 Resistance for disease management

Varietal resistance will remain the single most important and effective technology for disease management for many years to come. Since other control strategies are not always effective, and sometimes very costly, resistance should be the long-term goal to control crop diseases caused by *Xanthomonas* spp. Resistance is not, of course, perfect if pathogen variation is considered. Durable resistance (Johnson, 1981) is a retrospective concept. Resistance of a cultivar can only be called durable if the resistance of one cultivar remains effective while another succumbs to an epidemic after being planted to a large area for many years. To identify sources of durable resistance and to utilize these sources in resistance breeding is the goal of crop improvement programmes. Breeding for durable resistance always requires continuous field testing for extended periods at diverse geographical locations. The level of durability is different for different diseases and for the same disease under different environments (Mew, 1991).

Until recently, major genes have been used because they are easy to recognize and to incorporate into breeding lines. This is not to say that major genes for resistance to bacterial diseases caused by *Xanthomonas* spp. will not be durable. In the case of rice, the major gene, *Xa-4*, for bacterial blight resistance seems to be durable (Mew *et al.*, 1992). Other examples can be listed. The key issues in durable resistance depend largely on the types of diseases, the type of host–pathogen interaction and the ecosystems involved in the crop and xanthomonad pathogens.

Resistance breeding of cotton to *X. c.* pv. *malvacearum* first started in Sudan in the 1950s. Since then, breeding for resistance has been emphasized in the USA and other parts of the world. There are 18 R-genes identified with measurable effects on resistance. Both minor and major genes are used in order to obtain highly resistant or immune cultivars (Chase and Poole, 1987). Cotyledon, leaf, stem (black arm), and the boll infection are associated with high susceptibility to bacterial blight. Resistance seems to be associated with different species of *Gossypium*. Brinkerhoff (1970) summarized resistance in the different species following Knight's (1948) data. *G. barbadense* is highly susceptible. Some tolerance exists in the annual Sea Island cultivars which appear to be polygenic. The cultivated diploid cottons are resistant to angular leaf spot in India. The cultivars grown in the USA belong to *G. hirsutum* with low levels of tolerance. Brinkerhoff and Hussain (1978) attributed this tolerance to the restriction of the severity of the disease. Its resistance is also polygenic. Immunity is reported only in cultivars of *G. arboreum* and *G. berbaceum*. The major genes identified from these species have great effect. Genes with less effect were also identified and associated with each source of resistance. These genes were designated as minor or polygenes. New races of the pathogen build up within a single season on blight-resistant breeding lines with the B genes in a susceptible background. The durability of resistance appears to be associated with different combinations of major and minor genes. On the other hand, the bacterial pathogen also varied widely in virulence on the same host species cultivars and also on host species/cultivars with different levels of resistance.

Expression of resistance to crop diseases caused by xanthomonad pathogens is influenced by the environment. The bean cultivars Jules and the PI 207272 line are moderately resistant to *X. c.* pv. *phaseoli* in the temperate zone but are susceptible in field evaluation in the tropics (Webster *et al.*, 1983). Susceptibility of these cultivars to bacterial blight was associated with the effect of high temperature on the growth of the cultivars. The two cultivars mature early and growth was poor in the tropics. When bean is grown normally, it is likely to be resistant to *X. c.* pv. *phaseoli*.

Resistance to *X. c.* pv. *vesicatoria*, causing leaf spot of pepper and tomato, depends on the presence of two different loci: a resistance locus in a particular cultivar of pepper and an avirulence locus in a particular race of the pathogen. Resistance to *X. c.* pv. *vesicatoria* has also been reported in a number of cultivars of *Capsicum annuum*. Hypersensitivity and limited bacterial multiplication occurred following injection of a high inoculum dose into resistant plants selected from PI

173192. This type of resistance is simply inherited (Cook and Stall, 1973).

In the USA, the first cabbage truly resistant to *X. c.* pv. *campestris* was released to growers only recently (Williams, 1980). The source of resistance was derived from a Japanese cabbage cultivar, Early Fuji. The resistance is broad against many strains in many countries. The need to find new sources of resistance is pressing. Resistant cabbage is needed for the tropics and subtropics as *X. c.* pv. *campestris* has become endemic in these regions.

Although resistance appears to be the desirable technology to control crop diseases, resistant cultivars are not necessarily a demand commodity in the commercial market if other agronomic qualities are not taken into account. For example, the demand for a specific eating quality of a vegetable or fruit crop may be an essential consideration in resistance breeding. Snap bean cultivars grown in the USA, in general, are very susceptible to bacterial blight (Webster *et al.*, 1983a). Since current control measures are not adequate, the control can be improved by developing resistant cultivars. To achieve this, however, information on the level of resistance necessary for effective control is needed. It appears (Webster *et al.*, 1983a,b) that commerically acceptable cultivars are all very susceptible to bacterial blight.

7.6 CONCLUSION

In view of the above discussion, it is obvious that there is no single control strategy which is effective for all crop diseases caused by *Xanthomonas* spp. With a better understanding of the agroecosystems, and with the advancements in biotechnology, many disease management strategies can be improved, while also being ecologically sound and environmentally safe.

REFERENCES

Adaskaveg, J.E. and Hine, R.B. (1985) Copper tolerance and zinc sensitivity of Mexican strains of *Xanthomonas campestris* pv. *vesicatoria*, causal agent of bacterial spot of pepper. *Plant Disease*, **70**, 993–7.

Bender, C.L., Malvick, K.K., Conway, K.E., George, S. and Pratt, P. (1990) Characterization of pXV10A, a copper resistance plasmid in *Xanthomonas campestris* pv. *vesicatoria*. *Applied and Environmental Microbiology*, **57**, 170–5.

Brinkerhoff, L.A. (1970) Variation in *Xanthomonas malvacearum* and its relation to control. *Annual Review of Phytopathology*, **8**, 85–110.

Brinkerhoff, L.A. and Hussain, T. (1978) Evaluation of American and African bacterial blight resistant upland cotton lines in Pakistan in 1977. *Plant Disease Reporter*, **72**, 1082–5.

Chase, A.R. (1988) Chemical and nutritional aspects of controlling *Xanthomonas* diseases on Florida ornamentals. *Proceedings of the First Anthurium Blight*

Conference, 4 February 1988, Hawaii Institute of Tropical Agriculture and Human Resources.

Chase, A.R. (1989) Effect of nitrogen and potassium fertilizer on severity of *Xanthomonas* blight of *Syngonium podophyllum*. *Plant Disease*, **73(12)**, 972–5.

Chase, A.R. and Poole, R.T. (1987) Effects of nutrition on growth and susceptibility of *Anthurium scherzeranum* to *Xanthomonas* leaf spot. *AREC Apopka Research Report*, RH 87-4.

Chew, C.F. and Booth, J.A. (1974) Report of Bacterial Blight Committee – 1973. *Proceedings of the Cotton Disease Council*, **34**, 6–7.

Cook, A.A. and Stall, R.E. (1973) Inheritance of resistance in pepper to bacterial spot. *Phytopathology*, **53**, 1070–2.

Cook, A.A. and Stall, R.E. (1979) Differentiation of pathotypes among isolates of *Xanthomonas vesicatoria*. *Plant Disease Reporter*, **53(8)**, 717–19.

Crill, P., Jones, J.P. and Burgis, D.S. (1972) Relative susceptibility of some tomato genotypes to bacterial spot. *Plant Disease Reporter*, **57(7)**, 504–7.

Ekpo, J.A. and Saettler, A.W. (1977) Pathogenic variation in *Xanthomonas phaseoli* and *X. phaseoli* var. *fuscans*. *Plant Disease Reporter*, **70(1)**, 80–3.

Falico de Alcaraz, G. (1982) Control of citrus canker (*Xanthomonas citri* Hasse Dow.) on grapefruits. *Fitopatologia*, **17(1)**, 48–53.

Fang, C.T., Chu, C.L. and Lu, Q. (1982) Disinfection of the rice seed carrying the bacterial leaf blight organism. *Acta Phytophyla. Sinica*, **9(2)**, 109–12.

Fang, C.T., Lin, C.F. and Chu, C.L. (1957) A preliminary study on the disease cycle of the bacterial leaf blight of rice. *Acta Phytopathologica*, **2**, 173–85.

Fang, C.T., Xu, Z.G., Chu, C.L. and Lu, Q. (1982) Seedborne nature of the rice bacterial leaf blight organism and its inspection. *Acta Phytophyla. Sinica*, **9(2)**, 103–8.

Forster, R.L. and Schaad, N.W. (1988) Control of black chaff of wheat with seed treatment and foundation seed health program. *Plant Disease*, **72**, 935–8.

Fourest, E., Rehms, L.D., Sands, D.C., Bjarko, M. and Lund, R.E. (1990) Eradication of *Xanthomonas campestris* pv. *translucens* from barley seed with dry heat treatment. *Plant Disease*, **74**, 817–18.

Fry, W.E. (1982) *Principles of Plant Disease Management*, Academic Press, New York.

Gaur, R.B. and Ahmed, S.R. (1983) Studies in the chemical control of bacterial blight of moong incited by *Xanthomonas phaseoli* (Smith) Dowson. *Pesticides*, **17**, 22–3.

Gitaitis, R.D. (1983) Two resistant responses in cowpea induced by different strains of *Xanthomonas campestris* pv. *vignicola*. *Plant Disease*, **77**, 1025–8.

Gonzales, J.A. (1987) A procedure for the establishment of seed plants of cassava free of bacterial blight (*Xanthomonas campestris* pv. *manihotis*). *Agronomia Tropical*, **37(1–3)**, 75–84.

Goto, M. (1972) The significance of the vegetation for the survival of plant pathogenic bacteria, in *Plant Pathogenic Bacteria 1971, Proceedings of the Third International Conference on Plant Pathogenic Bacteria* (ed. H.P. Maas Geesteranus), PUDOC, Wageningen, pp. 39–53.

Graham, J.H., Gottwald, T.R., Civerolo, E.L. and McGuire, R.G. (1989) Population dynamics and survival of *Xanthomonas campestris* pv. *citri* in soil in citrus nurseries in Maryland and Argentina. *Plant Disease*, **73**, 423–7.

Groth, D.E. and Braun, E.J. (1989) Survival, seed transmission, and epiphytic development of *Xanthomonas campestris* pv. *glycines* in the north-central United States. *Plant Disease*, **73**, 327–30.

Gupta, P.P. and Chauhan, M.S. (1988) Control of bacterial blight of cotton through stable bleaching powder in Harvana. *Journal of the Indian Society for Cotton Improvement*, **13(1)**, 47–53.

Harman, E.G., Norton, J.M., Stasz, T.E. and Humaydon, H.S. (1987) Nyolate seed treatment of *Brassica* spp. to eradicate or reduce black rot caused by *Xanthomonas campestris* pv. *campestris*. *Plant Disease*, **71**, 27–30.

Hartman, G.L., Yang, C.H. and Wang, J.F. (1990) Races of *Xanthomonas campestris* pv. *vesicatoria* in Taiwan and an inoculation technique to evaluate host resistance. *Plant Protection Bulletin*, **32(2)**, 112–24.

Hunter, R.E., Brinkerhoff, L.A. and Bird, L.S. (1968) The development of a set of upland cotton lines for differentiating races of *Xanthomonas malvacearum*. *Phytopathology*, **58**, 830–2.

Hussain, T. and Brinkerhoff, L.A. (1978) Race 18 of the cotton bacterial blight pathogen, *Xanthomonas malvacearum*, identified in Pakistan in 1977. *Plant Disease Reporter*, **72**, 1085–7.

Johnson, R. (1981) Durable resistance: definition of genetic control and attainment in plant breeding. *Phytopathology*, **71**, 567–8.

Knight, R.L. (1948) The genetics of blackarm resistance. VI. Transference of resistance from *Gossypium arboreum* to *G. barbadense*. *Journal of Genetics*, **48**, 359–79.

Kuan, T.L. and Minsavage, G.V. (1985) Detection of *Xanthomonas campestris* pv. *carotae* in carrot seed. *Plant Disease*, **79**, 758–70.

Laurence, J.A. and Reynolds, K.L. (1986) The joint action of hydrogen fluoride and sulfur dioxide on the development of common blight of red kidney bean. *Phytopathology*, **76**, 514–17.

Lozano, C. and Wholey, D.W. (1974) The production of bacteria-free planting stock of cassava, Manihot seculenta Crantz. *World Crops*, **27**, 115–17.

Lozano, J.C. (1987) Cassava bacterial blight: a manageable disease. *Plant Disease*, **70**, 1089–93.

McGuire, R.G., Jones, J.B. and Sasser, M. (1987) Tween media for semi-selective isolation of *Xanthomonas campestris* pv. *vesicatoria* from soil and plant material. *Plant Disease*, **70**, 887–91.

Medina-Urrutia, V.M. and Stapleton, J.J. (1987) Control of Mexican lime bacteriosis with copper-based products. *Proceedings of the Florida State Horticultural Society*, **99**, 22–5.

Mew, T.W. (1987) Current status and future prospects of research on bacterial blight of rice. *Annual Review of Phytopathology*, **25**, 359–82.

Mew, T.W. (1991) Disease management in rice, in *CRC Handbook of Pest Management in Agriculture*, 2nd edn (ed. D. Pimentel), CRC Press, Boca Raton, FL, pp. 279–99.

Mew, T.W., Reyes, R.C. and Vera Cruz, C.M. (1989) Screening for bacterial blight resistance in rice, in *Methods in Phytobacteriology* (eds Z. Klement, K. Rudolph and D.C. Sands), Akademiae Kiado, Budapest, pp. 338–41.

Mew, T.W., Vera Cruz, C.M. and Medalla, E.S. (1992) Changes in race frequency of *Xanthomonas oryzae* pv. *oryzae* in response to the planting of rice cultivars in the Philippines. *Plant Disease*, in press.

Moffet, M.J. and Croft, B.J. (1983) *Xanthomonas*, in *Plant Bacterial Diseases: a diagnostic guide* (eds P.C. Fahy and G.J. Persley), Academic Press, New York, pp. 189–228.

Mount, M.S. and Lacy, G.H. (1982) *Phytopathogenic Prokaryotes*, Vol. 2, Academic Press, New York.

Nayudu, M.V. (1974) Variation in *Xanthomonas malvacearum*. *Indian Cotton Growers Review*, **18**, 350–5.

Nelson, R., Ardales, E.Y., Vera Cruz, C.M. and Mew, T.W. (1991) Spatial scale of variation of the bacterial blight pathogens of rice. *Phytopathology* (abstr. 457).

Nishijima, W.T. (1988) Anthurium blight: an overview, in *Proceedings of the First Anthurium Blight Conference*, 23–25 February 1988, Hawaii Institute of Tropical Agriculture and Human Resources, Hilo, Hawaii.

Nishijima, W.T. (1989) Current Anthurium blight control recommendations. *Proceedings of the Second Anthurium Blight Conference*, February 1989, Hilo, Hawaii.

Persley, G.J. (1979) Studies on the survival and transmission of *Xanthomonas manihotis* on cassava seed. *Annals of Applied Biology*, **93**, 159–77.

Persley, G.J. (1987) Cassava diseases and their control. *Proceedings of the 11th International Congress on Plant Protection*, 5–7 October 1987, Manila, Philippines.

Pohronezny, K., Moss, M.A., Dankers, W. and Schenk, J. (1990) Dispersal and management of *Xanthomonas campestris* pv. *vesicatoria* during thinning of direct-seeded tomato. *Plant Disease*, **74**, 800–5.

Qui, C.H. and Ni, B.Q. (1988) A preliminary test on the control of citrus canker with mono-molecular forming substance (MMFS) and other chemicals. *Fujian Agricultural Science and Technology*, **1**, 19.

Raju, K.S., Rao, G.S. and Subbayya, J. (1980) Chemical control of bacterial leaf spot of chili. *Madras Agricultural Journal*, **67**, 825–7.

Ritchie, D.F. and Dittapongpitch, V. (1991) Copper- and streptomycin-resistant strains and host differentiated races of *Xanthomonas campestris* pv. *vesicatoria* in North Carolina. *Plant Disease*, **75**, 733–6.

Robeson, D.J., Bretschneider, K.E. and Gonella, M.P. (1989) A hydathode inoculation technique for the simulation of natural black rot infection of cabbage by *Xanthomonas campestris*. pv. *campestris*. *Annals of Applied Biology*, **115**, 455–9.

Ryan, C.C. and Lopez, L.J. (1988) Control of leaf scald by cane stalk length treatment. *Sugarcane*, No. 6, p. 13.

Schaad, N.W., Gabrielson, R.L. and Mulanax, M.W. (1980) Hot acidified cupric acetate soaks for eradication of *Xanthomonas campestris* from crucifer seeds. *Applied and Environmental Microbiology*, **39**, 803–97.

Schaad, N.W., Sitterly, W.R. and Humaydan, H. (1980) Relationship of incidence of seedborne *Xanthomonas campestris* to black rot of crucifers. *Plant Disease*, **64**, 91–2.

Schaad, N.W. and White, W.C. (1974) A selective medium for soil isolation and enumeration of *Xanthomonas campestris*. *Phytopathology*, **64**, 876–80.

Schultz, T., Gabrielson, R.L. and Olson, S. (1986) Control of *Xanthomonas campestris* pv. *campestris* in crucifer seed with slurry treatment of calcium hypochlorite. *Plant Disease*, **70**, 1027–30.

Sharma, B.K. and Chauhan, M.S. (1985) Studies on the chemical control of foliar diseases of cotton in Haryana State. *Agricultural Science Digest*, **5(3)**, 153–6.

Singh, H. and Singh, N. (1988) Evaluation of some chemicals for the control of bacterial blight of cotton. *Pesticides*, **22**, 13–14.

Singh, J. and Swarup, J. (1986) Control of bacterial blight of cluster bean by seed treatment and foliar spray. *Pesticides*, **20**, 23–4.

Singh, R.B. and Jain, J.P. (1988) Chemical control of bacterial pustule of soybean. *Journal of Turkish Phytopathology*, **17**, 31–6.

Singh, R.P., Chowdhury, H.D. and Verma, J.P. (1977) Races of *Xanthomonas*

malvacearum. *Indian Phytopathology*, **30**, 572–3.
Srivastava, D.N. and Rao, Y.P. (1964) Seed transmission and epidemiology of the bacterial blight disease of rice in Northern India. *Indian Phytopathology*, **17**, 77–8.
Srivastava, S.S.L. and Bais, B.S. (1985) Chemical control of bacterial leaf pustule of soybean. *Indian Phytopathology*, **38**, 351–3.
Stephens, C.T. and Tuinier, J. (1989) Disease symptomatology and variation in susceptibility of seed-propagated hybrid geranium varieties to *Xanthomonas campestris* pv. *pelargonii*. *Plant Disease*, **73**, 559–62.
Strider, D.L. (1980) Control of bacterial leaf spot of zinnia with Captan. *Plant Disease*, **64**, 920–2.
Sun, Fuzai and He, Liyuan (1988) Studies on control of effects of Yukunin (fungicides) to black chaff (*Xanthomonas campestris* pv. *translucens*) of wheat. *Acta Phytopathol. Sinica*, **15**, 73–83.
Swanepoel, A. (1990) Bacterial blight in cotton. *Farming in South Africa*, G. 7, 1 p.
Tuinier, J.E. and Stephens, C.T. (1989) Use of serology to detect *Xanthomonas campestris* pv. *pelargonii* in aqueous extracts of geranium plants. *Plant Disease*, **73**, 875–8.
Verma, J.P. and Singh, R.P. (1974) Studies on the distribution of races of *Xanthomonas malvacearum* in India. *Indian Phytopathology*, **28**, 459–63.
Webster, D.M., Atkin, J.D. and Cross, J.E. (1983a) Bacterial blights of snap beans and their control. *Plant Disease*, **67**, 935–40.
Webster, D.M., Temple, S.R. and Galvex, G. (1983b) Expression of resistance to *Xanthomonas campestris* pv. *phaseoli* in *Phaseolus vulgaris* under tropical conditions. *Plant Disease*, **67**, 394–6.
Williams, P.H. (1980) Black rot: a continuing threat to world crucifers. *Plant Disease*, **64**, 736–42.
Young, J.M. (1978) Survival of bacteria on *Prunus* leaves. *Proceedings of the Fourth International Conference on Plant Pathogenic Bacteria*, **2**, 779–86.

8
Xanthan

I.W. Sutherland
Institute of Cell and Molecular Biology,
Division of Biology, Edinburgh University

8.1 INTRODUCTION

Xanthan is typically produced copiously as extracellular slime polysaccharide by the various pathovars of *X. campestris* and by some other *Xanthomonas* species. The production of this exopolysaccharide is manifested in the large mucoid colonies of *Xanthomonas* species on solid media and in the viscosity of liquid cultures. In the laboratory, the polysaccharide can be readily prepared in high yield then separated from the bacterial cells by high speed centrifugation. It is also an industrial product manufactured on a large scale by batch fermentation, the current estimated world production being at least 20 000 tonnes per annum.

There has been considerable work on the physiology of xanthan production by *X. campestris* and recently a number of the genes involved in xanthan synthesis have been identified. Because of its industrial applications and the very unusual properties of the polysaccharide in aqueous solution, xanthan has been the subject of a large number of structural and physicochemical studies.

8.2 STRUCTURE OF XANTHAN

The polymer can be readily precipitated from culture broths with polar solvents such as ethanol or isopropanol. Subsequent purification can

Xanthomonas.
Edited by J.G. Swings and E.L. Civerolo.
Published in 1993 by Chapman & Hall, London.
ISBN 0 412 434202

be achieved by a variety of techniques including fractional precipitation, chromatography and precipitation with quaternary ammonium compounds. The purified xanthan material can then be subjected to various analytical procedures to verify the composition and to standard techniques for the determination of polysaccharide structure. In procedures requiring prior acid hydrolysis to the monomeric constituents, problems may be encountered due to the relative stability of certain of the glycosidic linkages and the lability (and hence degradation) of some of the components. Acid hydrolysates of xanthan contain D-glucose, D-mannose and D-glucuronic acid in the molar ratio 2:2:1. Although other components have been reported in polysaccharides from bacterial isolates described as '*Xanthomonas*' species, the strains involved have generally been poorly characterized. Isolated polysaccharide from several of the other *Xanthomonas* species seems to share the same composition as that from pathovars of *X. campestris*. The primary structure of the polysaccharide is composed of a pentasaccharide repeat unit. This is effectively a main cellulose chain to which trisaccharide side-chains are attached at the C-3 position on alternate D-glucosyl residues (Jansson *et al.*, 1975; Melton *et al.*, 1976). Depending on the bacterial strain and on the physiological conditions for bacterial growth, the polysaccharide may carry varying amounts of O-acetyl groups on the C-6 position of the internal α-D-mannosyl residue and of 4,6-carboxyethylidene (pyruvate ketal) on the side-chain terminal β-D-mannosyl residue, respectively. Material from some strains, especially those in the *X. oryzae* group, carries two acetyl groups on the internal mannosyl residue and there is some evidence to suggest that certain polysaccharide preparations may even contain in excess of two moles of acetate per repeat unit. The term **xanthan** can thus be applied to polymers composed of a range of pentasaccharide repeating unit structures, all of which can be

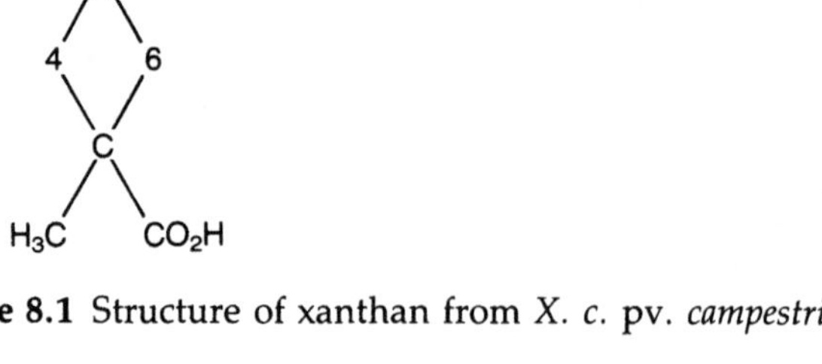

Figure 8.1 Structure of xanthan from *X. c.* pv. *campestris*.

represented as cellobiose carrying, at the non-reducing glucose residue, a trisaccharide side-chain of varying extent of acylation (Figure 8.1). A further variation in composition of xanthan occurs in the products from certain mutant strains in which the side-chains are truncated (Betlach *et al.*, 1987; Tait and Sutherland, 1989). These all lack pyruvate, but in addition may have lost the β-D-mannose **or** both the terminal D-mannosyl residue **and** the subterminal D-glucuronic acid (Figure 8.2). There is some indirect evidence to suggest that in some xanthan preparations there may be a small number of defective side-chains missing, thus possibly rendering portions of the molecule more susceptible to acid or to enzymic hydrolysis.

Xanthans are the substrates for a range of enzymes which have been termed **xanthanases**. There are mainly enzymes derived from mixed bacterial cultures, although some pure cultures of enzyme-synthesizing bacterial strains have been purified and identified as *Bacillus* species. The enzymes are induced following growth of the bacteria on xanthan as the sole carbon and energy source. Most enzyme preparations contain at least two different types of enzyme activity, although there may also be associated glycosidases (Figure 8.3). The name xanthanase should probably be used to describe enzymes with endo-1,4-β-glucanase activity cleaving the main chain of the polysaccharide. These enzymes will also degrade carboxymethylcellulose and other cellulose derivatives and

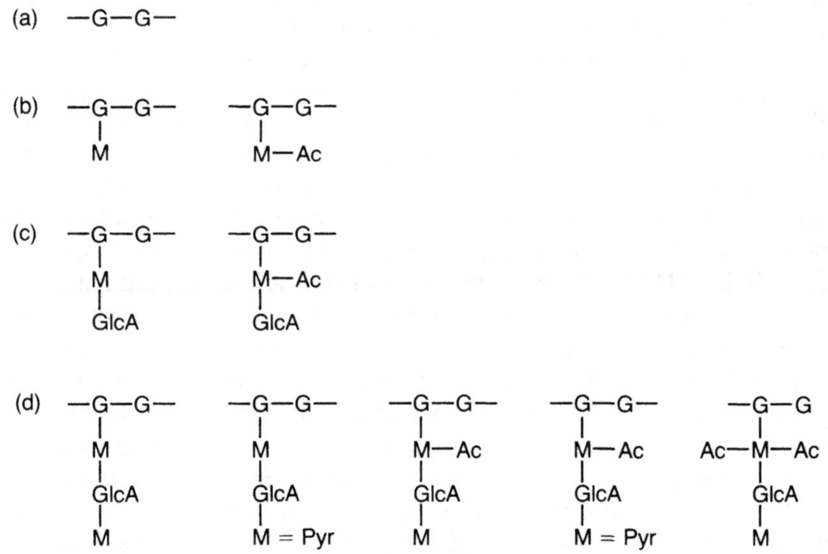

Figure 8.2 Structures of the repeat units occurring in preparations of xanthan from mutant (a–c) and wild-type (d) bacteria. M, D-mannose; G, D = glucose; Ac, acetyl; Pyr, pyruvate.

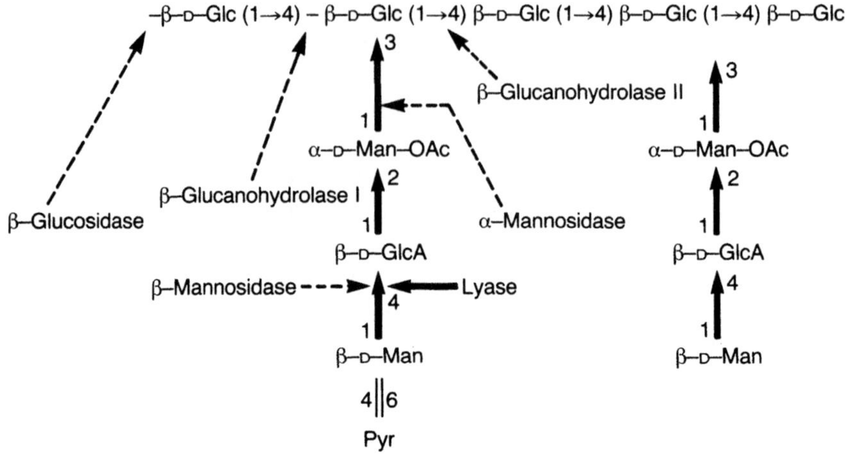

Figure 8.3 Enzymatic degradation of xanthan.

strictly speaking are therefore a type of 'cellulase'. The substrates are rapidly cleaved at random to yield oligosaccharides of different size; there is accompanying loss of solution viscosity. However, most of the commercial cellulase preparations derived from *Aspergillus* or *Trichoderma* species are devoid of activity against xanthan when the bacterial exopolysaccharide is in solution in the ordered form at ambient temperature. If the polysaccharide is maintained in the disordered form in ion-free water at temperatures near 50°C, limited degradation of the xanthan can be demonstrated by some of the heat-stable cellulases yielding free glucose and some oligomers of the repeat unit. As many of the xanthan molecules remain at least partially in the ordered form, they remain unhydrolysed. The second type of enzyme found associated with the bacterial preparations is a xanthan lyase (4,5-transeliminase) which cleaves the β-D-mannosyl-D-glucuronic acid linkage of the trisaccharide side-chains. This enzyme is unusual in that almost all enzymes of the lyase type which have been described so far cleave the **main chain** of their polysaccharide substrate. Also, the activity of the xanthan lyase does not appear to be greatly influenced by the presence or absence of pyruvate groups on the terminal mannose or of ester-linked acetyl groups on the internal mannose. The products of the **combined** action of the endoglucanase and the lyase are oligosaccharides containing the molar ratio D-glucose:D-mannose 2:1 and terminating in an unsaturated uronic acid residue (Sutherland, 1987). The only substrates which have been identified for the lyases are xanthans or high-molecular-weight fragments derived from them by main-chain cleavage. There is no action against closely related polysaccharide structures such as acetan from *Acetobacter xylinum* strains or the polymer from *Klebsiella aerogenes* type 5. Xanthan in which the D-glucuronic acid has been chemically converted to D-glucose is also resistant to hydrolysis by the enzymes, but this may be due to its very low water solubility rather than the altered structure. As so few of the enzymes have been purified, it is not yet clear whether the main chain is cleaved, then the oligosaccharides so formed are subject to the lyase action, or whether the lyase first acts on the xanthan molecule, making it perhaps more susceptible to cleavage by the endoglucanase. One complex enzyme preparation described by Cadmus *et al.* (1982) did cleave the side-chains from xanthan, but this material was not reported to contain any lyase activity.

The enzymes degrading xanthan are also useful in determining some of the structurally subtle, acyl modifications found on many xanthan molecules. The fragments generated are more suitable than undegraded polysaccharide for structural studies using fast atom bombardment spectroscopy. They have also proved useful in studies of monoclonal antibody specificity (Haaheim and Sutherland, 1989).

8.3 SYNTHESIS

8.3.1 Batch culture

It is possible to produce xanthan in the laboratory without difficulty using either batch or continuous fermentation. Growth of *X. campestris* under a wide range of cultural conditions is accompanied by copious production of exopolysaccharide. Indeed it has been said that in bacterial systems, 'exopolysaccharide production is a very efficient process' (Linton, 1990). Most, if not all, commercial production of xanthan is achieved in batch culture. Studies of this mode of growth have been made in complex media, thereby obscuring the effects of growth conditions on the composition and properties of the polymer formed. *X. campestris* will grow and produce exopolysaccharide on a wide range of carbon substrates, including amino acids, citric acid cycle intermediates and carbohydrates. The bacteria will use either ammonium salts or amino acids as sources of nitrogen. Although the structure of the polysaccharide formed is essentially independent of the substrate used, there may be some differences in acyl substituents and the physical properties of the product as well as in actual yields. Various ions are of course necessary for growth of the bacteria and equally for polysaccharide synthesis. Any limitation of ions needed for substrate uptake or for polymer synthesis can affect the yield and properties of the exopolysaccharide. Xanthan formation by *X. campestris* resembles several other bacterial polysaccharide-producing systems in that polymer production is favoured by a high ratio of carbon source/limiting nutrient such as nitrogen. Typically, media for laboratory synthesis of xanthan contain 0.1–0.2% ammonium salt and 2–3% glucose. Xanthan can indeed be produced in fairly good yield when *X. campestris* is grown in a simple synthetic medium composed of glucose, ammonium sulphate and salts. The quality and the final yield of xanthan may be enhanced by the addition of small amounts of organic acids or of citric acid cycle intermediates such as α-ketoglutaric acid. It has been suggested that their presence may lead to an improved metabolic balance between carbon flow from the carbohydrate substrate through the hexose monophosphate and Entner–Doudoroff pathways and the oxidation of the substrate through the tricarboxylic acid cycle. The energy required for xanthan synthesis is obtained from the oxidation of the carbon substrate, either through complete oxidation to carbon dioxide or through partial oxidation to products such as acetate or pyruvate which are components of normal xanthan. According to Jarman and Pace (1984), the proportion of the energy derived from partial oxidation of the carbon substrate depends both on the composition of the xanthan and on the ATP/O quotient. As *X. campestris* is a strict aerobe, oxygen is required both for growth and for xanthan production. Because of the increasing viscosity of the culture as xanthan is formed, oxygen may rapidly become limiting in shaken cultures. Thus, any laboratory culture has to be shaken at adequate speed if it is not to become oxygen-limited. Cultivation in a fermenter also has to be performed under conditions of vigorous aeration and the fermentation vessel has to be designed to ensure that there is a minimum dead space which would otherwise lead to stagnant areas of culture. Because of the effect of oxygen limitation on xanthan production, it is important that the efficiency of energy production and conservation is determined and that the true requirements of energy and oxygen for bacterial growth and for polysaccharide production are known. The process of xanthan production is complicated by mass transfer reactions in the high viscosity broths. Oxygen solubility decreases with increasing xanthan concentration and the diffusion constant of oxygen in dilute solutions of the polysaccharide is reduced relative to water (Ho *et al.*, 1988). However, in more concentrated solutions, where the polymer concentration exceeds 0.5%, the oxygen diffusion coefficients are higher than those obtained in water. The presence of polysaccharide in the fermentation broth may also affect the availability of carbon substrates such as glucose and sucrose by interaction with these carbohydrates.

In the design of the fermentation vessel, recent results from Nakajima *et al.* (1990) have suggested that the volume exposed to high shear is of critical importance. Thus, the specific rate of xanthan production depends on the volume of the high shear region. Under oxygen limitation, the specific rate of xanthan synthesis is a function of the oxygen

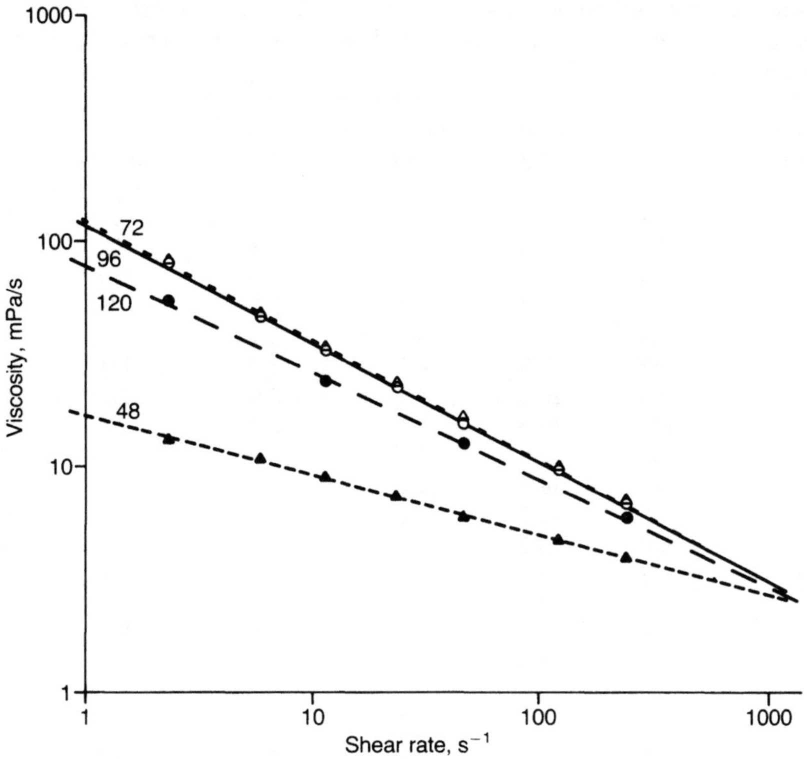

Figure 8.4 Viscosity of xanthan obtained from sulphate-limited batch culture at different times during growth.

transfer rate. There is also an effect on molecular weight which, under oxygen limitation, is reported to decrease linearly with the specific oxygen uptake rate (Suh *et al.*, 1990).

Xanthan is produced throughout growth of *X. campestris* as well as in the stationary phase. The specific rate of xanthan synthesis by *X. campestris* is closely related to the growth rate in batch culture; it is maximal during exponential growth and minimal during the stationary phase. In this respect, xanthan production resembles the synthesis of several other bacterial exopolysaccharides. However, differences can be seen in both the viscosity (Figure 8.4) and the acylation (Table 8.1) of xanthan synthesized in nutrient-limited media during various phases of the growth cycle. Although the reasons for these variations are not clear, it may be surmized that, at certain stages of growth, there may be only a limited availability of certain of the activated precursors required for polysaccharide synthesis. Acetyl CoA and phosphoenolpyruvate may not be readily available during certain stages of growth to permit the complete acylation of the xanthan repeating units. The pyruvate

Table 8.1 Composition and viscosity[a] of xanthan from sulphur-, ammonium-, phosphorus-, magnesium-, glucose- and iron-deficient batch cultures of *X. campestris* S459

Medium lacking	*Time (h)*	*GlcA (%)*	*Acetate (%)*	*Pyruvate (%)*	k^b *(mPa/s)*	n^c
S	48	14.2	2.9	3.3	17	0.73
	72	15.4	4.1	3.8	122	0.47
	96	14.5	4.0	3.3	119	0.47
	120	18.4	4.1	2.7	78	0.52
NH_4	48	14.5	3.7	2.9	21	0.69
	72	15.9	4.5	3.0	134	0.45
	96	16.6	4.4	2.6	113	0.49
	120	15.3	4.4	2.2	76	0.54
P	48	12.9	2.7	–	9	0.79
	72	14.4	4.0	–	76	0.52
	96	16.7	4.0	3.6	82	0.52
	120	16.3	4.1	2.8	122	0.46
Mg	48	9.8	1.9	–	5	0.83
	72	11.9	4.0	–	94	0.51
	96	16.1	4.1	3.8	139	0.44
	120	15.6	4.0	3.2	96	0.49
Glc	48	–	–	–	–	–
	72	14.3	3.5	–	109	0.47
	96	16.4	4.2	3.8	128	0.45
	120	15.2	4.0	3.1	62	0.55
Fe	48	–	–	–	–	–
	72	14.9	4.2	–	120	0.46
	96	14.1	4.3	4.0	161	0.42
	120	14.8	3.9	3.0	101	0.49

[a] 0.1% (w/v) Sodium-form xanthan in 9.6 mM NaCl (25°C).
[b] *k*, Consistency index.
[c] *n*, Pseudoplasticity index (relative units).

and acetate contents of xanthan (Table 8.1) peak in polysaccharide synthesized immediately after the end of exponential growth (72 h), whereas acyl contents are generally lower during exponential phase growth. Consequently, the xanthan synthesized in batch culture represents the products of **all** phases of growth and is likely to be a mixture of molecular types. The differences in the polysaccharide may well be accentuated by different strains as well as by different physiological conditions.

In batch culture exopolysaccharide production was greatest in ammonium-deficient medium and lower in sulphur- or magnesium-deficient media (Tait *et al.*, 1986). Similar results were obtained in an earlier study (Davidson, 1978). Less polymer was formed using phosphorus or iron deficiency and least of all under carbon-deficient conditions.

8.3.2 Continuous culture

In the research laboratory, continuous culture provides a number of advantages over batch culture for the study of exopolysaccharide production. It provides high productivity from the fermenter and the polymer formed can be obtained under highly standardized physiological conditions and should thus be much more uniform in its physical and chemical characteristics. One problem in using continuous culture for polysaccharide production is the inefficient stirring and mixing of the fermenter vessel. Any dead space behind baffles, etc., tends to accumulate very high polysaccharide concentrations and become increasingly stagnant. Sampling of the high viscosity culture and maintenance of culture purity may also be difficult. The capacity of the fermenter can be much smaller than would be required for a batch culture yielding the same amount of polysaccharide.

Various limiting nutrients have been used for the study of xanthan production in continuous culture, those most favoured being nitrogen, sulphur and magnesium. The effect of dilution (growth) rate on xanthan production in continuous culture has been examined for a number of strains and different limiting nutrients. Within the range $D = 0.03$–$0.07\,h^{-1}$, sulphur-limited cultures of *X. campestris* had very similar acyl contents, but the rheological characteristics differed with a change in dilution rate although the actual yield of polysaccharide was almost constant. Similarly in the range of specific growth rates from 0.05 to $0.2\,h^{-1}$, nitrogen-limited cultures differed little in their specific rate of xanthan synthesis. When complex media were used for continuous culture, the concentration of xanthan decreased with increasing dilution rate (Silman and Rogovin, 1972). In this study, pH was not controlled and increased with the dilution rate.

The conversion of substrate to xanthan in continuous culture can be very high. Values of 50–60% have been obtained in synthetic media under nitrogen limitation. Modification of such media by the addition of urea and distillers' solubles increased the conversion of substrate to xanthan up to values of 68% or higher. There are, however, marked differences between strains and the highest polysaccharide yields have been obtained in complex rather than synthetic media. It is also clear that the high specific rates of exopolysaccharide synthesis which are obtainable in continuous culture of *X. campestris* demand a considerable energy input. Jarman and Pace (1984) estimated that in sulphur-limited cultures, energy metabolism was inefficient and the bacteria exhibited either a low P/O ratio or a low Y_{ATP}. The maximum theoretical yield from glucose was 0.81–0.87 g per mol glucose for a P/O ratio of 1 to 3, whereas the actual yield of polysaccharide was 0.62. More recent studies on the growth efficiency of *X. campestris* in glucose-limited continuous culture (Rye *et al.*, 1988) produced molar growth yields of 53.6 g dry weight of bacteria per mol glucose and 26.4 g dry bacteria per mol oxygen. These values indicated an ATP/O quotient of approximately 1. Correction for polymer production only increased this value slightly.

Although single-stage continuous culture has been suggested in the patent literature and in various unpublished reports, it has apparently not been adopted for the industrial production of xanthan. Probable reasons for this include the complexity of equipment needed and the consequent high level of operator training, as well as the extensive testing of fermenter design and the problems of maintaining strain purity and culture sterility.

8.3.3 Industrial substrates

It should not be forgotten that in the laboratory, pure substrates such as glucose, sucrose, etc., are used whereas in the industrial situation different substrates are employed. These must be plentiful and cheap carbon sources which frequently include starch, starch hydrolysates, corn syrup, molasses, glucose and sucrose (derived from either sugar-beet or from sugarcane). Less pure products may be acceptable, in which case cruder, cheaper substrates can replace those listed above. Some may indeed be waste products from other industries. They include whey, cereal grain hydrolysates and dry milled corn starch. As the optimal synthesis of xanthan requires a balance between the utilizable carbon and nitrogen sources, care has to be taken to obtain consistency in yield and product quality when using some of these cruder substrates which may contain nitrogen in addition to carbohydrate. The nutritional versatility of *X. campestris* is clearly a major factor in favour of its use for commercial xanthan production. However, the **quality** of the xanthan produced from different substrates may vary considerably; the molecular weight and hence the rheological characteristics of xanthan synthesized from glucose or starch may well differ from that formed when whey or other proteinaceous material is employed. The nitrogen sources used for industrial production of xanthan may include yeast hydrolysate, soybean meal, cottonseed flour, distillers' solubles or casein hydrolysate. In batch culture for industrial production of the polysaccharide, there must be careful control of pH and of the aeration rate; indeed, because of the viscosity of the culture fluids, adequate oxygen transfer may be difficult to achieve unless the fermentation vessel has been carefully designed to ensure that mixing is optimal. The conditions used for polymer production and processing must be carefully standardized to ensure that product yield and quality are consistent.

Recovery of the polysaccharide from industrial cultures requires removal or destruction of the cells (if possible) followed by precipitation with polar organic solvents such as ethanol or isopropanol. The solvents are later recovered for reuse by distillation to reduce the cost of the process. Bacterial cells are difficult to remove from the highly viscous culture broth and it may, instead, be pasteurized to kill the bacteria and also to destroy enzymes such as cellulases, which might interfere with applications of the xanthan involving addition to or mixing with cellulosic compounds such as carboxymethylcellulose. The pasteurization process may also improve the subsequent separation of cells from polymer and the product quality. Polysaccharide recovery in the presence of polar solvents may be also increased or accelerated by the addition of electrolytes. After recovery the xanthan is spray dried and ground to yield a powder of uniform size and acceptable solubility. Alternatively, polymer concentrates may be prepared in liquid form by tangential flow filtration. These preparations containing about 12.5% dry weight of polymer are stabilized by the addition of appropriate antibacterial agents. They are reasonably stable if held at ambient temperature or below.

8.4 BIOSYNTHETIC PROCESS

Those bacterial polysaccharides found as cell wall components or extracellular products and composed of regular repeating units mainly share a common mode of biosynthesis. A series of membrane-bound enzymes utilizes various activated carbohydrate donors to form the polysaccharide on an acceptor molecule. The oligosaccharide repeat units of xanthan are formed by the sequential addition of monosaccharides from the energy-rich sugar nucleotides (usually nucleoside diphosphate sugars) to a C_{55} **isoprenoid lipid** acceptor molecule. At the same time, acyl adornments are added from appropriate activated donors. Thus, the xanthan backbone is formed by the sequential addition of D-glucose-1-phosphate and D-glucose, respectively, from two moles of UDP-D-glucose. Thereafter, D-mannose and D-glucuronic acid are added from GDP-mannose and UDP-glucuronic acid, respectively. Each step requires a specific enzyme and a specific substrate. Absence of the enzyme (or the substrate) inhibits synthesis of the polysaccharide. Depending on the strain used and the physiological conditions under which the bacteria have been grown, and hence on the exact structure of the polymer formed, O-acetyl groups are transferred from acetyl CoA to the internal mannose residue and pyruvate, from phospho-

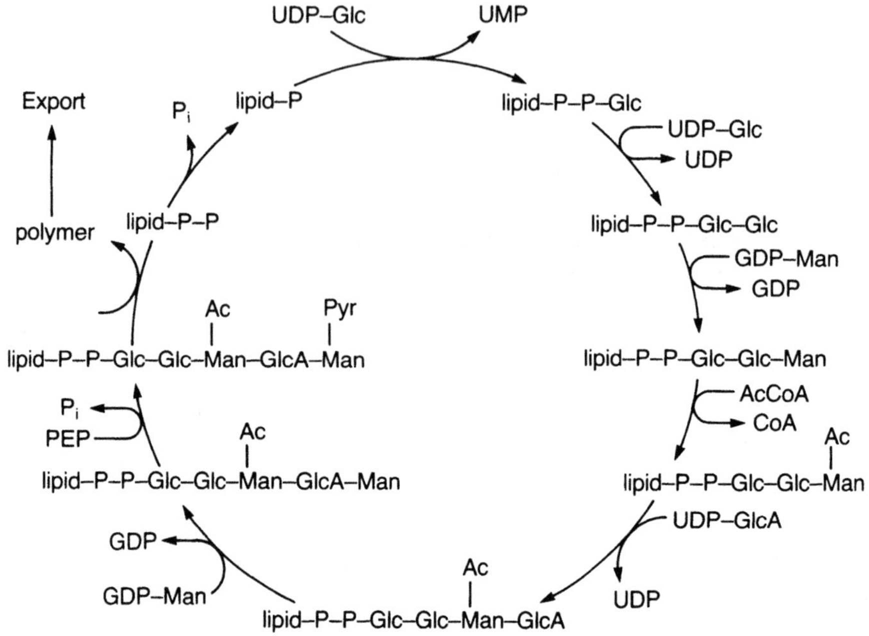

Figure 8.5 Biosynthesis of xanthan from activated precursors of carbohydrates and acyl groups (results of Ielpi *et al.*, 1981, 1983).

enolpyruvate (PEP), is added to the terminal mannose. This sequence of reactions was elegantly demonstrated by Ielpi and his colleagues (1981, 1983) using permeabilized cells and radioactively labelled precursors and is indicated in Figure 8.5.

In the initial reaction, the transfer of the sugar-1-phosphate from UDP-glucose, UMP is inhibitory, whereas the transfer of the second sugar can be inhibited by high concentrations of UDP. After the pentasaccharide repeat unit has been formed, oligomers are produced by transfer to other lipid intermediates, gradually increasing the size of the carbohydrate chain. This stage of biosynthesis has been examined in other bacteria such as *Klebsiella aerogenes*, but not in *X. campestris*. It normally involves addition of the longer oligosaccharide sequence to the **non-reducing terminus** of a single repeat unit attached to the isoprenoid lipid, simultaneously releasing a mole of isoprenoid lipid diphosphate. The inactive lipid carrier is dephosphorylated to yield isoprenyl phosphate which can then re-enter the biosynthetic sequence. Although the structure of the repeating unit is determined by the sequential transfer of the different monosaccharides and acyl groups from their respective donors by highly specific sugar **transferases**, the **polymerase** enzyme responsible for polymerization of the pentasaccharides into a macromolecule of $M_r > 10^6$ is less specific. This is clear from the ability of mutants to form xanthans defective in their trisaccharide side-chains. These mutants do not, however, yield as much polysaccharide as the wild-type bacteria. Such a lack of absolute specificity may also account for the natural evolution of bacterial exopolysaccharides such as xanthan and for the existence of families of structurally related but non-identical microbial polysaccharides. Examples are to be found in the polymers from certain strains of *Acetobacter xylinum* which share a tetrasaccharide structure of their repeating unit with xanthan and which are synthesized as alternatives to the bacterial cellulose normally produced by these bacteria. The side-chains of the exopolysaccharide from *A. xylinum* are larger than the trisaccharides of xanthan and differ in the terminal disaccharide. Whether there are other bacterial polysaccharides of the same general structural type as the xanthan 'family' remains to be seen.

The final stages of exopolysaccharide secretion from the cytoplasmic membrane, passage across the periplasm and outer membrane and finally excretion into the extracellular environment, are much less well defined than the earlier biosynthetic steps. Mechanisms must exist in all bacterial exopolysaccharide-synthesizing systems for releasing polymer from the isoprenoid lipid prior to transport to its final destination. The process undoubtedly requires an energy source and may well be analogous to the export of lipopolysaccharide to the outer membrane, in which ATP provides the energy. Recent studies with *X. campestris* and other bacterial systems have revealed mutants which are unable to excrete polysaccharide even though the biosynthetic pathway is complete. In such mutants, the polymer may be found in the periplasm or in the cytoplasm of the bacteria – inevitably with lethal results. There is, thus, a well-organized transport system in the original bacteria, capable of ensuring that the polymer is released from the carrier lipid, transported directly across the outer membrane and excreted into the extracellular environment. Several mutations can effectively disrupt this process and lead to internalization of the polysaccharide. The genetic system controlling xanthan production in *X. campestris* has been elegantly elucidated by Vanderslice *et al.* (1989). It differs in some respects from systems recently described in other bacteria (Boulnois and Jann, 1989). A linear sequence of 12 genes responsible for xanthan production was found. This included seven gene products needed for monosaccharide transfer and for acylation at the lipid intermediate level to form the completely acylated repeat unit (Figure 8.6 and Table 8.2). Interestingly, two acetyl transferases were found, one of which appears to function when the pyruvate transferase is defective. The region of the chromosome controlling xanthan synthesis comprised 13 kb. This is slightly smaller than the regions, each of *ca* 17 kb, re-

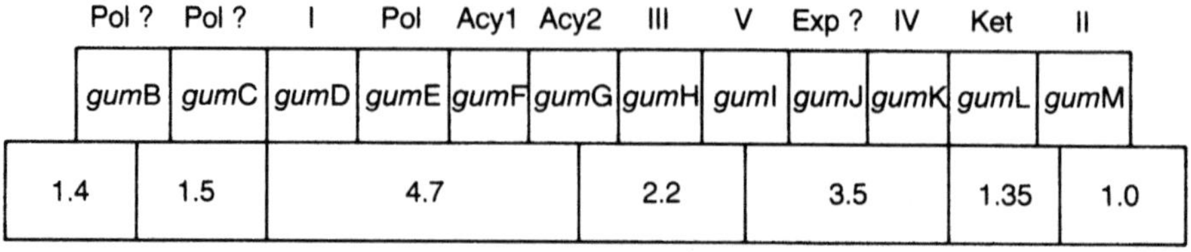

Figure 8.6 Genetic map of the genome segment of *X. campestris* involved in xanthan production. The *Bam*HI restriction map (bottom) indicates the order and the approximate size (in kilobases) of the fragments of the 16 kb of DNA of the xanthan gene cluster. The genetic map (centre) indicates the 12 separate xanthan genes, designated *gum*B to *gum*M. The biochemical functions (top) indicate the enzymic activity identified for each gene: I–V, transferases I–V; Acy, acetylases (1 and 2); Ket, ketalase; Pol, polymerase; Pol?, possible involvement with polymerization or export; Exp, post-polymerization processing or export.

Table 8.2 The functions of the *X. campestris* genes controlling xanthan synthesis

Enzyme activity	*Gene*	*Polymer synthesis*	*Lipid fraction*
Glc transferase I	D	0	No sugar
Glc transferase II	M	0	Glucose
Mannose transferase I	H	0	Cellobiose
GlcA transferase	K	+	Trimer
Mannose transferase II	I	+	None detected
Polymerase	E	–	Pentasaccharide
Wild type		+	None detected

sponsible for exopolysaccharide synthesis in *Escherichia coli* and in *Haemophilus influenzae*, respectively (Boulnois and Jann, 1989). However, other genes, responsible for less well-defined functions, may yet be identified in *X. campestris*.

The genetic control of the later stages of xanthan biosynthesis and export has not yet been elucidated, but it presumably includes features common to those exopolysaccharide-producing bacteria which have been studied in detail. Thus, in *E. coli*, a region of the chromosome extending to 11.6 kb codes for five proteins, expression of which appears to be necessary for exopolysaccharide production. These proteins **may** form part of a multicomponent system responsible for exporting the polymer from the site of polymerization at the cytoplasmic membrane to the exterior of the bacterial cell. One of the proteins involved does appear to have attributes common to other ATP-utilizing systems and it may be concerned with excretion. Certainly, lipopolysaccharide, which is synthesized by similar mechanisms to exopolysaccharide, does require ATP for its export and for the translocation of core material to the outer membrane.

Equally, **regulation** of xanthan synthesis is not yet well defined. Many of the enzymes needed for the formation of precursors not specifically associated with exopolysaccharide production appear to be under separate control. In contrast, the gene products needed solely for xanthan manufacture by the bacterium occur in the linked group of genes shown in Figure 8.6. In *X. campestris*, there are no significant competing demands for substrate as any carbohydrate supplied to the bacteria is metabolized to provide energy, or converted either to essential cell components or wall material or to exopolysaccharide with high efficiency. There are effectively no intracellular storage polymers competing with exopolysaccharide in their synthetic demands on substrate. Also, *X. campestris* only produces xanthan, unlike *Rhizobium* and other bacterial species where two or even three extracellular polymers, together with intracellular oligosaccharides, may be produced by a single microbial cell type.

Recent studies by Tang *et al.* (1991) report the existence of a region of the genome containing a cluster of genes in which mutation leads to failure to produce xanthan **and** a number of enzymes involved in pathogenesis of *X. campestris*. The genes were designated *rpf* (regulation of pathogenicity factors) and control various extracellular enzymes including cellulase (endo-1,4-β-glucanase), polygalacturonate lyase, amylase and protease. The gene cluster probably contains seven genes involved in the **positive** regulation of these factors. Another **negative** system coordinately repressing xanthan and extracellular enzyme synthesis has also been reported (Tang *et al.*, 1990).

8.5 PHYSICAL PROPERTIES OF XANTHAN

The industrial uses of xanthan all derive from the physical properties of the polysaccharide. The techniques applied to determine these physical attributes, together with a thorough knowledge of the chemical structure of the exopolysaccharide, have enabled the assembly of much information on structure/function relationships pertaining to xanthan and provided a model for other microbial exopolysaccharides. Electron microscopy has indicated the persistence length of the molecule as well as information as to whether the polysaccharide in solution exists in a single-stranded or double-stranded form. Values for the persistence length obtained using other techniques such as light scattering and viscometry in dilute solution range from 110 to 150 nm. There are, however, limitations on the interpretation of the data obtained from

Table 8.3 Molecular weight of xanthan. Many of these results represent reports from various authors. Light scattering has been the most widely used method

A. Application of different methods

Method	*Value obtained*
Contour length (electron microscopy)	$4–20 \times 10^6$
Diffusion and sedimentation	1.4×10^6
Intrinsic viscosity	62×10^6
Light scattering	$2.9–7.0 \times 10^6$
	(extremes of 1.1×10^6 and 47×10^6)
Low angle light scattering	$1.5–5.2 \times 10^6$
Sedimentation coefficient	14.8×10^6

B. Static light scattering measurements

Strain	*Molecular weight* $\times 10^6$
X. campestris pv. *campestris* 646	0.9–1.2
X. campestris pv. *phaseoli* 1128	1.27
X. campestris pv. *phaseoli* 556	1.48
X. oryzae pv. *oryzae* PXO61	1.60
Results of Shatwell *et al.* (1989)	
Translational diffusion coefficient (and intrinsic viscosity)	2.16×10^6

electron microscopy. Artifacts may be introduced either during the initial recovery of the polymer, or during sample preparation.

A wide range of values has been proposed for the molecular weight of xanthan. This has resulted partly from the application of different techniques and also from the tendency of the polysaccharide strands to aggregate in solution. The most consistent results have been obtained using light scattering techniques (Table 8.3). In a recent study, Shatwell (1990) noted a range of $0.9–1.6 \times 10^6$ daltons for four xanthans with differing acyl substitution from different pathovars prepared under standard cultural conditions. Others have obtained estimated results between the extremes of 1.1 and 47×10^6 daltons; clearly these very high values must be of dubious accuracy. Measurements of contour length similarly produced a wide range of estimates ($4–120 \times 10^6$).

X-ray fibre diffraction studies on xanthan indicate that it forms aligned helices in which there is poor lateral packing. This is consistent with an absence of crystallization and of gelation. The polysaccharide adopts a right-handed helical conformation with five-fold symmetry in which the helix pitch is 4.7 nm (Moorhouse *et al.*, 1977). Raising of the temperature of aqueous solutions of xanthan eventually produces a partial melting of the double helix. Xanthan is considered to adopt an ordered double helical conformation when in solution at lower temperatures, although this view is still not unanimous. This conformation causes the apparent 'stiffness' of the polysaccharide at low temperature or in high ionic strength. This stiffness is higher than that estimated for most other macromolecules and is comparable to that of fungal β-D-glucans which exist as triple helices (Norisuye *et al.*, 1980). The conformation is markedly affected by the presence of ions; salt increases the melting temperature and reduces the $[\,[\,]_{365}$ in the low temperature plateau region of the curve in relation to the charge carried by the polysaccharide. The conformational change, the shift from order to disorder, can be displaced to higher temperatures by the presence of increasing amounts of ions (Figure 8.7; Shatwell *et al.*, 1990). Acyl groups also have a noticeable influence on the conformational changes but results from light scattering experiments indicate that the inherent stiffness of the polysaccharide molecule is not greatly influenced by the pattern of acylation. If acyl groups are present, there is a significant correlation with the mid-point temperature of the conformational transition. The presence of acetyl groups on the xanthan structure has a stabilizing effect on the ordered form of the polysaccharide whereas pyruvate groups have the opposite effect. This can be attributed to intramolecular electrostatic repulsion between pyruvate groups. Conversely, the apolar interactions of the methyl groups of the acetate effect stabilization.

Aqueous solutions of xanthan are highly pseudoplastic. This is true whether or not acetyl and ketal substituents are present. Indeed, the absence of the acyl groups does not normally affect the solubility of xanthans although in an ion-free solution xanthan which is pyruvylated but not acetylated is poorly soluble. The viscosity, the degree of pseudoplasticity and the value for the transition from soft gel to pseudoplastic behaviour are all directly related to the polysaccharide concentration. The effect of addition of salts to the solution is also dependent on polymer concentration. At *ca* 0.3% (w/v) xanthan, addition of salts has practically no measurable effect but at higher concentrations increased viscosity may be observed when salts are present, presumably due to alteration of the macromolecular association. In dilute solutions, viscosity decreases in the presence of salts. This loss in viscosity may be due to decreased hydrodynamic volume resulting from charge shielding. There is probably no significant difference between the viscosity and the degree of acylation. Acetyl and pyruvate contents appear to have no influence either on dilute solution viscosity of xanthan or on the intrinsic viscosity at a given molecular weight, although some studies have suggested that under certain specific experimental conditions acylation may affect these values. Care must be taken in the interpretation of results as different xanthan samples may vary considerably in their

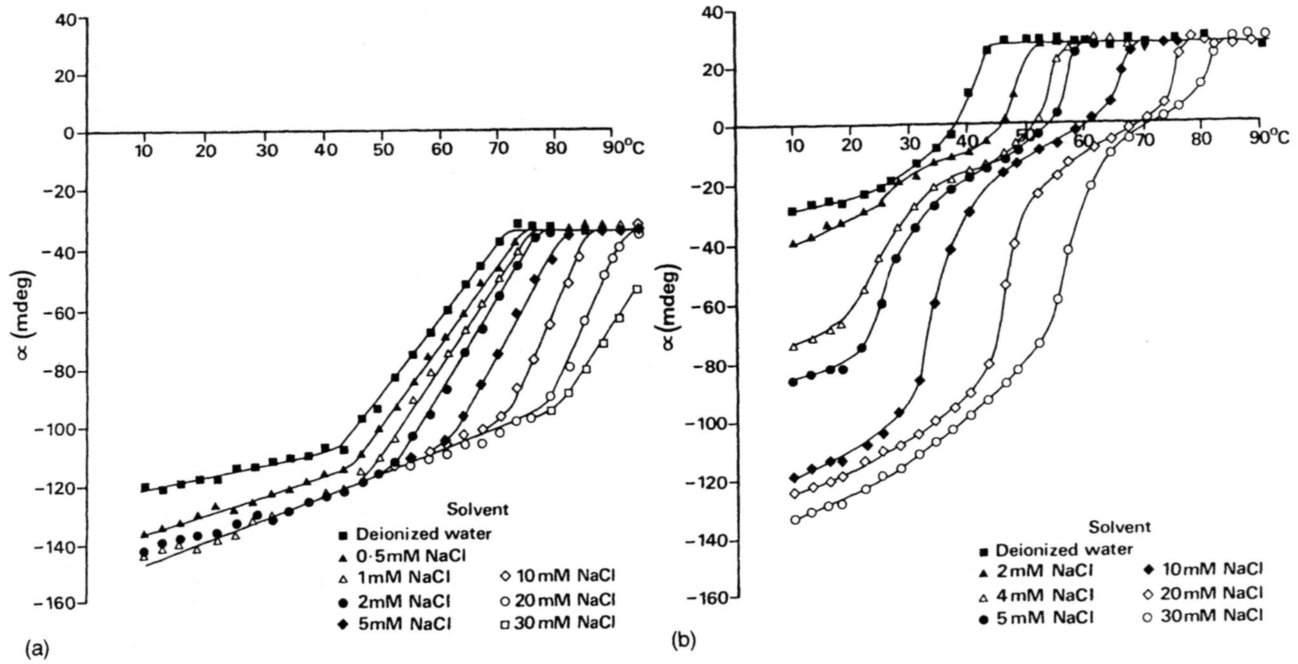

Figure 8.7 Effects of salts on the transition of xanthan from the ordered to the disordered state. Results for two different xanthans (61-acetylated, non-pyruvylated (a); 556-non-acetylated, pyruvylated (b)) monitored by optical rotation measurement (results of Shatwell, 1989).

molecular weight. The viscosity of truncated xanthans in which the terminal mannosyl residue is missing is much lower than the wild-type material, but, surprisingly, the 'polytrimer' in which the terminal disaccharide is missing is more viscous than the original xanthan (Betlach *et al.*, 1987; Tait and Sutherland, 1989).

The very high stability of xanthan molecules even when exposed to relatively high temperatures is especially important in the proposed usage of the polysaccharide for enhanced oil recovery. When stored at high temperature under conditions where oxygen is excluded, the xanthan solution gradually loses both viscosity and carbohydrate material. These losses increase with increased temperature, the main mechanism of degradation being apparently through random cleavage of the double-stranded xanthan chain. The acyl groups are lost very rapidly at temperatures of 60°C and above (Kierulf and Sutherland, 1988). Electron microscopy shows that heating at 80°C leads to dissolution of aggregates and this probably accounts for the **increase** in viscosity seen in some preparations. If the samples are held at the same temperature for longer periods of time or are kept at higher temperatures, the average contour length decreases (Table 8.4). When oxygen is present, the main mechanism of destruction of the xanthan molecule is probably through free radical attack. The products of such treatment are not recognizable oligosaccharides or monosaccharides but are all molecules of low molecular weight. This is in contrast to mild acid hydrolysis or to acetolysis, when the constituent monosaccharides together with a series of mannose-containing oligosaccharides and cellobiose are released (Lawson and Symes, 1977).

Aqueous solutions of xanthan when mixed with certain plant gluco- or galactomannans can be induced to form gels, whereas neither of

Table 8.4 Contour lengths of heated xanthan (results of Stokke *et al.*, 1989)

Treatment	*L_n (nm)*		*L_w (nm)*	
None	1285[a]	–[b]	2021[a]	–[b]
1 month, 80°C	–	462	–	876
5 months, 80°C	792	–	1119	–
12 months, 80°C	488	–	916	–
11 months, 90°C	–	508	–	932
2 months, 100°C	277	–	464	–
5 months, 100°C	–	432	–	709

L_n = number average contour length.
L_w = weight average contour length.
[a] Flocon; [b] xanthan 646.

these components will gel on its own. The mixed solution must be heated above the transition temperature of the xanthan in order to denature the xanthan helix and then allowed to cool. X-ray fibre diffraction studies on the mixed gels (Morris and Miles, 1986) reveal unique fibre patterns. Previously, gelation of the polymer mixtures had been attributed to the interaction between the helix of the microbial polysaccharide and unsubstituted regions of the plant glycomannan backbone. These results indicate that intermolecular binding could be due to co-crystallization resulting from the stereochemical compatibility between the cellulosic and mannan backbones. Acetyl groups play an important role on the xanthan molecules in inhibiting gelation of the polysaccharide mixtures and increasing the polymer concentrations required (Shatwell *et al.*, 1990, 1991).

8.6 INDUSTRIAL USAGE OF XANTHAN

Most of the industrial applications of xanthan derive from its ability to dissolve in hot or cold water to yield high-viscosity, pseudoplastic solutions, even at low polysaccharide concentrations. The xanthan solutions are compatible with acids, bases and salts and can remain stable for long periods at ambient temperatures.

Xanthan is used in the oil industry as a lubricant in oil drilling 'muds' where its capacity for suspending barytes and its viscosity regain after shearing is particularly valued. The pseudoplasticity of the xanthan solution provides low viscosity at the high shear region of the drill bit together with high viscosity in the annulus where the shear is low (Kang and Cottrell, 1979). One major potential application for xanthan is in **enhanced oil recovery**. As the reservoir pressure falls, water may be injected to displace the residual oil. However, the viscosity of the oil may be greater than water, leading to 'fingering' and failure to displace significant amounts of oil. This can be overcome by increasing the viscosity of the water with xanthan or other agents. Despite its relatively high cost and the consequent additional expense per barrel of oil produced, xanthan has the advantage of compatibility with high concentrations of total dissolved salts, conditions which are unfavourable to the use of synthetic polymers such as polyacrylamides. It can also produce solutions of relatively high viscosity at low polysaccharide concentration. Further, within certain limits, the viscosity is relatively insensitive to salts, pH and temperature. Ideally, xanthan for this purpose should be of high viscosity and free from bacterial cells which would otherwise block the rock pores of the oil reservoir. In practice, this is difficult to achieve and injectivity may be less than desired, but fermentation conditions and strains which yield small bacterial cells can be chosen. Biocides are necessary to ensure that biological degradation of the xanthan does not occur either at the surface or underground after injection. In the latter case, the relatively shallow, low-temperature reservoirs of the USA and other areas present this specific problem. In the deep, high-temperature, UK and Norwegian offshore oil reservoirs, the major difficulty is one of temperatures in excess of 80°C in which chemical degradation of polysaccharides presents a greater problem.

Another use for xanthan in the oil industry is in the process of selective pore blockage. The ability to inject xanthan solutions into the reservoir rocks, to place them selectively into zones of high permeability and then to cross-link them with Cr^{3+} is utilized. Once in place, the gels have a long stability. If there is later need to reopen the structure, the xanthan gel can be destroyed in a controlled manner with oxidizing agents (e.g. sodium hypochlorite) or acids. The gels are compatible with a wide range of salinities, over a pH range from 5 to 8 and at temperatures up to or even in excess of 65°C. The temperature at which the gels remain stable will depend on the salinity of the environment in which they are placed. When sheared, the gels will break and re-form. In this role as a gellant, the solubility of the polysaccharide in freshwater is very important, allowing it to be selectively placed initially. The rate of gellation must be controlled carefully. Once the gel has been formed, connate water or sea-water can be used. The aim is usually not to block the structure entirely, but to retain some residual permeability. It has been estimated that one well may require up to

Table 8.5 Industrial applications of xanthan

Physical properties	*Usage*
Suspension	Laundry chemicals
	Herbicides
	Liquid feed supplements
	Paper coatings
	Flowable pesticides
	Agrichemical sprays
Flocculant	Water clarification
	Ore extraction
Shear thinning/viscosity control	Oil drilling muds
Viscosity/cross-linking	Hydraulic fracturing
Viscosity control	Abrasives
	Jet printing
Stabilization	Thixotrophic paints
Foam stabilization	Fire-fighting fluids
Gelling agent	Explosives
	Blockage of permeable zones in oil reservoirs

1.35 mkg of polymer. Originally, it was thought that the demand for xanthan from the oil industry would reach 60 000 tonnes per annum by the mid 1990s, but this forecast has been proved to be highly optimistic. As a result, several of the commercial producers have ceased production in the last few years.

There are, however, many other industries which make use of the physicochemical properties of xanthan (Table 8.5). It has been incorporated as a stabilizer into emulsion paints and laundry starch as well as a carrier for agrochemicals such as herbicides and fertilizers. Xanthan can be incorporated into clay coatings to provide high-quality finish on papers; it is also a very good thickening agent for textile dyes in which it is incorporated to control rheological properties during textile printing. There are also numerous patents relating to derivatives of xanthan, but it is not clear whether this area has indeed provided a major commercial outlet for the polysaccharide.

8.7 FOOD USES OF XANTHAN

Many of the industrial applications of xanthan were originally in the food industry. It received approval for food use from the US FDA in 1969 and about 60% of the xanthan currently produced is probably food-grade polymer. Polysaccharides are incorporated into foods in order to alter the rheological properties of the water present. The texture of the product can therefore be changed. Thus, xanthan has found applications utilizing many of its physical properties (Table 8.6). In many foodstuffs, xanthan possesses further useful attributes, including rapid flavour release, good mouthfeel and compatibility with other food ingredients including proteins, lipids and polysaccharides.

Table 8.6 Different properties of xanthan used in foods

Function	*Application*
Adhesive	Icings and glazes
Binding agent	Pet foods
Coating	Confectionery
Emulsifying agent	Salad dressings
Encapsulation	Powdered flavours
Film formation	Protective coatings, sausage casings
Foam stabilizer	Beer
Stabilizer	Ice cream, salad dressings
Swelling agent	Processed meat products
Syneresis inhibitor	Cheeses, frozen foods
Thickening agent	Jams, sauces, syrups and pie fillings

It has to be remembered that many foodstuffs already contain polysaccharides such as starch or pectin in addition to proteins and lipids and it is particularly important that any added polymer such as xanthan should be totally compatible with them. Some of the foodstuffs to which polysaccharides are added are of a low pH. These include salad dressings, relishes and yoghurts. Evidently, any polysaccharide which is incorporated into foods of these types must be acid-stable. Xanthan, which is stable over a wide range of pH values, is thus very well suited to such applications.

Polysaccharides are added to foods to provide suitable rheological properties during processing and a final appearance which will enhance consumer appeal during purchase and consumption. Many polysaccharides are employed to provide gelation of foodstuffs. Starch is widely used because of its low cost and its compatibility with many food ingredients, but is not suitable for all food applications. Xanthan on its own only forms very weak gels but when mixed with various plant galactomannans or glucomannans, heated and cooled, it forms thermoreversible gels. Although these gels are opaque, this is still satisfactory for many food applications where clarity is not required. **Synergistic gels** of this type are employed in a range of foods including spreads, cream cheeses, etc. (Table 8.7).

When employed as a food additive, xanthan must conform to the definition provided in the National Formulary (UK) of '. . . a high molecular weight polysaccharide gum produced by a pure culture fermentation of a carbohydrate with *X. campestris* . . .'. The food grade polysaccharide must also meet the specifications listed in the Food Chemicals Codex (UK). These include absence of *E. coli* and *Salmonella*, and limits on arsenic (<3 ppm), heavy metals (<30 ppm) and isopropanol (<750 ppm).

Table 8.7 Usage of synergistic xanthan/galactomannan mixtures in foods

Gel formation and gel stabilization	Ice cream Cheese and cream cheese Dessert gels Milk shakes, milk drinks Puddings and pie fillings
Viscosity control	Ice cream Instant soups Chocolate drinks Cottage cheese dressings Milk shakes

Xanthan has many industrial applications unrelated to foods and beverages as can be seen in Table 8.5. It is also of value in the pharmaceutical and cosmetic industries where the same physical attibutes are used as in foods. An example is in the formulation of toothpastes, where the pseudoplastic behaviour of xanthan allows the product to flow freely when a tube is squeezed, but regains its viscosity immediately after the shear stress is removed.

Because of its valuable physical properties and the absence of competing microbial products, xanthan has a wide range of industrial applications. It could be used in others, but it suffers from a price disadvantage when compared with starch or some synthetic chemical polymers. The long and expensive testing programme needed before regulatory bodies will approve a microbial product means that it is unlikely that the overall industrial demand for xanthan will greatly diminish in the near future. Some applications such as enhanced oil recovery, if proved successful, will require very large quantities of the polysaccharide. There is probably adequate production capacity to meet such demand.

REFERENCES

Betlach, M.R., Capage, M.A., Doherty, D.H., *et al.* (1987) Genetically engineered polymers: manipulation of xanthan biosynthesis, in *Industrial Polysaccharides* (ed. M. Yalpani), Elsevier, Amsterdam, pp. 145–56.

Boulnois, G.J. (1989) Genetics of capsular production in bacteria. *Society for Experimental Biology Symposium*, **63**, 417–22.

Boulnois, G.J. and Jann, K. (1989) Bacterial polysaccharide capsule synthesis, export and evolution of structural diversity. *Molecular Microbiology*, **3**, 1819–23.

Cadmus, M.C., Jackson, L.K., Burton, K.A., Plattner, R.D. and Slodki, M.E. (1982) Biodegradation of xanthan gum by *Bacillus* sp. *Applied and Environmental Microbiology*, **44**, 5–11.

Davidson, I.W. (1978) Production of polysaccharide by *Xanthomonas campestris* in continuous culture. *FEMS Microbiology Letters*, **3**, 347–9.

Haaheim, L. and Sutherland, I.W. (1989) Monoclonal antibodies reacting with the exopolysaccharide from *Xanthomonas campestris*. *Journal of General Microbiology*, **135**, 605–12.

Ho, C.S., Ju, L.-K. and Baddour, R.F. (1988) The anomaly of oxygen diffusion in aqueous xanthan solutions. *Biotechnology and Bioengineering*, **32**, 8–17.

Ielpi, L., Couso, R. and Dankert, M. (1981) Lipid-linked intermediates in the biosynthesis of xanthan gum. *FEBS Letters*, **130**, 253–6.

Ielpi, L., Couso, R.O. and Dankert, M. (1983) Xanthan gum biosynthesis acetylation occurs at the prenyl-phospho-sugar stage. *Biochemical International*, **6**, 323–33.

Jansson, P.-E., Kenne, L. and Lindberg, B. (1975) Structure of the extracellular polysaccharide from *Xanthomonas campestris*. *Carbohydrate Research*, **45**, 275–82.

Jarman, T.R. and Pace, G.W. (1984) Energy requirements for microbial exopolysaccharide synthesis. *Archives of Microbiology*, **137**, 231–5.

Kang, K.S. and Cottrell, I.W. (1979) Polysaccharides, in *Microbial Technology*,

Vol. I (eds H.J. Peppler and D. Perlman), Academic Press, New York, pp. 417–81.

Kierulf, C. and Sutherland, I.W. (1988) Thermal stability of xanthan preparations. *Carbohydrate Polymers*, **9**, 185–94.

Lawson, C.J. and Symes, K.C. (1977) Oligosaccharides produced by partial acetolysis of xanthan gum. *Carbohydrate Research*, **58**, 433–8.

Linton, J.D. (1990) The relationship between metabolite production and the growth efficiency of the producing organism. *FEMS Microbiological Review*, **75**, 1–18.

Melton, L.D., Mindt, L., Rees, D.A. and Sanderson, G.R. (1976) Covalent structure of the extracellular polysaccharide from *Xanthomonas campestris*. *Carbohydrate Research*, **46**, 245–57.

Moorhouse, R., Walkinshaw, M.D. and Arnott, S. (1977) Xanthan gum – molecular conformation and interactions. *ACS Symposium*, **45**, 90–102.

Morris, V.J. and Miles, M. (1986) Effect of natural modifications on the functional properties of extracellular bacterial polysaccharides. *International Journal of Biol. Macromol.*, **8**, 342–8.

Nakajima, S., Funahashi, H. and Yoshida, T. (1990) Xanthan gum production in a fermentor with twin impellors. *Journal of Fermentation and Bioengineering*, **70**, 392–7.

Norisuye, T., Yanaki, T. and Fujita, H. (1980) Triple helix of a *Schizophyllum commune* polysaccharide in aqueous solution. *Journal of Polymer Science*, **18**, 547–8.

Rye, A.J., Drozd, J.W., Jones, C.W. and Linton, J.D. (1988) Growth efficiency of *Xanthomonas campestris* in continuous culture. *Journal of General Microbiology*, **134**, 1055–61.

Shatwell, K.P. (1989) The influence of acetyl and pyruvic acid substituents on the solution and interaction properties of xanthan. PhD thesis, Edinburgh University.

Shatwell, K.P., Sutherland, I.W., Dea, I.C.M. and Ross-Murphy, S.B. (1990) The influence of acetyl and pyruvate substituents on the solution properties of xanthan polysaccharide. *Carbohydrate Research*, **206**, 87–103.

Shatwell, K.P., Sutherland, I.W., Ross-Murphy, S.B. and Dea, I.C.M. (1991) Influence of the acetyl substituent on the interaction of xanthan with plant polysaccharides. I. Xanthan-locust bean gum systems. *Carbohydrate Polymers*, **14**, 29–51.

Silman, R.W. and Rogovin, P. (1972) Continuous fermentation to produce xanthan biopolymer: effect of dilution rate. *Biotechnol. Bioengin.*, **14**, 23–31.

Stokke, B.T., Foss, P., Christensen, B.E., Kierulf, C. and Sutherland, I.W. (1989) Long-term storage of xanthan in seawater at elevated temperature: physical dimensions and chemical composition of degradation products. *International Journal of Biol. Macromol.*, **11**, 137–44.

Suh, I.-S., Herbst, H., Schumpe, A. and Deckwer, W.-D. (1990) The molecular weight of xanthan polysaccharide produced under oxygen limitation. *Biotechnology Letters*, **12**, 201–6.

Sutherland, I.W. (1987) Xanthan lyases – novel enzymes found in various bacterial species. *Journal of General Microbiology*, **133**, 3129–34.

Tait, M.I. and Sutherland, I.W. (1989) Synthesis and properties of a mutant type of xanthan. *Journal of Applied Bacteriology*, **66**, 457–60.

Tait, M.I., Sutherland, I.W. and Clarke-Sturman, A.J. (1986) Effect of growth conditions on the production, composition and viscosity of *Xanthomonas campestris* exopolysaccharide. *Journal of General Microbiology*, **132**, 1483–92.

Tang, J.-L., Gough, C.L. and Daniels, M.J. (1990) Cloning of genes involved in negative regulation of production of extracellular enzymes and polysaccharide of *Xanthomonas campestris* pathovar *campestris*. *Molecular and General Genetics*, **222**, 157–60.

Tang, J.-L., Liu, Y.-N., Barber, C.E., Dow, J.M., Wootton, J.C. and Daniels, M.J. (1991) Genetic and molecular analysis of a cluster of *rpf* genes involved in positive regulation of synthesis of extracellular enzymes and polysaccharide in *Xanthomonas campestris* pathovar *campestris*. *Molecular and General Genetics*, **226**, 409–17.

Vanderslice, R.W., Doherty, D.H., Capage, M., *et al*. (1989) Genetic engineering of polysaccharide structure in *Xanthomonas campestris*, in *Biomedical and Biotechnological Advances in Industrial Polysaccharides* (eds V. Crescenzi, I.C.M. Dea, S. Paoletti, S.S. Stivala and I.W. Sutherland), Gordon and Breach, New York, pp. 145–56.

Index

GPSR Compliance

The European Union's (EU) General Product Safety Regulation (GPSR) is a set of rules that requires consumer products to be safe and our obligations to ensure this.

If you have any concerns about our products, you can contact us on ProductSafety@springernature.com

In case Publisher is established outside the EU, the EU authorized representative is:

Springer Nature Customer Service Center GmbH
Europaplatz 3
69115 Heidelberg, Germany

Batch number: 10815974

Printed by Printforce, the Netherlands